High-Performance Automation Methods for Computational Intelligent Systems

Computational methods are necessary for the proper execution of the applications for the benefit of society and technological development. Technological development makes life easier by constructing powerful systems with the help of computational methods. The nature of computational methods changes from time to time and retains only efficient applicable theories. Researchers take the idea of existing computational systems and advance them as per possible future needs. Efficient computational methods solve complex problems and also help to make the system more intelligent. Automation process requires decision-making computational systems. A more intelligent system contains an efficient computational method, which is described by powerful algorithm development. The aim of this book is to identify the technological development for future computational systems, which ultimately reflects the more intelligent system. Automation has become the need of the hour in today's world, and hence, the computational systems need to be upgraded to that level to perform the required tasks. The most efficient computational algorithm acts like a human being and offers a full sense of intelligent automation.

This book:

- Presents the latest research trends for the upcoming computational intelligent systems in a comprehensive manner.
- Focuses on the integration of multi-purpose and multi-dimension natural language into intelligent systems.
- Elaborates on nature-inspired and intelligent behaviour-based computational methods to deal with the observation of nature.
- Illustrates applications of quantum cellular energy-efficient computing methods for automation and applications of genetic algorithms in multidisciplinary fields.
- Discusses aspects of intelligent automation like technology-based, architecture-based, logic implementation-based, and the different algorithms-based concepts.

It is primarily written for senior undergraduates, graduate students, and academic researchers in the fields of electrical engineering, electronics and communications engineering, computer science and engineering, and information technology.

Computational Intelligence in Signal Processing and Automation

Rajeev Arya and Brijesh Iyer

Computational Intelligence in Signal Processing and Automation is the need of the hour for next-generation intelligent system design and deployment. It is forecasted that automation techniques will be the backbone of next-generation intelligent systems. The three automation variants, namely fixed automation, programmable automation, and flexible automation, will play a vital role in all design and development activities. By enhancing and utilizing signal processing algorithms in the field of biomedical engineering amongst myriad of other domains, the series shall cover the most comprehensive range of computationally intelligent signal processing paradigms. The series will incorporate topics of multidisciplinary nature such as expert systems, dynamic neural networks, probabilistic reasoning, fuzzy logic, genetic algorithms, multi-agent systems, distributed non-linear computing, and pattern recognition.

High-Performance Automation Methods for Computational Intelligent Systems: Design and Enabling Technologies
Vijay Kumar Sharma and Sachin Kumar Gupta

High-Performance Automation Methods for Computational Intelligent Systems

Design and Enabling Technologies

Edited by

Vijay Kumar Sharma and

Sachin Kumar Gupta

CRC Press
Taylor & Francis Group
Boca Raton London New York

CRC Press is an imprint of the
Taylor & Francis Group, an informa business

First edition published 2026
by CRC Press
2385 NW Executive Center Drive, Suite 320, Boca Raton FL 33431

and by CRC Press
4 Park Square, Milton Park, Abingdon, Oxon, OX14 4RN

CRC Press is an imprint of Taylor & Francis Group, LLC

ISBN: 978-1-032-74063-8 (hbk)
ISBN: 978-1-041-08070-1 (pbk)
ISBN: 978-1-003-64360-9 (ebk)

DOI: 10.1201/9781003643609

Typeset in Sabon
by codeMantra

Contents

About the editors

Vijay Kumar Sharma received his M.Tech. degree in VLSI Design from the National Institute of Technology, Hamirpur (H.P.), India, in 2009 and his Ph.D. in Low-Power VLSI Design from the ABV Indian Institute of Information Technology & Management, Gwalior (M.P.), India, in 2015. Currently, he is an Associate Professor in the Department of Electronics and Communication Engineering, Madan Mohan Malaviya University of Technology, Gorakhpur, India, since January 2025. He is also associated with the School of Electronics & Communication Engineering, Shri Mata Vaishno Devi University, Katra, India. His research interests are in the area of low-power circuit design. He has published widely in international journals and conferences of repute. He is also listed in list of world's top 2% scientist, published by Stanford University. He has authored many book chapters for reputed publishers and has many patents to his credit. He is a lifetime member of various societies. He has organized many short-term programmes, workshops, faculty development programmes, seminars, and conferences. He participates actively in all related events.

Sachin Kumar Gupta received his M.Tech. and Ph.D. from the Indian Institute of Technology (Banaras Hindu University), Varanasi (Uttar Pradesh), India, in 2011 and 2016, respectively. Currently, he is an Associate Professor in the Department of Electronics and Communication Engineering, Central University of Jammu, Samba, India, since September 2023. His research and teaching interest includes computer systems and networking, cryptography and network security, cyber security, UAV, SDN, and IoT. He has published widely in reputed international journals and prestigious conference proceedings, and is an author of many book chapters as well. He has also been granted a number of patents from the German patent and trademark office, and Office of the Controller General of Patents Designs & Trademarks, Govt. of India, South Africa, etc. He has organized a number of faculty development programmes, short-term courses, workshops, conference, etc. in the capacity of coordinator and organizing secretary.

Contributors

Kavya Varshini Anburasu
Department of Computing Technologies
SRM Institute of Science and Technology
Kattankulathur, India

Md. Waseem Akram
Department of Electronics and Communication Engineering
Jamia Millia Islamia
Delhi, India

Nazida Ansari
Department of Electronics and Communication Engineering
Jamia Millia Islamia
Delhi, India

Rakesh Kumar Bajaj
Department of Mathematics
Jaypee University of Information Technology
Solan, India

Shashi Bala
Department of ECE
Chandigarh Engineering College
Mohali, India

Abhay Bhatia
Department of Computer Science & Engineering
RIT
Roorkee, India

Jayaashri Chezhian
Department of Computing Technologies
SRM Institute of Science and Technology
Kattankulathur, India

Muskan Dixit
School of Management North Campus
IIT Mandi
Mandi, India

Mageshkumar Gunasekar
Department of Electronics and Communication Engineering
Kongu Engineering College
Erode, India

Murugesan Govindasamy
Department of Electronics and Communication Engineering
Kongu Engineering College
Erode, India

Sachin Kumar Gupta
Department of Electronics and Communication Engineering
Central University of Jammu
Samba, India

Sumeet Gupta
School of Electronics and Communication Engineering
Shri Mata Vaishno Devi University
Katra, India

Syed Mohd Hussain
Department of Electronics and Communication Engineering
Jamia Millia Islamia
Delhi, India

Tamilselvan Kumaravel Subramaniam
Department of Electronics and Communication Engineering
KPR Institute of Engineering and Technology
Arasur, India

Isha Kansal
Chitkara University Institute of Engineering and Technology
Chitkara University
Punjab, India

Amina Khan
Department of Electronics and Communication Engineering
PES University
Bengaluru, India

Anil Kumar
Department of Computer Science & Engineering
RIT
Roorkee, India

Arun Kumar
Department of ECE
GITAM University
Bangalore, India

Ashish Kumar
Chitkara University Institute of Engineering and Technology
Chitkara University
Punjab, India

Sunil Kumar
Department of IT
AKGEC
Ghaziabad, India

Anmol Madan
Chitkara University Institute of Engineering and Technology
Chitkara University
Punjab, India

Monika
Department of Mathematics
Jaypee University of Information Technology
Solan, India

Umayia Mushtaq
Department of Electronics and Communication Engineering
Jamia Millia Islamia
Delhi, India

Preeti
Department of Computer Science and Applications
Maharshi Dayanand University
Rohtak, India

Balwant Raj
Department of ECE
UIET Panjab University
Hoshiarpur, India

Balwinder Raj
Department of ECE
NIT
Jalandhar, India

Lalithraj Subramaniyan
Department of Electronics and Communication Engineering
Kongu Engineering College
Erode, India

Padmini Sankaramurthy
Department of Computing Technologies
SRM Institute of Science and Technology
Kattankulathur, India

Suthagar Subramaniyan
Department of Electronics and Communication Engineering
Kongu Engineering College
Erode, India

Indu Saini
Department of Electronics and Communication Engineering
Dr B R Ambedkar National Institute of Technology
Jalandhar, India

Aum Shah
Department of Computing Technologies
SRM Institute of Science and Technology
Kattankulathur, India

Aman Sharma
Department of Computer Science and Engineering
Jaypee University of Information Technology
Solan, India

Priti Sharma
Department of Computer Science and Applications
Maharshi Dayanand University
Rohtak, India

Yogesh Shastri
School of Electronics and Communication Engineering
Shri Mata Vaishno Devi University
Katra, India

Animesh Singh
Department of Computing Technologies
SRM Institute of Science and Technology
Kattankulathur, India

Mayank Kumar Singh
Department of Electronics and Communication Engineering
Dr B R Ambedkar National Institute of Technology
Jalandhar, India

Surender Singh
Department of Electrical Engineering
UIET, MDU
Rohtak, India

Jyoti Snehi
Department of CSE
Punjabi University
Patiala, India

Neetu Sood
Department of Electronics and Communication Engineering
Dr B R Ambedkar National Institute of Technology
Jalandhar, India

Manish Verma
Department of Computer Science & Engineering
AKGEC
Ghaziabad, India

Acknowledgements

This book contains the chapters related to the advancement in the domain of computational intelligent for the development of future systems. The various methods have been included in the book to deal the several issues.

We are grateful to the editorial team of CRC press for regular support at each stage of the preparation and outcome of this book.

We gratefully acknowledge the our organizations; Madan Mohan Malaviya University of Technology, Gorakhpur, India, Shri Mata Vaishno Devi University, Katra, India, and Central University of Jammu, Samba, India, for providing all necessary supports during the editing of this book.

The chapters presented in the book would not have been possible without the tremendous contribution of the authors.

This book would be the great resource for the researchers and academicians, who are working in the field of artificial intelligence, machine learning, deep learning, genetic algorithm, fuzzy logic, automation approach, natural language processing, cybersecurity, drone, memory, etc.

Vijay Kumar Sharma

Madan Mohan Malaviya University of Technology, Gorakhpur, India
Shri Mata Vaishno Devi University, Katra, India

Sachin Kumar Gupta
Central University of Jammu, Samba, India

Chapter 1

Fuzzy similarity measures for clustering validation

Monika, Rakesh Kumar Bajaj, and Aman Sharma

1.1 INTRODUCTION TO CLUSTERING VALIDATION

Every day, people collect and store huge amounts of information as data for analysis and management. Classifying and grouping data into categories or clusters is crucial for effective management [1]. The most beneficial task in data mining is clustering, in which data are labeled by groups and finding patterns and dissemination in the raw data [2]. Clustering means the data with similar properties are grouped into one cluster and the data with dissimilar properties are grouped into different clusters [3]. It is extensively utilized in machine learning, [4] data mining, pattern recognition, decision-making, and several other applications. An example is online app data in which items selected by customers to buy are gathered in one group and the customers who want to buy in the future are gathered in another group, named a wish list. We can also see another example of medical stores where there are different groups or patterns used to store all types of medicine because of the large data. Therefore, the main purpose of the clustering techniques is to identify how patterns are grouped into the right groups so that we can recognize their similarities and differences and draw tangible outcomes from them [5]. Clustering plays a crucial role in numerous data analysis and machine learning applications, including data mining, regression, prediction, and more. Engineering, technology and science, the humanities, medical science, and everyday life all require the grouping of objects for many reasons. Clustering falls under unsupervised machine learning in which we can draw references from datasets that contain input data but no labeled responses, i.e. data are not pre-classified. For this reason clustering is the part of learning by experience but not by examples. It is essential to understand the difference between clustering as an unsupervised classification and discriminant analysis [2] as a supervised classification. In the case of supervised classification data are provided with labeled means of pre-classified patterns. With labeled data, we easily differentiate the objects or items in different clusters. Several research communities use the term "clustering" to describe methods of grouping unlabeled data.

DOI: 10.1201/9781003643609-1

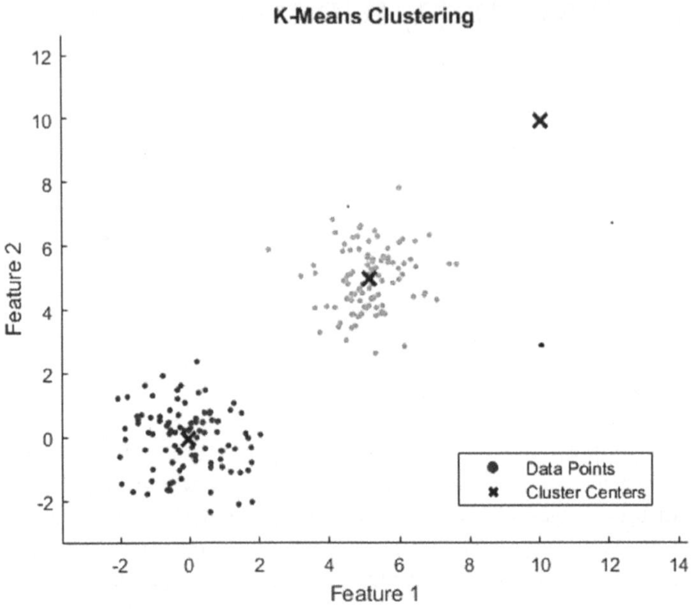

Figure 1.1 Results based on K-means clustering.

In Figure 1.1, there are easily identifiable clusters of data points, and each cluster is represented by a different color.

In 1999, Jain et al. [2] identified the following steps for clustering: pattern representation, i.e. selecting and extracting features, similarity calculation, grouping, and cluster visualization. According to Liao [6], time series clustering involves three key components: the clustering method, the similarity/dissimilarity measure, and the assessment of performance. Further, four primary feedback steps make up the elements of a clustering task as described by Xu and Wunsch [7]. The steps included selecting and extracting features, designing and selecting clustering algorithms, validating clusters, and interpreting results. To address the imprecise nature of clustering results, it's important to evaluate the performance of clustering methods used during the process. In Figure 1.2, we have presented the clustering activity steps based on these observations, and these steps are summarized as follows [8]:

Pattern representation: It is the number of classes, patterns, and features attainable for clustering, which includes feature extraction or selection. Feature selection refers to noticing the most beneficial subset of original features for clustering, whereas extraction of feature is the process of applying several transformations to the initial features to generate new, important characteristics for data clustering.

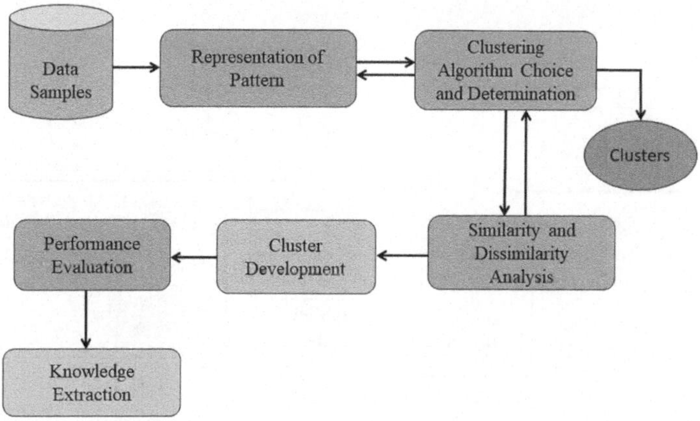

Figure 1.2 Detailed step in the clustering process.

Grouping process: It is a process of separating into distinct groups or groups with varying degrees of membership. Clustering algorithms depended on the specific objective and application in addition to the kind of data that were available. Similarity/dissimilarity computation involves measuring the similarity of two data types, like raw data, matrices and feature pairs, for comparison.

Performance evaluation: Performance evaluation is the assessment of a system's effectiveness, efficiency, accuracy, or quality. Clustering evaluates an algorithm's ability to accurately identify patterns or structures within data, to group data points that are comparable together. These are the most useful steps which we have discussed here.

We discuss various clustering techniques that are useful in multiple types of circumstances. It is divided into two parts: hard clustering and soft clustering. In hard clustering, each object falls into only one cluster, but in the case of soft clustering, objects could fall into more than one cluster. Several other methods exist for clustering that are widely used in different approaches to machine learning, named as partitioning, hierarchical, density-based, distribution model-based, and fuzzy clustering. Fraley et al. [9] proposed a clustering approach in two categories: partitioning and hierarchical techniques. After that another additional method of clustering was developed, i.e. density-based, grid-based methods, and model-based techniques. Figure 1.3 presents several kinds of clustering algorithm types, which are explained here.

A brief explanation of these techniques or algorithms is presented here based on the existing research [10].

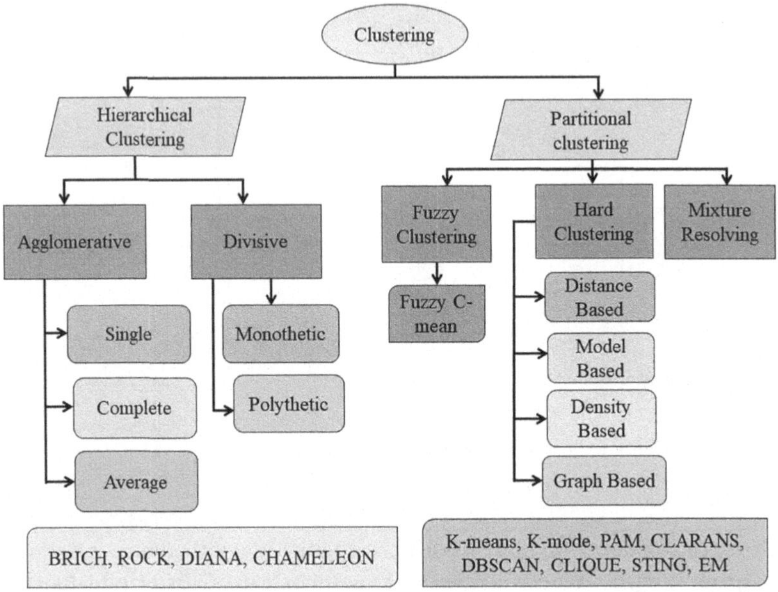

Figure 1.3 General classification of clustering algorithm.

Partitional clustering: It divides the dataset into distinct clusters. The objective is to find the optimal number of partitions for a specific criterion function.

Hierarchical clustering: It blends smaller clusters into larger ones or breaks larger clusters sequentially. This algorithm generates a cluster tree, known as a dendrogram, indicating their relationship. It is separated into two parts: divisive hierarchical clustering and agglomerative. The agglomerative takes a bottom-up approach, starting with a single item and merging smaller atomic groups into bigger ones until certain termination scenarios are met. The divisive hierarchical clustering approach works from the top-down, breaking up a group of all objects into smaller groups until each item forms a group or meets certain termination criteria.

Density-based clustering: The main concept behind this kind of clustering is to use density conditions to arrange nearby objects of a set of data into groups.

Grid-based clustering: This type of algorithm is primarily designed for geographic data mining. It quantizes the area through a finite amount of cells and performs all operations on them. The main benefits of this approach include its rapid processing time, without the need for distance calculations, and ease of identifying neighboring clusters.

Statistical: It depends on the analysis of statistics concepts and utilizes similarity measures to separate objects with limited numerical data.

Conceptual: Objects are clustered based on their associated concepts.

There is another classification in the case of clustering to handle the ambiguity in terms of overlapping clusters, named fuzzy clustering. In fuzzy clustering, objects can be classified into more than one cluster by using fuzzy techniques. These algorithms create clustering methods that can handle the uncertainty of real data, making them suitable for everyday use. In 1984, Bezdek et al. [11] proposed the most essential fuzzy clustering algorithm, the Fuzzy c-means (FCM) clustering algorithm. As a result, in the field of clustering, several types of algorithms are developed for handling large amounts of data in everyday life. Clustering helps uncover hidden patterns, structures, or natural groupings in the data, which can be useful for exploratory data analysis, pattern recognition, and decision-making. It is widely used in many areas. Here we summarize some applications in which clustering is utilized [8].

Image segmentation: A crucial step in the processing of images is image segmentation. It can be completed through hierarchical clustering and K-means can also be utilized for segmentation procedures. Magnetic resonance imaging (MRI) can visualize the internal parts of objects and living things. Since MRI images have a higher variance than CT scans, the majority of research on medical image segmentation uses MRI data.

Data mining: It is the technique of researching as well as evaluating large amounts of raw information to find patterns and extract useful knowledge. It can be used with relational, transactional, and relative databases, as well as large unstructured data collections like the global web. Recently, a number of articles on data mining have appeared in special issues.

Data reduction: It is one of the significant tasks for dealing with huge amounts of knowledge, and its analysis becomes extremely challenging. Clustering can compress data by grouping it into a variety of relevant clusters. After sorting through the various clusters, we can select the relevant information or dataset. In addition to performing data reduction, this procedure will save processing time for data.

Big data mining: This is a developing challenge. It is the processing of data volumes that are larger than what can be handled by traditional database management systems. An enormous quantity of data are constantly generated due to the utilization of different social media platforms, traveling, digital governance, and other procedures. Clustering regarding data can aid in collecting similar data from unformatted databases. It plays an important role in cybercrime and social networks like WhatsApp and Facebook.

Business: Clustering can help marketers identify and categorize significant customer groups according to their purchasing patterns.
The field of biology: In biological processes, it can help define taxonomies, categorize similar genes, and understand population structures.
Spatial analysis of data: Clustering techniques were used to analyze large datasets of geographical information systems and satellite images.

There are several other applications of clustering including sequence analysis, clustering of human genes given by Kaplan and Winther [12], mobile banking, and knowledge-based system addressed by Motiwalla et al. [13], educational data mining [14], climatology [15], field robotics [16], etc. in this study.

The overall structure of the chapter is as follows:

- The introduction covered the importance and challenges of clustering validation, as well as the implementation of fuzzy similarity measures in clustering.
- Section 1.2 introduces fuzzy and traditional similarity measures, as well as discusses their degree of similarity and dissimilarity. Section 1.3 discusses the limitations of traditional clustering validation and provides a numerical example of how intuitionistic fuzzy similarity measures can be applied to cluster analysis.
- In Section 1.4, the challenges of real-world data and the robustness of fuzzy similarity measures are discussed. Section 1.5 explains partial membership and multi-clustering. In Section 1.6, selection of appropriate similarity measures and the clustering objective are explained.
- Section 1.7 outlines the current state of the field, potential challenges, and research directions. Finally, the chapter is appropriately concluded, with future scope.

1.1.1 The significance of clustering validation

This section presents various clustering approaches that involve fundamental techniques. In data clustering, finding the optimum amount of clusters is a major output and it has remained an active issue for researchers. The motive for having distinct clustering approaches to various techniques is that there is no such specific definition of "cluster" [17]. As a result, various clustering methods have been suggested based on a distinct inclusion principle. According to Saxena et al. [8], user control over the number of clustering results can be achieved through trial and error, heuristics, or evolutionary procedures. The latest study by Hannah et al. [18] and Hu et al. [19] aims to improve the quality of clustering outcomes obtained through conventional algorithms and variants. Several clustering methods have been developed to deal with issues. Cluster validation is a technique for determining the effectiveness of clustering. It is an important task to incorporate applications. It assists users in selecting appropriate clustering variables, including the number of clusters and the suitable clustering model and algorithm. Cluster validity indices (CVIs) are frequently utilized to measure the quality of clustering in a quantitative way. Numerous CVIs have been suggested and used successfully. According to Wu et al. [20] clustering validity is generally divided into two categories: internal CVIs and external CVIs. They differ in terms of labeled and unlabeled data. For labeled data, the ground truth partition can be used in conjunction with external CVI to investigate the correlation between prospective clustering and the ground truth partition. In the case of unlabeled information, a significant post-clustering question is employed to evaluate various candidate partitions that meet internal cluster validity. Further, we provide an overview of these two types of validation measures [5,21].

External clustering: External CVIs are analyzed as measures of similarity between the fundamental truth and potential partitions. The fundamental truth partition, which is usually created by an expert in the data domain, determines the main structure of interest for the expert. Furthermore, we present an essential concept that is the basis for external CVIs [5,21].

Using a clustering algorithm, a derived clustering format P is connected to dataset E, while a prescribed clustering format S is connected to prior dataset information.

Summary of selected external indices: There are four groups depending on the number of object pairs that belong between R and S.

w, the number of pairs of objects which lies in E is the component of the same clusters in P and S.

x, the number of pairs of objects which lies in E is the component of the same clusters in P and distinct cluster in S.

y, the number of pairs of objects which lies in E is the component of the distinct clusters in P and same cluster in S.

z, the number of pairs of objects which lies in E is the component of the distinct clusters in P and S.

T, total number of object pairs with n number of dataset in E s.t., $n(n-1)/2$.

Now, we present four popular pair-counting-based CVIs.

Rand index: It is a popular pair-counting-based CVI which is utilized for comparing similarity between two groups and range from 0 to 1

$$\boldsymbol{R'} = \frac{w+z}{T}. \tag{1.1}$$

Jaccard index: It is similar to rand index but ignores one term, i.e. neutral term:

$$J' = \frac{w}{w+x+y}. \tag{1.2}$$

Fowlkes and Mallows index:

$$F' = \sqrt{\frac{a}{a+b}\frac{a}{a+c}}. \tag{1.3}$$

Internal clustering: It addresses the accuracy of clustering based on the data itself. Several internal clustering indices have been developed by researchers. Mostly, it builds up the index with two factors: one is cluster cohesion and another is cluster separation. There are various popular internal CVIs, some of which are discussed here [5,21].

Consider n data element in E (dataset) along with M number of partitions indexed from i from 1 to M, where

n_i = Number of data objects that correspond to cluster R_i,
c_i = Centroid linked to the cluster R_j,
c = Total number of centroid vectors

a_i = Average error for the group R_j, and,

a_j = Average error for the group $R_j d(R_i, R_j)$ = Distance between R_i and R_j in dataset.

Calinski and Harabasz index:

$$H' = \frac{\rho_r(\delta_e)}{M-1} / \frac{\rho_r(\delta_f)}{n-M}. \tag{1.4}$$

where

$\rho_r(\delta_e) = \Sigma_{i=1}^{M} n_i \| c_i - c \|^2$, Trace of the cluster scatter matrix.

$\rho_r(\delta_f) = \Sigma_{i=}^{M} \Sigma_{j=1}^{n_i} \| e_j - c_i \|^2$, Trace through a cluster scatter matrix.

The higher the value of the Calinski and Harabasz index, the better the quality of the clustering solution.

Dunn index: Utilizes the maximum diameter across all clusters to represent intra-cluster compactness and the smallest pairwise distance between objects in distinct clusters to represent inter-cluster dispersion [5,21]. It uses the ratio of separation under cohesion.

$$B' = \min_{i=1,\ldots,M}\left(\min_{j=1,\ldots,M,j\neq i}\left(\min_{i\neq j}\frac{d(R_i,R_j)}{\max\limits_{l=1,\ldots,M}\varepsilon(R_l)}\right)\right) \tag{1.5}$$

where

$$d(R_i,R_j)=\min_{r_1\in R_i,r_2\in R_j} d(r_1,r_2)$$

and

$$\varepsilon(R_l)=\max_{r_1,r_2\in R_l} dr_1,r_2\Big).$$

The higher the Dunn index, the better the estimation of partition.

Davies–Bouldin index:

$$D'=\frac{1}{M}\sum\nolimits_{i=1}^{M} K_i \tag{1.6}$$

where

$$K_i=\max_{j,j\neq i}\left(\frac{a_i+a_j}{c_i-c^2}\right).$$

The smallest value of Davies–Bouldin index indicates the potential of M in the dataset.

These are some internal or external cluster validity, which can also be used in generating multiple clustering [22] or an ensemble clustering. Also they can be utilized to select the best clustering technique along with the optimum number of clusters.

1.1.2 Challenges in clustering validation

Cluster analysis requires evaluating results to determine the most effective partitioning for the data. This is the primary focus of cluster validity. There are several challenges in this field, including [23,24]:

Quality of the cluster: Cluster validation faces a significant challenge in assessing the quality and meaningfulness of the clusters, which effectively represent and convey information. There are distinct methods to measure the quality of the cluster like external validation, internal validation, and relative validation. In the case of external data, results are compared with some external indication or ground truth, but it also affects the quality of the data. There are also inconsistencies between objective and criteria. Internal validation depends on data itself, there is no need for any label data, but it can be favored by the selection of distance or similarity measures, cluster number.

In the case of relative validation for finding the best solution it compares with different clustering situations but it may be costly to compute and sensitive with outliers and noise.

Stability of the cluster: Cluster validation examines the reliability of the clusters, such as how robust they are to modifications in the data or clustering variables. It may be affected by various factors, such as sampling, feature selection, normalization, dimensionality reduction, and cluster algorithm, etc.

Dimensionality of data: The application of certain analysis, such as K-means, maximum-margin grouping, and information-theoretic clustering, is difficult to apply to high-dimensional data. This type of problem is solved by presenting raw data on the smallest dimensional subsystem and then categorizing based on feature embedding, such as sparse codes.

Data object representation: It is another issue in clustering. It is represented differently in various application fields. Some data components are expressed as feature vectors, whereas others are indicated as object-similarity graphs. Identifying an effective method to convey data in clustering enhances performance while reducing computational complexity.

Data streaming: Clustering necessitates additional effort compared to clustering static information due to its distinct characteristics. Kokate et al. [25] recognized some challenges associated with data stream clustering methods. The increasing volume of data streams from different sources, such as social networks, can boost computational power and optimize storage capacity for clustering algorithms. More research is needed to create adaptable methods for clustering evolving data streams and to improve existing contextual adaptive clustering methods.

To overcome these challenges, careful consideration of dataset characteristics is required, as are the selection of appropriate algorithms and the use of preprocessing methods or algorithmic modifications. To ensure the accuracy and validity of the clustering results, they should be evaluated and validated using domain knowledge and further analyses.

1.1.3 The role of fuzzy similarity measures

Similarity measure plays a vital role in cluster analysis for estimating the degrees of relationship between two or more objects. Those objects that are similar to each other lie in the identical cluster and dissimilar objects lie in separate clusters. In 1999, Jain et al. [2] identified several clustering algorithms that utilized distance measures to examine the similarity measures within a combination of elements and they also defined the situation for any acceptable distance measure. Researchers point out the significant role of the similarity of items between groups in a clustering procedure. In traditional clustering, an element of the dataset either links to a group/cluster or not, such as K-means clustering. However, in numerous practical situations, data points may link to more than one group. Therefore, fuzzy clustering algorithms such as FCM clustering, utilize fuzzy similarity measures to examine degree of membership to each data point in a group. Fuzzy similarity measures play an essential role in many fields like clustering, pattern recognition, and in the decision-making process. These measures are useful when dealing with imprecise or incomplete data; they allow you to determine the degree of similarity among objects in an ambiguous environment. Providing the degree of similarity between cluster data and centroids, fuzzy clustering algorithms can handle noisy or outlier data [26] that would interrupt the clustering process using traditional methods. Yu et al. [27] proposed the fuzzy rough c-means clustering approach, effectively solving the fuzzy boundary problem between items and cluster centers. This approach increases clustering results and interpretability while achieving higher repetition efficiency. The robustness and adaptability of clustering algorithms can be increased by integrating fuzzy similarity measures with fuzzy logic-based techniques. For clustering challenges in which the data distribution is complex or unclear, fuzzy similarity measures are appropriate since they take non-linear, irregular, or interconnecting cluster boundaries into account.

1.1.4 Foundations of fuzzy similarity measures

Real-world data are full of vagueness, inconsistency, and inexactness. There is no crisp boundary between the data. So for dealing with uncertainty, Zadeh [28] developed the fuzzy sets (FSs) theory in 1965, which balances the space between real-world problems and mathematics. This theory is represented by the membership degree, which lies between [0, 1]. Here, zero means "no" and 1 means "yes", with partial membership in between and there is one condition: the sum of all membership values is less than or equal to one. After developing this theory, many researchers extended it to include intuitionistic fuzzy set (IFS) theory [29], picture fuzzy set (PFS) theory [30], and many more. The purpose of using fuzzy theory in clustering is that data objects can fall into multiple clusters, i.e. it handles incomplete data, so this theory is useful for improving the clustering accuracy. Firstly, Wang [31] introduced the idea of similarity measures for FSs in 1982. Further, many similarity measures are defined on a FS [32,33]. Several investigations have expanded the entropy, distance, and similarity measures of FS to include PFS [34] and many more. Next, Majumdar and

Samanta introduced the similarity measures for soft sets (SSs) [35] and fuzzy soft sets (FSSs) [36] for computing the degree of similarity. Further, in the environment of the Pythagorean fuzzy soft set (PFSS), Athira et al. developed new similarity measures for cluster analysis. Furthermore, in order to cluster heterogeneous components into a homogeneous form, new distance measures for clustering techniques have been employed. Next, Saracoglu et al. [37] proposed a fast and high-quality method for locating similar documents in large datasets using fuzzy clustering. Zeng et al. [38] combined the implication function with hesitant fuzzy sets (HFSs) to establish new formulas for computing the similarity measure of HFSs and also created a hesitant fuzzy clustering algorithm based on the transitive closure. Kamran and Hassan [39] introduced new distance and similarity measures in the context of HFSs and implemented them for hierarchical clustering. Further, S. Singh and Ganie [40] introduced new similarity measures for PFSs, and employed these measures in the field of pattern recognition, clustering, and multi-attribute decision-making (MADM) problems. Orthogonal distance and similarity measures for single-valued neutrosophic FSSs were presented by Al-Sharqi et al. [41], and they also developed new algorithms based on these measures to address decision-making challenges. As a result, there are several extensions available on fuzzy similarity measures which are utilized in many decision-making problems.

1.1.5 Traditional similarity metrics

The conventional similarity metrics are mathematical measures which are applied to quantify the similarity between two data objects. They applied in numerous fields such as pattern recognition, machine learning, etc. Such measurements are necessary for understanding data relationships, identifying structures, making predictions, and gaining meaningful insights. Some similarity metrics are discussed here [42].

Minkowski distance: This distance works properly when the dataset collection is isolated or compressed; if this condition is not met, the large-scale attributes perform better than the others. It includes Manhattan and Euclidean distance which are special cases.

$$\mathrm{MD} = \left(\sum_{i=1}^{n} |r_i - s_i|^p\right)^{\frac{1}{p}}, \quad p \geq 1 \tag{1.7}$$

where p is positive real number, and r_i and s_i are two vectors in n -dimensional space.

Manhattan distance: The difference between two points is calculated by adding the absolute differences between their coordinates

$$\mathrm{MH} = \Sigma_{i=1}^{n} \left| r_i - s_i \right|. \tag{1.8}$$

Euclidean distance: It is a straight line distance among two points in Euclidean space

$$\mathrm{ED} = \sqrt{\Sigma_{i=1}^{n} \left(r_i - s_i \right)^2}. \tag{1.9}$$

Jaccard similarity: The size of the intersection between two finite sets is compared to the size of the union to determine their similarity

$$\mathrm{JC} = \frac{\left| A_1 \cap A_2 \right|}{\left| A_1 \cup A_2 \right|}. \tag{1.10}$$

Pearson correlation: It is commonly utilized in clustering gene expression data. This computes the similarity between the structures of two gene function patterns

$$\mathrm{PC} = \frac{\Sigma_{i=1}^{n} \left(r_i - u_r \right)\left(s_i - u_s \right)}{\sqrt{\Sigma_{i=1}^{n} \left(r_i - s_i \right)^2} \sqrt{\Sigma_{i=1}^{n} \left(r_i - s_i \right)^2}}. \tag{1.11}$$

Cosine similarity: It is defined as the angle between two points

$$\mathrm{CS} = \frac{\Sigma_{i=1}^{n} r_i \cdot s_i}{\sqrt{\Sigma_{i-1}^{n} r_i^2} \sqrt{\Sigma_{i-1}^{n} s_i^2}}. \tag{1.12}$$

where u_r and u_s represent the mean of the r_i and s_i. It has a drawback in that it does not deal with outliers.

Traditional similarity metrics have limitations in that they can't deal with incomplete or uncertain data. Therefore, researchers extended these measures in the areas of FS theory, which are useful for several decision-making problems.

1.1.6 Extending similarity metrics with fuzziness

Extending similarity metrics with fuzziness involves integrating the concept of fuzziness or ambiguity into traditional similarity measures. Fuzziness describes the degree of membership or uncertainty within a set, which allows a more flexible and refined understanding of similarity. Various clustering techniques utilized distance measures to establish the similarity and dissimilarity between pairs of components. To compute the distance between two elements, there are different types of distance measures available in the literature.

The distance measure satisfies the following properties [43]:

Non-negativity: distance between two elements r_1 and r_2 is always greater than or equal to 0, i.e. $d(r_1,r_2)\geq 0$ and $d(r_1,r_2)=0$ iff $r_1=r_2$. Symmetric: distance measure should be symmetric, i.e. $d(r_1,r_2)=d(r_2,r_1)$. Transitivity: $d(r_1,r_2)\leq d(r_1,r_3)+d(r_3,r_2)$.

Fundamental mathematical concepts of some of these measures as presented by Xu and Wunsch in 2010 [7] are given below.

Minkowski distance: Consider a collection of dataset denoted by R, with m data objects $n$number of features, then the Minkowski distance [7] is defined as

$$d(r_i,r_j)=\left(\sum_{i=1}^{n}\left|r_{il}-r_{jl}\right|^q\right)^{\frac{1}{q}}. \tag{1.13}$$

where q is the generic number.

Cosine similarity:

$$S(r_i,r_j)=\cos\alpha=\frac{r_i^T r_j}{\|r_i\|\|r_j\|}. \tag{1.14}$$

Extended Jaccard measure:

$$J(r_i,r_j)=\frac{r_i^T r_j}{\|r_i\|^2+\|r_j\|^2-r_i^T r_j}. \tag{1.15}$$

The mathematical definition of similarity measures is defined as

Definition 1.1

Consider $\varphi(R)$ is a collection of all FSs over the dataset R. Then a mapping $\Psi:\varphi(R)\times\varphi(R)\rightarrow[0,1]$ is defined as similarity measures for any sets A_1, A_2, $A_3\in\varphi(R)$, with the following axioms:

1. $0\leq\varphi(A_1,A_2)\leq 1$;
2. $\varphi(A_1,A_2)=0$ iff $A_1=A_2$;
3. $\varphi(A_1,A_2)=\varphi(A_2,A_1)$;
4. If $A_1\subset A_2\subset A_3$ then $\varphi(A_1,A_2)\leq\varphi(A_1,A_3)$ and $\varphi(A_2,A_3)\leq\varphi(A_1,A_3)$.

In Table 1.1, we present the mathematical notion of some existing similarity measures that are utilized in decision-making challenges.

Table 1.1 Existing similarities measures

Authors	*Similarity measures*
Chen [43]	$S'_l(A,B)=1-\frac{1}{2n}\sum_{i=1}^{n}\left\lvert\mu_A(r_i)-\sigma_A(r_i)-\left(u_A(r_i)-\sigma_A(r_i)\right)\right\rvert$ where μ and σ represent the member and non-membership degree lies in [0, 1], and n is the total number of element in a set A and B.
Li et al. [44]	$S'_\varepsilon(A,B)=1-\sqrt{\frac{1}{2n}\sum_{i=1}^{n}\left(\mu_A(r_i)-\mu_B(r_i)\right)^2+\left(\sigma_A(r_i)-\sigma_B(r_i)\right)^2}$
Ye [45]	$S'_\vartheta(A,B)=\frac{1}{n}\sum_{i=1}^{n}\frac{\mu_A(r_i)\mu_B(r_i)+\sigma_A(r_i)\sigma_B(r_i)}{\sqrt{\mu^2{}_A(r_i)+\sigma^2{}_A(r_i)}\sqrt{\mu^2{}_B(r_i)+\sigma^2{}_B(r_i)}}$
Muthukumar et al. [46]	$S'_\gamma(A,B)=\frac{\sum_{i=1}^{n}\left(\mu_A(r_i)\cdot\mu_B(r_i)+\sigma_A(r_i)\cdot\sigma_B(r_i)\right)}{\sum_{i=1}^{n}\left(\left(\mu_A(r_i)^2\vee\mu_B(r_i)^2\right)+\left(\sigma_A(r_i)^2\vee\sigma_B(r_i)^2\right)\right)}$
Zhang [47]	$S'_\beta(A,B)=\frac{1}{n}\sum_{i=1}^{n}\frac{\left\lvert\mu_A^2(r_i)-\sigma_B^2(r_i)\right\rvert+\left\lvert\sigma_A^2(r_i)-\mu_B^2(r_i)\right\rvert+\left\lvert\pi_A^2(r_i)-\pi_B^2(r_i)\right\rvert}{\left\lvert\mu_A^2(r_i)-\mu_B^2(r_i)\right\rvert+\left\lvert\sigma_A^2(r_i)-\sigma_B^2(r_i)\right\rvert+\left\lvert\pi_A^2(r_i)-\pi_B^2(r_i)\right\rvert+\left\lvert\mu_A^2(r_i)-\sigma_B^2(r_i)\right\rvert+\left\lvert\sigma_A^2(r_i)-\mu_B^2(r_i)\right\rvert+\left\lvert\pi_A^2(r_i)-\pi_B^2(r_i)\right\rvert}$ where π represents the degree of indeterminacy between 0 and 1.

1.1.7 Degrees of similarity and dissimilarity

The concept of similarity, as well as its inverse, dissimilarity, is one of the most fundamental concepts in science. It is found everywhere in mathematics, as well as in the most distinct and remote interpretations of mathematics as the descriptors of physical reality. In group technology (GT), a similarity or dissimilarity coefficient is utilized to assess cluster grouping ability. One significant and often applied method for creating cellular manufacturing systems is GT [48]. It organizes components or machines with similar functional characteristics into circles or cells. Components with comparable processing needs are categorized into part families and machines specifically designed to process them into machine cells according to clustering techniques. The measures of similarity coefficients employed in GT, firstly the Jaccard coefficient, was introduced by McAuley [49] for producing cells using single linkage clustering techniques. After that, Mosier and Taube [50] extended this coefficient into weighted similarity coefficient. Dissimilarity coefficients are inversely related to similar coefficients, which means that when similarity between pairs of data objects increases, dissimilarity decreases. When comparing two entities or phenomena, degrees of similarity and dissimilarity are used to measure how comparable or distinct they are in terms of their traits, qualities, or attributes. In clustering algorithms, dissimilarity metrics are commonly employed to group similar entities together while exceeding the dissimilarity between clusters. Currently, a number of types of similarity and dissimilarity measures have been introduced [51] for calculating the degree of similarity between different types of FSs.

1.1.8 Assessing clustering quality with fuzzy similarity

As clustering is an unsupervised process, the outcomes of the clustering process must be evaluated using validity indices, which are outside standards. These indices are crucial in identifying the number of clusters in a dataset. Minimizing compactness within a cluster and maximizing the distance between clusters is the general idea behind these indices. Existing indices typically perform well for datasets with well-separated clusters, but have trouble with complex datasets with overlapping clusters [52]. Clustering quality with fuzzy similarity can be measured by evaluating how properly the clusters formed via a clustering algorithm representing the fundamental structure of the data while taking into account the uncertainty or vagueness associated with the data point's memberships in multiple groups. Fuzzy clustering approaches, such as FCM [11] and Possibilistic c-means (PCM) [53], allow data points to be assigned to multiple clusters to varying degrees, rather than a single cluster. In fuzzy clustering, a fuzzy partition matrix represents each data point's membership degree in every single cluster. These membership degrees express the degree to which an element is associated with each cluster. Utilize fuzzy similarity metrics to evaluate the clustering's quality. These measures account for the degree of overlap or similarity between clusters while also acknowledging the fuzziness of cluster assignments. Therefore, assessing the quality of the clustering outcomes is an essential concern in cluster analysis. It usually depends on a clustering validity index. Since it is an unsupervised process, its results must be evaluated using an external criterion. In the case of a low-dimensional set of data, humans can examine the clustering outcomes by visual observation, but in the case of high-dimensional datasets, there is no desired outcome criterion to examine the clustering outcomes; they are evaluated using a cluster validity index [54].

1.1.9 Limitations of traditional clustering validation

Clustering objective is to separate the dataset into the classes, so that dataset in the same bunch is similar, and dissimilar in separate classes. Several clustering techniques are developed for this purpose. As we discussed above, clustering algorithms can be organized into categories: hard clustering, where one data element links with one group, and soft clustering, where data elements can link with more than one group. In the beginning, several conventional measures utilize ground truth labels for comparison, which may not be available or may be subjective in real-world datasets, limiting their applicability. Moreover, finding the best clustering solution in complex datasets can be difficult because traditional validation measures are sensitive to factors like the number of clusters and cluster shape.

In addition, traditional validation techniques can have difficulties with high-dimensional data that may not be resistant to noise or outliers, reducing the accuracy of the evaluation. Also, these methods may not properly represent hierarchical cluster structures, limiting their application in datasets with hierarchical relationships.

In the case of partitional clustering, a single cluster must include at least one data point or each data point must be allocated to at least one cluster, for example, K-means clustering [55] in which each cluster is expressed by a centroid and each object is allocated to the closest centroid during clustering. It obtains this by optimizing the total of squared error (SSE) objective:

$$O = \sum_{i=1}^{k} \sum_{r \in \mu_i} \|r - c_i\|^2 \tag{1.16}$$

where μ_i is the ith cluster and c_i is the cluster center of μ_i. It is a 50-year-old technique that is widely utilized because it is simple, easy to implement, efficient, and empirically successful [56]. But still, it has some limitations which are that it is sensitive to centroid initialization and not effective in identifying clusters with non-spherical shapes, i.e. data with an irregular shape of cluster. K-means is effective on well-separated clusters rather than overlapping clusters and is also sensitive to outliers. In spite of these drawbacks, clustering is still a useful technique for segmentation, pattern recognition, and exploratory data analysis across a range of industries. Researchers are constantly developing new algorithms and techniques to address these challenges while also improving the effectiveness and robustness of clustering methods.

1.1.10 Fuzzy similarity measures: a solution

Fuzzy similarity measures are useful in many fields, involving pattern recognition, data mining, image processing, and artificial intelligence. They provide a flexible structure for comparing objects or data points in scenarios where conventional binary logic may not be sufficient. As we discussed above Wang introduced similarity measures for FSs. Following that, researchers extended this theory to different sets in different applications for dealing with uncertain or complex data. Based on fuzzy logic Santini and Jain [57] suggested similarity measures. There are many extensions in the area of FSs and similarity measures were also defined for these extensions with their applications. Jay Chen and Ming Chen [58] proposed a novel approach for fuzzy risk analysis based on similarity measures of extended fuzzy numbers. Shishavan et al. [59] developed new information measures for spherical FSs, including Jaccard, exponential, and square root cosine similarity measures. Further, Khan and Lohani [60] gave the similarity measures for IFSs with application in cluster analysis. Following that, Garg et al. [61] presented correlation coefficients for T-spherical FSs for use in clustering and MADM problems. Next, similarity measures for PFSSs were suggested by Athira et al. and applied in the area of clustering. A parametric similarity measure under spherical FSs, introduced by Sarfraz and Pamucar [62], has been applied to the selection of medical equipment. In a subsequent study, Rani et al. [63] proposed a MADM approach based on similarity measures between PFSs. They also developed an alternative method for measuring alternatives and ranking them using the MARCOS method (Measurement of Alternatives and Ranking according to Compromise Solution). Patel et al. [64] proposed similarity measures for intuitionistic sets and developed a MADM algorithm, which has been employed in face recognition and software quality analysis. In general, the evolution of fuzzy similarity measures has been driven by advances in theory, algorithms, and practical use across multiple disciplines. As the need for dealing with uncertainty and imprecision in data develops, fuzzy similarity measures are going to remain an important tool in the frameworks of both researchers and practitioners. By using the normalized hamming distance of the IFSs, which was provided by Khan and Lohani [60], we represent the clustering algorithm as an application of the intuitionistic fuzzy similarity measures with the aid of a numerical example.

1.1.11 Case studies illustrating fuzzy similarity in action

Clustering analysis is widely studied and applied in diverse fields. Using distance measurements, researchers have proposed a number of similarity metrics to quantify the similarity between two Atanassov Intuitionistic Fuzzy Sets (AIFSs). We compute a numerical example using the normalized hamming distance.

The normalized hamming distance [60] for two IFSs A and B in R is

$$h(A,B)=\frac{1}{2n}\sum_{i=1}^{n}\left|\mu_A(r_i)-\mu_B(r_i)\right|+\left|\pi_A(r_i)-\pi_B(r_i)\right|+\left|\sigma_A(r_i)-\sigma_B(r_i)\right|. \quad (1.17)$$

On the basis of this distance measure, the similarity measure [60] is defined as

$$S'(A,B)=1-\frac{1}{2n}\sum_{i=1}^{n}\left|\mu_A(r_i)-\mu_B(r_i)\right|+\left|\pi_A(r_i)-\pi_B(r_i)\right|+\left|\sigma_A(r_i)-\sigma_B(r_i)\right|. \quad (1.18)$$

where μ, π, and σ represent the membership degree, indeterminacy degree, and non-membership degree, respectively, r_i is the element of R and n is the number of elements in the set.

Now, some helpful definitions are included before proceeding.

Definition 1.2

Consider A_n be n number of IFSs over R and a matrix $\mathfrak{D}=\left[\mathfrak{D}_{ij}\right]_{n\times n}$, where $\mathfrak{D}_{ij}=S'\left(L_i,L_j\right), i,j=1,2,\ldots,n$ known as a similarity matrix between the IFSs satisfying the following properties:

1. $0\le\mathfrak{D}_{ij}\le 1$;
2. $\mathfrak{D}_{ij}=1$;
3. $\mathfrak{D}_{ij}=\mathfrak{D}_{ji}$ where $i,j=1,2,\ldots,n$.

Definition 1.3

As per reference [65] consider $\mathfrak{D}=\left[\mathfrak{D}_{ij}\right]_{n\times n}$ be a similarity matrix, if $\mathfrak{D}^2=\mathfrak{D}*\mathfrak{D}$ be a composition matrix of $\mathfrak{D}$, where $\mathfrak{D}_{ij}=\max_k\left\{\min\left\{\mathfrak{D}_{ik},\mathfrak{D}_{kj}\right\}\right\}, k=1,2,\ldots,n$.

Definition 1.4

As per reference [65] for given similarity matrix $\mathfrak{D} = \left[\mathfrak{D}_{ij}\right]_{n\times n}$ and in the finite composition $\mathfrak{D} \to \mathfrak{D}^2 \to \mathfrak{D}^4 \to, \cdots, \mathfrak{D}^{2z} \to \cdots$, matrix there exists a number called an integer $z > 0$ s.t. $\mathfrak{D}^{2z} = \mathfrak{D}^{(2z+1)}$ and then $\mathfrak{D}^{2z}$ is called equivalent similarity matrix (ESM).

Definition 1.5

As per reference [65] for considering equivalent similarity matrix $\mathfrak{D} = \left[\mathfrak{D}_{ij}\right]_{n\times n}$ α-cutting matrix of $\mathfrak{D}$ is $\mathfrak{D}^{\alpha} = \left[\mathfrak{D}_{ij}^{\alpha}\right]_{m\times m}$, where

$$\mathfrak{D}_{ij}^{\alpha} = \begin{cases} 0; \mathfrak{D}_{ij}^{\alpha} < \alpha \\ 1; \mathfrak{D}_{ij}^{\alpha} \geq \alpha \end{cases}$$

and $\alpha \in [0,1]$ is known as the confidential interval.

The idea for clustering uses the IFSs form to express expert opinions on $n$alternatives, $\{w_1, w_2, \cdots, w_n\}$ corresponding to their m number of features, to group the given W_i' into the same groups by using the suggested steps for this task:

Step 1: Using Equation (1.18), create the similarity matrix $\mathfrak{D} = \left[\mathfrak{D}_{ij}\right]_{n\times n}$, where $\mathfrak{D}_{ij} = S'(L_i, L_j)$, $i, j = 1, 2, \ldots, n$.
Step 2: Find the ESM $\mathfrak{D}^{2z}$, where $\mathfrak{D} \to \mathfrak{D}^2 \to \mathfrak{D}^4 \to, \cdots, \mathfrak{D}^{2z} \to \cdots$, until $\mathfrak{D}^{2z} = \mathfrak{D}^{(2z+1)}$.
Step 3: Develop the α–cutting matrix $\mathfrak{D}^{\alpha} = \left[\mathfrak{D}_{ij}^{\alpha}\right]_{m\times m}$, using definition (1.5).
Step 4: Assign w_i and w_j, which are comparable to the same group.

1.1.12 Numerical illustration

Following the above-described clustering step, we computed a numerical example.

Consider top brands of cars $\{W_1, W_2, W_3\}$. Selection of the car is on the basis of their features $\{Q_1, Q_2, Q_3\}$ which symbolizes Q_1=Selling price, Q_2=Mileage, and Q_3=Performance. The objective is to classify the cars using these features. Each car is rated by an expert using the IFSs data, including indeterminacy degree. Table 1.2 provides a complete summary of their ratings. We are now utilizing the measures given by Khan and Lohani [60] W_i has the following steps in order to group the cars, W_i which has the following steps:

Table 1.2 Value assigned by experts to each object

	Q_1	Q_2	Q_3
W_1	(0.1,0.5,0.2)	(0.5,0.4,0.0)	(0.0,0.1,0.5)
W_2	(0.2,0.3,0.4)	(0.0,0.6,0.0)	(0.9,0.1,0.0)
W_3	(0.5,0.0,0.4)	(0.1,0.7,0.2)	(0.8,0.0,0.1)

Step 1: First, calculate the degree of similarity between IFSs, $S'(L_i, L_j)$, $i, j = 1, 2, \ldots, n$ using Equation (1.18):

$$\mathfrak{D} = \begin{bmatrix} 1.00 & 0.57 & 0.45 \\ 0.57 & 1.00 & 0.64 \\ 0.45 & 0.64 & 1.00 \end{bmatrix}.$$

Step 2: Compute $\mathfrak{D}^2 = \mathfrak{D} * \mathfrak{D}$ using definition (1.3):

$$\mathfrak{D}^2 = \begin{bmatrix} 1.00 & 0.57 & 0.57 \\ 0.57 & 1.00 & 0.64 \\ 0.57 & 0.64 & 1.00 \end{bmatrix}.$$

Therefore, $\mathfrak{D}^2 \neq \mathfrak{D}$, calculate the value of $\mathfrak{D}^4$ as $\mathfrak{D}^4 = \mathfrak{D}^2 * \mathfrak{D}^2$

$$\mathfrak{D}^4 = \begin{bmatrix} 1.00 & 0.57 & 0.57 \\ 0.57 & 1.00 & 0.64 \\ 0.57 & 0.64 & 1.00 \end{bmatrix}.$$

We obtain $\mathfrak{D}^4 = \mathfrak{D}^2$, so $\mathfrak{D}^2$ is classified as an equivalent matrix.

Step 3: Using the definition (1.5), compute the α–cutting matrix$\mathfrak{D}^\alpha$ with a confidential level $\alpha = 0.64$.

Step 4: Organize W_i into two groups using the $\mathfrak{D}^\alpha$ matrix: $\{W_1\}$ and $\{W_2, W_3\}$.

From improving document search technologies to refining customer segmentation approaches and assisting medical diagnosis, the easy implementation of fuzzy similarity measures illustrates their transformative impact across multiple domains, highlighting new pathways for dealing with inconsistencies and optimizing decision-making procedures.

1.2 HANDLING IMPRECISE AND AMBIGUOUS DATA

In recent years, fuzzy data analysis and its study methods have grown in popularity. However, in numerous instances, existing methods for accurate non-fuzzy facts are improved and broadened to the fuzzy situation, without carefully considering how a FS is interpreted for data modeling. Learning models from ambiguous data, which includes interval data or, more broadly, data modeled as fuzzy subsets of an underpinning reference space, has become prominent in recent times [66]. Furthermore, while difficulties such as fuzzy regression evaluation [67] have been investigated for a long time, the study is now expanding, both in the context of the challenges addressed in classification, clustering, and ranking, as well as the uncertainty formalisms employed. Therefore, Learning from uncertain and imprecise data necessitates an extension of the equivalent learning algorithm. Ultimately, this occurs without outlining the specific significance of an uncertain assessment, even though representations like intervals or FSs can be determined in a variety of ways. Consequently, handling imprecise and ambiguous data is an issue that arises in many fields, including natural language processing, decision-making systems, and machine learning. Fuzzy logic gives a mathematical framework for dealing with uncertainty by representing imprecise information with FSs and fuzzy logic operators such as And, OR, and Not, allowing the changing of FSs and fuzzy information. These operators enable the blend of imprecise information from multiple sources as well as the utilization of fuzzy rules to make decisions in the face of ambiguity.

1.2.1 Challenges posed by real-world data

Human decisions and reasoning are often ambiguous and uncertain. However, traditional databases deal with data that is crisp, deterministic, and precise. If we intend the computer system to respond to actual conversation, we must incorporate vagueness into our databases through the use of fuzzy logic, which is an expansion of standard Boolean logic that represents aspect between absolute truth and absolute false. Conventional ways for precise data handling can cause two major complications. First, real-world situations are rarely crisp and determined, and they cannot be described specifically. For example, if we choose smart students in class, we can see that the word "smart" does not convey a clear message. Everyone has an opinion about this word; they select it based on marks or sincerity. Second, an in-depth account of the real thing would necessitate far more detailed information than a human being can detect and process at the same time. Because most traditional databases in use presently are built on the relational model, such an overview is hard to represent using the traditional relational framework. If we desire the computer technology to react to human word, we need to present uncertainty into our databases.

Clustering is a challenging task due to the complexity and diversity of real-world data. Some of the main challenges are as follows [68]:

Practical application datasets frequently contain noisy or irrelevant features, which obscure the underlying cluster structure and reduce clustering accuracy.

- Outliers can have a significant impact on clustering algorithm performance, causing unreliable clusters to form or cluster boundaries to be distorted.
- The size of real-world datasets presents scalability challenges for clustering algorithms, requiring efficient processing techniques to handle large amounts of data.

A lot of effort has been put into overcoming these types of challenges in this field. Wiroonsri [69] proposed a new correlation cluster validity index to improve the performance of clustering analysis in finding multiple optimal solutions. Next, Hong et al. [70] proposed a novel method for generating synthetic varied information networks that aids in identifying patterns in real-world datasets, summarizing graph statistics, and constructing a synthetic network with In-Cluster and Out-Cluster Merge components. A non-negative matrix factorization (NMF)-based approach for attributed graph clustering is proposed [71] to extract and integrate significant varied information from structure and attribute sources. Consequently, researchers are constantly developing new clustering algorithms and techniques to tackle complex data problems and improve clustering effectiveness on real-world datasets.

1.2.2 Fuzzy similarity measures robustness

Robustness of fuzzy similarity measures is defined as their ability to perform consistently and accurately across multiple datasets and conditions, even in the presence of noise or uncertainty. Fuzzy similarity measures are applied in fuzzy logic and FS theory to quantify the similarity of objects or sets with ambiguous boundaries. Robustness is an essential evaluation criterion for fuzzy reasoning approaches. It is a crucial characteristic for any similarity measure because the real world usually includes noise, incompleteness, or uncertainties. In the fuzzy literature, there are numerous methods to assess the robustness of fuzzy reasoning. Cai introduced [72] δ-equalities for mathematical operators, fuzzy relations, fuzzy modus ponens (FMP), and fuzzy modus tollens (FMT). The robustness problem of fuzzy reasoning in term of δ-equalities under s-norm and fuzzy is defined as:

Let R be the universal set and A' and B' be two FSs on R and $\mu_{A'}(r)$ and $\mu_{B'}(s)$ be their membership degree, respectively, then

$$\delta = \left(\inf_{r,s} \max\left(\mu_{A'}(r),\left(\mu_{B'}(s)\right)\right) * \min\left(\delta_1,\delta_2\right)\right)$$

Zang and Cai [73] gave the idea of optimal fuzzy reasoning by using a new optimization objective function and evaluated its robustness using Monte Carlo simulation in probabilistic or statistical contexts. Inspired by the ensemble method, authors introduced the ensemble loss function and showed that this function is Bayes consistent and robust and improves the performance of a simple regression in comparison with state-of-the-art methods.

Robust loss function $= \sum_{i=1}^{n} \theta\left(x_i - y_i \cdot w\right)$, where $y_i \cdot w$ denote the inner product of two vectors y and$w$. Furthermore, Hu et al. [74] discussed the sensitivity of the conventional fuzzy rough set model to noise, as well as how noisy samples influence fuzzy rough computation. After that, a new image retrieval scheme employed IFSs for color images and fuzzy concepts for similarity, as shown in large-scale experiments, for robustness and effectiveness. Next, Nik-Khorasani et al. [75] introduced robust loss functions to boost the hybrid learning effectiveness of adaptive neuro-fuzzy inference systems against noise and outliers. Robustness in fuzzy similarity measures requires balancing simplicity and expressiveness, considering computational efficiency and meaningful data similarities. Empirical validation and thorough sensitivity analysis are crucial for improving these measures.

1.2.3 Practical applications in data imperfections

Most machine learning applications must draw accurate conclusions from available data, even if some data are mistaken or important data are missing. Human learners can easily compensate for a lack of information. Even animals can learn through training in which they acquire inconsistent benefits. However, incorporating such credibility into machine learning systems appears to be difficult. Real-world datasets often include data imperfections such as missing values, outliers, noise, and inconsistencies. Despite their difficulties, there are several practical applications in which dealing with data imperfections is essential [76]:

- Electronic Health Records (EHRs) often contain missing or incomplete data, requiring techniques like imputation and data fusion to handle missing values and integrate data from multiple sources. Detecting outliers in medical data is crucial for diagnosis and treatment.
- Financial transaction datasets often include fraudulent activities, requiring robust detection algorithms to identify these outliers while minimizing false positives accurately. Such variations may also indicate accounting errors or inconsistencies that need further investigation.
- Customer databases often contain missing or incomplete information due to errors or privacy concerns. Techniques like data imputation

and probabilistic record linkage help businesses identify unusual buying patterns and fraud.
- Data cleansing techniques are used to detect and correct outliers or missing values in environmental sensor data, ensuring the accuracy of environmental monitoring systems.

A new outlier detection method proposed for data with imperfect labels incorporates limited abnormal examples into learning by assigning likelihood values to each input data. Ghorbel et al. [77] suggested a typology of imperfection for data specified by Alzheimer's patients or caregivers in these applications (mainly due to disease-induced memory discordance). Using Bayesian inference and the principle of likelihood [78], the article suggests a likelihood analysis method for making use of imperfect data from multiple sources for inference and decision-making. Next, to improve decision-making by lowering uncertainty and strengthening data imputation, the authors [79] suggested two techniques for mining partial data utilizing interval-valued FS theory and rough set theory, particularly for posture detection in fall detection systems. Despite increased missing data, these techniques retain stable performance. Handling data imperfections is important for providing dependable, effective solutions in various applications. Statistical analysis, machine learning, and domain-specific knowledge are all useful techniques for preprocessing and analyzing imperfect data. Furthermore, the current study continues to address the challenges presented by data imperfections in various domains, providing valuable insights and approaches for researchers and practitioners seeking to create effective data-driven solutions.

1.3 PARTIAL MEMBERSHIPS AND MULTIPLE CLUSTER ASSIGNMENTS

Partial memberships and multiple cluster assignments are scenarios in which data components can belong to multiple clusters or have partial membership in one cluster. This concept is especially important in clustering algorithms that allow for soft and fuzzy clustering, in which data components may belong to multiple clusters with varying degrees [80].

In the case of partial memberships: Crisp clustering algorithms like K-means assign data components to one cluster, while fuzzy clustering techniques like FCM assign partial membership to clusters, reflecting any uncertainty or ambiguity in their categorization, rather than making hard assignments.

In the case of multiple cluster tasks: Clustering algorithms like Gaussian Mixture Models (GMM) allow data components to belong to multiple clusters simultaneously, based on their characteristics or proximity to the border between clusters. Each cluster contributes probabilistically to the overall assignment. There are many applications of partial membership and multiple clusters such as image segmentation, market segmentation, and genetic analysis.

1.3.1 Fuzziness and partial memberships

FS theory is useful in the study of control, pattern recognition, and medical diagnosis, but classical probability theory is affected by uncertainties in natural language and machine learning, necessitating alternative approaches [80]. Fuzziness and partial memberships in fuzzy clustering techniques refer to the level of uncertainty in assigning data points to clusters, unlike traditional hard clustering methods.

- Fuzziness in clustering suggests data points may not belong to a single cluster but align with multiple clusters to varying degrees, with membership values ranging from 0 to 1, indicating strong or weak association.
- Partial memberships in fuzzy clustering method like FCM permit data points to have fractional memberships to multiple clusters simultaneously, extending the concept of fuzziness and allowing for more uncertain categorization.

Mathematically, partial membership is expressed by membership values where each element r_i is an associated membership vector $\mu_i = \mu_{i1}, \mu_{i2}, \ldots, \mu_{in}$ representing degree of belongingness to each of the n clusters. Here, each membership value μ_{ij} falls within the interval [0, 1] and the sum of all membership values for each element r_i is equal to one, i.e. $\sum_{j=1}^{n} \mu_{ij} = 1$.

The fuzzy membership function is an important component of FS theory, but its measurement is still discussed. Jixiang Deng and Yong Deng [81] suggested a new information volume for fuzzy membership functions, including first-order and higher-order information volumes, which can be calculated by continuously separating hesitancy degrees until convergence. Bose et al. [82] introduced entropy-based soft clustering for brain MRI, improving performance and reducing segmentation errors, and suggested a new fuzzy membership function with Shannon's entropy-based variation. Therefore, the concepts of fuzziness and partial membership have applications in a variety of fields where data relationships are complex and multifaceted.

1.3.2 Capturing data points' degrees of belonging

Degrees of belonging is a technique used in clustering algorithms to quantify the degree to which a data point belongs to a specific category or cluster, often based on similarity. Common approach included:

Membership functions, particularly in fuzzy logic-based approaches, assign membership values to data points to indicate their degree of belonging to a specific category or cluster. Distance-based clustering algorithms like DBSCAN or OPTICS indirectly assign degrees of belonging to data points through measures like reachability distance or density, determining data points' strength in dense regions.

GMM use probability density functions to capture the probability of a data point belonging to a specific cluster.

Take the degrees of belonging of data points; we can represent each data point r_i as a tuple $(r_i, \mu_A(r_i))$, where $\mu_A(r_i)$ denotes the membership degree of the component r_i to the set$A$. The degree of membership in a challenge can be determined by factors such as distance from cluster centroids, similarity to prototype vectors, and other pertinent metrics. Here we discuss this with the help of fuzzy clustering, i.e. FCM algorithm [11].

The objective function is

$$\sum_{i=1}^{n} \sum_{j=1}^{c} \mu_{ij}^{m} \left\| r_i - v_j \right\|^2 \tag{1.19}$$

where

μ_{ij} is the degree of membership of data element r_i to the cluster j.
m is a weighting exponent typically type 2, but it can vary.
v_j is the centroid of the cluster$j$.
$\left\| r_i - v_j \right\|^2$ represent Euclidean distance between data element r_i and centroid v_j.

The degrees of membership μ_{ij} are repeatedly updated until convergence, so that each data point has a degree of membership for each cluster, showing its degree of belonging to that cluster. Therefore, capturing degrees of belonging in data allows for more nuanced analysis and decision-making in applications like pattern recognition, classification, and segmentation, as data points may not neatly belong to a single category.

1.4 SELECTING THE RIGHT FUZZY SIMILARITY MEASURE

Fuzzy similarity measures build on traditional similarity metrics by including the concept of fuzziness or inconsistency. Unlike classical similarity measures, which are binary (i.e. objects are either identical or not), fuzzy similarity measures accept partial correlations with and can quantify the degree of similarity between objects. This is especially useful in domains where experts' qualitative descriptions may contain imprecise or subjective data. The right fuzzy similarity measure is determined by a variety of factors, including the specific problem domain, the nature of the data, and the application's requirements. Here we give some suggestions for selecting the right one:

- Different similarity metrics are appropriate for various types of data. For example, cosine similarity may be useful for text data, whereas Jaccard or Dice similarity may be more appropriate for set data.
- Certain similarity measures may be more robust to noise and uncertainty in data than others. If you have noisy or uncertain data, select a measure that performs well in the presence of noise.
- Consider the computational cost of calculating the similarity measure, especially if working with large datasets. Choose a measure that balances accuracy and computational effectiveness.

The text provides a list of commonly used fuzzy similarity measures, along with references for further reading [83].

Fuzzy cosine similarity: This measure extends traditional cosine similarity [83] to FSs by taking into account the degree of membership of elements in sets

$$S_{\rho}'^{(A,B)} = \frac{\sum_{i=1}^{n} \mu_A(r_i) \cdot \mu_B(r_i)}{\sqrt{\sum_{i=1}^{n} \mu_A(r_i)^2} \sqrt{\sum_{i=1}^{n} \mu_B(r_i)^2}} \tag{1.20}$$

where A and B are two FSs, r_i is the element of the universal set R, $\mu_A(r_i)$, and$\mu_B(r_i)$ are member degrees of r_i in A and B.

Fuzzy hamming distance: It calculates the minimum number of substitutions required to convert one FS to another. This measure [83] is useful for comparing sets that use binary membership functions

$$H(A,B) = \frac{1}{n} \sum_{i=1}^{n} \delta\left(\mu_A(r_i), \mu_B(r_i)\right) \tag{1.21}$$

where $\delta(x,y) = \begin{cases} 1, \text{ if } x = y \\ 0, \text{ if } x \neq y \end{cases}$ represent the Kronecker delta function.

Rani et al. [84] presented a multi-criteria decision-making method based on a new similarity measure between PFSs. With the application of face recognition and software equality evaluation, Patel et al. [85] proposed a similarity measure in the environment of IFSs. Similarity measures between IFS A and B are

$$I(A,B)=\frac{1}{2n}\sum_{i=1}^{n}\left[\begin{array}{c}\dfrac{2-\Delta\varepsilon_i-\Delta\varepsilon_{\max}}{2+\Delta\varepsilon_{\min}-\Delta\varepsilon_{\max}}(1-\Delta\varepsilon_i)\\+\dfrac{2-\Delta\eta_i-\Delta\eta_{\max}}{2+\Delta\eta_{\min}-\Delta\eta_{\max}}(1-\Delta\eta_i)\end{array}\right] \tag{1.22}$$

There are several similarity measures available and researchers have extended these measures to find better results when dealing with decision-making problems. Selecting a similarity measure for quantifying qualitative descriptions requires expertise, data understanding, and analysis requirements. This method aids in information retrieval, recommendation systems, and decision-making processes.

1.4.1 Clustering objectives and measure suitability

Clustering is a fundamental unsupervised learning process that divides a dataset into related groups. Its objectives vary depending on the context and application, and its suitability is assessed using a variety of criteria. Common objectives of the clustering are provided, as well as references to relevant literature [2].

Finding hidden patterns: Clustering can assist in locating hidden structures or patterns in data that may not be immediately known. Combining similar data points reveals natural clusters or segments within the dataset.

Anomaly identification: Another goal of clustering is to find outliers or irregularities in the dataset. Anomalies frequently occur in clusters with few data points or in clusters that differ significantly from the majority.

Feature instruction: Clustering can be used as a preliminary step in feature learning or extraction. By identifying clusters, it may become clear which features are most important for distinguishing between various groups of data points.

Data reduction: Clustering is a technique that reduces data complexity by summarizing data with cluster centroids or prototypes, while retaining crucial information.

The suitability of clustering algorithms is assessed using a variety of criteria, taking into account things like computational complexity, scalability, internal cohesion, and robustness to noise. The interpretability of clustering results is critical for sound decision-making. Recent research highlights the importance of evaluating algorithms to finding the optimum number of clusters, allowing for informed algorithm selection and application in specific tasks and domains.

1.4.2 A decision framework for fuzzy similarity measure selection

In numerous real-world scenarios, especially those involving uncertain or imprecise data, the concept of similarity is important. However, traditional similarity measures may fail to capture the variations of fuzzy or uncertain data. In such cases, using fuzzy similarity measures becomes vital to accurately modeling and analyzing complex systems. Selecting an appropriate fuzzy similarity measure is a difficult task that must take into account a number of factors, including the nature of the data, the desired properties of the measure, and computational analysis constraints. A decision framework can help with this process by providing a systematic approach for assessing and choosing the best fuzzy similarity measure for a specific problem domain [86].

Identifying the problem: Identify the problem's circumstances and objectives, as well as identify the type of fuzzy data present, such as linguistic or interval data.

Requirement evaluation: Determine the desired properties of the similarity measures such as reflexivity, symmetry, transitivity, and also examine the computational complexity considerations.

Selection criteria: Develop criteria for evaluating similarity measures, such as sensitivity to uncertainty, robustness to noise, and interpretability.

The applicant similarity measures: Determine potential similarity measures appropriate for fuzzy data (e.g. Jaccard index, cosine similarity, FS-based measures).

Mathematical formulation: The article explains the mathematical expression of each potential similarity measure, which incorporates FS theory, as well as the specifics of fuzzy membership functions and operations.

Analysis and comparison: Compare each similarity measure to the defined criteria and conducting experiments or simulations if necessary.

Selection decision: Choose the most suitable similarity measure based on the evaluation results.

Validation and sensitivity evaluation: Validate your chosen similarity measure with real-world data and conduct a sensitivity analysis to understand its behavior under different conditions.

At last, the selection process should be documented, along with the mathematical formulation and evaluation results of the chosen similarity measure.

The decision framework offers a structured approach to selecting fuzzy similarity measures, allowing participants to make more informed decisions, improve fuzzy modeling effectiveness, and advance computational intelligence, resulting in robust solutions in uncertain environments.

1.5 EMERGING TRENDS AND FUTURE CHALLENGES

The field of fuzzy clustering has advanced significantly in response to the growing demand for effective data analysis techniques, particularly in big data and the World Wide Web of Issues applications. To deal with uncertainty and ambiguity in data, researchers looked into various fuzzy clustering algorithms and validation measures. Recent studies provide insights into current research as well as potential future directions in this rapidly evolving field. The authors emphasized the importance of statistical analyses and comparative evaluations of various fuzzy similarity measures for clustering validation, as well as the development of standardized objectives and evaluation metrics. Yang et al. [87] present a comprehensive study on fuzzy clustering algorithms for big data analysis, highlighting various approaches and challenges associated with this method. The period of big data presents an enormous volume and high velocity of complex data generated by a variety of sources, including transactions, social networks, and daily life activities. Proper processing can lead to intelligent and relevant decision-making across a variety of industries. FSs have been used in big data processing because they can represent and quantify uncertainty. The authors [88] summarized fuzzy techniques that have emerged in big data problems, addressing challenges and providing potential for further development. A critical analysis of contributions and relevance to big data challenges has been presented. Future research directions may include creating adaptive and scalable fuzzy similarity measures, investigating novel validation metrics, and incorporating domain knowledge to improve the understanding of clustering results. Continued study and creativity in this field are essential for dealing with existing challenges and creating new opportunities for implementing fuzzy clustering techniques in real-world scenarios.

1.5.1 Current state of the field

Clustering is a crucial unsupervised learning method in data analysis that groups data objects into clusters, aiming to find the most similar and dissimilar data objects. However, data can be complex and multifaceted, leading to the need for multiple clustering solutions. Meta-clustering, an accepted technique, investigates multiple reasonable and separate clustering by examining a large number of base clustering. Nonetheless, low-quality and unnecessary base clustering can affect the generation of high-quality and diverse clustering views, and consequent clustering descriptions may not be entirely relevant. Additionally, checking all returned solutions can be time and energy-consuming for users. Through the years, one of the results of data clustering has been the classification of objects into distinct classes. Several authors, including Saxena et al. [8] and Baidari and Patil [89], have highlighted the challenge in identifying the optimal number of classes for multiple clustering problems. The researchers have described the matter as the subjectivity of clustering. According to Li et al. [90], subjectivity arises when determining the optimal partition or cluster. If a single clustering technique cannot solve all clustering problems, it's important to carefully choose clustering parameters that are appropriate for the user. Over the years, there has been a rise in curiosity about clustering algorithms and methods, with general reviews becoming more visible trends. Furthermore, researchers [91] have worked on reducing data dimensionality using the extraction of features, choosing methods, and subspace clustering. To identify and eliminate outliers from the data, Pereira and Frazzon [92] first normalized the data before applying a clustering algorithm. Ni et al. [93] noticed that deep neural networks (DNNs) perform better in speech and text modeling, as well as video and classification of pictures, due to less training creating and larger training datasets. Further, Jia et al. [94] developed similarity measures between Pythagorean FS on the basis of overlapping and grouping function. To overcome the limitation of FCM clustering algorithm including initial sensitivity, readily dropping into local minima, and influencing via noise, etc. an improved variant of FCM inside the context of an intuitionistic fuzzy environment and a metric to reduce the effect of noise were developed. The study of clustering using fuzzy information continues to grow with researchers addressing challenges and exploring new methods. The need for robust, scalable, and accessible methods grows by the complexity and volume of data. Efforts are underway to create algorithms with high-dimensional datasets to enhance interpretability. Interdisciplinary collaborations are also gaining traction, drawing on insights from optimizing, statistics, and computer vision. The future development of fuzzy clustering looks promising to tackle complex real-world problems.

1.5.2 Potential challenges and research direction

Clustering, an important part of data mining and machine learning, faces several potential challenges and also provides numerous opportunities for future research. Researchers carry out an in-depth examination of existing research on clustering algorithms as well as fuzzy similarity measures, with an emphasis on research articles, past progress, theoretical foundations, and practical implications. This helps in the identification of shortcomings, limitations, and future developments, as well as understanding researchers with standard datasets, metrics for evaluation, and methodologies for comparing various algorithms and similarity measurements. Streams and clustering are more closely related considering incoming data are unlabeled.

Researchers have concentrated on stream analysis because streams are generated by a wide range of stream applications, which include cell phones, GPS, automobiles, security sensors, plasma screens, social media, and more [95]. In many years of research, learning from imbalanced data has remained a major focus. The area of imbalanced learning faces numerous obstacles that necessitate investigation and intensive development, with many untested paths yet to be explored, making it unique and exciting in machine learning [96]. This can be explored for low-power designs [99–109].

Research is needed to create clustering algorithms that can manage diverse kinds of data, adjust to dynamic and evolving data, handle high-dimensional information, scalable large datasets, manage noise and outliers, handle privacy and security issues, and handle domain-specific difficulties in applications such as financial services, healthcare, and social networks. These algorithms will enhance the quality of outcomes from clustering and address a variety of issues.

1.6 CONCLUSION

The implementation of fuzzy similarity measures for clustering validation creates an effective structure for evaluating the quality and effectiveness of clustering algorithms. Through comprehensive analysis, these measures provide valuable insights into cluster separation, stability, and membership, allowing for a better understanding of the data's underlying structure. Furthermore, they make comparative evaluations easier, allowing for the selection of appropriate metrics adapted to specific datasets and clustering objectives. Furthermore, the interpretability and visualization abilities of fuzzy similarity measures improve everyone's understanding of clustering results, promoting informed decision-making. This study first addressed the importance and challenges of clustering validation, emphasizing the use of fuzzy similarity measures. It then examined both traditional and fuzzy similarity metrics, highlighting their limitations and practical applications in cluster analysis using numerical examples. It also investigated the complexities of real world data and the reliability of fuzzy similarity measures. Furthermore, it covered topics such as partial membership, multi-clustering, choosing appropriate similarity metrics, and defining clustering objectives. Finally, the current state of the field was discussed, along with possibilities for future research. While acknowledging the benefits, it is essential to recognize limitations and areas for improvement, providing a way for future studies to focus on improving existing measures and developing novel approaches to address the complexities of real-world datasets. Advanced measures are being developed to deal with high-dimensional datasets, leveraging machine learning and data mining techniques to improve clustering validation accuracy and robustness. The scalability of fuzzy similarity measures is critical, necessitating additional research into algorithms for large-scale dataset handling, particularly in the healthcare and finance domains, to identify new applications and challenges. Standardized comparative analysis frameworks and evaluation procedures enable equal evaluations of measures and algorithms, thereby promoting progress. Integrating fuzzy logic with deep learning techniques can improve clustering performance and interpretability. Therefore, the use of fuzzy similarity measures in clustering validation helps to advance the field of data clustering and its practical applications.

REFERENCES

1. Xu, Rui, and Donald Wunsch, "Survey of clustering algorithms." *IEEE Transactions on Neural Networks*, vol. 16, no. 3, pp. 645–678, 2005.
2. Jain, Anil K., M. Narasimha Murty, and Patrick J. Flynn, "Data clustering: A review." *ACM Computing Surveys (CSUR)*, vol. 31, no. 3, pp. 264–323, 1999.
3. Guha, Sudipto, Rajeev Rastogi, and Kyuseok Shim, "CURE: An efficient clustering algorithm for large databases." *ACM Sigmod Record*, vol. 27, no. 2, pp. 73–84, 1998.
4. Carpineto, Claudio, and Giovanni Romano, "A lattice conceptual clustering system and its application to browsing retrieval." *Machine Learning*, vol. 24, pp. 95–122, 1996.
5. Halkidi, Maria, Yannis Batistakis, and Michalis Vazirgiannis, "On clustering validation techniques." *Journal of Intelligent Information Systems*, vol. 17, pp. 107–145, 2001.
6. Liao, T. Warren, "Clustering of time series data-a survey." *Pattern Recognition*, vol. 38, no. 11, pp. 1857–1874, 2005.
7. Xu, Rui, and Donald C. Wunsch, "Clustering algorithms in biomedical research: A review." *IEEE Reviews in Biomedical Engineering*, vol. 3, pp. 120–154, 2010.
8. Saxena, Amit, Mukesh Prasad, Akshansh Gupta, Neha Bharill, Om Prakash Patel, Aruna Tiwari, Meng Joo Er, Weiping Ding, and Chin-Teng Lin, "A review of clustering techniques and development." *Neurocomputing*, vol. 267, pp. 664–681, 2017.
9. Fraley, Chris, and Adrian E. Raftery, "How many clusters? Which clustering method? Answers via model-based cluster analysis." *The Computer Journal*, vol. 41, no. 8, pp. 578–588, 1998.
10. Oyewole, Gbeminiyi John, and George Alex Thopil, "Data clustering: Application and trends." *Artificial Intelligence Review*, vol. 56, no. 7, pp. 6439–6475, 2023.
11. Bezdek, James C., Robert Ehrlich, and William Full, "FCM: The fuzzy c-means clustering algorithm." *Computers & Geosciences*, vol. 10, no. 2–3, pp. 191–203, 1984.
12. Kaplan, Jonathan Michael, and Rasmus Gronfeldt Winthe, "Prisoners of abstraction? The theory and measure of genetic variation, and the very concept of "race"." *Biological Theory*, vol. 7, pp. 401–412, 2013.
13. Motiwalla, Luvai F., Mousa Albashrawi, and Hasan B. Kartal, "Uncovering unobserved heterogeneity bias: Measuring mobile banking system success." *International Journal of Information Management*, vol. 49, pp. 439–451, 2019.
14. Baker, Ryan Shaun, "Data mining for education." *International Encyclopedia of Education*, vol. 7, no. 3, pp. 112–118, 2010.

15. Huth, Radan, Christoph Beck, Andreas Philipp, Matthias Demuzere, Zbigniew Ustrnul, Monika Cahynova, Jan Kysely, and Ole Einar Tveito, "Classifications of atmospheric circulation patterns: Recent advances and applications." *Annals of the New York Academy of Sciences*, vol. 1146, no. 1, pp. 105–152, 2008.
16. Bewley, Alex, Rajiv Shekhar, Sam Leonard, Ben Upcroft, and Paul Lever, "Real-time volume estimation of a dragline payload." In *2011 IEEE International Conference on Robotics and Automation*, IEEE, Shanghai, 2011.
17. Maimon, Oded, and Lior Rokach, eds. *Data Mining and Knowledge Discovery Handbook*. Springer, New York, 2005.
18. Wayment-Steele, Hannah K., Adedolapo Ojoawo, Renee Otten, Julia M. Apitz, Warintra Pitsawong, Marc Hömberger, Sergey Ovchinnikov, Lucy Colwell, and Dorothee Kern., "Predicting multiple conformations via sequence clustering and AlphaFold2." *Nature*, vol. 625, no. 7996, pp. 832–839, 2024.
19. Hu, Lun, Yue Yang, Zehai Tang, Yizhou He, and Xin Luo, "FCAN-MOPSO: An improved fuzzy-based graph clustering algorithm for complex networks with multi-objective particle swarm optimization." *IEEE Transactions on Fuzzy Systems*, vol. 31, no. 10, pp. 3470–3484, 2023.
20. Wu, Junjie, Hui Xiong, and Jian Chen, "Adapting the right measures for k-means clustering." In *Proceedings of the 15th ACM SIGKDD International Conference on Knowledge Discovery and Data Mining*, Paris, 2009.
21. Oyewole, Gbeminiyi John, and George Alex Thopil, "Data clustering: Application and trends." *Artificial Intelligence Review*, vol. 56, no. 7, pp. 6439–6475, 2023.
22. Liu, Yanchi, Zhongmou Li, Hui Xiong, Xuedong Gao, and Junjie Wu, "Understanding of internal clustering validation measures." *2010 IEEE International Conference on Data Mining*, IEEE, Sydney, 2010.
23. Ezugwu, Absalom E., Abiodun M. Ikotun, Olaide O. Oyelade, Laith Abualigah, Jeffery O. Agushaka, Christopher I. Eke, and Andronicus A. Akinyelu, "A comprehensive survey of clustering algorithms: State-of-the-art machine learning applications, taxonomy, challenges, and future research prospects." *Engineering Applications of Artificial Intelligence*, vol. 110, p. 104743, 2022.
24. Wang, Zhangyang, Shiyu Chang, Jiayu Zhou, Meng Wang, and Thomas S. Huang, "Learning a task-specific deep architecture for clustering." In *Proceedings of the 2016 SIAM International Conference on Data Mining. Society for Industrial and Applied Mathematics*, Miami, FL, 2016.
25. Kokate, Umesh, Arvind Deshpande, Parikshit Mahalle, and Pramod Patil, "Data stream clustering techniques, applications, and models: Comparative analysis and discussion." *Big Data and Cognitive Computing*, vol. 2, no. 4, p. 32, 2018.
26. Askari, Salar, "Fuzzy C-Means clustering algorithm for data with unequal cluster sizes and contaminated with noise and outliers: Review and development." *Expert Systems with Applications*, vol. 165, p. 113856, 2021.
27. Yu, Bin, Zijian Zheng, Mingjie Cai, Witold Pedrycz, and Weiping Ding, "FRCM: A fuzzy rough c-means clustering method." *Fuzzy Sets and Systems*, vol. 480, p. 108860, 2024.

28. Zadeh, Lotfi A, "Fuzzy sets." *Information and Control*, vol. 8, no. 3, pp. 338–353, 1965.
29. Atanassov, Krassimir T., and Krassimir T. Atanassov, *Intuitionistic Fuzzy Sets*. Physica, Heidelberg, pp. 1–137, 1999.
30. Cuong Bui Cong, "Picture fuzzy sets." *Journal of Computer Science and Cybernetics*, vol. 30, no. 4, pp. 409–420, 2014.
31. Wang, Pei Zhuang, *Theory of Fuzzy Sets and Their Applications*. Shanghai Science and Technology Publishing House, Shanghai, 2014.
32. Hong, Dug Hun, and Seok Yoon Hwang, "A note on the value similarity of fuzzy systems variables." *Fuzzy Sets and Systems*, vol. 66, no. 3, pp. 383–386, 1994.
33. Liang, Zhizhen, and Pengfei Shi, "Similarity measures on intuitionistic fuzzy sets." *Pattern Recognition Letters*, vol. 24, no. 15, pp. 2687–2693, 2003.
34. Peng, Xindong, and Harish Garg, "Multiparametric similarity measures on Pythagorean fuzzy sets with applications to pattern recognition." *Applied Intelligence*, vol. 49, pp. 1–39, 2019.
35. Majumdar, Pinaki, and Syamal Kumar Samanta, "Similarity measure of soft sets." *New Mathematics and Natural Computation*, vol. 4, no. 1, pp. 1–12, 2008.
36. Majumdar, Pinaki, and S. K. Samanta, "On similarity measures of fuzzy soft sets." *International Journal of Advance Soft Computing and Applications*, vol. 3, no. 2, pp. 1–8, 2011.
37. Saracoglu, Rıdvan, Kemal Tutuncu, and Novruz Allahverdi, "A fuzzy clustering approach for finding similar documents using a novel similarity measure." *Expert Systems with Applications*, vol. 33, no. 3, pp. 600–605, 2007.
38. Zeng, Wenyi, Rong Ma, Qian Yin, and Zeshui Xu, "Similarity measure of hesitant fuzzy sets based on implication function and clustering analysis." *IEEE Access*, vol. 8, pp. 119995–120008, 2020.
39. Rezaei, Kamran, and Hassan Rezaei, "New distance and similarity measures for hesitant fuzzy sets and their application in hierarchical clustering." *Journal of Intelligent & Fuzzy Systems*, vol. 39, no. 3, pp. 4349–4360, 2020.
40. Singh, Surender, and Abdul Haseeb Ganie, "Applications of picture fuzzy similarity measures in pattern recognition, clustering, and MADM." *Expert Systems with Applications*, vol. 168, p. 114264, 2021.
41. Al-Sharqi, Faisal, Ashraf Al-Quran, Hamiden Abd El-Wahed Khalifa, Haifa Alqahtani, Badria A. Ali Yousif, A. Rawan, and Mona Aladil, "Orthogonal distance and similarity for single-valued neutrosophic fuzzy soft expert environment and its application in decision-making." *Full Length Article*, vol. 23, no. 4, p. 08-22, 2024.
42. Shirkhorshidi, Ali Seyed, Saeed Aghabozorgi, and Teh Ying Wah, "A comparison study on similarity and dissimilarity measures in clustering continuous data." *PLoS One*, vol. 10, no. 12, p. e0144059, 2015.
43. Chen, Shyi-Ming, "Similarity measures between vague sets and between elements." *IEEE Transactions on Systems, Man, and Cybernetics, Part B (Cybernetics)*, vol. 27, no. 1, pp. 153–158, 1997.
44. Li, Yanhong, David L. Olson, and Zheng Qin, "Similarity measures between intuitionistic fuzzy (vague) sets: A comparative analysis." *Pattern Recognition Letters*, vol. 28 no. 2, pp. 278–285, 2007.

45. Ye, Jun, "Cosine similarity measures for intuitionistic fuzzy sets and their applications." *Mathematical and Computer Modelling*, vol. 53, no. 1–2, pp. 91–97, 2011.
46. Muthukumar, Pachaiyappan, and G. Sai Sundara Krishnan, "A similarity measure of intuitionistic fuzzy soft sets and its application in medical diagnosis." *Applied Soft Computing*, vol. 41, pp. 148–156, 2016.
47. Zhang, Xiaolu, "A novel approach based on similarity measure for Pythagorean fuzzy multiple criteria group decision making." *International Journal of Intelligent Systems*, vol. 31, no. 6, pp. 593–611, 2016.
48. Sarker, Bhaba R., and Khan M. Saiful Islam, "Relative performances of similarity and dissimilarity measures." *Computers & Industrial Engineering*, vol. 37, no. 4, pp. 769–807, 1999.
49. McAuley, John, "Machine grouping for efficient production." *Production Engineer*, vol. 51, no. 2, p. 53, 1972.
50. Mosier, Charles, and Larry Taube, "Weighted similarity measure heuristics for the group technology machine clustering problem." *Omega*, vol. 13, no. 6, pp. 577–579, 1985.
51. Al-Sharqi, Faisal, Yousef Al-Qudah, and Naif Alotaibi, "Decision-making techniques based on similarity measures of possibility neutrosophic soft expert sets." *Neutrosophic Sets and Systems*, vol. 55, no. 1, p. 22, 2023.
52. Sassi, Minyar, Amel Grissa Touzi, and Habib Ounelli, "Clustering quality evaluation based on fuzzy FCA." In *Database and Expert Systems Applications: 18th International Conference, DEXA 2007*, Regensburg, Germany, September 3–7, 2007. Proceedings 18. Springer, Berlin, Heidelberg, 2007.
53. Krishnapuram, Raghuram, and James M. Keller, "The possibilistic c-means algorithm: Insights and recommendations." *IEEE Transactions on Fuzzy Systems*, vol. 4, no. 3, pp. 385–393, 1996.
54. Bouveyron, Charles, Stéphane Girard, and Cordelia Schmid, "High-dimensional data clustering." *Computational Statistics & Data Analysis*, vol. 52, no. 1, pp. 502–519, 2007.
55. Lloyd, Stuart, "Least squares quantization in PCM." *IEEE Transactions on Information Theory*, vol. 28, no. 2, pp. 129–137, 1982.
56. Jain, Anil K., "Data clustering: 50 years beyond K-means." *Pattern Recognition Letters*, vol. 31, no. 8, pp. 651–666, 2010.
57. Santini, Simone, and Ramesh Jain, "Similarity measures." *IEEE Transactions on Pattern Analysis and Machine Intelligence*, vol. 21, no. 9, pp. 871–883, 1999.
58. Chen, Shi-Jay, and Shyi-Ming Chen, "Fuzzy risk analysis based on similarity measures of generalized fuzzy numbers." *IEEE Transactions on Fuzzy Systems*, vol. 11, no. 1, pp. 45–56, 2003.
59. Shishavan, Seyed Amin Seyfi, Fatma Kutlu Gundogdu, Elmira Farrokhizadeh, Yaser Donyatalab, and Cengiz Kahraman, "Novel similarity measures in spherical fuzzy environment and their applications." *Engineering Applications of Artificial Intelligence*, vol. 94, p. 103837, 2020.
60. Khan, Mohd Shoaib, and QM Danish Lohani, "A similarity measure for Atanassov intuitionistic fuzzy sets and its application to clustering." In *2016 International Workshop on Computational Intelligence (IWCI)*. IEEE, Dhaka, 2016.

61. Ullah, Kifayat, Harish Garg, Tahir Mahmood, Naeem Jan, and Zeeshan Ali, "Correlation coefficients for T-spherical fuzzy sets and their applications in clustering and multi-attribute decision making." *Soft Computing*, vol. 24, pp. 1647–1659, 2020.
62. Sarfraz, Mehwish, and Dragan Pamucar, "A parametric similarity measure for spherical fuzzy sets and its applications in medical equipment selection." *Journal of Engineering Management Systems Engineering*, vol. 3, no. 1, pp. 38–52, 2024.
63. Rani, Pratibha, Shyi-Ming Chen, and Arunodaya Raj Mishra, "Multi-attribute decision-making based on similarity measure between picture fuzzy sets and the MARCOS method." *Information Sciences*, vol. 658, p. 119990, 2024.
64. Patel, Anjali, Subhankar Jana, and Juthika Mahanta, "Construction of similarity measure for intuitionistic fuzzy sets and its application in face recognition and software quality evaluation." *Expert Systems with Applications*, vol. 237, p. 121491, 2024.
65. Xu, Zeshui, Jian Chen, and Junjie Wu, "Clustering algorithm for intuitionistic fuzzy sets." *Information Sciences*, vol. 178, no. 19, pp. 3775–3790, 2008.
66. Hüllermeier, Eyke, "Learning from imprecise and fuzzy observations: Data disambiguation through generalized loss minimization." *International Journal of Approximate Reasoning* vol. 55, no. 7, pp. 1519–1534, 2014.
67. Ferraro, Maria Brigida, Renato Coppi, G. Gonzalez Rodriguez, and Ana Colubi, "A linear regression model for imprecise response." *International Journal of Approximate Reasoning*, vol. 51, no. 7, pp. 759–770, 2010.
68. Han, Jiawei, Micheline Kamber, and Data Mining, "Concepts and techniques." *Morgan Kaufmann*, vol. 340, pp. 94104–3205, 2006.
69. Wiroonsri, Nathakhun, "Clustering performance analysis using a new correlation-based cluster validity index." *Pattern Recognition*, vol. 145, p. 109910, 2024.
70. Hong, Ming-Yi, Yi-Hsiang Huang, You-Chen Teng, Chih-Yu Wang, and Che Lin, "SynHIN: generating synthetic heterogeneous information network for explainable AI." *arXiv preprint arXiv*, vol. 2401, p. 04133, 2024.
71. Jannesari, Vahid, Maryam Keshvari, and Kamal Berahmand, "A novel nonnegative matrix factorization-based model for attributed graph clustering by incorporating complementary information." *Expert Systems with Applications*, vol. 242, p. 122799, 2024.
72. Cai, Kai-Yuan, "Robustness of fuzzy reasoning and/spl delta/-equalities of fuzzy sets." *IEEE Transactions on Fuzzy Systems*, vol. 9, no. 5, pp. 738–750, 2001.
73. Zhang, Lei, and Kai-Yuan Cai, "Optimal fuzzy reasoning and its robustness analysis." *International Journal of Intelligent Systems*, vol. 19, no. 11, pp. 1033–1049, 2004.
74. Hu, Qinghua, Lei Zhang, Shuang An, David Zhang, and Daren Yu, "On robust fuzzy rough set models." *IEEE Transactions on Fuzzy Systems*, vol. 20, no. 4, pp. 636–651, 2011.
75. Nik-Khorasani, Ali, Hadi Sadoghi-Yazdi, and Ali Mehrizi, "Robust hybrid learning approach for adaptive neuro-fuzzy inference systems." *Fuzzy Sets and Systems*, vol. 481, p. 108890, 2024.
76. Parsons, Simon, "Current approaches to handling imperfect information in data and knowledge bases." *IEEE Transactions on Knowledge and Data Engineering*, vol. 8, no. 3, pp. 353–372, 1996.

77. Ghorbel, Fatma, Fayçal Hamdi, Nassira Achich, and Elisabeth Metais, "Handling data imperfection—False data inputs in applications for Alzheimer's patients." *Data & Knowledge Engineering*, vol. 130, p. 101864, 2020.
78. Yang, Jian-Bo, Dong-Ling Xu, Xiaobin Xu, and Chao Fu, "Likelihood analysis of imperfect data." IEEE Transactions on Systems, Man, and Cybernetics: Systems, vol. 53, no. 8, pp. 5046–5057, 2023.
79. Mroczek, Teresa, Dorota Gil, and Barbara Pękala, "Fuzzy and rough approach to the problem of missing data in fall detection system." *Fuzzy Sets and Systems*, vol. 480, p. 108868, 2024.
80. Liu, Zhe, and Sukumar Letchmunan, "Enhanced fuzzy clustering for incomplete instance with evidence combination." *ACM Transactions on Knowledge Discovery from Data*, vol. 18, no. 3, pp. 1–20, 2024.
81. Deng, Jixiang, and Yong Deng, "Information volume of fuzzy membership function." *International Journal of Computers Communications & Control*, vol. 16, no. 1, pp. 1–14, 2021. DOI:10.15837/ijccc.2021.1.4106
82. Bose, Ananya, Ujjwal Maulik, and Anasua Sarkar, "An entropy-based membership approach on type-II fuzzy set (EMT2FCM) for biomedical image segmentation." *Engineering Applications of Artificial Intelligence*, vol. 127, p. 107267, 2024.
83. Landau, Gad M., Abraham Bookstein, Shmuel Tomi Klein, and Timo Raita, "Fuzzy hamming distance: A new dissimilarity measure." In *Combinatorial Pattern Matching: 12th Annual Symposium, CPM 2001* Jerusalem, Israel, July 1–4, 2001 *Proceedings 12*. Springer, Berlin, Heidelberg, 2001.
84. Rani, Pratibha, Shyi-Ming Chen, and Arunodaya Raj Mishra, "Multi-attribute decision-making based on similarity measure between picture fuzzy sets and the MARCOS method." *Information Sciences*, vol. 658, p. 119990, 2024.
85. Patel, Anjali, Subhankar Jana, and Juthika Mahanta, "Construction of similarity measure for intuitionistic fuzzy sets and its application in face recognition and software quality evaluation." *Expert Systems with Applications*, vol. 237, p. 121491, 2024.
86. Al-Sharqi, Faisal, Ashraf Al-Quran, and Mamika Ujianita Romdhini, "Decision-making techniques based on similarity measures of possibility interval fuzzy soft environment." *Iraqi Journal for Computer Science and Mathematics*, vol. 4, no. 4, pp. 18–29, 2023.
87. Yan, Hang, Nan Yang, Yi Peng, and Yitian Ren, "Data mining in the construction industry: Present status, opportunities, and future trends." *Automation in Construction*, vol. 119, p. 103331, 2020.
88. Wang, Hai, Zeshui Xu, and Witold Pedrycz, "An overview on the roles of fuzzy set techniques in big data processing: Trends, challenges and opportunities." *Knowledge-Based Systems*, vol. 118, pp. 15–30, 2017.
89. Baidari, Ishwar, and Channamma Patil, "A criterion for deciding the number of clusters in a dataset based on data depth." *Vietnam Journal of Computer Science*, vol. 7, no. 04, pp. 417–431, 2020.
90. Li, Qi, Shuliang Wang, Chuanfeng Zhao, Boxiang Zhao, Xin Yue, and Jing Geng, "HIBOG: Improving the clustering accuracy by ameliorating dataset with gravitation." *Information Sciences*, vol. 550, pp. 41–56, 2021.
91. Solorio-Fernandez, Saul, J. Ariel Carrasco-Ochoa, and José Fco Martínez-Trinidad, "A review of unsupervised feature selection methods." *Artificial Intelligence Review*, vol. 53, no. 2, pp. 907–948, 2020.

92. Pereira, Marina Meireles, and Enzo Morosini Frazzon, "A data-driven approach to adaptive synchronization of demand and supply in omni-channel retail supply chains." *International Journal of Information Management*, vol. 57, p. 102165, 2021.
93. Ni, Jinjie, Tom Young, Vlad Pandelea, Fuzhao Xue, and Erik Cambria, "Recent advances in deep learning based dialogue systems: A systematic survey." *Artificial Intelligence Review*, vol. 56, no. 4, pp. 3055–3155, 2023.
94. Jia, Zihang, Junsheng Qiao, and Minghao Chen, "On similarity measures between Pythagorean fuzzy sets derived from overlap and grouping functions." *International Journal of Fuzzy Systems*, vol. 25, no. 6, pp. 1–17, 2023.
95. Tiwari, Suman R., and Kaushik K. Rana, "Challenges and future research directions for stream clustering." In *2023 7th International Conference on Trends in Electronics and* Informatics *(ICOEI)*. IEEE, Tirunelveli, pp. 525–531, 2023.
96. Krawczyk, Bartosz, "Learning from imbalanced data: Open challenges and future directions." *Progress in Artificial Intelligence*, vol. 5, no. 4, pp. 221–232, 2016.
97. Sharma, Vijay Kumar, "CNTFET circuit-based wide fan-in domino logic for low power applications." *Journal of Circuits Systems Computers*, vol. 31, no. 2, p. 2250036, 2022.
98. Sharma, Vijay Kumar, and Pattanaik Manisha, "A reliable ground bounce noise reduction technique for nanoscale CMOS circuits." *International Journal of Electronics*, vol. 102, no. 11, pp. 1852–1866, 2015.
99. Sharma, Vijay Kumar, Patel Sumit, and Pattanaik Manisha, "High performance process variations aware technique for sub-threshold 8T-SRAM cell." *Wireless Personal Communications*, vol. 78, no. 1, pp. 57–68, 2014.
100. Sharma, Vijay Kumar, and Pattanaik Manisha, "Techniques for low leakage nanoscale VLSI circuits: A comparative study." *Journal of Circuits Systems Computers*, vol. 23, no. 5, p. 1450061, 2014.
101. Sharma, Vijay Kumar, "A survey of leakage reduction techniques in CMOS digital circuits for nanoscale regime." *Australian Journal of Electrical and Electronics Engineering*, vol. 18, no. 4, pp. 217–236, 2021.
102. Kumar, Chanchal, Avinash Sharan Mishra, and Vijay Kumar Sharma, "Leakage power reduction in CMOS logic circuits using stack ONOFIC technique." In *2018 Second International Conference on Intelligent Computing and Control Systems (ICICCS)*, Madurai, 2018.
103. Sharma, Vijay Kumar, and Pattanaik Manisha, "Design of low leakage variability aware ONOFIC CMOS standard cell library." *Journal of Circuits Systems and Computers*, vol. 25, no. 11, p. 1650134, 2016.
104. Naz, Syed F., Sadat Riyaz, and Vijay Kumar Sharma, "A review of QCA nanotechnology as an alternate to CMOS." *Current Nanoscience*, vol. 18, no. 1, pp. 18–30, 2022.
105. Riyaz, Sadat and Vijay Kumar Sharma, "Design of reversible Feynman and double Feynman gates in quantum-dot cellular automata nanotechnology." *Circuit World*, vol. 49, no. 1, pp. 28–37, 2023.
106. Raina, Bhoomika, Chetna Verma, Mehak Gupta, and Vijay Kumar Sharma, "Binary coded decimal (BCD) seven segment circuit designing using quantum-dot cellular automata (QCA)." In *2021 5th International Conference on Trends in Electronics and Informatics (ICOEI)*, Tirunelveli, 2021.

107. Mushtaq, Umayia, and Vijay Kumar Sharma, "Performance analysis for reliable nanoscaled FinFET logic circuits." *Analog Integrated Circuits Signal Processing*, vol. 107, no. 3, pp. 671–682, 2021.
108. Kajal, and Vijay Kumar Sharma, "Design and simulation of FinFET circuits at different technologies." In *2021 6th International Conference on Inventive Computation Technologies (ICICT)*, 2021.
109. Sharma, Vijay Kumar, "Designing of face recognition system." In *2019 International Conference on Intelligent Computing and Control Systems (ICCS)*, Madurai, 2019.

Chapter 2

Determination and classification of human blood contamination diseases and Rh factor using ensemble algorithm

Suthagar Subramaniyan, Mageshkumar Gunasekar, Tamilselvan Kumaravel Subramaniam, Murugesan Govindasamy, and Lalithraj Subramaniyan

2.1 INTRODUCTION

Blood is a major requirement for human survival and is made up of white blood cells (WBCs), red blood cells (RBCs), plasma, and platelets. This may additionally be divided into: platelets and cells make up approximately 45% of human blood [1], and another 55% are packed with plasma. Those additives and their bodily residences, which include size, shape, shade, and matter in all blood modifications, because of the intrusion of any foreign substance or microorganism, can lead to any sort of sickness. There are extraordinary forms of illness. In lots of instances, small pixels play a critical role in detecting and predicting abnormalities and the emergence of illnesses in the frame. Typically, a person's fitness is judged by analyzing the various functions of blood cells and counting them [2].

2.1.1 Blood components and hematological parameters

Blood is a body fluid that provides the body with vitamins, oxygen, and waste merchandise. The common adult has five liters, or six quarts, of blood in their frame. Red blood cells (RBCs), white blood cells (WBCs), platelets, and blood plasma all carry out vital bodily processes. In addition to preventing excessive blood loss through blood clotting, these elements aid in the transportation of oxygen and nutrients to tissues and the lungs. Platelets aid in the creation of clots, whereas red blood cells transport oxygen. By generating antibodies and battling infections, white blood cells aid in the body's defense. Along with regulating body temperature to maintain stable internal conditions, blood also collaborates closely with organs such as the liver and kidneys to filter waste and purify the body. Furthermore, the rule of contamination highlights how crucial it is to keep areas clean in order to prevent infection, particularly in healthcare settings. Blood shifting within

DOI: 10.1201/9781003643609-2

the arteries, blood vessels, and capillaries, called complete blood, carries 50%–55% of the plasma issue [3], which is a part of blood plasma, a mixture of water, protein, sugar, and salt.

2.1.2 Red blood cells

RBCs, or crimson blood cells, are very massive cells that bring oxygen to the tissues and make up 40%–45% of the blood volume [4,5]. Pink blood cells (RBCs) make up most blood cells (99%) on a microscopic scale. The crimson blood cell is in the form of a hermaphrodite with a flat center. RBC interruptions are adjustments that have an effect on RBCs. There is a ramification of RBC problems, which include anemia, pink cell enzyme deficiency, hemolytic anemia, purple cellular membrane disorder, hemoglobinopathy, heme manufacturing sickness, nutritional anemia, polycythemia, and hemochromatosis.

2.1.3 White blood cells

WBCs, or leukocytes, are part of the immune system that fights infections. WBCs form in the marrow and are found within the blood and lymphatic tissues. Unique forms of WBCs are granulocytes (neutrophils, eosinophils, and basophils), monocytes, lymphocytes (T and B cells), and macrophages. Platelets, or thrombocytes, are unusual mobile structures that help the blood clot. Hemoglobin is a protein that includes oxygen in RBCs, which efficiently includes oxygen from the lungs to frame tissues and enables transport of carbon dioxide and hydrogen ions returned to the lungs. The shade of hemoglobin gives it a reddish brown coloration [6]. A boom or decrease within the wide variety of blood cells suggests disorders that include leukopenia, leukocytosis, neutropenia, neutrophil cytosis, lymphocytopenia, and lymphocytic leukocytosis. Interruption of neutrophils and lymphocytes is commonplace, monocytes and eosinophils are rare, and disturbances associated with basophils are very uncommon. Mobile disruption and hematological changes are the result of many conditions within the body, including high stress or emotional strain, smoking, being pregnant, or excessive exercising. One of the most dangerous compounds in smoke is nicotine, which can alter the everyday variety of those parameters, along with RBC, WBC, and hemoglobin counts. Enhancing the parameters will increase the hazard of coronary heart disorder, cancer, polycythemia, continual obstructive pulmonary disease, and heart sickness [7].

2.2 LITERATURE REVIEW

Er. Nirpjeet Kaur et al. [1] have presented a topic on the identification of blood groups by using laser technology. Images are inspected, and numerical capacities are discredited. The purpose of computerized image processing is to improve the character of pictorial data and stimulate machine explanation through programming. A computerized imaging framework should have basic components for image acquisition, extraordinary equipment for supporting image applications, and a large amount of memory for capacity and information about yield devices. Picture segmentation is an area that has been extensively studied, particularly in a variety of restorative applications, and it continues to provide specialists with a variety of challenges. In computerized mammograms, segmentation is a crucial task for identifying suspicious tumor regions. Each image has its own set of edges and different levels of constraint. The most widely used approach for extracting articles from a photograph is "thresholding" in image processing.

Fabien Picot et al. [2] analyzed that clinicians often need to perform a brain biopsy to get histological evidence during the diagnosis and remedy of mind cancer. This method, however, incorporates the following risks: The samples received aren't necessarily typical of the tumor, and while the pattern is taken, there is a danger of blood vessel rupture. This sort of bleeding occurs in 0% (no bleeding after biopsy procedure for one group), 3% (partial bleeding after biopsy procedure for other group), and 59.8% (partial bleeding after biopsy procedure for yet other group), with a fatality rate of up to 9%. A diffuse reflectance spectroscopic imaging system integrated into a mind biopsy needle is supplied right here to help surgeons during needle biopsy tactics. With a purpose to lessen the risks of the technique, the imaging era uses 18 optical fibers to collect 81 reflectance spectra in an unmarried acquisition.

Ferraz et al. [3] have provided an algorithm to determine the blood type. With the use of image processing by a computerized system even before giving a blood transfusion, it is important to recognize your blood type, even if it's an emergency. These exams are currently executed manually by technicians, which can lead to human blunders. Various strategies to automate those checks were created; however, none of them can execute the evaluation in time for emergency conditions.

Khurram Khirshid et al. [4] have designed a system on the basis of an image segmentation algorithm. In this system, called "NICK," a new sliding window-based local thresholding methodology is compared to some current sliding window-based thresholding techniques. The suggested method seeks to improve binarization outcomes for old document images in particular. When tested on low-quality ancient document images, NICK is inspired by Niblack's binarization method and demonstrates its robustness and effectiveness.

Mehdi Hasan Talukder et al. [5] have designed an algorithm to improve the accuracy of human blood type. In an emergency case, determining human blood types is critical. However, the existing system's detection technique is quite slow. Human blood types are now identified manually using a palette process. It entails collecting blood and combining it with certain reagents to determine blood agglutination. The results are examined under a microscope. The major goal of this study is to offer a method for employing image processing techniques to determine human blood types.

Neha Srivathsa et al. [6] achieved a task using simple morphological photoprocessing algorithms and a primarily material strip-based short diagnostic check. They describe a completely unique methodology for computerized ABO Rh-D blood typing. Photographs of the material strip after testing are intended for low-cost cellular phones, and the proposed gadget strategizes the images using noise reduction, variety filtering, and empirically determined heuristics to automatically pick out the blood type. The purpose is to develop a simple cell phone application with a view to permitting automatic, brief, and easy blood testing.

Priyadarshini R et al. [7] have proposed that the idea of this research article is to update guide work in clinical laboratories for blood group identification. No matter the fact that automation plays an increasingly critical role in scientific packages, no device exists to automatically determine blood types. Commonly, blood institution analysis is accomplished through laboratory technicians. They examine blood groups using the following methods: ABO ahead typing, ABO reverse typing, and Rhesus type trying out. ABO typing and Rhesus typing are the most widely used of the three procedures in all laboratories.

Qingqian Yang et al. [8] have presented the design of image segmentation as the foundation of feature extraction and pattern recognition, and it is one of the most fundamental methodologies in digital image processing. The primary picture segmentation algorithms are listed and reviewed in this work, followed by fundamental assessment methods and a discussion of the future of image segmentation. Based on a vast number of comparative tests, some important properties of picture segmentation emerge.

Ryszard S. et al. [9] have designed an algorithm called LASER that is used to detect blood groups. In this approach, the transmitter is utilized to create pulses with a frequency of 10 kHz. The light-emitting diode converts electrical modifications into optical versions and then receives those pulses. The optical alerts were then transmitted through the fiber to the blood sample, which is then received by a detector that converts the optical versions back into electric versions.

Satoh et al. [10] researched and described successful polymerase chain reaction (PCR)-based procedures for forensic ABO typing, including the use of collection-specific primers with positive controls and the application of two pairs of primers analyzed using fragment analysis on an ABI PRISM 3100 genetic analyzer. The technique offers testing of well-known base alterations at four nucleotide positions: 261, 796, 802, and 803, allowing for strong group prediction based on the presence of three usual alleles like A, B, and O. As a result, the algorithms recognized well all unusual alleles within the database.

Tejaswini et al. [11] have designed an algorithm for the determination and classification of blood types using photoprocessing. Basically, identifying blood types is crucial before giving a blood transfusion in an emergency state of affairs. Those assessments are presently carried out manually with the aid of technicians, which could lead to human mistakes.

U.R. Brown et al. [12] have done work in creating an antibody identification summary form that necessitates input from a large number of people and must be adapted to the needs of a certain institution. This form, which collects a lot of information, should be as complete and feasible in terms of patient history, serological results, and transfusion implications. This well-organized, comprehensive source of patient data makes serological resolution for a patient's antibody easier and reduces the risk of organizational and communication problems.

Edeh et al. [13] provide an effective argument for the application of ensemble learning models in hepatitis C illness prediction. The study offers important new information for clinical practice and public health by showing how artificial intelligence can improve diagnosis accuracy. Although the results show promise, more testing and improvement are required to fully realize the model's potential in many real-world circumstances.

Ykhlef et al.'s [14] study promises supervised learning techniques in anticipating blood product requirements for patients with hematologic illnesses. Through the utilization of sophisticated machine learning algorithms and multi-center data, the research offers a useful instrument for enhancing patient care and blood resource management. The study's encouraging findings support more research and implementation of AI in healthcare logistics.

Using blood test data, Tutsoy et al. [15] offer a substantial breakthrough in health risk classification through creative feature selection strategies and deep self-supervised machine learning. The study's potential influence on clinical diagnostics and health risk management is highlighted by its high classification accuracy and efficient feature removal approach. Further efforts to evaluate and incorporate these models into clinical practice may improve their usefulness and practicality.

2.3 EXISTING METHOD

At present, human blood groups are determined by plate test manually. Most of the blood assessments now do not need unique situations like fasting for 8–12 hours before a blood draw or stopping certain medicines. By testing the fluid, various parameters within the blood may be measured. The outcomes help to perceive fitness problems much earlier or may additionally predict sicknesses. The system comprises of picture acquisition followed by numerous image processing operations that might be median filtering, segmentation, erosion, area detection, and counting of cells. The counting of RBCs could be done based on the edge detection.

2.3.1 Convolutional neural network

Existing cell detection algorithms use convolutional neural networks to detect malignant cells, which is consistent with training with datasets for disease detection. Convolution, activation functions, and max pooling operation are used to train the model. The model is trained by varying batch sizes and epoch values.

2.3.2 Disadvantages

- An increasing computational expense.
- Time complexity.
- Human errors lead to misinterpretation of results.
- Difficult to get consistency in results from visual inspection of cells, because the cell structure is complex and overlap occurs.
- If you just want a good pre-trained model, the network is little too slow and complicated.

2.4 PROPOSED METHOD

Figure 2.1 shows a methodical procedure for applying image processing and machine learning approaches to classify different blood-related illnesses. Below is a thorough breakdown of every step:

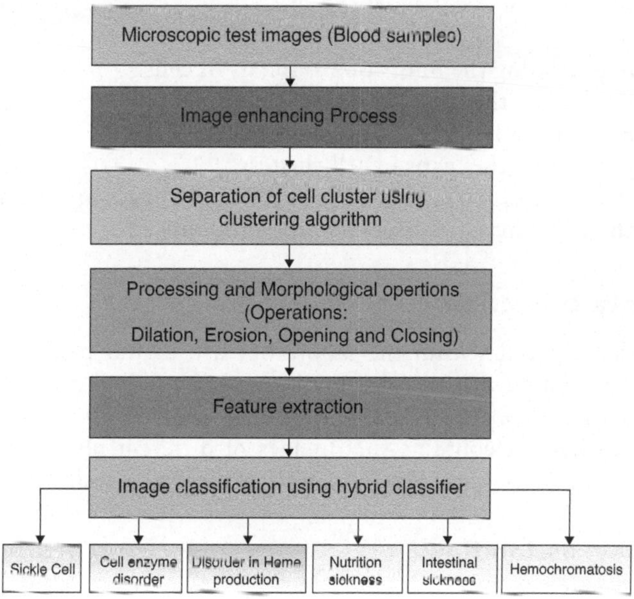

Figure 2.1 Flowchart of proposed method.

2.4.1 Blood samples used in microscopic test images

Microscopic pictures of blood samples are taken at the start of the procedure. These pictures were probably taken in lab experiments where microscopes were used to get close-up images of blood cells.

2.4.2 Image improvement methods

To raise the quality of the raw microscopic images, augmentation is used. This could entail applying methods like sharpening, contrast modification, and noise reduction to enhance visibility and simplify the analysis of blood cell properties.

2.4.3 Cell cluster separation using cluster algorithm

Enhanced photos are segmented and comparable cells are grouped together with the use of clustering techniques. This stage aids in the separation of various blood cell types and the detection of groupings that might point to anomalies.

2.4.4 Morphological operations (such as dilation, erosion, opening, and closing) and processing

To improve the cell clusters, morphological manipulations are performed on the segmented pictures. Dilation, erosion, opening, and closing are examples of operations that assist in:

Dilation: extending the limits of groupings of cells.
Erosion: reducing the limits in order to eliminate minute sounds.
Opening: clearing the foreground of any little things.
Closing: plugging tiny gaps in cell clusters.

Through these techniques, the segmented images are cleaned up, and the characteristics become more pronounced.

2.4.5 Data collection

Photographs received within the laboratory are digital pictures saved in JPEG format. Those pictures are pre-processed using a colored aircraft. There are various websites available for public analysis of very small blood images. This dataset contains tiny images of different blood cell components, including WBCs and RBCs, as well as their subspecies.

2.4.6 Progression flow

In our proposed system, reagents are mixed with three blood samples. After a while, agglutination may additionally or may not arise. After agglutination is created, the slide is taken as an image and allowed to be processed within the MATLAB picture processing toolbox. With the use of this software, the chances of human blunders can be decreased.

In our proposed method, various pre-processing strategies have been used, along with gray conversion. Pre-image processing can substantially increase the reliability of visible recognition. Many filtering features that enhance or reduce certain image information enable smooth and easier testing. Users are able to enhance the digital camera photograph with just a few clicks. Filters contain many photo filters for photograph enhancement, combined filters for part enhancement, noise discount, individual conversion, and many others. Image processing includes a few picture processing capabilities, multiplied comparisons with twin or dynamic switches, viewing tables, or split aircraft pictures.

2.4.7 Gray level co-occurrence matrices

The mathematical distribution of the marked stiffness compounds in the specified areas related to the image is used to calculate the texture characteristics in the mathematical analysis. Statistics are divided into consecutive, and higher numbers based on the number of intensity points (pixels) in each compound. The Gray Level Co-occurrence Matrix (GLCM) method is an initial method for extracting the mathematical properties of the second system.

2.4.8 Classification technique

Supportive vector machines are supervised learning models with integrated reading algorithms for analyzing and retrospecting data and detecting patterns. The support vector machine (SVM) training algorithm creates a model that provides new models in one of the two categories based on a set of training examples, each identified as one of two phases, making it an unparalleled binary line divider. The SVM model is a model of examples such as points in space, mapped in such a way that the examples of different categories are separated by a large distance. There are new ones. Then, depending on which side of the distance they fell, they were drawn to the same place and predicted to be part of the group. SVMs can perform non-linear and linear partitions by automatically mapping their input into high-resolution areas, known as kernel tricks.

2.4.9 Advantages of the proposed system

- Provides accurate results in predicting stem cell transplantation and provides a recommendation to the tech coordinator.
- Cost savings and of short term.
- Identifies hematological parameters such as polycythemia, dengue, and more than 20 disorders.
- Indicates that there is alcohol in the blood before transfusion.

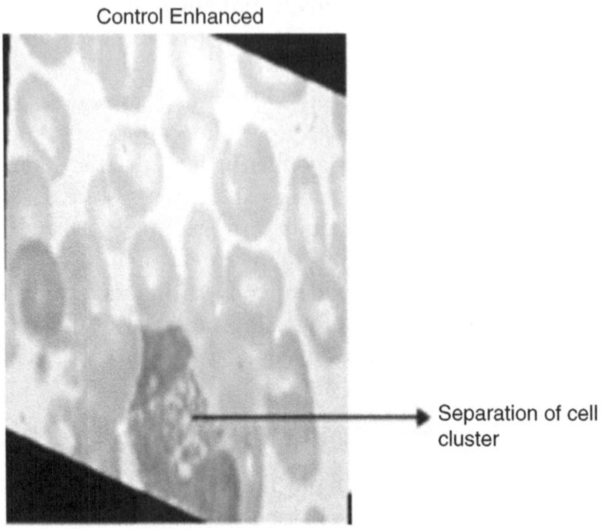

Figure 2.2 Enhanced image technique to remove noise.

- Contaminates blood, such as toxic contamination due to high blood pressure and rupture, anemia, and fetal erythroblastosis.

2.5 RESULTS AND DISCUSSION

2.5.1 Enhanced image

This method changed digital images in this step so that the findings are better suited for display or further image analysis, as shown in Figure 2.2. It would eliminate noise, sharpen, or brighten an image in this image to make it easier to recognize key features.

2.5.2 Cluster segmentation

Using similarities in pixel intensity, color, texture, or other characteristics, an image is divided into sections (clusters) by a process known as cluster segmentation. Cluster segmentation classifies the given image into three divisions. Division one represents the group of defected regions or cells called clusters, which shows the tumor region from the input dataset. Divisions two and three depict the chromatic layers of a and b formats, as shown in Figure 2.3. Cluster evaluation, frequently referred to as clustering, is a method that requires unsupervised machine learning. It consists of locating herbal groupings in records automatically. Clustering algorithms,

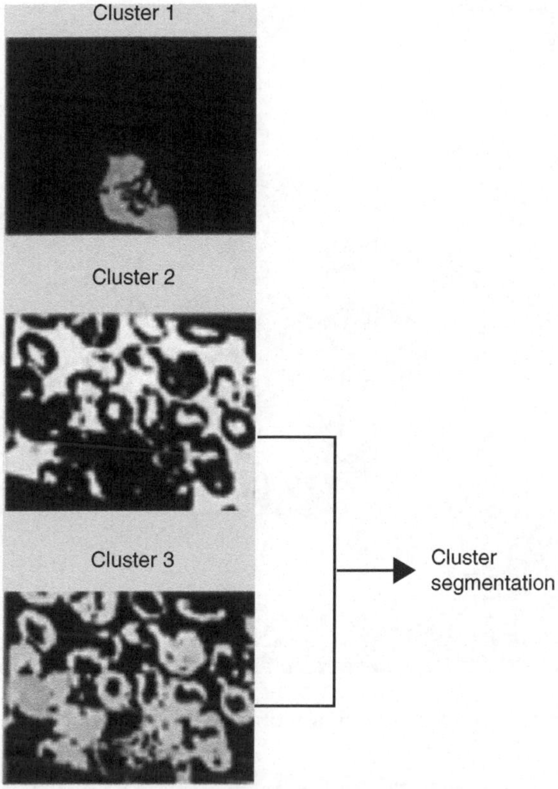

Figure 2.3 Separation of cluster chromatic layer using clustering algorithm.

unlike supervised studying (which consists of predictive modeling), simply compare the incoming facts and look for herbal corporations or clusters in the function area.

2.5.3 Morphology operation

A group of image processing methods known as morphological operations manipulate pictures according to their morphologies. They create the same-sized output picture by applying a structural element to the input image. Only when the test is successful at that position in the input image does a morphological operation in a binary image result in a new binary image with a non-zero value for the pixel. Although they may also be applied to grayscale pictures, these techniques are specifically utilized for binary images. This operation includes dilation, erosion, opening, and closing processes. Dilation makes the foreground item larger or the white

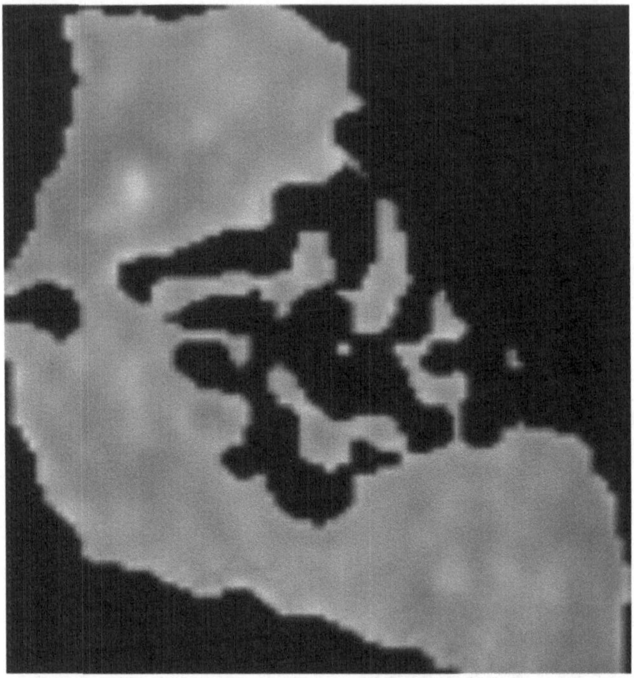

Figure 2.4 Separation of cell cluster using morphological operation.

area in the image larger. The object borders are expanded by pixels with this procedure. Deterioration reduces the foreground object's size, or the white area in the picture. Pixels on the edges of the object are eliminated by this technique. Noise cancellation is achieved by opening operations; it is described as dilation that is followed by erosion. Small dark spots or holes inside foreground objects are sealed using a closing technique. It's described as an erosion that happens after a dilation. The morphological operation is done to remove imperfections from the image shown in Figure 2.4.

After the separation of cluster, the region of imperfections is removed and further clarifications are shown in Figure 2.5.

2.5.4 Malignant cell

A malignant cell was identified in the K-bounding prop region, as shown in Figure 2.6. K-bounding prop region generally describes the procedure that uses linked component analysis and morphological procedures to locate and extract properties from "K" distinct regions (or blobs) in a picture. In this case, "K" denotes the total number of regions of interest, while "bounding prop" most likely refers to these areas' centroid, area, bounding boxes, and other characteristics. A tumor is a clump of cells that has grown out of control. Tumors are classified as benign or

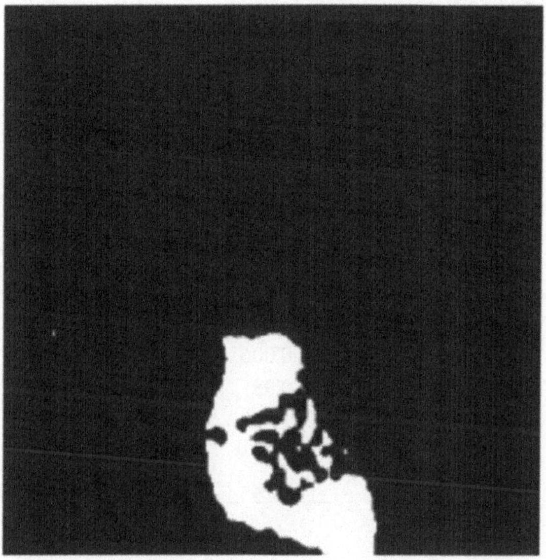

Figure 2.5 Bounding formation stage.

Figure 2.6 K-bounding prop region.

malignant. Benign tumors are cell fragments that grow locally and do not spread. Tumors that are malignant can spread and invade other tissues. Figure 2.6 shows the depiction of cells for visible clarity and the identification of cells in that cluster.

Table 2.1 Parameters analysis between datasets

Parameters	*Test dataset*	*Training dataset*
Contrast	0.24	0.22
Correlation	0.20	0.17
Energy	0.84	0.83
Homogeneity	0.95	0.95
Mean	0.0037	0.0035
Standard deviation	0.081	0.080
Entropy	1.9	2.0
RMS	0.08	0.08
Variance	0.006	0.006
Smoothness	0.95	0.97
Kurtosis	19.40	15.54
Skewness	1.50	1.20
Invariance	0.12	0.20

2.5.5 Analysis between the test vs training dataset

The analysis is done between the test and training dataset to absorb the statistical feature co-relating with both datasets in real-time performance as shown in Table 2.1.

2.5.6 Confirmation message of result

Figure 2.7 shows the results of identifying the types of Rh factors in the input image and also predicting the diseases based on threshold values. The prediction is based on 13 selected extractions from the GLCM features, and then the confidence score is an analysis of the blood cell range and conditions related to healthy situations.

The results of prediction are based on cell classification; in this lymphocyte cell classification, there are no diseases identified. Hence, it is termed as healthy or no disease found in that classification shown in Figure 2.8.

2.5.7 Disease analysis

The demographics of the collected datasets are shown in Table 2.2. It has various age groups from South India, among whom are 21 males at the ages of 0–15, 87 males at the ages of 16–30, 97 males at the ages of 31–45, 88 males at the ages of 46–60, and 41 males at the ages of 61–80, for a total of 334. It also has female data, with 17 at the ages of 0–15, 107 at the ages of 16–30, 102 at the ages of 31–45, 71 at the ages of 46–60, and 26 at the ages of 61–80, for a total of 323. A total of 22 transgender data points were taken with various age groups.

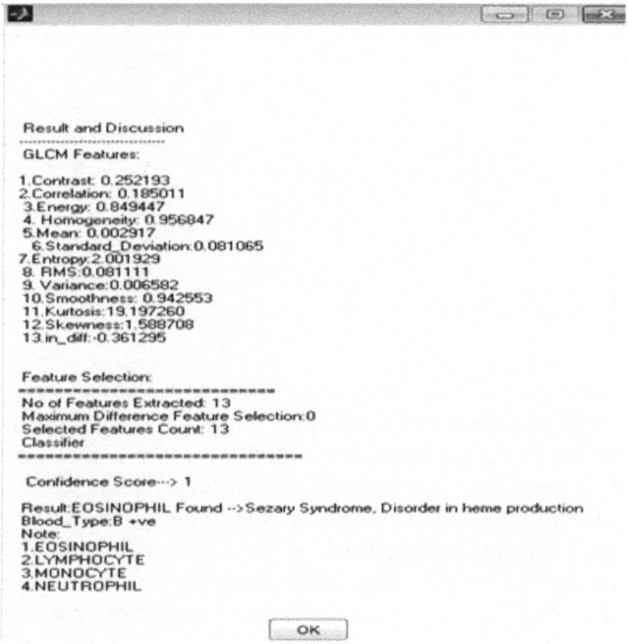

Figure 2.7 Result predicted in dataset.

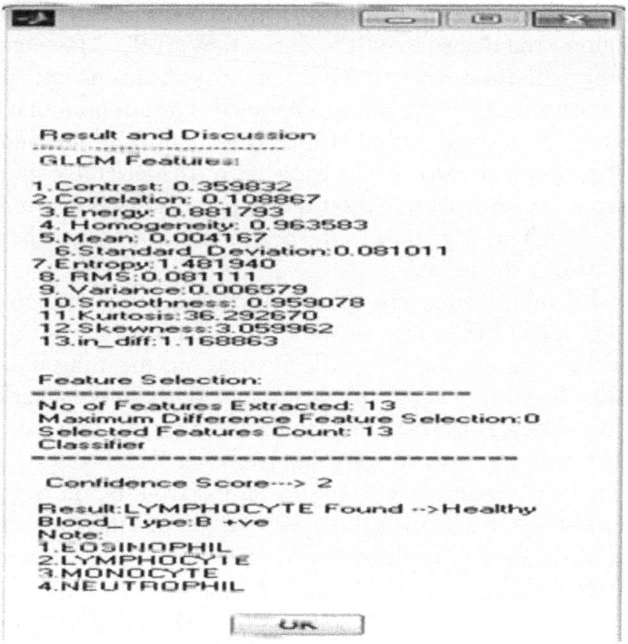

Figure 2.8 Lymphocyte disease predicted in dataset.

Table 2.2 Demographics details of the collected data

	Male			*Female*			*Transgender*			
Age	*R*	*SU*	*U*	*R*	*SU*	*U*	*R*	*SU*	*U*	*Origin*
0–15	11	8	3	9	7	1	0	0	0	South Indian
16–30	51	23	13	53	39	15	4	5	2	South Indian
31–45	49	31	17	47	34	21	2	1	2	South Indian
46–60	43	26	19	37	21	13	3	0	3	South Indian
61–80	18	17	6	11	9	6	0	0	0	South Indian

R, Rural; SU, Semiurban; U, Urban.

The disease analyzed is based on threshold values in the prediction ranges, and the variation in the threshold diseases can be precisely identified and displayed as shown in Table 2.3. Among the 679 records, a hybrid classifier identifies 120 as having sickle cell disease, 80 as having hemolytic enzyme disorder, 77 as having dysregulation in heme production and nutrition sickness, 55 as having intestinal sickness, 150 as having hemochromatosis disease, and 87 records show healthier reports. The proposed method of determining disease is compared with the physical method, and it shows that the ensemble algorithm performs better than the physical method with less processing time and an accuracy of 95.16%.

Figure 2.9 demonstrates how the ensemble algorithm's classifications across six distinct blood-related illnesses are distributed graphically. With a 25% classification rate, hemochromatosis has the greatest rate of identification among the disorders in the dataset. With 9% classification rate, intestinal illness is the least common. The classification outcomes of the ensemble algorithm exhibit a heterogeneous distribution among the many illnesses, which is indicative of its utilization in a multi-dimensional health risk categorization framework. The capacity of the ensemble model to categorize diverse health risks based on blood test results is demonstrated by this graphic, which clearly conveys the proportions of various health disorders recognized by the model.

Figure 2.10 summarizes the physical classification system used for various blood-related illnesses. Using a classification rate of 25%, hemochromatosis has the highest rate of identification, meaning it is the most common disorder found using this technique. With a classification rate of 9%, intestinal illness is the least common. The distribution across different disorders demonstrates how the physical technique can be used to diagnose a range of medical conditions using information from blood tests. For each of the six illnesses, the categorization percentages produced by the physical technique and the ensemble algorithm are similar. The same percentage of cases are attributed by each technique to hemochromatosis, sickle cell disease, cell enzyme disorders, heme production disorders, digestive illnesses, and nutrition sickness. The same outcomes imply that both approaches are equally successful in categorizing particular blood-related illnesses. But the

Table 2.3 Accuracy of disease determination and analysis for the given datasets

S. No.	Disease	*Determination of disease using ensemble algorithm (proposed method)*	*Determination of disease in physical method*	*Determination of diseases in given dataset in %*	*Error between physical method and proposed method in %*	*Accuracy of proposed method in %*
1.	Sickle cell	120	116	20	3.44	96.66
2.	Cell enzyme disorder	80	78	13.3	2.50	97.50
3.	Disorder in heme production	77	76	12.8	1.30	98.70
4.	Nutrition sickness	110	105	18.3	4.55	95.45
5.	Intestinal sickness	55	53	9.16	3.64	96.36
6	Hemochromatosis	150	143	25	4.67	95.33

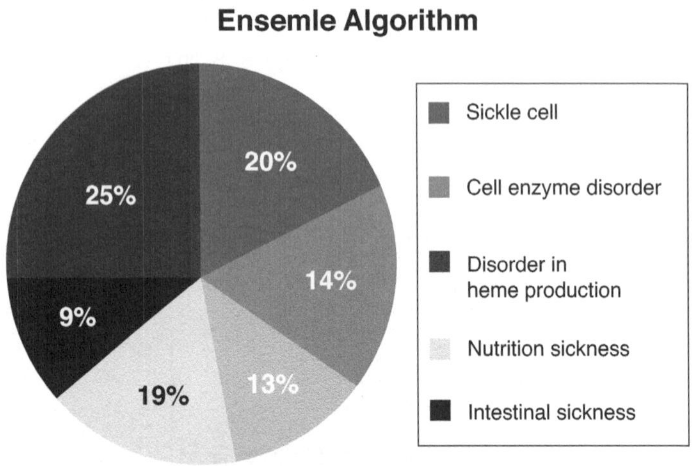

Figure 2.9 Determination of disease using ensemble algorithm.

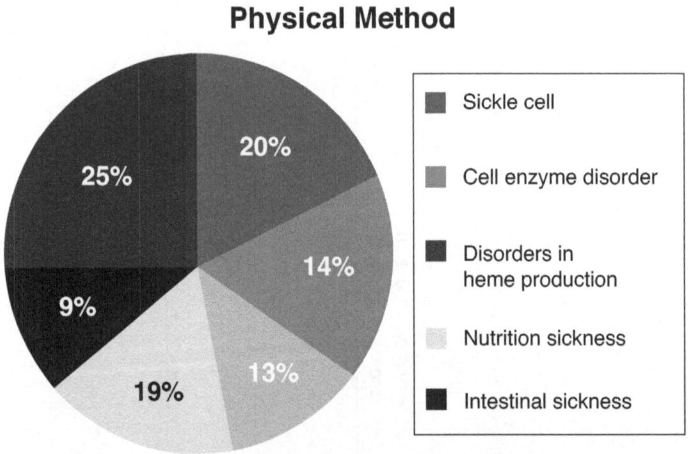

Figure 2.10 Determination of disease using physical (conventional) method.

proposed method has good computational effectiveness, simplicity of use, cost, and producing the output with less time for the patient, as shown in Figure 2.11.

2.5.8 Analysis of ensemble classifier accuracy

Figure 2.12 shows the analysis of proposed method with comparison of existing method, whereas neural network has an accuracy of 85% and ensemble classifier has an accuracy of 95%. The accuracy is calculated

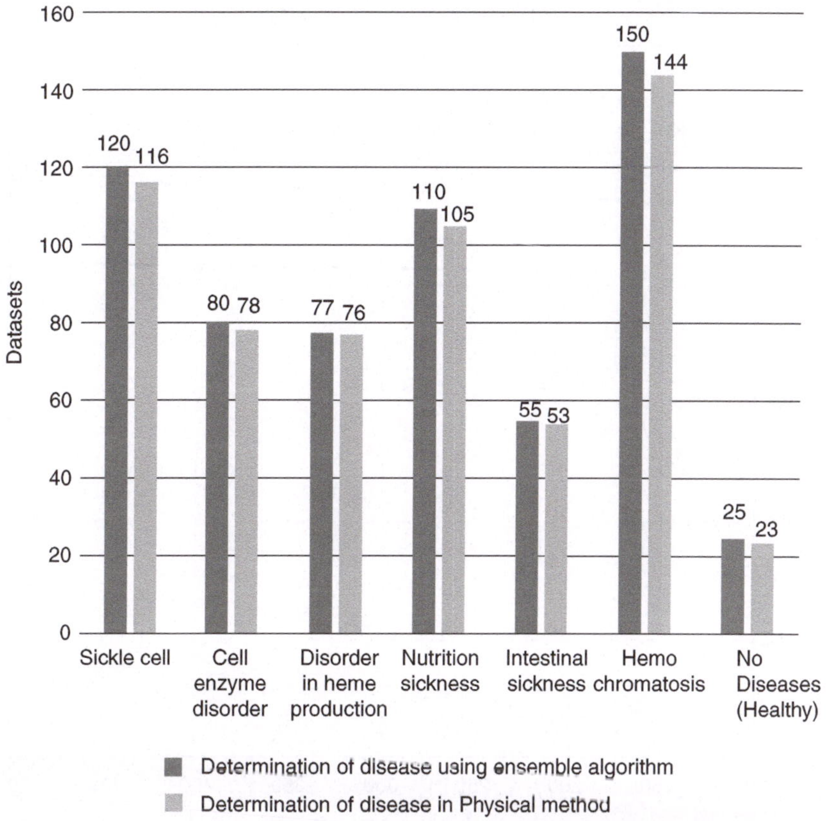

Figure 2.11 Performance comparison of diseases determination by using ensemble algorithm and conventional method.

based on the parameters value with respect to true positive, true negative, false positive, and false negative as shown in Equation 2.1:

$$\text{Accuracy} = (TP + TN) / (TP + FP + TN + FN) \tag{2.1}$$

Here among this analysis the proposed method of ensemble classifier has highest level of accuracy. This shows ensemble classifier has more efficiency than the other classifiers [16–28].

2.6 CONCLUSION

Machine learning has become an essential tool for modeling human processes in a range of fields, including medicine, because of the amount of data. Blood analysis is one of the most important illness detectors because it contains

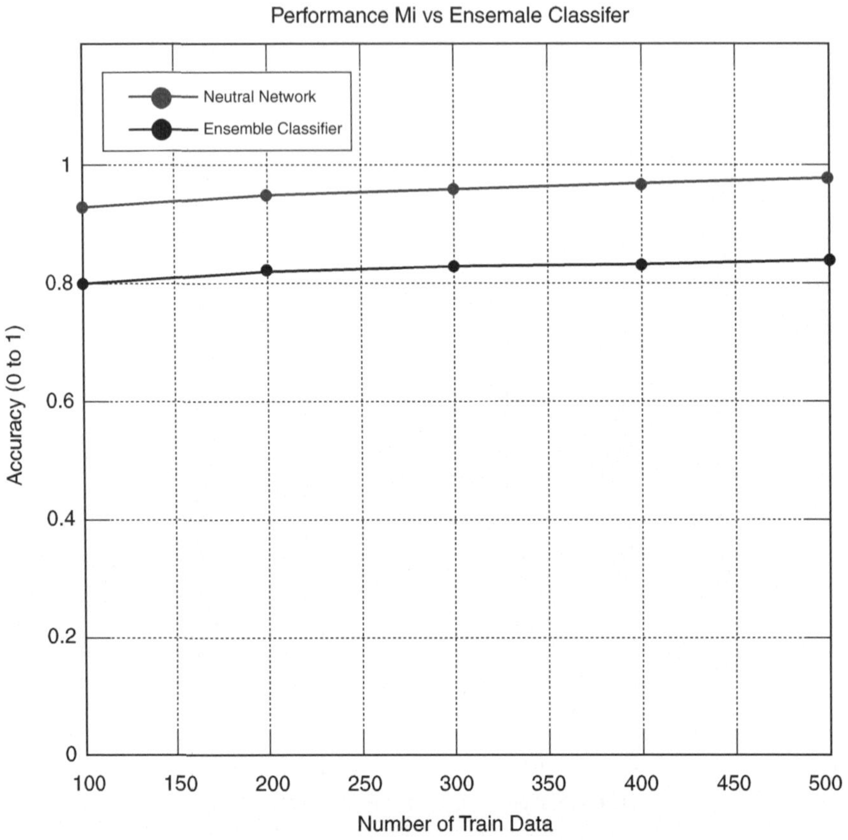

Figure 2.12 Analysis and comparison of accuracy of ensemble algorithm with neural network.

many characteristics with varying values, indicating definite proof of the diseases' existence. The devised approach demonstrates that it is an effective and efficient way for detecting and precisely determining the patient's blood type. The application of image processing technique allows for the automation of the detection of agglutination and the determination of the patients' blood type in a short period of time. In an emergency, the method is appropriate and useful. A high-quality dataset is obtained and assessed by physicians who possess expertise in the field, as this determines the accuracy of the machine learning algorithm. Classifiers are trained on this dataset in order to maximize their accuracy. In order to help doctors forecast blood illnesses based on standard blood tests, the researchers evaluated many classifiers and attained accuracy of up to 95.16%. The proposed dataset will be tested using various deep learning algorithms in the future to compare traditional and deep learning approaches in this study field. In addition, to gather and test more blood data, an online Internet of Things (IoT) is developed.

REFERENCES

1. Kaur, Nirpjeet, and Rajpreet Kaur, "A review on various methods of image thresholding," *International Journal on Computer Science and Engineering*, vol. 3, no. 10, p. 3441, 2019.
2. Picot, Fabien, Julien Pichette, Joannie Desroches, Andréanne Goyette, Marie-Andrée Tremblay, Yasmine Ben-Mansour, Frédéric Leblond, Gilles Soulez, and Brian C. Wilson. "Imaging system based on diffusive reflectance spectroscopy for blood vessels detection during brain biopsy procedure," in *2016 Photonics North (PN)*, 2016, doi: 10.1109/PN.2016.7537924.
3. Ferraz, Ana, "Automatic system for determination of blood types using image processing techniques," in *2013 IEEE 3rd Portuguese Meeting in Bioengineering (ENBENG)*, pp. 1–6, 2013. IEEE.
4. Khurshid, Khurram, Imran Siddiqi, Claudie Faure, and Nicole Vincent, "Comparison of Niblack inspired binarization methods for ancient documents," *Proc. SPIE*, vol. 7247, no. 5, pp. 1–10, 2017.
5. HasanTalukder, Mehedi, Md Mahfuz Reza, Mahbuba Begum, Md Rabiul Islam, and Md Mahmudul Hasan, "Improvement of accuracy of human blood groups determination using image processing techniques," *International Journal of Advanced Research in Computer and Communication*, vol. 4, no. 10, pp. 110–124, 2018.
6. Srivathsa, Neha, and Dhananjaya Dendukuri, "Automated ABO Rh-D blood type detection using smartphone imaging for point-of-care medical diagnostics," in *IEEE Conference Publications*, 2016, doi: 10.1109/EMBC.2016.7591689.
7. Priyadharshini, Ravi, Saravanan Ramya, Subramaniyan Kalaiyarasi, Devi Subramaniyam Kalpana, and Vanitha Suthathira, "A novel approach in identification of blood group using laser technology," *International Journal of Research in Engineering and Technology*, vol. 85, no. 10, pp. 2319–1163, 2014.
8. Yang, Qingqiang, and Wenxiong Kang, "General research on image segmentation algorithms," *International Journal of Image, Graphics and Signal Processing*, vol. 5, no. 2, pp. 1–8, 2019
9. Romaniuk, Ryszard S., and Jerzy Gajda, "Laser technology and application," *International Journal of Electronics and Telecommunications*, vol. 59, no. 12, pp. 195–202, 2013.
10. Satoh, Ken, and Yukio Itoh, "Forensic ABO blood grouping by 4 SNPs analyses using an ABIPRISM 3100 genetic analyzer," in *International Congress Series*, vol. 1288, no. 18, pp. 49–51, 2006.
11. Tejaswini, Hassan Vasudev, and Mallikarjuna Swamy, "Determination and classification of blood types using image processing techniques," *ITSI Transactions on Electrical and Electronics Engineering*, vol. 2, no. 2, pp. 29–33, 2014.
12. Brown, Michelle R., and Peggy Crim, "Organizing the antibody identification process," *Clinical Laboratory Science*, vol. 20, no. 2, pp. 122–126, 2017.
13. Edeh, Michael Onyema, Surjeet Dalal, Imed Ben Dhaou, Charles Chuka Agubosim, Chukwudum Collins Umoke, Nneka Ernestina Richard-Nnabu, and Neeraj Dahiya, "Artificial intelligence-based ensemble learning model for prediction of hepatitis C disease," *Frontiers in Public Health*, vol. 10, p. 892371, 2022.

14. Ykhlef, Amel, Nedjla Selma Labri, and Menaouer Brahami, "Supervised learning techniques for blood product prediction in patients with hematologic diseases: A multi-centre study in Western Algeria," *International Journal of Information Technology*, vol. 01, pp. 1–12, 2024.
15. Tutsoy, Onder, and Gizem Gul Koç, "Deep self-supervised machine learning algorithms with a novel feature elimination and selection approaches for blood test-based multi-dimensional health risks classification," *BMC Bioinformatics*, vol. 25, no. 1, p. 103, 2024.
16. Sharma, Vijay Kumar, "CNTFET circuit-based wide fan-in domino logic for low power applications," *Journal of Circuits Systems Computers*, vol. 31, no. 2, p. 2250036, 2022.
17. Sharma, Vijay Kumar, and Manisha Pattanaik, "A reliable ground bounce noise reduction technique for nanoscale CMOS circuits," *International Journal of Electronics*, vol. 102, no. 11, pp. 1852–1866, 2015.
18. Sharma, Vijay Kumar, Sumit Patel, and Manisha Pattanaik, "High performance process variations aware technique for sub-threshold 8T-SRAM cell," *Wireless Personal Communications*, vol. 78, no. 1, pp. 57–68, 2014.
19. Sharma, Vijay Kumar, and Manisha Pattanaik, "Techniques for low leakage nanoscale VLSI circuits: A comparative study," *Journal of Circuits Systems and Computers*, vol. 23, no. 5, p. 1450061, 2014.
20. Sharma, Vijay Kumar, "A survey of leakage reduction techniques in CMOS digital circuits for nanoscale regime," *Australian Journal of Electrical and Electronics Engineering*, vol. 18, no. 4, pp. 217–236, 2021.
21. Kumar, Chanchal, Avinash Sharan Mishra, and Vijay Kumar Sharma, "Leakage power reduction in CMOS logic circuits using stack ONOFIC technique," in *2018 Second International Conference on Intelligent Computing and Control Systems (ICICCS)*, 2018.
22. Sharma, Vijay Kumar, and Manisha Pattanaik, "Design of low leakage variability aware ONOFIC CMOS standard cell library," *Journal of Circuits Systems and Computers*, vol. 25, no. 11, p. 1650134, 2016.
23. Naz, Syed F., Sadat Riyaz, and Vijay Kumar Sharma, "A review of QCA nanotechnology as an alternate to CMOS," *Current Nanoscience*, vol. 18, no. 1, pp. 18–30, 2022.
24. Riyaz, Sadat, and Vijay Kumar Sharma, "Design of reversible Feynman and double Feynman gates in quantum-dot cellular automata nanotechnology," *Circuit World*, vol. 49, no. 1, pp. 28–37, 2023.
25. Raina, Bhoomika, Chetna Verma, Mehak Gupta, and Vijay Kumar Sharma, "Binary coded decimal (BCD) seven segment circuit designing using quantum-dot cellular automata (QCA)," in *2021 5th International Conference on Trends in Electronics and Informatics (ICOEI)*, 2021.
26. Mushtaq, Umayia, and Vijay Kumar Sharma, "Performance analysis for reliable nanoscaled FinFET logic circuits," *Analog Integrated Circuits and Signal Processing*, vol. 107, no. 3, pp. 671–682, 2021.
27. Kajal, and Vijay Kumar Sharma, "Design and simulation of FinFET circuits at different technologies," in *2021 6th International Conference on Inventive Computation Technologies (ICICT)*, 2021.
28. Sharma, Vijay Kumar, "Designing of face recognition system," in *2019 International Conference on Intelligent Computing and Control Systems (ICCS)*, 2019.

Chapter 3

Deep learning technique for Dark Web threat detection and analysis on social media

Anil Kumar, Abhay Bhatia, Manish Verma, and Sunil Kumar

3.1 INTRODUCTION

WWW is a worldwide network of computers and networks that share information. The Dark Web, also known as the Darknet or the Deep Web, is a decentralized, encrypted network that allows its users to conceal their identities. Without worrying about being tracked or caught, criminals may use the Tor network and protocol to spread propaganda, advertise illicit goods, and recruit new members. The rise and fall of the Silk Road marketplace in 2011 and 2013, respectively, helped to promote underground networks. According to Cherqi et al. [13], law enforcement is cracking down on the Dark Web because it provides a haven for criminals. In this chapter, the author explores the relationship between criminal activity on the Dark Web and the reaction from law enforcement.

Surface web: The part of the World Wide Web accessible to the general public and searchable by standard web search engines is known as the Surface Web, Visible Web, Indexed Web, Indexable Web, or Light net. The Surface Web may only include 10% of all internet material. The Surface Web comprises all publicly available web pages that may be found using any search engine.

Deep web: Parts of the World Wide Web that are not crawled and cataloged by standard web search engines are known as the "Deep Web," "invisible web," or "hidden web." The Deep Web are services that require users to log in before accessing their content, such as webmail, online banking, restricted access to social media pages and profiles, specific web forums and programming languages, and pay walls for content like video-on-demand and select online periodicals. Access to the Deep Web's content is possible through a specific URL or IP address, but it may be protected behind a login screen or other security measure.

Dark Web: The Dark Web refers to information found on darknets, which are internet based "overlay networks" that need special client software, network settings, or authentication (Almukaynizi et al., 2018). A user's

DOI: 10.1201/9781003643609-3

location, name, and email address are all kept secret when they chat and transact business on hidden computer networks over the Dark Web (Miloshevska et al.). The Deep Web is part of the internet that is not crawled by search engines, while the term is often incorrectly used to refer just to the Dark Web. The Dark Web comprises many kinds of darkness, such as smaller, more intimate peer-to-peer networks and more prominent, well-known networks like Tor, Freenet, I2P, and Riffle, which are all run by the public and private sectors. Sites accessible through the Tor browser on the Dark Web are often labeled with the "onion" extension (Figure 3.1).

Tor: A Tor browser ensures the user's privacy when searching the Dark Web by encrypting all of the user's entry points and paths. Many layers of encryption conceal users' identities and whereabouts on the Dark Web. In order to conceal the users' true identities and ensure their anonymity, the darknet employs encryption technology that reroutes their data via an extensive network of intermediary servers. The information being broadcast can only be decrypted by a node further down in the chain, which the final node can only decrypt. Because of the system's complexity, reproducing the node route and decrypting the information layer by layer is almost hard. The high degree of encryption prevents the website owner and the visitor from learning the other's physical location or Internet Protocol address. Because of this robust encryption, darknet users may have private conversations, publish blogs, and transfer information without fear of being tracked (Figure 3.2).

Cybercrime: An example of a crime that may be committed using a computer and network is known as cybercrime. It's possible the computer was either a tool or an object of criminal activity. An individual's safety and financial well-being are only two of the many things that

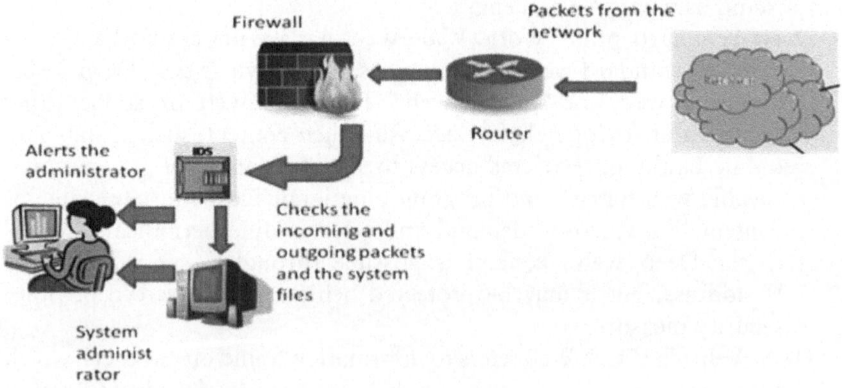

Figure 3.1 Dark net.

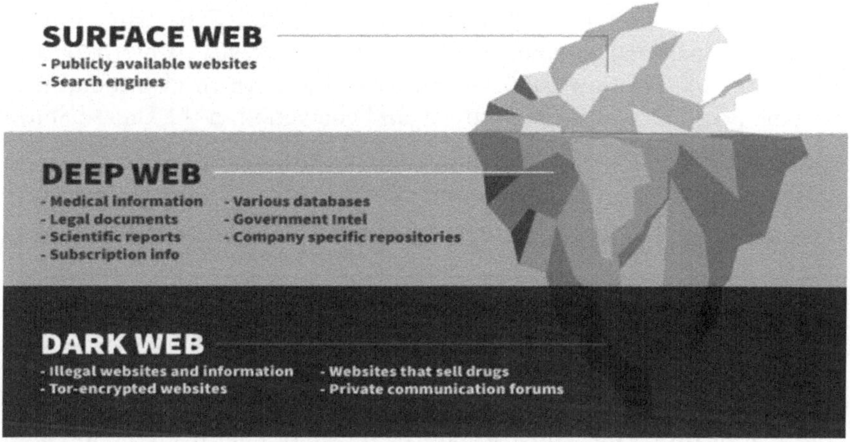

Figure 3.2 Encryption techniques in TOR.

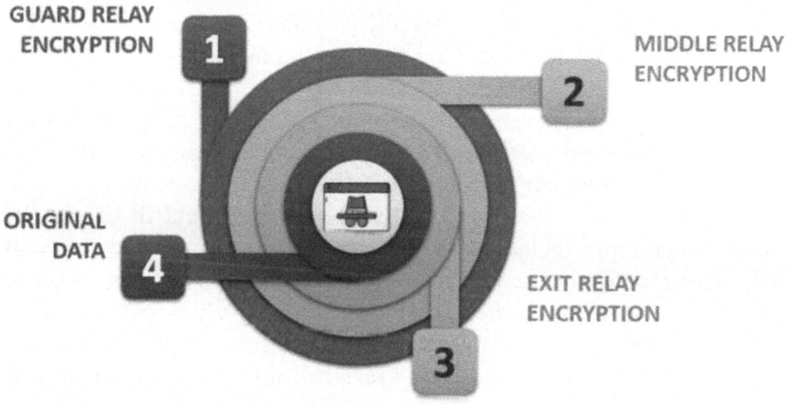

Figure 3.3 Cyber crime.

cybercrime may threaten. When private information is intercepted or released, whether legally or illegally, it raises serious privacy issues and cybercrime is no exception. Cybercrime, which includes espionage, money theft, and other transnational crimes, is committed on a global scale by both state and non-state actors. Cyber warfare is a term used to describe cybercrimes that span national boundaries and include the involvement of at least one nation-state. Cybercrime, in Warren Buffett's estimation, is the world's biggest issue and a major threat to civilization (Figure 3.3).

The yearly cost to the world economy was estimated at $445 billion in a paper released in 2014. Online theft of credit and debit cards costs Americans about $1.5 billion in 2012. Cybercrime is estimated to cost the world economy up to $600 billion each year, according to a study done by CSIS and McAfee in 2018. Cybercrime groups are cooperating to commit crimes online, as revealed by the World Economic Forum's 2020 Global Risk study, which also found that the possibility of their discovery and conviction in the United States is less than 1%.

3.2 BACKGROUND

In 2019, Tounsi et al. [1] proposed the electronic system and the internet, together known as the "digital world," have become an indispensable part of our daily lives. With the advent of digitization, it is now possible to complete any job quickly, which increases the number of possible interruptions. An intrusion detection system (IDS) employs hardware or software to monitor potentially dangerous activity or infringements on a system or framework. SIEM frameworks (security data and event management) are often used to trace out any malicious activity or violation. Many data sources are merged in a SIEM system, and alert separation methods are utilized to discern between hostile behavior and false alarms.

From 2017 to 2020, Dutta, Mador, Queiroz, Leukfeldt, Shakarian, Tounsi, Wagner, Miloshevska, and Chayal et al. [2–10] proposed postmortem analysis looks on methods to minimize such dangers in the future. The detection component is crucial to all three components and is given more weight owing to the proved efficacy of the concept of online threat detection. Networks include a massive quantity of data, and this collection focuses mostly on the application of information mining in the field of interruption detection. In information mining, order is a particularly efficient way for spotting disruptions. To represent categorization operations, IF-THEN rules and decision trees are utilized. Classifying online dangers is required in order to detect outliers using algorithmic approaches.

In 2018, according to Mishra et al. [36], VMI can identify attacks on virtual computers at the fine granular level. To detect attacks, the authors keep tabs on and analyze Tenant Virtual Machines (TVM) down to the level of individual processes and system calls. Security safeguards in the virtual machine (VM), for example, might be removed, and genuine applications can be altered with to get access to sensitive data. A Random Forest classifier was used to create consistent behavior over a wide range of TVM incursions, with features retrieved and chosen at the VMM layer utilizing the introspection feature. VM Guard is the name of the architecture. In 2018, Sklavounos et al. [37] developed a novel NIDS for detecting DOS using the CUSUM and EWMA charts in their experimental dataset NSL-KDD.

In 2020, according to Liggett et al. [12], in a large dataset, using support vector machines for classification and the Chi-selector for feature selection, a model for detecting intrusions was developed. The suggested model was tested on the KDD dataset. There are some studies that have made mention of PIDS (Perimeter Intrusion Detection Systems):

Scientists proposed detecting jamming attacks in a LoRaWAN network. The LoRaWAN protocol was used to build the system. ID's architecture for IoT-based M2M collaborates across nodes in the Internet of Things (IoT) to deliver precise identification with low impact on communication, computation, or energy (Arshad et al. [38]). Combining data from two different sources is a good idea, according to this paradigm, because of the network's flexibility, resource limits, and collaborative nature.

In 2017, Subba et al. [24] attempted to improve the computing efficiency of host-based intrusion detection system (HIDS). This system call underwent dimensionality reduction to shrink the feature vectors down to a more manageable size. Using classifiers such as MLPs, C4.5 decision trees, and Naive Bayes, intrusive processes in the feature vector may be discovered. By comparing it to the ADFA-LD, the suggested technique was put to the test.

In 2017, Parvat et al. [39] employed deep learning to propose NIDS. The suggested strategy, which divides and conquers, employs an ensemble of binary classifiers. The system's performance was assessed using the NSL-KDD dataset. The usage of IDS in a 5G wireless communication network was investigated by Devi et al. (2017) using data from KDD cup 99 and Adaptive's Neuro-Fuzzy Inference System (ANFIS) [40]. For wireless networks, Kolias et al. [41] suggested a decentralized IDS in 2017. To detect intrusions in the data stream, the system utilizes categorization rule induction and swarm intelligence methodologies. The proposed method was validated using the Aegean wireless intrusion dataset version 2. Using game theory in 2017, Gupta et al. [42] found and evaluated a 5G wireless small cell access point bandwidth spoofing assault. The researchers also presented an adaptive IDS based on a hidden Markov Model to detect intrusions into 5G wireless networks. P. Herrmann developed and tested a distributed IDS approach for the IoT that is suitable for tiny devices. The researchers utilized a trust management technique to keep track of their neighbors' reputation. Using the suggested technology, malware may be identified in an energy- and processing-efficient manner.

The tactics outlined above have aided in the accurate detection of IDS incursions, resulting in enhanced IDS security. IDS research is underway, but there are a number of issues that must be solved. A serious problem that may have a major effect on IDS performance is the bias in the training datasets. Compared to the negative class, the positive class has a relatively low number of examples. It's possible that a classifier that was trained with biased data might give more weight to false positives. We've chosen to adjust the class imbalance in order to avoid this situation. The initial stage in a

Figure 3.4 Web threat/intrusion detection system.

proposed strategy to minimize class imbalance is resampling datasets and then employing classifier ensembles to improve classification performance.

From 2018 to 2019, Marin et al. [22,31,32,35] devised a system for detecting key participants and determining distinguishing traits of cybercriminals. To create interaction graphs and identify groups of hackers, a mix of SNA (Social Network Analysis) and seniority analysis was utilized (Figure 3.4).

3.2.1 Cyber defense

Cyber defense is the strategic execution of activities in the cyber domain to achieve mission goals. When comparing network assurance (security) with mission assurance, the key difference is the priority placed on preventing cyber-attacks. The four pillars of cyber defense are detection, identification, orientation, and engagement with the adversary to guarantee mission success and outmaneuver them. When the focus moves from security to defense, intelligence and reconnaissance must be given more weight, and staff functions including intelligence, operations, communications, and planning must be coordinated with one another. As a matter of mission assurance, organizations and their associated electronic information and information infrastructures need defensive cyber operations to be carried out on or via the global information infrastructure. In most cases, a defensive cyber strategy will not include attacking the enemy. It is possible to launch offensive cyber operations against the world's computer networks in an effort to sabotage the resources, plans, and operations of a hostile nation or terrorist organization. Active cyber defense comprises pursuit operations aimed at the attacker (Figure 3.5).

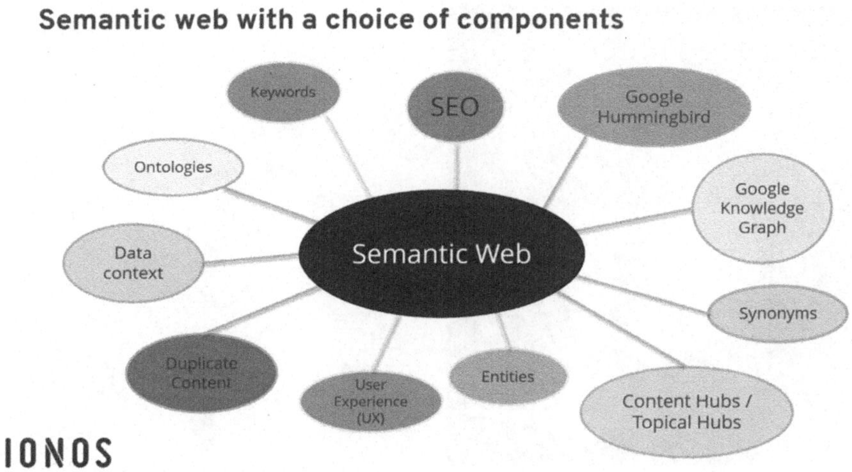

Figure 3.5 Cyber defense.

3.2.1.1 Proactive cyber defense

Anticipating and countering an attack in the cyber and mental worlds is the essence of proactive cyber defense. Taking preventative steps in cyberspace may be seen as a compromise between being too aggressive and being too passive. As a preventative measure or a means of self-defense, this might mean stopping an assault in its tracks or preventing an attacker from gathering his or her forces. Common approaches include attribution, threat hunting, aggressive pursuit, and cyber deception. The employment of psychological operations, controlled broadcast of information, precise targeting, information warfare operations, computer network exploitation, and other active measures is an objective of both preventive and offensive operations. As well as improving operational preparedness in the actual or virtual fighting field, the goal of proactive defensive measures is to improve information gathering by stimulating the responses of the threat agents. Cyber defense may be pre-emptive in nature, meaning that it aims to stop a potential cyber assault before it happens, or it can be defensive in nature, meaning that it seeks for the source of an oncoming cyber operation in order to stop it.

3.2.2 Semantic web

The World Wide Web Consortium (W3C) has established a set of standards for the development of the Semantic Web, often known as Web 3.0. (W3C). The Semantic Web's mission is to make the information found on the internet accessible to computers. Technologies like Resource Description Framework (RDF) and Web Ontology Language (OWL) are used to allow

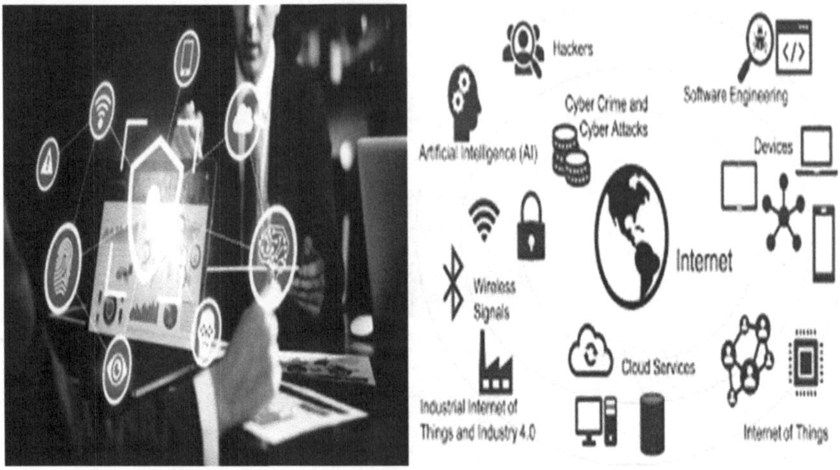

Figure 3.6 Semanticweb.(https://www.ionos.com/digitalguide/online-marketing/search-engine-marketing/semantic-web/)

for the encoding of semantics with the data. Technologies like this are used to officially represent metadata. Concepts, entities, and connections among them are all describable by ontology. These embedded semantics provide several benefits, including the ability to reason about data and use a variety of different types of data. In essence, the RDF is the backbone of these standards since it promotes widespread use of the Web's common data formats and exchange methods (Figure 3.6).

3.2.2.1 Semantic Web classification

The Semantic Web employs the same document classification mechanisms as the rest of the internet. In addition to their regular purposes, annotations may be used to improve or organize documents. Categorization may be enhanced by inferring more information about texts using knowledge in the form of ontology. To classify documents using this approach, you must combine classification learning with previous knowledge and feature development. Document classes may be included when annotating documents. For categorization, predefined portions of documents should be employed. The use of non-textual data, such as images, is a current and future research topic. R.-H. Ferguson, the study's author, presents an application used by Western senators to analyze news and media information.

V. M. Vilic's study looks at a syntactic-based approach of word extraction and inquiry that employs an information dictionary and is employed in the actual world. By abusing syntactic examples, technical methodologies that depend on language structure often make incorrect assumptions about the meaning of opinion aims. Synaptic pattern based on direct reliance and the word identified by exact match, rather than other correlations and word conversions. Patterns of synaptic connections, the exact dictionary, and matching may be found and accomplished using this way. Straight dependencies are classified into two types. The main type shows that a single word is reliant on another word without the need of extra words to show their dependency. The second kind implies that two words rely on a third word. When dealing with small datasets, syntactic-based strategies perform well, but when used with large datasets, they exhibit a reduction in accuracy and the weakest recall when compared to other contemporary models. This is based on three different datasets. Furthermore, it was discovered that semantic terms could be recovered and additional connections and studies may benefit from the technique.

The Dark Web's hacker forums in 2019, according to Zenebe et al. [26], were a rich source of information that may be used to identify possible cyber threats and discover important participants. They used IBM Watson Analytics and WEKA ML algorithms to identify the most prevalent threats and important players in the hacking community.

As Marin and Almukaynizi and other academics have suggested, a temporal logic framework may be able to detect near-future threats (post conditions). In research, cyber-attacks, hacker activities, and technical features were examined as socio-personal variables.

In 2021, Huang et al. [29] created the hybrid CA/SNA method Hacker Rank to identify renowned hackers. CA is used to extract members' interests in topics, while SNA is used to build a network portraying their connections and identifying notable hackers as the next step. Key hackers are individuals with the highest Hacker Rank rankings, which assesses members' ranks. Members' interactions (responds) and the number of responses a user posts and replies are used to assess how active they are in social network graphs.

3.2.3 Hacking

The CTI framework created by Samtani and colleagues is used to examine hackers' assets in forum discussions. Hacking tools are disseminated via classification and TM. Trending tools are found by combining existing metadata with the text of individual posts. The framework employs bipartite SNA to identify the community's most hazardous hackers. Each problem is represented by a bipartite network that depicts the interaction between hackers and threads of certain asset types.

In 2017, the CTI tool created by Grisham et al. [21] can read hacker forums on the Dark Web, where malware is discussed in a multitude of languages, and identify the authors and developers of specific mobile spyware. Malware was detected using a neural network and recurrent neural network (RNN) architecture, and the most prolific malware producers were tracked using SNA. The application extracts textual information such as sub-forum names, subject titles, post content, and attachment names before classifying it as harmful software. It identifies popular and pervasive malware by propagating the discovered infection during the publication time. In order to find co-occurrences between hackers and authors, an actor's bipartite author-thread network was projected onto the writers' network.

In 2018, Pastrana et al. [22] created a technique with two goals: preventing young people from becoming cybercriminals and supporting security companies in reacting effectively to cyber-attacks. They use a variety of techniques to better comprehend the behaviors of online criminals. The method detects important players and evaluates the temporal growth of their interests and skills to anticipate which users will become big actors in the future by modeling the common elements of their forum activity. The system we've envisioned so far includes ML, natural language processing (NLP), SNA, and TM. Furthermore, the actors' interactions were represented via a directed network of answers and citations.

From 2017 to 2018, Biswas et al. [23,24] analyzed hacker forums to find crucial characteristics for recognizing notable hackers. Every member of a hacker community has a distinct function in the overall success of the organization. To generate predictions and develop a model to categories hackers, text mining and sentiment analysis approaches were applied. Four hypotheses were determined to be extremely significant. The number of replies in each thread, the number of messages submitted, and the average message size are all criteria that may be used to categorize hackers based on their degree of expertise. The third and fourth theories have little to no support. Botnet traffic-based intrusion detection and prevention systems (TIDS) for infected devices known as "bots" were allegedly distributed via the bot master workstation. Randomized data partitioned learning model (RDPLM) and packet header analysis are used in the proposed approach to detect attacks. The authors used new feature selection approaches in order to more correctly detect intrusions. There is evidence that this method may enhance detection accuracy while using less computer resources and scaling to big networks.

When looking at machine learning techniques for intrusion detection, Buczak and Guven considered a wide range of elements, such as algorithm complexity and security concerns. When determining the optimum intrusion detection approach, the authors considered a number of aspects, including accuracy and algorithm complexity.

In 2012, cascaded classifiers, as proposed by Khor et al. [45], increased the detection rates of the IDS for atypical attacks. To allow each expert to focus on a reduced number of categories, the proposed method first separates the uncommon intrusions from the non-rare intrusion category. Because of the dominant intrusion category, detection rates for rare intrusions have increased, which this method helps to reduce. Furthermore, the detection rates of the technique are enhanced while the computational costs are lowered due to the employment of two network traffic filters. Aburomman and

Reaz presented classifier ensembles to increase the accuracy of the IDS. The authors built an ensemble for this investigation and evaluated it against the Weighted Majority Algorithm (WMA). The weights provided by LUS IDS's performance has improved as a result of Meta tuning.

In 2016, Qassim et al. [30] investigated the characteristics that are most effective in identifying a broad variety of anomalies in network data. They developed A-IDS, an alert classifier that detects and categorizes abnormalities in packet headers. The flow of network traffic is watched to determine if an attack is taking place, and characteristics are chosen and compared to current data. By fusing radial basis with support vector machines, Govindarajan et al. [33] developed a groundbreaking hybrid IDS. According to a number of data sources, heterogeneous models are more successful than homogeneous models for detecting intrusions.

In 2016–2018, Liu et al. [34] were able to increase the number of flows in the minority class while simultaneously enhancing the generalization of the classifier by using resampling and a variety of ensemble approaches. Adding weights to flows that have been erroneously classified may assist. In 2018, Marteau et al. [35] presented a new similarity measure for symbolic sequential data to detect an unknown attack. The authors applied the SC4ID approach (Sequence Covering for Intrusion Detection) to the best of their abilities while trying to cover a certain sequence of sequences. SC4ID was evaluated using well-known system call datasets including the UNM and ADFA-LD.

They employed the reputation systems of hacker forums as a source of truth data for each of the three experiments. An approach is provided to illustrate how a model derived from one topic may be stretched to identify prominent players in other forums. Scientists believe that hackers' usage of new tools might give them a notion of what they'll be releasing in the future. Adoption behavior may be better understood by seeing how the hacker community adopts and publishes on topics that affect other members of the community. A method known as sequential rule mining analyzes the sequences of posts made over certain time periods to determine members' posting tendencies (day and hour). These criteria are applied to the dataset in order to anticipate future postings.

In 2018, A. Deb et al. [25] proved that sentiment analysis might aid in the prediction of cyber-attacks. By examining the main ideas stated in internet forums, scientists hope to learn more about hacker behavior over time.

3.2.4 Cyber threats

In 2017, Sapienza et al. [14] facilitated the merging of social networks across the open, encrypted, and hidden webs. Cyber security professionals use Twitter and Dark Web forums established by hacker organizations to keep the public informed about possible cyber risks. Text mining techniques are used in Dark Web hacker forums to detect significant phrases and how often they occur. Words having these keywords in their contexts are included in the system's extensive vocabulary, enabling it to better understand and enhance the warnings, as well as follow and watch the progression of actions linked with previously found terms on the Dark Web. Predictions about hackers may eventually be generated utilizing data from the white-hat community, vulnerability researchers, and Deep Web sites. The technology may send out timely notifications to cyber security firms utilizing the found correlations. Rules are learnt by using concepts from logical programming (the Point Frequent Function in particular) and causal inference (PFR).

In 2018, Williams et al. [15] created an incremental technique for crawling and categorizing cyber hazards for the first time. In the classification step, all of the most current hacking tools and attachments, as well as the most recent trends in threats, are classified.

In 2018, Narayanan et al. [16] proposed an ontology-based system with semantically improved knowledge representation and reasoning for early detection of cyber-attacks. Because information is scarce, security blogs and Dark Web conversations are sometimes tailored to a particular target audience, such as like-minded individuals or fellow experts. Its reasoning is based on previous assaults, the instruments used in them, and a range of similar indicators. Complex patterns in data streams may also be discovered using association rules and clustering algorithms.

In 2018, Tavabi et al. [17] used neural language modeling to create Dark Embed, a platform for predicting vulnerability exploitability. Low-dimensional vector space embedding is used to represent data from Deep and Dark Web forums, with the goal of discovering grammatical, contextual, and semantic links between words. After that, we utilize the distributed representations as criteria for categorization. Looking at content from the Dark Web to learn about cyber dangers, such as information on dangerous tools and services, the individuals use and spread them and hack data. Text feature analysis is a technique that combines data from Dark Web forums and markets with audio recordings of actors chatting and engaging (breached data). Key businesses publish a broad variety of content, including information about their goods and services, as well as information about the market in which they operate.

The authors were able to verify the necessity of considering forum data due to their wide usage of language. In 2010, according to Ampel et al. [18], deep transfer learning was used to categorize attack source code. Expert labels from public exploit repositories are utilized in the Deep Transfer Learning for Exploit Labeling (DTL-EL) application to do a more extensive examination of hacker forums. Predefined categories enable cyber security businesses to take proactive measures based on information gathered by the system. Because of the large quantity of information and categories available on the Dark Web, this method might be employed there.

For example, the INTIME program offers a structure for analyzing and recognizing cyber threats in the context of unique cyber security concerns, such as the IoT (Koloveas et al. [19,20]). Rather than depending just on publicly accessible data, this technique incorporates information obtained from the Deep and Dark Webs, as well as confidential databases collected by intelligence agencies. Information is gathered, its utility is assessed, threats are recognized, and the resulting cyber threat information is disseminated through technology. Crawlers and social network monitors, as well as NLP and machine learning algorithms, are employed for these operations.

In 2019, T. D. Wagner, K. Mahbub, et al. [8] enhanced message structure that may assist multicast key management systems. Users in the subgroups are assigned one-way tasks to transmit fresh important material in the new message structure. Using this method instead of the previous one, just the sections of the message intended for normal users will be conveyed, avoiding the need for further communications. According to T. Miloshevska, the system provides a decentralized access control framework for cloud storage techniques by using the secret key of privacy preserving extraction. This approach eliminates the requirement for authentication and synchronization across many authorities. The envelope rules-based insensitive assurance and Pedersen assurance are the most effective techniques for addressing cryptographic challenges. As a result, the passwords for the components of legal authentication are provided to the users.

N. M. Chayal and N. P. Patel add identity-based user revocation to the ABE distribution mechanism. In addition, the approach generates a significant number of independent attribute authorities. According to S. Samtani and M. Kantarcioglu, "Only a Genuine User" may access the encrypted data. The ability of this method to evade replay attacks may aid in the prevention of eavesdropping. Only an authorized person may see the data decryption. It is feasible to verify the user's authorization to deliver a message stored in the cloud using F's approach (R. Montasari et al. [11]) in 2021. It also meets the user's requirements for secrecy and consent.

In 2000, Hwang et al. showed that if the password table is compromised, the whole system is jeopardized. A new smart card-based authentication technique was then suggested. Despite this, their system is vulnerable to an impersonation attack, in which a user assumes the identity of another user and uses their passwords without knowing the secret key. Despite the fact that the preceding data's policy is inaccurate, utilize the previous write of sale data to replace the new information. Authentication and security methods are also reviewed. This chapter covers how users in a distributed system might be permitted to access cloud data. Repetition attacks may be avoided, and user withdrawal can be treated.

According to Rescorla, two users may exchange sensitive keys over an insecure network by using the Diffie-Hellman protocol Man-in-the-Middle. Because of this method's ability to communicate user credentials over the network, middle attacks are conceivable. If the user's credentials are not safeguarded, their privacy may be jeopardized. Password-Authenticated Key Exchange (PAKE) is the outcome of combining key exchange methods with password-based authentication. This method utilizes a one-way function to generate the verifier, and the secret key is computed by all parties involved. In this method, however, a secret password is still necessary to safeguard the verifier from a malicious attacker.

3.2.5 Cyber security

To prevent unauthorized users from gaining access to or destroying information stored on a computer system or over a network, IT security measures, programs, or hardware must be put in place and also unauthorized modification of data and unauthorized diversion of services. The growing use of "smart" devices like smart phones, smart TVs, and the other gadgets that make up the "IoT" have all contributed to the field's increased visibility and importance. Due to the political use and technological complexity of information systems, cybersecurity has emerged as one of the world's most pressing problems in the modern era. Assuring the system's stability, integrity, and data protection is its top priority.

3.2.5.1 Cyber security techniques

We have relied heavily on the concept of a user name and password to keep our data safe. It's possible that this is the first step in ensuring the safety of our digital infrastructure. All papers must be checked for validity before download to guarantee they are from a trustworthy source and have not been altered. Commonly, anti-virus software on the devices verifies the legitimacy of the documents. It is essential that you have reliable anti-virus software installed on all of your devices to protect them from malicious code.

Malware scanners are software that scans a computer's files and documents for dangerous code or viruses. Malicious software is an umbrella word for many different types of potentially destructive computer programs, such as viruses, worms, and Trojan horses.

A firewall may help shield your computer from intruders like hackers, viruses, and worms that try to access your system through the internet. If a communication is not secure enough, the firewall will not allow it to enter or leave the network. Firewalls are essential for protecting against malware. Anti-virus software is designed to identify potentially harmful software on a computer system, such as viruses and worms, and to stop them from doing damage. The auto-update function included in all current anti-virus programs ensures that the software always has the most up-to-date viral signature database. It goes without saying that in this day and age of computers, you need some kind of anti-virus software.

3.2.6 Cyber threat intelligence (CTI)

CTI relies heavily on the Dark Web for information. In the first half of this section, we'll go through all of CTI's many components. Following that, I'll provide a quick summary of the most current study, including its aims, techniques, instruments, and findings. I'll also go through any potential limits.

Companies in both the public and commercial sectors may utilize CTI to keep an eye out for and deal with cyber threats. This data might provide light on the tactics, methods, and procedures (TTPs) used by adversaries. Based on the kind and purpose of a CTI system, CTI security alerts and recommended settings are also offered. CTI can answer often asked questions such as what, who, where, when, and how. CTI has access to several data sources. If the data comes from an internal or external source, it may be feasible to extract information from log files, firewall logs, alarms, historical event responses, malware used in attacks, and network traffic.

You should have high-quality customer relationship intelligence (CRI) on hand when you need it. Based on the organization's aims and roots, a CTI-specific subgroup may be identified. It is possible to create several types of intelligence by combining open data sources and social media with human ingenuity and technological capacity, including but not limited to technical intelligence (TECHINT). Sari had something to say about each of the CTIs. Alsmadi claims that COMINT, DARK WEB Information, SIGINT, and Geospatial Intelligence are all additional forms of intelligence gathering (GEOINT). Alsmadi identifies the following as the three tiers of cyber intelligence.

To identify possible dangers demands understanding their sources, aims, and likely results. Strategic Cyber Intelligence (SCI), for example, provides information on the resources and capabilities of possible attackers and may predict the pathways that various players would take to accomplish their goals. As a subset of Cyber Intelligence, "Tactical" or "Technical" Intelligence informs organizations about the methods and tools used in cyber-attacks, as well as how they might defend themselves. Information obtained from the Dark Web might be used to prevent and repair intrusions. Several factors must be considered while conducting this kind of inquiry, including maintaining track of hacker activities in organized organizations and reviewing previous attacks.

Shakarian divides CTI into four phases, with the highest levels focusing on long-term issues:

- Situational awareness,
- Impending threats,
- Capability knowledge, and
- Community understanding.

Data-driven CTI necessitates multiple techniques for data gathering, processing, and analysis in order to fulfill an organization's security goals. Examine the existing security domain, examine current cyber risks, monitor your own cyber assets, and predict potential future threats to get a sense of the intelligence required by a corporation.

3.2.7 Machine learning for cyber defense

In 2017, Gupta et al. [42] demonstrated that global research has been conducted to tackle the issues associated with growing user numbers, such as interference control, load sharing, and greater capacity. 5G is emerging as a promising potential for meeting demand and addressing these obstacles. New technologies, such as relays in device-to-device communication and tiny cell access points, have been developed. This chapter focuses on 5G wireless communication network security challenges.

In 2017, Deepak et al. [43] said that nowadays, more people are using various web services for activities such as online shopping and payment operations. IDS is one of the top systems used by numerous businesses to identify these attacks. The distributed IDS is proposed in this work for efficient distributed denial-of-service (DDoS) assaults. The suggested intrusion detection uses server IDS and client IDS installed on a central server and individual clients.

In 2019, Zenebe et al. [26] said that IBM Watson Analytics and the WEKA machine learning techniques were employed to analyze cyber threat information from darknet forums. CTI is information or knowledge about dangers that may aid in preventing cyber security breaches. The findings revealed that the most popular tools on the darknet were Crypter, Password cracker, RATs (Remote Administration Tools), buffer overflow attack tools, and keylogger system exploits. As a result, studying darknet forum postings may yield actionable information, and machine learning helps create classifiers for attack type prediction. CTI researchers uncovered cyber threat information in darknet hacker forum postings. Hackers' most often used tools include Crypter, Password cracker, RATs, buffer overflow exploit tools, and keylogger programs. Knowledge of these tools will assist security experts in determining which tools to monitor and optimize appropriately. The study used descriptive analytics to identify patterns in attack type and prominent participants in darknet forums.

In 2019, Marin et al. [27] said that AI program learns rules that link harmful hacking behavior with real-world cyber occurrences through a temporal logical framework. The framework analyzes hackers and their techniques when they organize cyber offensives online, taking into account socio-personal and technological signs of corporate assaults. Our findings show that the suggested technique is viable, outperforming baseline systems by an average F1 score increase of 138%, 71%, and 17%. The waste of time, money, and resources used by enterprises to combat unrelated cyber threats may lead to them becoming criminal hackers' next target.

In 2018, Sarkar et al. [28] clarified that with the surge in security breaches in recent years, there has been a growing need to mine information from social media platforms to raise alarms of potential assaults. The authors used SNA techniques to examine forum reply networks in order to better comprehend the structure of underground message boards. We conclude from our experiments that the Deep Web may be a reliable informational resource for predicting external company threats. Using machine learning models to forecast security risks opens up several research options, such as predicting if vulnerability would be exploited via Dark Web sources. Several efforts have been made to forecast real-world cyber-attacks using external social media data sources.

3.3 PROPOSED METHODOLOGY

"Deep Learning Technique for Dark Web Threat Detection and Analysis on Social Media" is entirely based on Dark Web threat monitoring. The growth of Online Social Networks in recent years has spread social news such as commercial advertisements, political news, and celebrity information. Some social networks, such as Facebook, Instagram, and Twitter, have been impacted by false news. Recent information arriving on an online social network is suspect, and in many circumstances, it misleads other network members. We deployed a pre-trained model on our dataset and discovered that RNN outperforms other known deep learning approaches in identifying social media threats. Social networks on the Deep and Dark Webs are essential study and practice grounds for hackers. Some hacking-related things exchanged, discussed, and offered for sale on such networks include leaked information, stolen credit card data, and system vulnerabilities. It is critical to develop efficient strategies for detecting them early. This research presents a unique deep neural network for the early detection of threats.

3.3.1 Methodologies to be used

In this part, we will first explain the system's overall structure. Following that, we will go through each stage of the proposed framework and the dataset utilized in the experiments. The following phases comprise the general structure of the proposed system:

3.3.1.1 RNN architecture

Studies using sequence information, such as word, audio, and video, have made substantial use of RNNs. Sigma cells and tanh cells, the building blocks of RNNs, are unable to glean enough information from the input data when the input gap is large. By incorporating gate functions into the cell structure, long short-term memory (LSTM) may be able to successfully tackle the problem of long-term dependency. The authors provide a road map for the future of LSTM network research. To investigate the LSTM cell's learning potential, we evaluate the LSTM cell and its modifications [44].

Recurrent cells in RNNs' recurrent layers (sometimes called hidden layers) include feedback connections that cause their states to alter based on both historical data and current input. A wide variety of RNNs may be created using different structures for the recurrent layers. Therefore, the structure of the RNNs' recurrent cells and networks is what sets them apart from conventional neural networks. RNNs may have varying capabilities because of differences in cells and inner connections. This section initially provides a quick overview of the LSTM cell and its variations before delving into the construction of LSTM networks. Recurrent networks have the benefit of sharing weights for each location of the input vector. Another benefit is that it decreases the number of parameters the network must learn (Figure 3.7).

RNNs are a kind of artificial neural network. They are employed in a variety of applications where data containing sequences is involved, like predicting the following word in a sentence. Unlike feed-forward networks, which influence all brain layers, RNNs only affect the layer they are applied to. Consider networks with recurrent layers to better grasp how one

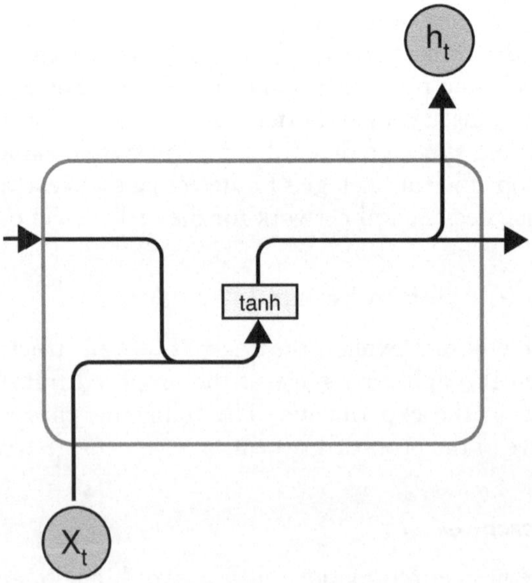

Figure 3.7 The general schema of an RNN unit. (Adapted from Olah 2015.)

performs after being adequately trained. At each iteration of the timeline, we provide the model the next piece of information. This instance may be represented as a feed-forward network if its lifetime is known in advance (let's say, *t* time steps). The RNN may be trained using gradient-based methods, depending on the unrolled model. All back-propagation techniques used for feed-forward networks may also be employed for RNN training. Our unrolled network consists of clusters of connections that all point back to the same node in the raw network. It allows us to accurately drive the network by applying an error derivative that takes into account all the variables that impact a link's weight. Therefore, the state of a recurrent network is represented by the activations of its neurons (Figure 3.8).

Every input and every output in a convolutional neural network (CNN) is completely autonomous. Predicting the next word in a sentence, for example, requires knowledge of the previous words in the phrase, so it's helpful to keep them in mind. Therefore, RNN was created, and it addressed the problem by adding a Hidden Layer. One of the most important features of an RNN is its hidden state, which remembers specific details about a sequence. RNNs are equipped with a memory unit to store the results of all their calculations. It utilizes the same parameters for each input since it always produces the same result by performing the same operation on all inputs or hidden layers.

3.3.2 Long short-term memory

According to Hochreiter and Schmidhuber (1997), the LSTM cell tackles the problem of "long-term dependency." By inserting a "gate" into the cell, they increased the usual recurrent cell remembering capabilities. Since this groundbreaking study, several scholars have modified and popularized LSTMs. There are many variants of LSTM, such as those with a forget gate, without a forget gate, and with a peephole connection. When people talk about LSTM cells, they typically mean LSTMs that have a forget gate [44].

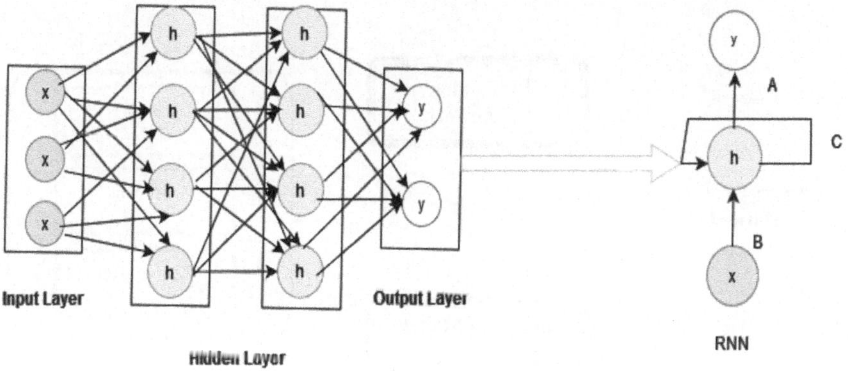

Figure 3.8 Simple recurrent neural network

3.3.3 Pre-trained models: the scenario of new age transfer learning

A comparable problem has previously been addressed by a pre-trained model. Instead of proposing a model to deal with a comparable problem, we might use the model trained on the prior issue as a starting place. Assume we want to develop a self-learning car. We could spend years constructing a solid image recognition algorithm from the ground up, or we could utilize Google's inception model (a pre-trained model based on ImageNet data) to find pictures in those photographs. Although a pre-trained model's accuracy may be compromised in practice, using one eliminates the need to start from scratch. Keras Applications are deep learning models that come with pre-trained weights. These models may be used to forecast, extract features, and fine-tune them.

3.3.4 High-level view of proposed experimental diagram

The high-level framework represents the block diagram of the suggested experimental framework. It demonstrates that we may utilize any approaches to complete our job, including the idea of transfer learning by applying any pre-trained model or tailoring it to our dataset. It depicts how our dataset may be fed into any pre-trained model or improved to meet our needs. We will generate findings, analyze them, and have further conversations using specific detecting techniques.

3.3.5 Operational view of proposed experimental diagram

The proposed framework's operational perspective is a detailed high-level view. Figure 3.9 depicts the entire flow of it, as the dataset is now expanded through a series of steps such as data collection and data

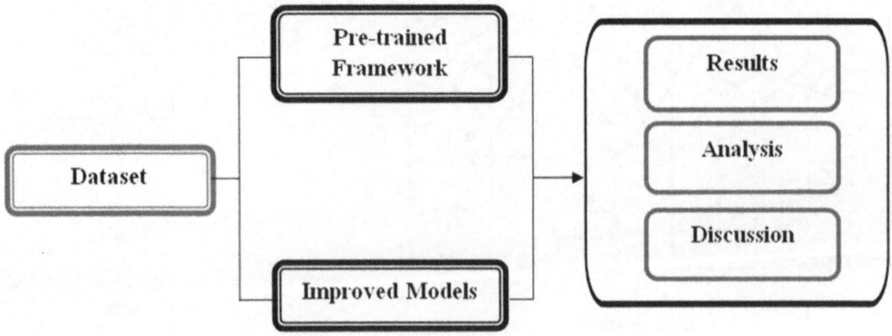

Figure 3.9 High-level view of proposed experimental diagram.

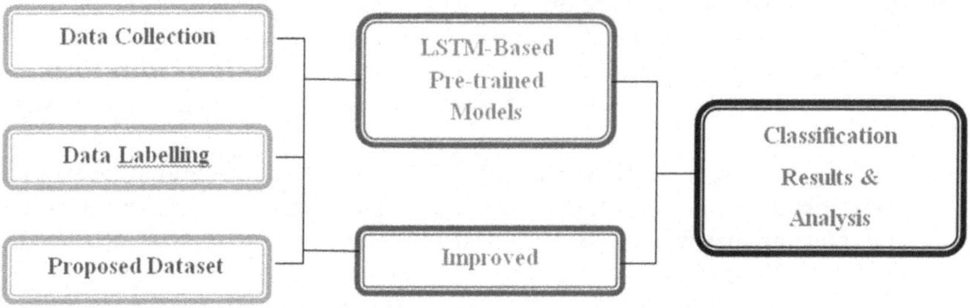

Figure 3.10 Operational view of proposed experimental diagram.

labeling (image labeling) to achieve the final proposed dataset, which is then transferred to pre-trained models or some LSTM-based pre-trained models, or we can customize them, and we can get some meaningful results [46–58] (Figure 3.10).

3.4 CONCLUSION

LSTM is being selected and implemented in order to identify and evaluate attacks that take advantage of security issues using machine learning methods. The framework established will give society the necessary wide knowledge of cybercrimes, allowing them to analyze the danger of such assaults and prevent cybercrime offenses. The framework built saves time and eliminates the need for manual reporting. It assists in determining the number of cases filed by occurrence and by location. This is important for classifying instances and taking precautions against filing cybercrime complaints in recognized hotspot locations. The proposed strategy as opposed to other conventional methods, NS-NN, reveals the utility of parameter computation. Future work would have a smart interface implementation with more parameters for reference.

REFERENCES

1. W. Tounsi, "What is cyber threat intelligence and how is it evolving?," in *Cyber-Vigilance and Digital Trust: Cyber Security in the Era of Cloud Computing and IoT*, W. Tounsi, Ed., ISTE Ltd, Washington, NJ, 2019.
2. A. Dutta, and S. Kant, "An overview of cyber threat intelligence platform and role of artificial intelligence and machine learning," in *Proceedings of the 16th International Conference, ICISS 2020, Lecture Notes in Computer Science Book Series (LNCS)*, pp. 81–86, Jammu, India, December 2020.
3. Z. Mador, "Keep the dark web close and your cyber security tighter," *Computer Fraud & Security*, vol. 2021, no. 1, pp. 6–8, 2021.

4. A. L. Queiroz, and B. Keegan, "Challenges of using machine learning algorithms for cybersecurity: a study of threat classification models applied to social media communication data," in *Cyber Influence and Cognitive Threats*, J. M. Vladlena Benson, Ed., Academic Press, Elsevier Inc., Cambridge, MA, 2020.
5. E. R. Leukfeldt, E. R. Kleemans, and W. P. Stol, "Cybercriminal networks, social ties and online forums: social ties versus digital ties within phishing and malware networks," *British Journal of Criminology*, vol. 57, no. 3, pp. 704–722, 2017.
6. P. Shakarian, "Dark-web cyber threat intelligence: from data to intelligence to prediction," *Information*, vol. 9, no. 12, p. 305, 2018.
7. W. Tounsi, and H. Rais, "A survey on technical threat intelligence in the age of sophisticated cyber attacks," *Computers & Security*, vol. 72, pp. 212–233, 2018.
8. T. D. Wagner, K. Mahbub, E. Palomar, and A. E. Abdallah, "Cyber threat intelligence sharing: survey and research directions," *Computers & Security*, vol. 87, p. 101589, 2019.
9. T. Miloshevska, "Dark web as a contemporary challenge to cyber security," *KriminalističkeTeme*, vol. 5, pp. 117–128, 2019.
10. N. M. Chayal, and N. P. Patel, "Review of machine learning and data mining methods to predict different cyberattacks," in *Data Science and Intelligent Applications, Proceedings of ICDSIA 2020,* Springer, Gujarat, India, 2020.
11. R. Montasari, F. Carroll, S. Macdonald, H. Jahankhani, A. Hosseinian-Far, and A. Daneshkhah, "Application of artificial intelligence and machine learning in producing actionable cyber threat intelligence," in *Digital Forensic Investigation of Internet of Things (IoT) Devices*, R. Montasari, H. Jahankhani, R. Hill, and S. Parkinson, Eds., Springer, Berlin, Germany, 2021.
12. R. Liggett, J. R. Lee, A. L. Roddy, and M. A. Wallin, "The dark web as a platform for crime: an exploration of illicit drug, firearm, CSAM, and cybercrime markets," in *'e Palgrave Handbook of International Cybercrime and Cyberdeviance*, T. J. Holt, and A. M. Bossler, Eds., Palgrave Macmillan, London, 2020.
13. O. Cherqi, G. Mezzour, M. Ghogho, and M. E. Koutbi, "Analysis of hacking related trade in the darkweb," in *Proceedings of the 2018 IEEE International Conference on Intelligence and Security Informatics (ISI)*, Miami, FL, USA, November 2018.
14. A. Sapienza, A. Bessi, S. Damodaran, P. Shakarian, K. Lerman, and E. Ferrara, "Early warnings of cyber threats in online discussions," in *Proceedings of the 2017 IEEE International Conference on Data Mining Workshops (ICDMW)*, New Orleans, LA, USA, November 2017.
15. R. Williams, S. Samtani, M. Patton, and H. Chen, "Incremental hacker forum exploit collection and classification for proactive cyber threat intelligence: an exploratory study," in *Proceedings of the 2018 IEEE International Conference on Intelligence and Security Informatics (ISI)*, Miami, FL, USA, November 2018.

16. S. Narayanan, A. Ganesan, K. Joshi, T. Oates, A. Joshi, and T. Finin, "Early detection of cybersecurity threats using collaborative cognition," in *Proceedings of the 2018 IEEE 4th International Conference on Collaboration and Internet Computing (CIC)*, pp. 354–363, Philadelphia, PA, USA, October 2018.
17. N. Tavabi, P. Goyal, M. Almukaynizi, P. Shakarian, and K. Lerman, "DarkEmbed: exploit prediction with Neural Language models," in *Proceedings of the AAAI Conference on Artificial Intelligence*, pp. 7849–7854, New Orleans, LA, USA, February 2018.
18. B. Ampel, S. Samtani, H. Zhu, S. Ullman, and H. Chen, "Labeling hacker exploits for proactive cyber threat intelligence: a deep transfer learning approach," in *Proceedings of the 2020 IEEE International Conference on Intelligence and Security Informatics (ISI)*, pp. 1–6, Arlington, VA, USA, November 2020.
19. P. Koloveas, T. Chantzios, S. Alevizopoulou, S. Tryfonopoulos, and C. Tryfonopoulos,"INTIME: a machine learning-based framework for gathering and leveraging web data to cyber-threat intelligence," *Electronics*, vol. 10, no. 7, p. 818, 2021.
20. P. Koloveas, T. Chantzios, C. Tryfonopoulos, and S. Skiadopoulos, "A crawler architecture for harvesting the clear, social, and dark web for IoT-related cyber-threat intelligence," in *Proceedings of the 2019 IEEE World Congress on Services*, pp. 3–8, Milan, Italy, July 2019.
21. J. Grisham, S. Samtani, M. Patton, and H. Chen, "Identifying mobile malware and key threat actors in online hacker forums for proactive cyber threat intelligence," in *Proceedings of the 2017 IEEE International Conference on Intelligence and Security Informatics (ISI)*, pp. 13–18, Beijing, China, November 2017.
22. S. Pastrana, A. Hutchings, A. Caines, and P. Buttery, "Characterizing eve: analysing cybercrime actors in a large underground forum," in *Research in Attacks, Intrusions, and Defenses*, M. Bailey, T. Holz, M. Stamatogiannakis, and S. Ioannidis, Eds., RAIDSpringer, Berlin, Germany, 2018.
23. B. Biswas, A. Mukhopadhyay, and G. Gupta, ""Leadership in action: how top hackers behave": a big-data approach with text-mining and sentiment analysis," in *Proceedings of the 51st Hawaii International Conference on System Sciences*, pp. 1752–1761, Big Island, HI, USA, January 2018.
24. B. Subba, S. Biswas, and S. Karmakar, "Host based intrusion detection system using frequency analysis of n-gram terms," in *Region 10 Conference, TENCON 2017-2017 IEEE*. IEEE, Penang, 2017.
25. A. Deb, K. Lerman, and E. Ferrara, "Predicting cyber-events by leveraging hacker sentiment," *Information*, vol. 9, no. 11, p. 280, 2018.
26. A. Zenebe, M. Shumba, A. Carillo, and S. Cuenca, "Cyber threat discovery from dark web," in *Proceedings of the 28th International Conference on Software Engineering and Data Engineering*, Toulouse, France, October 2019.
27. E. Marin, M. Almukaynizi, and P. Shakarian, "Reasoning about future cyber-attacks through socio-technical hacking information," in *Proceedings of the 2019 IEEE 31st International Conference on Tools with Artificial Intelligence (ICTAI)*, pp. 157–164, Portland, OR, USA, November 2019.

28. S. Sarkar, M. Almukaynizi, J. Shakarian, and P. Shakarian, "Predicting enterprise cyber incidents using social network analysis on dark web hacker forums," in *Proceedings of the Cyber Defense Review, Special Edition: International Conference on Cyber Conflict (CYCON U.S.): Cyber Conflict During Competition*, pp. 87–102, Washington, DC, USA, November 2018.
29. C. Huang, Y. Guo, W. Guo, and Y. Li, "HackerRank: identifying key hackers in underground forums," *International Journal of Distributed Sensor Networks*, vol. 17, no. 5, p. 15501477211015145, 2021.
30. Q. S. Qassim, A. M. Zin, and M. J. Ab Aziz, "Anomalies classification approach for network—based intrusion detection system," *International Journal of Network Security*, vol. 18, pp. 1159–1171, 2016.
31. E. Marin, J. Shakarian, and P. Shakarian, "Mining key-hackers on darkweb forums," in *Proceedings of the 2018 1st International Conference on Data Intelligence and Security (ICDIS)*, pp. 73–80, South Padre Island, TX, USA, April 2018.
32. E. Marin, M. Almukaynizi, E. Nunes, J. Shakarian, and P. Shakarian, "Predicting hacker adoption on darkweb forums using sequential rule mining," in *Proceedings of the 2018 IEEE Intl Conf on Parallel & Distributed Processing with Applications*, pp. 1183–1190, Melbourne, Australia, December 2018.
33. M. Govindarajan, "Evaluation of ensemble classifiers for intrusion detection," *World Academy of Science, Engineering and Technology, International Journal of Computer, Electrical, Automation, Control and Information Engineering*, vol. 10, no. 6, pp. 876–884, 2016.
34. Z. Liu, R. Wang, and M. Tao, "SmoteAda NL: a learning method for network traffic classification," *Journal of Ambient Intelligence and Humanized Computing*, vol. 7, no. 1, pp. 121–130, 2016.
35. P.-F. Marteau, "Sequence covering for efficient host-based intrusion detection," *IEEE Transactions on Information Forensics and Security*, vol. 14, no. 4, pp. 994–1006, 2018.
36. P. Mishra, et al., "VMGuard: a VMI-based security architecture for intrusion detection in cloud environment," *IEEE Transactions on Cloud Computing*, vol. 8, no. 3, pp. 957–971, 2018.
37. D. Sklavounos, A. Edoh, and G. Paraskevopoulos, "Utilization of statistical control charts for DoS network intrusion detection," *International Journal of Cyber-Security and Digital Forensics (IJCSDF)*, vol. 7, no. 2, pp. 166–174, 2018.
38. Arshad, J., et al., "A novel framework for collaborative intrusion detection for M2M networks," in *International Conference on Information and Communication Systems (ICICS)*, IEEE, Irbid, 2018.
39. A. Parvat, et al., "Network intrusion detection system using ensemble of binary deep learning classifiers," in *International Conference on Smart Trends for Information Technology and Computer Communications*, Springer, Pune, 2017.
40. R. Devi, et al., "Implementation of intrusion detection system using adaptive neuro-fuzzy inference system for 5G wireless communication network," *AEU-International Journal of Electronics and Communications*, vol. 74, pp. 94–106, 2017.

41. C. Kolias, V. Kolias, and G. Kambourakis, “Term ID: a distributed swarm intelligence-based approach for wireless intrusion detection,” *International Journal of Information Security*, vol. 16, no. 4, pp. 401–416, 2017.
42. A. Gupta, R. K. Jha, and S. Jain, “Attack modeling and intrusion detection system for 5G wireless communication networks,” *International Journal of Communication Systems*, vol. 30, no. 10, p. e3237, 2017.
43. K. Deepak, et al., “Distributed intrusion detection system for TCP flood attack,” in *Proceeding of International Conference on Intelligent Communication, Control and Devices*, Singapore, Springer Singapore, 2017.
44. Y. Yu, X. Si, C. Hu, and J. Zhang, “A review of recurrent neural networks: LSTM cells and network architectures,” *Neural Computation*, vol. 31, no. 7, pp. 1235–1270, Jul. 2019, doi: 10.1162/neco_a_01199.
45. A. Kumar, P. Verma, M. Kumar, and J. Kumar, “Cyber threat: a review on dark sides of dark web,” in *2023 3rd International Conference on Advancement in Electronics & Communication Engineering (AECE)*, pp. 306–310, Ghaziabad, India, 2023, doi: 10.1109/AECE59614.2023.10428580.
46. V. K. Sharma, “CNTFET circuit-based wide fan-in domino logic for low power applications,” *Journal of Circuits, Systems and Computers*, vol. 31, no. 2, p. 2250036, 2022.
47. V. K. Sharma, and M. Pattanaik, “A reliable ground bounce noise reduction technique for nanoscale CMOS circuits,” *International Journal of Electronics*, vol. 102, no. 11, pp. 1852–1866, 2015.
48. V. K. Sharma, S. Patel, and M. Pattanaik, “High performance process variations aware technique for sub-threshold 8T-SRAM cell,” *Wireless Personal Communications*, vol. 78, no. 1, pp. 57–68, 2014.
49. V. K. Sharma, and M. Pattanaik, “Techniques for low leakage nanoscale VLSI circuits: a comparative study,” *Journal of Circuits Systems and Computers*, vol. 23, no. 5, p. 1450061, 2014.
50. V. K. Sharma, “A survey of leakage reduction techniques in CMOS digital circuits for nanoscale regime,” *Australian Journal of Electrical and Electronics Engineering*, vol. 18, no. 4, pp. 217–236, 2021.
51. C. Kumar, A. S. Mishra, and V. K. Sharma, “Leakage power reduction in CMOS logic circuits using stack ONOFIC technique,” in *2018 Second International Conference on Intelligent Computing and Control Systems (ICICCS)*, Madurai, 2018.
52. V. K. Sharma, and M. Pattanaik, “Design of low leakage variability aware ONOFIC CMOS standard cell library,” *Journal of Circuits Systems and Computers*, vol. 25, no. 11, p. 1650134, 2016.
53. S. F. Naz, S. Riyaz, and V. K. Sharma, “A review of QCA nanotechnology as an alternate to CMOS,” *Current Nanoscience*, vol. 18, no. 1, pp. 18–30, 2022.
54. S. Riyaz, and V. K. Sharma, “Design of reversible Feynman and double Feynman gates in quantum-dot cellular automata nanotechnology,” *Circuit World*, vol. 49, no. 1, pp. 28–37, 2023.
55. B. Raina, C. Verma, M. Gupta, and V. K. Sharma, “Binary coded decimal (BCD) seven segment circuit designing using quantum-dot cellular automata (QCA),” in *2021 5th International Conference on Trends in Electronics and Informatics (ICOEI)*, Tirunelveli, 2021.

56. U. Mushtaq, and V. K. Sharma, "Performance analysis for reliable nanoscaled FinFET logic circuits," *Analog Integrated Circuits Signal Processing.*, vol. 107, no. 3, pp. 671–682, 2021.
57. Kajal and V. K. Sharma, "Design and simulation of FinFET circuits at different technologies," in *2021 6th International Conference on Inventive Computation Technologies (ICICT)*, Coimbatore, 2021.
58. V. K. Sharma, "Designing of face recognition system," in *2019 International Conference on Intelligent Computing and Control Systems (ICCS)*, Madurai, 2019.

Chapter 4

Energy-efficient hybrid algorithm for reconfiguration in radial distribution system

Surender Singh

4.1 INTRODUCTION

Electrical demand is rapidly increasing, depending on the various sectors such as domestic, industrial, defense, transportation, educational institutions, etc. The growing energy demand in different power sectors, generally in transportation, is met either by generation from nonrenewable/renewable energy sources or by optimal placement of Electric Vehicle Charging Stations (EVCS) in existing radial distribution systems (RDSs). Power engineers focus on providing consumers with reliable, efficient power in quality and economical energy. The requisite demands of the consumers are estimated using load flow within the constraints of line flows in RDS. Furthermore, it can be understood using the well-known load flow backward/Forward sweep method. Many researchers [1–22] have reported the load flow analysis based on backward/Forward sweep method in RDS. The voltage stability is an essential parameter for the system stability, which operates closer to its operating limit. Due to this, the system is severely loaded. IEEE defines voltage stability (VI) as the ability of a system to maintain its voltage stability when the line parameter and line flow are subjected to change. The voltage stability by the CIGRE/IEEE joint task force is defined as the "ability of the power system to maintain the voltage within limits after being subjected to unhealthy condition from a given healthy condition" [23–26]. The primary cause of blackouts in the world is voltage instability. From 1965 to 2005, statistical data has been observed that approximately 12 blackouts out of these seven are responsible for voltage instability [27]. The hybrid algorithm using reconfiguration in RDS is applied in this chapter to achieve the energy-efficient objective with a new voltage stability index (VSI). The term "reconfiguration" in RDS refers to the change process in the existing system's topology, which is done by simultaneously switching sectionalized and tie-line switches. Many researchers [28–112] have reported that optimal reconfiguration is divided into three approaches—heuristic-based, mathematical optimization, and meta-heuristic—that are distinguished in the RDS reconfiguration. In real time, the load flow analysis can be done in

DOI: 10.1201/9781003643609-4

3-phase instead of 1-phase system using the reconfiguration in RDS [113]. The 3-phase system is unbalanced due to unequal loading and impedance in each phase. The unbalancing causes circulating current in the neutral point of the 3-phase system.

The distribution system delivers power to different loads, such as commercial, industrial, and residential, which usually involve dynamic changes in loading with a wide range. The loading patterns are a typical combination of these loads during peak time, leading to the RDS (substation and specific feeders/lines) becoming heavily loaded. Due to this, considerable energy loss will occur, which may lead to instability in the system. Therefore, this chapter implements an energy-efficient hybrid algorithm in MATLAB on IEEE 33-bus and 69-bus using reconfiguration in RDS. The most essential aspect of designing RDS is using reconfiguration algorithms with the help of Genetic Algorithm (GA), which is an appropriate step that guarantees the system stays radial during the implementation. The selection operator, such as the initial population, fitness function, and mating pool, forms the evolution, and genetic operators, such as crossover, mutation, and inversion, form reproduction at every stage of the genetic evolution. Particle Swarm Optimization (PSO) is a random-based search optimization technique that helps to find the optimal solution, to make the hybrid algorithm blend the GA and PSO to get an optimal solution using reconfiguration in RDS. This chapter is divided into seven sections, and the following work is included in each section: An overview of the research proposal and the researchers' stated investigations are provided in the first section. It also covers extensive literature on RDS reconfiguration and voltage stability. The problem formulation to accomplish the proposed investigation is included in the second section. This section's hierarchy outlines the critical characteristics of RDS that must be met before formulating the objective function. The objectives combine with minimizing power loss and maximizing the new VSI. The equality and inequality constraints are also taken into consideration. It also contains the logically deduced equation for the two-bus system of the suggested new VSI. A brief explanation of the PSO principle, together with a flowchart, pseudo-code, and algorithm implementation, is provided in the third section. A quick description of GA and its flowchart is given in the fourth section. A hybrid method with its pseudo-code is described in the fifth section, and its implementation is on two test systems. The result and discussion of the proposed research hybrid algorithm are explained in the sixth section. The hybrid algorithm is implemented on IEEE 33-bus and 69-bus to check the effectiveness of the proposed algorithm. This section also presents a demonstration for comparing the voltage profile, new VSI, and losses with existing methods. The seventh section concludes by implementing a hybrid algorithm using reconfiguration in RDS.

4.2 PROBLEM FORMULATION

This chapter develops a basic algorithm based on circuit theory and fundamental systems analysis techniques. The main aim is to provide an improved calculation model for the RDS, which is computationally quick. The following important aspects are considered to be obtained using the suggested approach just by solving recursive algebraic equations:

- The system is capable of connecting the remote site loads.
- Each segment's line losses and the total power losses are calculated.
- Rerouting or reconfiguring of the system is required for the power distribution to maintain a better voltage profile at each node.

The load flow backward/Forward sweep is an essential method for implementing a hybrid algorithm. The complex power at a particular node is estimated as "the apparent power at a particular node is the sum of its connected load and the power of all the nodes connected beyond to that particular node."

The formulation of a problem is categorized into two steps. In the first step, the total power loss is minimized, and in the succeeding step, the new VSI of the distribution system is maximized.

$$\text{Obj}(F) = \max\left[w_1 \times F_1 + w_2 \times F_{2\,\text{prime}}\right] \tag{4.1}$$

where

F_1 = Real power loss in kW
F_2 = New VSI in p.u.
w_1 and w_2 = Corresponding weight constant
$F_{2\,\text{prime}} = [1/F_2]$

4.2.1 Minimization of total power loss

Power loss combines active and reactive losses due to the resistive and reactive parts of passive branch elements between the nodes. It is well known that active power is dissipated in resistance, and the reactive power is consumed by inductive reactance and generated by the capacitive part of the reactance. In this chapter, the research proposal focuses on minimizing the total active power losses and improving the voltage profile with new voltage stability indices using the proposed algorithm. The total active power loss in RDS is as follows:

$$F_2 = \sum_{j=1}^{Nb} I_j^2 \times R_j \tag{4.2}$$

4.2.2 New voltage stability index

Automation in distribution systems is one of the most extensive challenges faced by power engineers. The distribution system is generally radial in topology. The RDS topology is reconfigured to enhance the operational environment. The system is adversely affected under peak load situations as the loading patterns significantly impact its voltage stability. Therefore, it makes sense that the distribution system might encounter voltage failure or instability in high-loading situations. For every node in RDS, a new VSI is considered together with its voltage angle, which promotes better outcomes. The proposed new VSI is estimated from an equivalent two-bus system. Consider a standard RDS with several buses like IEEE 33-bus and 69-bus, etc. An equivalent circuit of RDS in a two-bus system is shown in Figure 4.1 [115].

The mathematical formulation for new VSI using a two-bus equivalent circuit in RDS is as under [120]:

$$I = \frac{|S|\angle -\phi}{V_2\angle -\delta_2} \tag{4.3}$$

$$V_1\angle\delta_1 - V_2\angle\delta_2 = I\,|Z|\angle\theta \tag{4.4}$$

$$\left(V_1 \times V_2\angle\delta_1 - \delta_2\right) - V_2^2\angle 0 = |S| \times |Z|\angle(\theta - \phi) \tag{4.5}$$

$$V_1^2 - 4\times|S|\times|Z|\cos(\theta-\phi) \geq 0 \tag{4.6}$$

$$\text{new VSI} = V_1^2 - 4\times|S|\times|Z|\cos(\theta-\phi) \tag{4.7}$$

$$f_2 = \left[\left|V_1^2\right| - (4\times|S|\times|Z|\times\cos(\theta-\phi))\right] \tag{4.8}$$

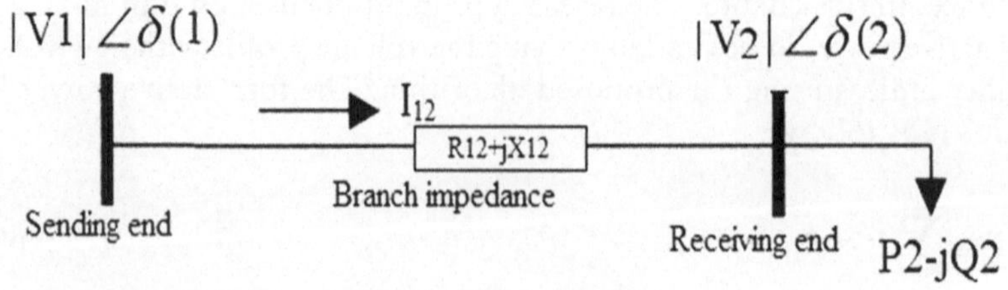

Figure 4.1 A two-bus system equivalent circuit in RDS.

where $\theta = \tan^{-1}\left(\frac{X}{R}\right), \phi = \tan^{-1}\left(\frac{Q}{P}\right)$

V_1 is the sending node voltage, S is the power at a particular node, I_j specifies the branch current, and Z represents the impedance between sending and receiving nodes.

4.2.3 Constraints

The equality and inequality constraints in RDS are as follows:

$$P_{min} \leq P \leq P_{max}, \quad V_{min} \leq V \leq V_{max} \tag{4.9}$$

$$P_G = P_D + P_L \tag{4.10}$$

4.3 PARTICLE SWARM OPTIMIZATION

PSO is a novel heuristic technique developed in 1995 by J. Kennedy and R. Eberhart [112]. Their research principle's objective is to visualize the behavior of fish or birds in a flock. Adjusting their social behavior model can serve as an effective optimizer. The creation of a random population based on a non-linear continuous fitness function served as the objective function for their research. The fundamental PSO algorithm is as follows:

- There are initially multiple potential solutions in the random population. The input data includes the acceleration coefficient C_1, C_2, inertia weight, and minimum/maximum values for position and velocity.

```
position(j,k)=pmin+(pmax-pmin)*position(j,k)
velocity(j,k)=vmin+(vmax-vmin)*velocity(j,k)
```

- Next, the equality and inequality constraints are applied to the position and velocity. Estimate the local fitness and corresponding position.
- The objective function for the best global fitness and corresponding position is calculated using the min/max function.

```
fit(k)=(position(m,k))*(position(m,k))
```

- Now, the iteration for optimizing the objective function starts and computes the updated velocity and position.

```
New veloc(j,k)=w*velocity(j,k)+c1*rand(1)*(l_b(j,k)-posi
tion(j,k))+c2*rand(1)*(g_b(j,i-1)-position(j,k))
```

- Apply the equality and inequality constraints again to assess the local and global optimum fitness/position.

```
if(nv(j,k)<vmin)
nv(j,k)=vmin
position(j,k)=position(j,k)+nv(j,k)
if(position(j,k)>pmax)
position(j,k)=pmax
if(position(j,k)<pmin)
position(j,k)=pmin
```

- Update the position and velocity recommended locally and globally.
- When optimizing, use the convergence criteria. Print the optimized results if the condition has been met; otherwise, return to the beginning of the iteration.

The flowchart of fundamental PSO is shown in Figure 4.2.

The implementation of PSO in MATLAB is as under:

```
%pso main program
% it- no of iterations
clc;
clear all;
close all;
population=input(' Enter the population size ');
variable=input(' Enter the number of variable ');
it=1000;
C1=C2=2;
wmin=0.4;wmax=0.9;
position=rand(population,variable);
velocity=rand(population,variable);
pmin=-10;
pmax=10;
for j=1:population
    for k=1:variable
        position(j,k)=pmin+(pmax-pmin)*position(j,k);
    end
end
vmin=pmin;
vmax=pmax;
for j=1:population
    for k=1:variable
        velocity(j,k)=vmin+(vmax-vmin)*velocity(j,k);
    end
end
position
velocity
```

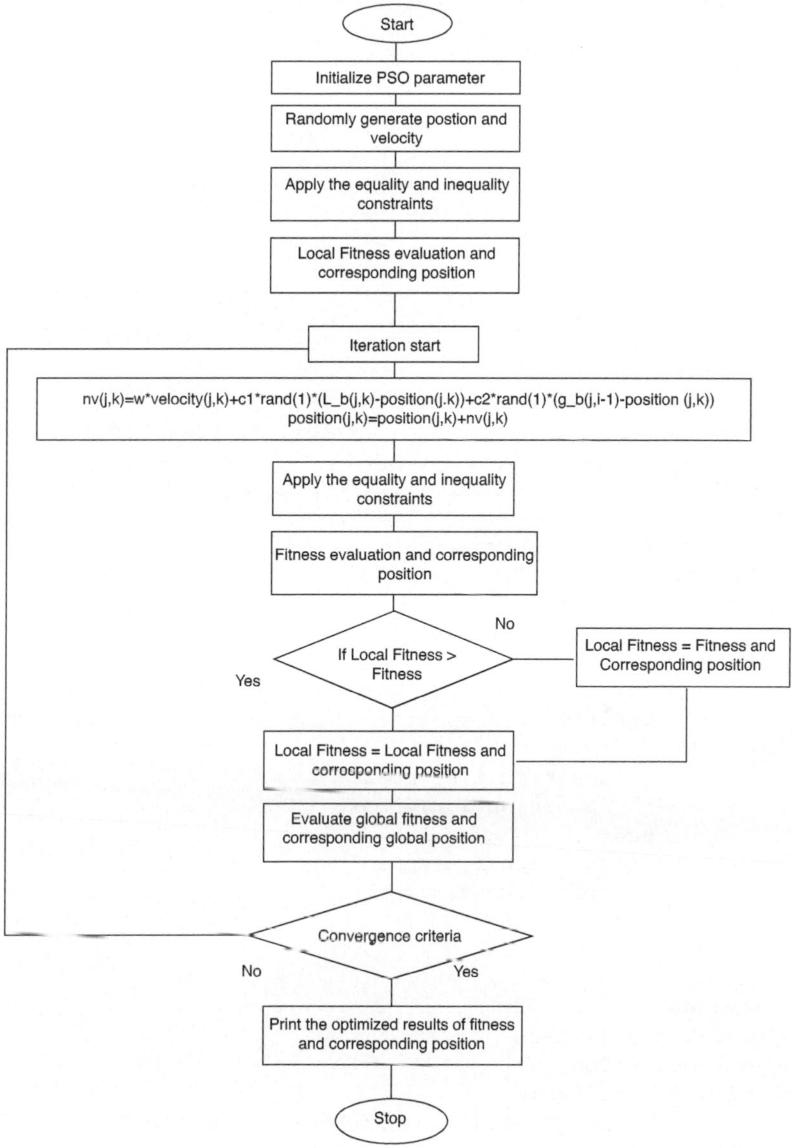

Figure 4.2 Fundamental flowchart of PSO.

```
%change
%calculate fitness
for n=1:population
for k=1:variable
    fit(n,k)=(position(n,k))*(position(n,k));        %x²;
end
```

```
end
%fit
%find global and local best
lb=position;
lf=fit;
[c,d]=min(fit);
gf(1)=c;
%change
g_b(:,1)=position(:,b);
%g_b=position(b);
%iterations begin
for j=2:it
    w=wmax-(wmax-wmin)*j/it;
    for i=1:population
        for k=1:variable
nv(i,k)=w*velocity(i,k)+C1*rand(1)*(lb(i,k)-position(i,k))+C2
*rand(1)*(g_b(i,j-1)-position(i,k));
            %nv(i,k)=velocity(i,k)+C2*rand(1)*(g_b-position
(i,k));
            if(nv(i,k)>vmax)
nv(i,k)=vmax;
            end
            if(nv(i,k)<vmin)
nv(i,k)=vmin;
            end
            position(i,k)=position(i,k)+nv(i,k);  %position
update
            if(position(i,k)>pmax)
                position(i,k)=pmax;
            end
            if(position(i,k)<pmin)
                position(i,k)=pmin;
            end
        end
    end
    %change
    %calculate fitness
for m=1:population
      for k=1:variable
            fit(m,k)=(position(m,k))*(position(m,k));        %x2
    end
end
    %find global and local best
    for m=1:variable
        if(fit(m)<l_f(m));
l_f(m)=fit(m);
l_b(:,m)=position(:,m);
        end
    end
```

```
    [c,d]=min(l_f);
g_f(i)=c;
g_b(:,j)=lb(:,d);
    %g_b=lb(d);
    velocity=nv;     % update the initial velocity with
updated velocity
end
g_b
g_f
```

This method originally emerged for non-linear functions, but it may now be modified to address complex objective functions in the sciences. It is a population-based heuristic optimization technique that derives inspiration from nature. It is superior to earlier optimization techniques in numerous aspects.

4.4 GENETIC ALGORITHM

It is based on Darwin's theory of fittest natural selection, and genetic mechanisms that mimic evolution to successive generations are better and better, like the Hill climbing technique. John Holland, with his students and colleagues at the University of Michigan—most notably David E, who originated the idea in 1965 and published it in 1975—developed GA. Goldberg successfully applied it to several optimization problems. This specific type of evolutionary algorithm uses crossover, also known as recombination, inheritance, mutation, and selection—all of which are influenced by evolutionary biology. It belongs to the class of heuristics for global searches. In general, evolution starts over multiple generations with a population of randomly created individuals. Each generation evaluates each member of the population's fitness, chooses many individuals from the current population (based on fitness), and then modifies the group to establish a new one. The fundamental biological process of GA is Genetics and Evolution, which is briefly explained below, and its implementation in the proposed hybrid algorithm on 33-bus and 69-bus using reconfiguration in RDS.

4.4.1 Evolution

It will be further subcategorized into two parts: initialization and selection, which are briefly narrated as follows.

4.4.1.1 Initialization

The potential solutions usually comprise the population, which varies depending on the nature of the objective function. Generally, the initial population is generated randomly, allowing for the whole search space range. Sometimes, the sample probability distribution is adjusted to concentrate on regions of higher interest, or the solutions are "seeded" in places where optimal solutions are probably to be discovered.

4.4.1.2 Selection

A subset of the existing population is chosen to procreate for the benefit of future generations during each one. Better solutions usually have a higher chance of being chosen individually in a fitness-based process. Specific selection techniques assign a fitness value to each solution and select the best ones first. Some methods rate a representative population sample because the former can take a long period.

4.4.2 Genetics

It originated in 1865 by Gregor Johann Mendel and was based on genes and their characteristics. Our body-building block is a cell with a fixed number of chromosomes for particular living beings. The cell's nucleus is located in its middle. The genetic code is located in the nucleus. The chromosomes hold genetic information. DNA makes up each chromosome, and genes are stored inside them. Each gene is located at a distinct location on a chromosome. The essential schematic representation of genetics (reproduction) is shown in Figure 4.3.

One of the most common methods is to use roulette wheel selection to copy the string to the mating pool. To produce offspring of the next generation, the parents reproduce themselves from the pool. Each string's fitness is expressed by the area on the roulette wheel that each string contributes to the total fitness. For each new solution developed, two-parent solutions are selected for breeding from the previously selected pool. A "child" solution is created by applying the crossover and mutation processes previously discussed. This new solution often has numerous traits of its "parents." For

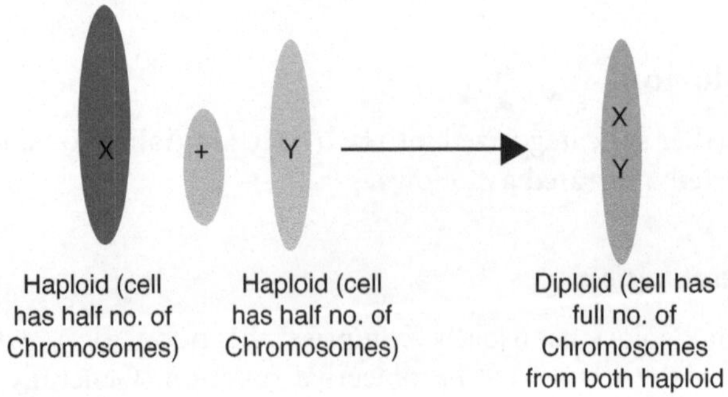

Figure 4.3 Basic schematic representation of genetics.

every additional kid, a fresh set of parents is chosen, and so on, until a fresh population of solutions of a suitable size is generated.

It comprises three basic building blocks: Crossover, Mutation, and Inversion, which are individually explained below:

4.4.2.1 Crossover

Before pairing strings together to cross, use a preset probability to decide whether or not to do the crossover for each pair. The crossover positions are taken at random if a crossover is selected. The flowchart of fundamental GA is shown in Figure 4.4.

4.4.2.2 Mutation

Random changes to string settings are carried out using this low-probability genetic operator. It will slightly alter a string from 0 to 1 and vice versa, and as a result, it diversifies the search and eliminates convergence.

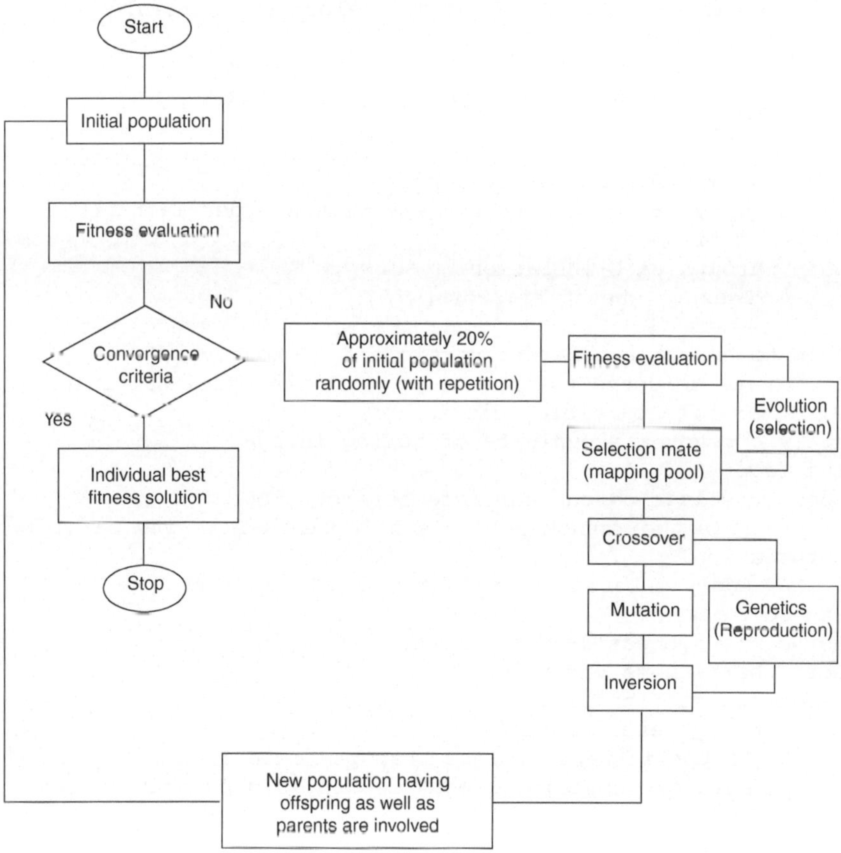

Figure 4.4 Fundamental flowchart of GA.

4.4.2.3 Inversion

As in jumble mutation, we choose a subset of genes for inversion alterations, but instead of rearranging the selection, that may flip every string within the subgroup.

4.5 PROPOSED HYBRID ALGORITHM

The focus of the proposed algorithm is to minimize the total power loss and maximize the new VSI in RDS. The pseudo-code for the proposed hybrid algorithm is as follows:

```
Declare & initialize PSO (pop size), C1=C2=2, min./max.
(position & velocity), random no., etc

Declare & initialize particle with an empty array for cost,
position &velocity
Loop through each particle
      Initialize position, cost, velocity & particle
position & cost
      If  best fitness < global best fitness
            Then Global best = best fitness
      End if
End loop
Store & sort fitness array & pop
Store empty array to best fitness values & pop last fitness
as worst fitness
Loop through each iteration
      P=exp. (-beta* fit/worst fit)
      P=sum (P)
Loop to select two-parent particle for crossover
I1& I2 are selected parents using Roulette wheel selection ()
P1& P2 are the particle information
Apply crossover & evaluate offspring in popc
End loop
Loop to select parent particle for mutation
      I is the random particle selected & P is its particle
information
      Apply mutation & evaluate mutants in popm
End loop
Merge all fitness from crossover & mutation
Sort the fitness variable
Update worst fitness & best solution
Loop through each particle
      If pop fitness is less than particle fitness
      Particle position & cost is replaced by that of pop
      End if
```

```
Particle best fitness is stored in (x)
End loop
Update global best fitness & position
Loop through each particle
       Update velocity
       Apply velocity limits
       Update position & apply position limits
       Evaluate fitness
       If particle fitness < best fitness
       Particle best position & fitness is equal to particle
position & fitness
       If particle best fitness < global best fitness
              Update both global best position and fitness
       End if
       End if
End loop
Loop through each particle
       If particle fitness is less than pop fitness
       Update pop position & fitness
       End if
Store pop fitness in (x)
End loop
Update global best fitness & position
Store best fitness as best fitness of PSO
Update w
End loop
Store global best in the best solution
```

4.6 RESULTS AND DISCUSSION

In RDS, a single source feeds through one end, creating no closed paths but connecting all nodes. This connection approach is commonly used worldwide and makes it inexpensive and easy. Sectionalized and tie-line switches are the two types of switches used in RDS. Sectionalized and tie-line switches operate simultaneously to keep the radial topology, termed a process known as reconfiguration. In this chapter, two test systems, the IEEE-33 bus and the IEEE-69 bus, are examined. Using a forward-backward sweep approach, both test systems estimate the voltage profile, losses, and new VSI. The system's reliability becomes compromised whenever unhealthy conditions arise as nodes beyond the defective node are not supplied. Tie-line and sectionalized switches must alternatively operate to enhance reliability, optimize voltage profile, and reduce line losses. The proposed methodology is categorized into two categories: before and after reconfiguration. The comparative results are illustrated on 33-bus and 69-bus in RDS.

4.6.1 Before reconfiguration

4.6.1.1 33-bus system

This system has three lateral and five tie-lines sending and receiving nodes as 25–29, 18–33, 9–15, 8–21, and 11–22. Figure 4.5 shows the base case of the IEEE 33-bus RDS. Table 4.1 shows the load and line data for the 33-bus [114].

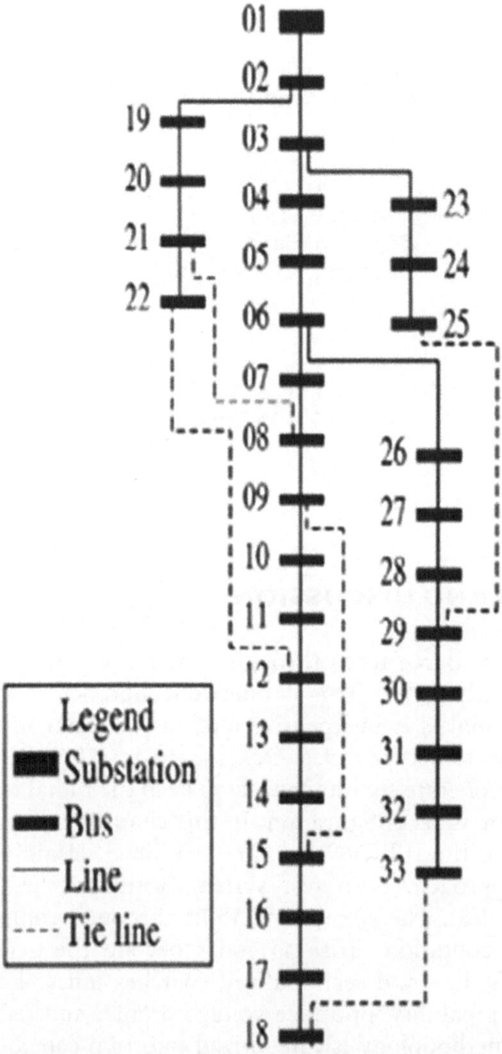

Figure 4.5 Before reconfiguration of IEEE 33-bus RDS.

Table 4.1 Load and line data for the 33-bus [114]

Branch no.	*Sending node*	*Receiving node*	*R (ohm)*	*X (ohm)*	*Real load*	*Reactive load*
1	1	2	0.0922	0.0477	100	60
2	2	3	0.493	0.2511	90	40
3	3	4	0.366	0.1864	120	80
4	4	5	0.3811	0.1941	60	30
5	5	6	0.819	0.707	60	20
6	6	7	0.1872	0.6188	200	100
7	7	8	1.7114	1.2351	200	100
8	8	9	1.03	0.74	60	20
9	9	10	1.044	0.74	60	20
10	10	11	0.1966	0.065	45	30
11	11	12	0.3744	0.1238	60	35
12	12	13	1.468	1.155	60	35
13	13	14	0.5416	0.7129	120	80
14	14	15	0.591	0.526	60	10
15	15	16	0.7463	0.545	60	20
16	16	17	1.289	1.721	60	20
17	17	18	0.732	0.574	90	40
18	2	19	0.164	0.1565	90	40
19	19	20	1.5042	1.3554	90	40
20	20	21	0.4095	0.4784	90	40
21	21	22	0.7089	0.9373	90	40
22	3	23	0.4512	0.3083	90	50
23	23	24	0.898	0.7091	420	200
24	24	25	0.896	0.7011	420	200
25	6	26	0.203	0.1034	60	25
26	26	27	0.2842	0.1447	60	25
27	27	28	1.059	0.9337	60	20
28	28	29	0.8042	0.7006	120	70
29	29	30	0.5075	0.2585	200	600
30	30	31	0.9744	0.963	150	70
31	31	32	0.3105	0.3619	210	100
32	32	33	0.341	0.5302	60	40
	Tie-sectionalize switch data					
33	21	8	2	2		
34	9	14	2	2		
35	12	22	2	2		
36	18	33	0.5	0.5		
37	25	29	0.5	0.5		

4.6.1.2 69-bus system

The load and line data of the distribution system for the proposed methodology are taken from Ranjan and Das [114]. Table 4.2 shows the load and line data for the IEEE 69-bus. Additionally, it is noted that the sending and receiving nodes' connect the opening/closing of the tie-line and the sectionalized switches alternatively. An optimal solution is found when the proposed approach is implemented into a realistic problem, but the primary objective is a combination of minimizing losses and maximizing new VSI. The system topology for the 69-bus as per Table 4.2 data according to sending and receiving nodes with incorporation of sectionalized switches before reconfiguration is shown in Figure 4.6.

4.6.2 After reconfiguration

Reconfiguration in the distribution system is a process of change in system topology by simultaneously altering the position of switches. The radial nature of the distribution system must maintain the optimal status of switches by using a hybrid algorithm. The methodology for the objective function is tested on 33-bus and 69-bus in RDS. First, the results are shown for the 33-bus and then for the 69-bus in RDS.

4.6.2.1 33-Bus system

The reconfigured topology for 33-bus in RDS by applying the proposed algorithm is shown in Figure 4.7. It is observed that the tie-line switches are now connected, and corresponding optimal sectionalized switches are open to maintain the radial topology after reconfiguration using a hybrid algorithm.

The proposed algorithm evaluated active power loss using a reconfiguration of 128.1 kW. However, in the base case, the actual power loss is 210 kW due to a decrease in 78 kW total active power losses and the corresponding increase in minimum voltage from 0.90376 to 0.9390 p.u. The reduction in power loss also indirectly improves the voltage profile at each node. Table 4.3 illustrates the comparative results for reducing active power loss and enhancing minimum voltage after implementing the proposed approach using reconfiguration in RDS.

The results obtained by applying GA are satisfactory, but the computational efficiency is improved by applying PSO to get near the optimal solution. For the execution time utilizing hybrid and GA on 33-bus, the computing efficiency is fast and expressed in seconds, as shown in Table 4.4.

The evaluation of new VSI before and after reconfiguration using Equation (4.7) of the proposed approach is illustrated in Figure 4.8. The new VSI after reconfiguration is observed to improve using a hybrid algorithm.

Table 4.2 Load and line data for the 69-bus

Branch no.	Sending node	Receiving node	R (ohm)	X (ohm)	Real load	Reactive load	Branch no.	Sending node	Receiving node	R (ohm)	X (ohm)	Real load	Reactive load
	1	2	0.0005	0.0005	0	0	38	38	39	0.0304	0.0355	24	17
2	2	3	0.0005	0.0012	0	0	39	39	40	0.0018	0.0021	24	17
3	3	4	0.0015	0.0036	0	0	40	40	41	0.7283	0.8509	1.2	1
4	4	5	0.0251	0.0294	0	0	41	41	42	0.31	0.3623	0	0
5	5	6	0.366	0.1864	2.6	2.2	42	42	43	0.041	0.0478	6	4.3
6	6	7	0.3811	0.1941	40.4	30	43	43	44	0.0092	0.0116	0	0
7	7	8	0.0922	0.047	75	54	44	44	45	0.1089	0.1373	39.22	26.3
8	8	9	0.0493	0.0251	30	22	45	45	46	9.00E-04	0.0012	39.22	26.3
9	9	10	0.819	0.2707	28	19	46	4	47	0.0034	0.0084	0	0
10	10	11	0.1872	0.0619	145	104	47	47	48	0.0851	0.2083	79	56.4
11	11	12	0.7114	0.2351	145	104	48	48	49	0.2898	0.7091	384.7	274.5
12	12	13	1.03	0.34	8	5.5	49	49	50	0.0822	0.2011	384.7	274.5
13	13	14	1.044	0.345	8	5.5	50	8	51	0.0928	0.0473	40.5	28.3
14	14	15	1.058	0.3496	0	0	51	51	52	0.3319	0.1114	3.6	2.7
15	15	16	0.1966	0.065	45.5	30	52	9	53	0.174	0.0886	4.35	3.5
16	16	17	0.3744	0.1238	60	35	53	53	54	0.203	0.1034	26.4	19
17	17	18	0.0047	0.0016	60	35	54	54	55	0.2842	0.1447	24	17.2
18	18	19	0.3276	0.1083	0	0	55	55	56	0.2813	0.1433	0	0
19	19	20	0.2106	0.0696	1	0.6	56	56	57	1.59	0.5337	0	0
20	20	21	0.3416	0.1129	114	81	57	57	58	0.7837	0.263	0	0
21	21	22	0.014	0.0046	5.3	3.5	58	58	59	0.3042	0.1006	100	72
22	22	23	0.1591	0.0526	0	0	59	59	60	0.3861	0.1172	0	0

(Continued)

Table 4.2 (Continued) Load and line data for the 69-bus

Branch no.	*Sending node*	*Receiving node*	*R (ohm)*	*X (ohm)*	*Real load*	*Reactive load*	*Branch no.*	*Sending node*	*Receiving node*	*R (ohm)*	*X (ohm)*	*Real load*	*Reactive load*
23	23	24	0.3463	0.1145	28	20	60	60	61	0.5075	0.2585	1244	888
24	24	25	0.7488	0.2475	0	0	61	61	62	0.0974	0.0496	32	23
25	25	26	0.3089	0.1021	14	10	62	62	63	0.145	0.0738	0	0
26	26	27	0.1732	0.0572	14	10	63	63	64	0.7105	0.3619	227	162
27	3	28	0.0044	0.0108	26	18.6	64	64	65	1.041	0.5302	59	42
28	28	29	0.064	0.1565	26	18.6	65	11	66	0.2012	0.0611	18	13
29	29	30	0.3978	0.1315	0	0	66	66	67	0.0047	0.0014	18	13
30	30	31	0.0702	0.0232	0	0	67	12	68	0.7394	0.2444	28	20
31	31	32	0.351	0.116	0	0	68	68	69	0.0047	0.0016	28	20
32	32	33	0.839	0.2816	14	10	Tie-sectionalize switch data						
33	33	34	1.708	0.5646	19.5	14	69	11	43	0.5	0.5		
34	34	35	1.474	0.4873	6	4	70	13	21	0.5	0.5		
35	3	36	0.0044	0.0108	26	18.55	71	15	46	1	0.5		
36	36	37	0.064	0.1565	26	18.55	72	50	59	2	1		
37	37	38	0.1053	0.123	0	0	73	27	65	1	0.5		

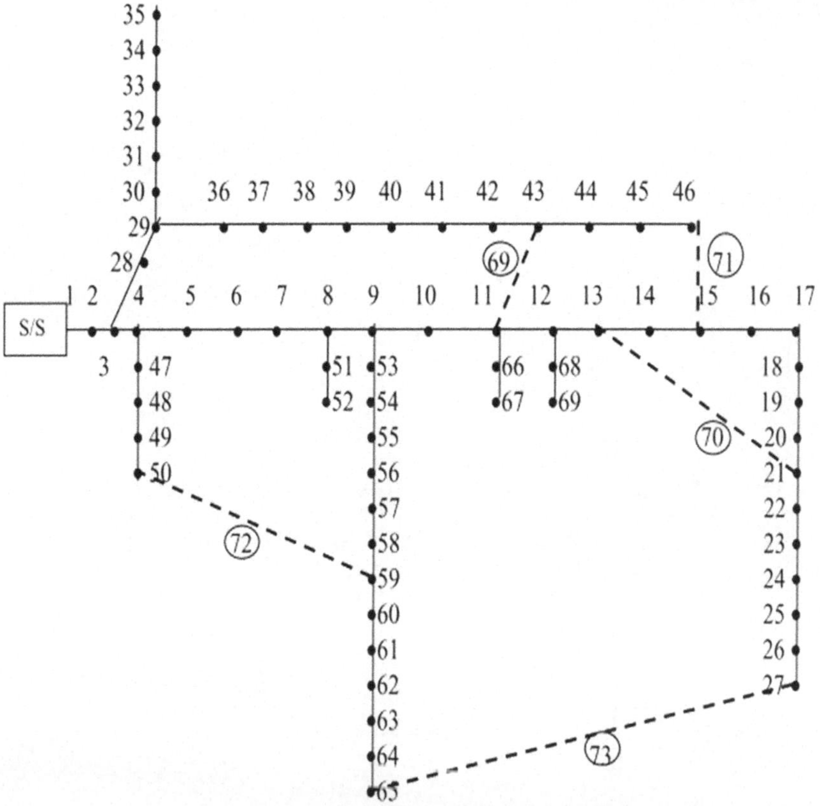

Figure 4.6 Before reconfiguration of IEEE 69-bus RDS.

The minimum voltage before reconfiguration at node 18 is 0.9026 p.u., whereas, after reconfiguration, the minimum voltage at node 32 is 0.9309 p.u. Figure 4.9 shows each node's voltage comparison (before and after reconfiguration).

4.6.2.2 69-Bus system

It is observed from the line and load data in Table 4.2 that the tie-line switches branch no. 69, 70, 71, 72, and 73 are connected after reconfiguration using a hybrid algorithm. The 69-bus topology in RDS becomes reconfigured using the proposed algorithm, as demonstrated in Figure 4.10.

A comparison of active power loss and minimum voltage with the reported method and the proposed method is shown in Table 4.5. The table shows that the reduction in active power loss and enhancement in minimum voltage are obtained using a proposed hybrid algorithm.

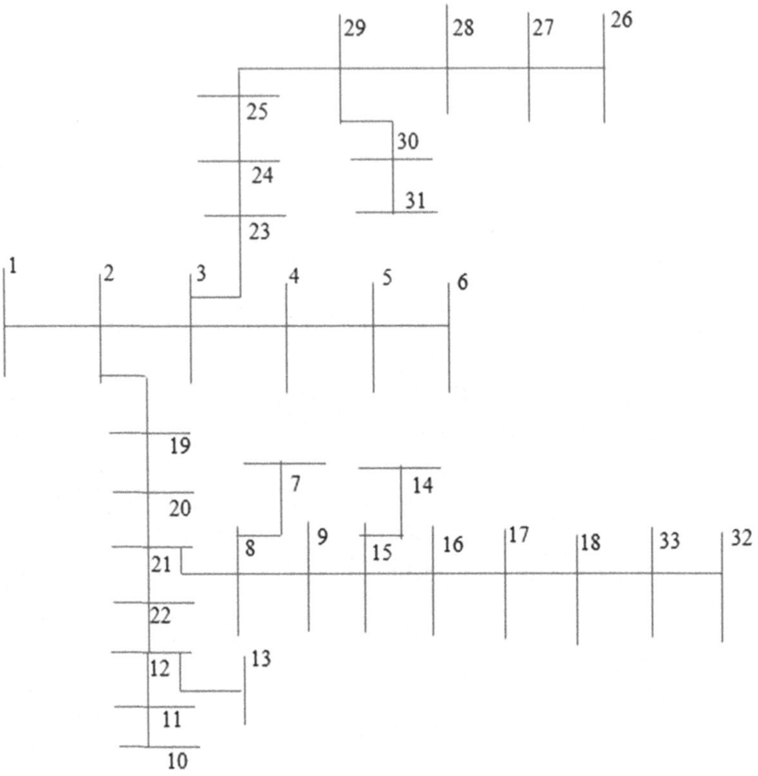

Figure 4.7 Reconfigured IEEE 33-bus RDS.

Table 4.3 Comparative results for reducing active power loss and enhancing minimum voltage

Method	*Line switches out*	*Total kW loss*	*Minimum voltage*	*Saving energy in kW*
Base case [33]	33-34-35-36-37	211.22	0.90376	–
Baran and Wu M1 [1]	35-11, 36-31, 37-28	167	–	44.22
Baran and Wu M2 [1]	33-6,35-11, 36-31	178.4	–	32.82
Baran and Wu M3 [1]	33-6, 35-11, 36-31	178.4	–	32.82
Shirmohammadi & Hong [35]	33-7, 34-10, 35-14, 36-32	158.2	–	53.02
Goswami and Basu [41]	33-7, 34-9, 35-14, 36-32	157.1	–	54.12

(Continued)

Table 4.3 (Continued) Comparative results for reducing active power loss and enhancing minimum voltage

Method	*Line switches out*	*Total kW loss*	*Minimum voltage*	*Saving energy in kW*
Essallah and Khedher [116]	33-7, 34-14, 35-11, 36-32, 37-28	155.1	–	56.12
Moghari et al. [122]	7-9-14-28-32	155.63	–	55.57
Venkatesh et al. [70]	6-14-9-32-37	128.26	0.9378	83.98
Venkatesh and Ranjan [68]	07-14-09-32-37	139.83	0.9378	73.39
Bezao Silveira et al. [121]	–	139.55	–	71.67
Merlin and Back [28]	06-14-09-32-37	128.28	0.9378	82.96
Proposed algorithm	As per Figure 4.3	128.1	0.939	83.12

Table 4.4 Comparison of execution time between hybrid and GA algorithm findings

Method	*Losses (kW)*	*Minimum voltage (p.u.)*	*Time (seconds)*
Genetic algorithm	128.1	0.9309	606
Hybrid algorithm	128.1	0.9309	145

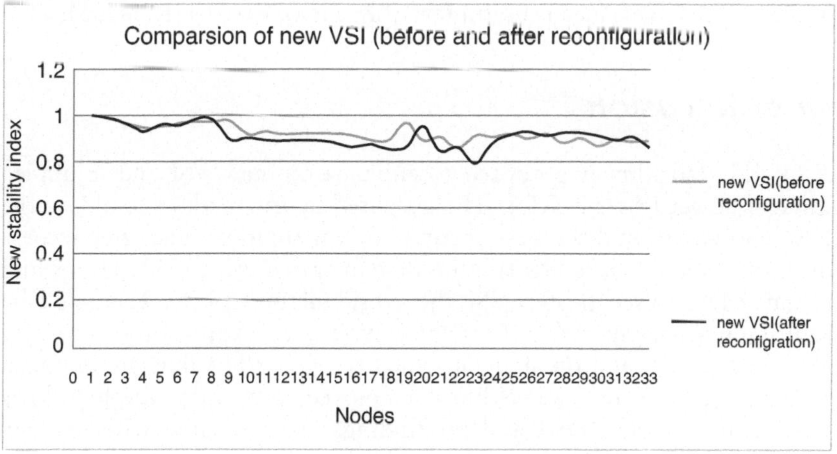

Figure 4.8 Comparison of new VSI before and after reconfiguration.

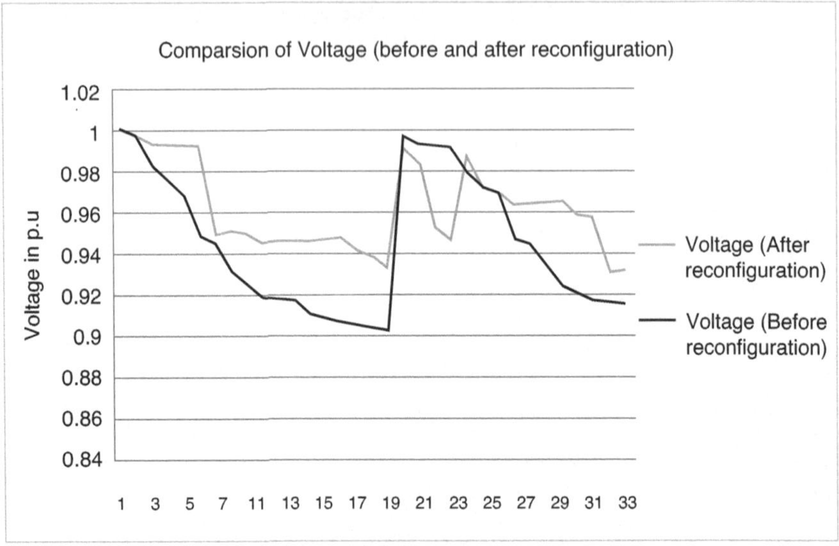

Figure 4.9 Comparison of voltage (before and after reconfiguration) at each node.

At node 61, the minimum voltage after reconfiguration is 0.9497 p.u., while at node 65, the minimum voltage before reconfiguration is 0.9080 p.u. Figure 4.11 demonstrates the voltage shifts before and after reconfiguration.

The minimum value of the new VSI after reconfiguration is 0.8352 p.u. at node 54, whereas for the before reconfiguration, it is 0.8154 p.u. at node 64. The variation of the new VSI before and after the reconfiguration is shown in Figure 4.12. This idea is useful for other circuit designs [123–135].

4.7 CONCLUSION

A hybrid algorithm is proposed to enhance the new VSI and minimize the real power losses in RDS. The proposed hybrid method combines the PSO and GA to find the best energy-efficient solution. The effectiveness of the suggested hybrid algorithm is demonstrated using IEEE 33-bus and 69-bus RDS. To see the new VSI, the hybrid method for the RDS line flow solution has been modified. For such a system, an expression for the new VSI is proposed, and the node at which the new VSI value is minimized is identified. A hybrid algorithm is employed to evaluate the line flows in several realistic RDSs, and the findings are compared with existing methods.

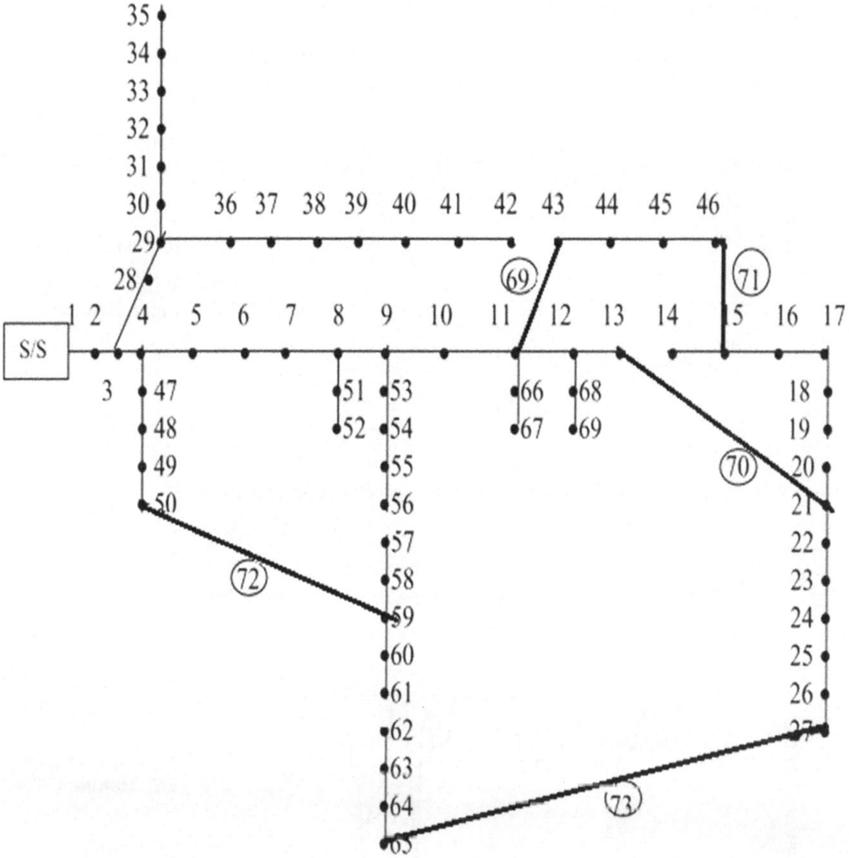

Figure 4.10 Reconfigured IEEE 69-bus RDS.

Table 4.5 Comparison of active power loss and minimum voltage

Method	*Active power loss (kW)*	*Minimum voltage (p.u.)*
Base case [33]	317.18	0.8754
Sivanagaraju et al. [109]	143.27	0.9303
MPSO [116] case 1	275.28	0.8993
Moghari et al. [122]	166.45	0.9470
FMOA [117]	125.3	0.9483
Meta-heuristic [118]	125.3	0.9428
Meta-heuristic H.A.S.	125.54	0.9428
GSA [119]	126.324	0.9495
Proposed hybrid algorithm	126.30	0.9497

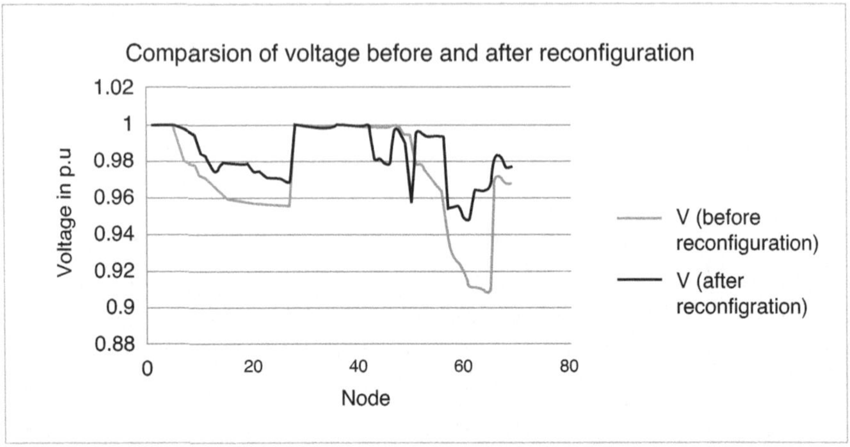

Figure 4.11 Comparison of voltage before and after reconfiguration.

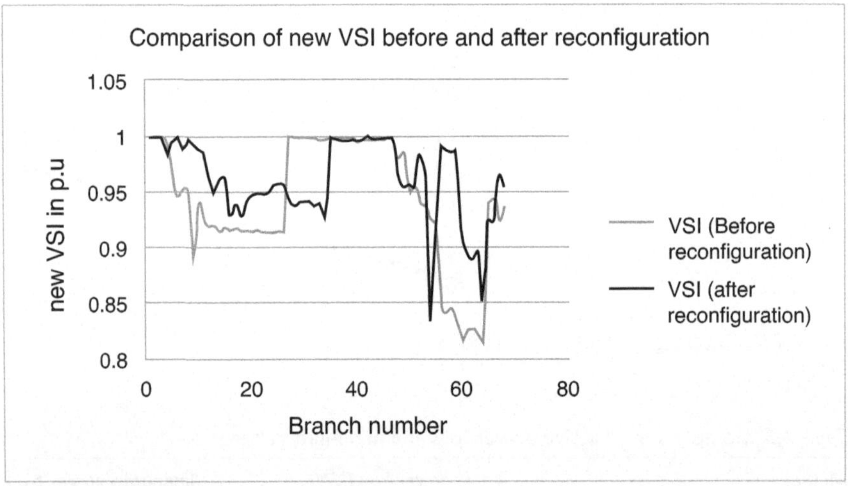

Figure 4.12 Comparison of new VSI before and after reconfiguration.

REFERENCES

1. H.-D. Chiang and M. E. Baran, "On the existence and uniqueness of load flow solution for radial distribution power networks," *IEEE Transactions on Circuits and Systems*, Vol. 37(3), 1990. https://doi.org/10.1109/31.52734
2. M. H. Haque, "Efficient load flow method for distribution systems with radial or mesh configuration," *I.E.E. Proceeding-Generation Transmission and Distribution*, Vol. 143(1), pp. 33–38, 1996. https://doi.org/10.1049/ip-gtd:19960045

3. J. Nanda, M. S. Srinivas, M. Sharma, S. S. Dey and L. L. Lai, "New findings on radial distribution system load flow algorithms," in *IEEE Power Engineering Society Winter Meeting. Conference Proceedings (Cat. No.00CH37077)*, Singapore, pp. 1157–1161, 2000. https://doi.org/10.1109/PESW.2000.850107
4. M. M. Medina, L. Qi and K. L. Butler-Purry, "A three phase load flow algorithm for shipboard power systems," in *2003 IEEE PES Transmission and Distribution Conference and Exposition (IEEE Cat. No. 03CH37495)*, Vol. 1, pp. 227–233, 2003. https://doi.org/10.1109/TDC.2003.1335220
5. B. Venkatesh and R. Ranjan, "Data structure for radial distribution system load flow analysis," *I.E.E. Proceedings - Generation Transmission and Distribution*, Vol. 150(1), pp. 101–106, 2003. https://doi.org/10.1049/ip-gid:20030013
6. M. P. Selvan and K. S. Swarup, "Distribution system load flow using object-oriented methodology," in *International Conference on Power System Technology, Power-Con*, Singapore, 2004. https://doi.org/10.1109/ICPST.2004.1460178
7. H. M. Khodr, L. Ocque, J. M. Yusta and M. A. Rosa, "New load flow method S-E oriented for large radial distribution networks," in *IEEE Conference*, Caracas, 2006. https://doi.org/10.1109/TDCLA.2006.311490
8. T. Thakur and J. Dhiman, "A new approach to load flow solutions for radial distribution system," in *IEEE/PES Transmission & Distribution Conference and Exposition* Latin America, pp. 1–6, 2006. https://doi.org/10.1109/TDCLA.2006.311634
9. G. W. Chang, S. Y. Chu and H. L. Wang, "An improved backward/forward sweep load flow algorithm for radial distribution systems" *IEEE Transactions on Power Systems*, Vol. 22(2), pp. 882–884, 2007. https://doi.org/10.1109/TPWRS.2007.894848
10. X. Yang, M. Choi, H. Lim and S. Lee, "Load flow analysis for distribution automation system based on distributed load modeling," *Journal of Electrical Engineering & Technology*, Vol. 2, pp. 329–334, 2007. https://doi.org/10.5370/JEET.2007.2.3.329
11. H. Shateri and S. Jamali, "Load flow method for distribution networks with multiple source nodes," in *IEEE Electrical Power & Energy Conference*, Vancouver, 2009. https://doi.org/10.1109/EPC.2008.4763361.
12. S. Ghosh, "A new technique for load-flow analysis of radial distribution networks," *International Journal of Engineering and Technology*, Vol. 1, pp. 1793–8236, 2009.
13. R. A. Jabr, "Radial distribution load flow using conic programming," *IEEE Transactions on Power Systems*, Vol. 21(3), pp. 1458–1459, 2006. https://doi.org/10.1109/TPWRS.2006.879234
14. S. Mishra and D. Das, "Evolution of distribution system load flow methods – a bibliographic review," *Journal Institution Engineers India Electrical Engineering Division*, Vol. 91, pp. 1–7, 2010.
15. A. Hamouda and K. Zehar, "Improved algorithm for radial distribution networks load flow solution," *International Journal of Electrical Power and Energy Systems*, Vol. 33(3), pp. 508–514, 2011. https://doi.org/10.1016/j.ijepes.2010.11.004

16. Z. Wang and F. L. Alvarado, "Interval arithmetic in power flow analysis," *IEEE Transactions on Power Systems*, Vol. 7(3), pp. 1341–1349, 1992. https://doi.org/10.1109/59.207353
17. D. Das, S. Ghosh and D. K. Sriniwas, "Fuzzy distribution load flow," *Electric Machines Power System*, Vol. 27, pp. 1215–1226, 2010. https://doi.org/10.1080/073135699268678
18. B. Das, "Radial distribution power flow using interval arithmetic," *International Journal of Electrical Power and Energy Systems*, Vol. 24(10), pp. 827–836, 2002. https://doi.org/10.1016/S0142-0615(01)00092-8
19. A. Boukaroura, L. Slimani and T. Bouktir, "Analysis of radial distribution system load flow under uncertainties with fuzzy arithmetic algorithm," in *3rd International Conference on Information Processing and Electrical Engineering*, Tebessa, 2014. https://doi.org/10.13140/2.1.3776.9924
20. A. R. Gupta and A. Kumar, "Comparison of deterministic and probabilistic radial distribution systems load flow," *Journal of The Institution of Engineers (India): Series B*, Vol. 98, pp. 547–556, 2017.
21. R. H. Zubo, G. Mokryani, H.-S. Rajamani, J. Aghaei, T. Niknam and P. Pillai, "Operation and planning of distribution networks with integration of renewable distributed generators considering uncertainties: A review," *Renewable and Sustainable Energy Reviews*, Vol. 72, pp. 1177–1198, 2016. https://doi.org/10.1016/j.rser.2016.10.036
22. F. J. Ruiz-Rodriguez, M. Gomez-Gonzalez and F. Jurado, "Improving voltage profile in radial distribution systems using binary particle swarm optimization and probabilistic load flow," in *Proceedings of the International Conference on Power Engineering, Energy and Electrical Drives*, Malaga, 2011
23. S. B. Crary, I. Herlitz and B. Favez, CIGRE SC32 Report: "System stability and voltage, power and frequency control," *C.I.G.R.E.*, Appendix 1, Rep. 347, 1948.
24. C.I.G.R.E. Report: "Definitions of general terms relating to the stability of the interconnected synchronous machine," *C.I.G.R.E.*, Paper No. 334a, 1966.
25. C. Barbier, L. Carpentier and F. Saccomanno, CIGRE SC32 Report: "Tentative classification and terminologies relating to stability problems of power systems," *E.L.E.C.T.R.A.*, No. 56, 1978.
26. IEEE TF Report: "Proposed terms and definitions for power system stability," *IEEE Transaction Power Apparatus and Systems*, Vol. PAS-101, pp. 1894–1897, 1982.
27. W. Lu, Y. Besanger, É. Zamaï and D. Radu, "Blackouts: Description, analysis and classification," in *International Conference on Power Systems*, pp. 429–434, 2006.
28. A. Merlin and H. Back, "Search for a minimal loss operating spanning tree configuration in an urban power distribution system," in *Proceedings of 5th Power System Computing Conference*, Cambridge, pp. 1–18, 1975.
29. C. H. Castro, J. B. Bunch and T. M. Topka, "Generalized algorithms for distribution feeder deployment and sectionalizing," *IEEE Transaction Power Apparatus Systems*, Vol. 99(2), pp. 549–557, 1980.

30. D. W. Ross, J. Paton, A. I. Cohen and M. Carson, "New methods for evaluating distribution automation and control system benefits," *IEEE Transaction Power Apparatus Systems*, Vol. 100(6), pp. 2978–2986, 1981.
31. K. Aoki, H. Kuwabara, T. Satoh and M. Kanezashi, "An efficient algorithm for load balancing for transformers and feeders by switch operation in large scale distribution systems," *IEEE Transaction Power Delivery*, Vol. 3(4), pp. 1865–1872, 1988.
32. S. Civanlar, J. J Grainger, H. Yin and S. S. H. Lee, "Distribution feeder reconfiguration for loss reduction," *IEEE Transaction Power Delivery*, Vol. 3(3), pp. 1217–1223, 1988.
33. M. E. Baran and F. F. Wu, "Network reconfiguration in distribution system for loss reduction and load balancing," *IEEE Transaction Power Delivery*, Vol. 4(2), pp. 1401–1407, 1989.
34. C. C. Liu, S. J. Lee and K. Vu, "Loss minimization of distribution feeders: Optimality and algorithms," *IEEE Transaction Power Delivery*, Vol. 4(2), pp. 1281–1289, 1989.
35. D. Shirmohammadi and H. W. Hong, "Reconfiguration of electric distribution on networks for resistive line loss reduction," *IEEE Transaction Power Delivery*, Vol. 4(1), pp. 1492–1498, 1989.
36. C. A. Castro and A. A. Watanabe, "An efficient reconfiguration algorithm for loss reduction of distribution systems," *Electrical Power System Research.* Vol. 19(1), pp. 137–144, 1990.
37. C. T. Huddleston, R. P. Broadwater and A. Chandrasekaran, "Reconfiguration algorithms for minimizing losses in radial electric distribution systems," *Electric Power System Research*, Vol. 18(1), pp. 57–66, 1990.
38. T. Taylor and D. Lubkeman, "Implementation of heuristic strategies for distribution feeder reconfiguration," *IEEE Transaction Power Delivery*, Vol. 5(1), pp. 239–246, 1990.
39. T. P. Wagner, A. Y. Chikhani and R. Hackam, "Feeder reconfiguration for loss reduction," *IEEE Transaction Power Delivery*, Vol. 6(4), pp. 1922–1933, 1991.
40. S. C. Chen and M. Y. Cho, "Determination of critical switches in distribution systems," *IEEE Transaction Power Delivery*, Vol. 7(3), pp. 1443–1449,1992.
41. S. K. Goswami and S. K. Basu, "A new algorithm for the reconfiguration of distribution feeders for loss minimization," *IEEE Transaction Power Delivery*, Vol. 7(3), pp. 1484–1491,1992.
42. G. J. Rosseti, E. J. Oliveira, L. W. Oliveira, I. C. Silva Jr. and W. Peres, "Optimal allocation of distributed generation with reconfiguration in electric distribution systems," *Electric Power System Research*, Vol. 103(3), pp. 178–183, 2013.
43. H. Ahmadi and J. R. Marti, "Minimum loss network reconfiguration: A minimum spanning tree problem sustain," *Energy Grids Network*, Vol. 1, pp. 1–9, 2014.
44. D. P. Bernardon, A. P. C. Mello, L. L. Pfitscher, L. N. Canha, A. R. Abaide and A. A. B. Ferreira, "Real time reconfiguration of distribution network with distributed generation," *Electric Power System Research*, Vol. 107(1), pp. 59–67, 2014.

45. F. Ding and K. A. Laparo, "Hierarchical decentralized network reconfiguration for smart distribution systems-part I: Problem formulation and algorithm development," *IEEE Transaction Power System*, Vol. 30(2), pp. 734–743, 2015.
46. S. Ghasemi and J. Moshtag, "A novel codification and modified heuristic approaches for optimal reconfiguration of distribution networks considering losses cost and cost-benefit from voltage profile improvement," *Applied Soft Computing*, Vol. 25, pp. 360–368, 2014.
47. E. J. Oliveira, G. J. Rosseti, L. W. Oliveira and F. V. Gomes, "New algorithm for reconfiguration and operating procedures in electric distribution systems," *International Journal Electrical Power Energy System*, Vol. 57(1), pp. 121–134, 2014.
48. K. Aoki, T. Ichimori and M. Kanezashi, "Normal state load allocation in distribution systems," *IEEE Transaction Power Delivery*, Vol. 2(1), pp. 147–155, 1987.
49. A. Augugliaro, L. Dusonchet and S. Mangione, "Optimal reconfiguration of distribution networks for loss reduction using nonlinear programming," *European Transaction Electrical Power*, Vol. 1(6), pp. 317–324,1991.
50. A. A. Abur, "Modified linear programming method for distribution system reconfiguration," *International Journal Electricity Power Energy System*, Vol. 18(7), pp. 469–474,1996.
51. A. B. Morton and I. M. Y. Mareels, "An efficient brute force solution to the network reconfiguration problem," *IEEE Transaction Power Delivery*, Vol. 15(3), pp. 996–1000, 2000.
52. E. R. Ramos, A. G. Exposito, J. R. Santos and F. L. Iborra, "Path-based distribution network modeling: Application to reconfiguration for loss reduction," *IEEE Transaction Power System*, Vol. 20(2), pp. 556–564, 2005.
53. H. P. Schmidt, N. Kagan and J. C. Guaraldo, "Fast reconfiguration of distribution systems considering loss minimization," *IEEE Transaction Power System*, Vol. 20(3), pp. 1311–1319, 2005.
54. R. El-Ramli, M. Awad and R. A. Jabr, "Ordinal optimization for dynamic network reconfiguration," *Electric Power Components System*, Vol. 39(16), pp. 1845–1857, 2011.
55. R. A. Jabr, R. Singh and B. C. Pal, "Minimum loss network reconfiguration using mixed integer convex programming," *IEEE Transaction Power System*, Vol. 27(2), pp. 1106–1115, 2012.
56. E. D. Anese and G. B. Giannakis, "Sparsity leveraging reconfiguration of smart distribution systems," *IEEE Transaction Power Delivery*, Vol. 29(3), pp. 1417–1426, 2014.
57. A. S. Deese, "Comparative study of accuracy and computation time for optimal network reconfiguration techniques via simulation," *International Journal Electric Power Energy System*, Vol. 63(1), pp. 394–400, 2014.
58. K. Nara, A. Shiose, M. Kitagawa and T. Ishihara, "Implementation of genetic algorithm for distribution system loss minimum reconfiguration," *IEEE Transaction Power System*, Vol. 7(3), pp. 1044–1051, 1992.
59. D. L. Duan, X. D. Ling, X. Y. Wu and B. Zhong, "Reconfiguration of distribution network for loss reduction and reliability improvement based on an enhanced genetic algorithm," *International Journal Electric Power Energy System*, Vol. 64(1), pp. 88–95, 2015.

60. D. J. Shin, J. O. Kim, T. K. Kim, J. B. Choo and C. Singh, "Optimal service restoration and reconfiguration using genetic-tabu algorithm," *Electric Power System Research*, Vol. 71(2), pp. 145–152, 2004.
61. Y. Y. Hong and S. Y. Ho, "Determination of network reconfiguration considering multi-objective in distribution systems using genetic algorithms," *IEEE Transaction Power System*, Vol. 20(2), pp. 1062–1069, 2005.
62. F. Mendoza, J. L. B. Agustin and J. A. D. Navarro, "N.S.GA and S.P.GA applied to multi objective design power distribution systems," *IEEE Transaction Power System*, Vol. 21(4), pp. 1938–1945, 2006.
63. A. Mazza, G. Chicco and A. Russo, "Optimal multi objective distribution system reconfiguration with multi criteria decision making based solution ranking and enhanced genetic operators," *International Journal Electric Power Energy System*, Vol. 54(1), pp. 255–267, 2014.
64. H. C. Chang and C. C. Kuo, "Network reconfiguration in distribution systems using simulated annealing," *Electric Power System Research*, Vol. 29(1), pp. 227–238, 1994.
65. C. T. Su and C. S. Lee, "Feeder reconfiguration and capacitor settings for loss reduction of distribution systems," *Electric Power System Research*, Vol. 58(1), pp. 97–102, 2001.
66. Y. J. Jeon and J. C. Kim, "Application of simulated annealing and tabu search for loss minimization in distribution systems," *International Journal Electric Power Energy System*, Vol. 26(1), pp. 9–18, 2004.
67. J. Chen, F. Zhang and Y. Zhang, "Distribution network reconfiguration using simulated annealing immune algorithm," *Energy Procedia*, Vol. 12, pp. 271–277, 2011.
68. B. Venkatesh and R. Ranjan, "Optimal radial distribution system reconfiguration using fuzzy adaption of evolutionary programming," *International Journal Electric Power Energy System*, Vol. 25(10), pp. 775–780, 2003.
69. Y. T. Hsiao, "Multi-objective evolution programming method for feeder reconfiguration," *IEEE Transaction Power System*, Vol. 19(1), pp. 594–599, 2004.
70. B. Venkatesh, R. Ranjan and H. B. Gooi, "Optimal reconfiguration of radial distribution systems to maximize loadability," *IEEE Transaction Power System*, Vol. 19(1), pp. 260–266, 2004.
71. A. C. B. Delbem, A. Carvalho and N. G. Bretas, "Main chain representation for evolutionary algorithms applied to distribution system reconfiguration," *IEEE Transaction Power System*, Vol. 20(1), pp. 425–436, 2005.
72. C. T. Su and C. S. Lee, "Network reconfiguration of distribution system using improved mixed integer hybrid differential evolution," *IEEE Transaction Power Delivery*, Vol. 18(3), pp. 1022–1027, 2003.
73. J. P. Chiou, C. F. Chung and C. T. Su, "Variable scaling hybrid differential evolution for solving network reconfiguration of distribution systems," *IEEE Transaction Power System*, Vol. 20(2), pp. 668–674, 2005.
74. S. Jazebi and B. Vahidi, "Reconfiguration of distribution networks to mitigate utilities power quality disturbances," *Electric Power System Research*, Vol. 91(1), pp. 9–17, 2012.
75. K. K. Kim and K. H. Hung, "Artificial neural network based feeder reconfiguration for loss reduction in distribution systems," *IEEE Transaction Power Delivery*, Vol. 8(3), pp. 1356–1366, 1997.

76. M. A. Kashem, G. B. Jasmon, A. Mohamed and M. Moghavvemi, "Artificial neural network approach to network reconfiguration for loss minimization in distribution networks," *International Journal Electric Power Energy System*, Vol. 20(4), pp. 247–258, 1998.
77. M. W. Siti, D. V. Nicolae, A. A. Jimoh and A. Ukil, "Reconfiguration and load balancing in the LV and MV distribution networks for performance," *IEEE Transaction Power Delivery*, Vol. 22(4), pp. 2534–2540, 2007.
78. A. Augugliaro, L. Dusonchet and E. R. Sanseverino, "Genetic, simulated annealing and tabu search algorithms: Three heuristics methods for optimal distribution network's reconfiguration and compensation," *European Transaction Electric Power*, Vol. 9(1), pp. 35–41, 1999.
79. K. K. Li, T. S. Chung, G. J. Chen and G. Q. Tang, "A tabu search approach to distribution network reconfiguration for loss reduction," *Electric Power Component System*, Vol. 32(6), pp. 571–585, 2004.
80. Y. Mishima, K. Nara, T. Satoh, T. Ito and H. Kaneda, "Method for minimum loss reconfiguration of distribution system by tabu search," *Electric Engineering Japan*, Vol. 152(2), pp. 1149–1155, 2005.
81. A. Y. Abdelaziz, F. M. Mohammed, S. F. Mekhamer and M. A. L. Badr, "Distribution system reconfiguration using a modified tabu search algorithm," *Electric Power System Research*, Vol. 80(8), pp. 943–953, 2010.
82. C. T. Su, C. F. Chang and J. P. Chiou, "Distribution network reconfiguration by ant colony search algorithm," *Electric Power System Research*, Vol. 75(2–3), pp. 190–199, 2005.
83. A. Ahuja, S. Das and A. Pahwa, "An AIS-ACO hybrid approach for multi objective distribution system reconfiguration," *IEEE Transaction Power System*, Vol. 22(3), pp. 1101–1111, 2007.
84. C. F. Chang, "Reconfiguration and capacitor placement for loss reduction of distribution systems by ant colony search algorithm," *IEEE Transaction Power System*, Vol. 23(4), pp. 1747–1755, 2008.
85. T. Niknam, "An efficient hybrid evolutionary algorithm based on PSO and A.C.O. for distribution feeder reconfiguration," *European Transaction Electric Power*, Vol. 20, pp. 575–590, 2010.
86. Y. K. Wu, C. Y. Lee, L. C. Liu and S. H. Tsai, "Study of reconfiguration for the distribution system with distributed generators," *IEEE Transaction Power System*, Vol. 25(3), pp. 1678–1685, 2010.
87. L. W. Oliveira, E. J. Oliveira, F. V. Gomes, I. C. Jr. Silva, A. L. M. Marcato and P. V. C. Resende, "Artificial immune systems applied to the reconfiguration of electrical power distribution networks for energy loss minimization," *International Journal Electric Power Energy System*, Vol. 56(1), pp. 64–74, 2014.
88. K. K. Satish and T. Jayabharthi, "Power system reconfiguration and loss minimization for a distribution system using bacterial foraging optimization algorithm," *International Journal Electric Power Energy System*, Vol. 36(1), pp. 13–17, 2012.
89. R. Hooshmand and S. H. Soltani, "Simultaneous optimization of phase balancing and reconfiguration in distribution networks using BF-NF algorithm," *International Journal Electric Power Energy System*, Vol. 41(1), pp. 76–86, 2012.

90. T. Niknam, "An efficient hybrid evolutionary algorithm based on PSO and H.B.M.O. algorithms for multi-objective distribution feeder reconfiguration," *Energy Convers Management*, Vol. 50(8), pp. 2074–2082, 2009.
91. T. Niknam, A. K. Fard and A. Seifi, "Distribution feeder reconfiguration considering fuel cell/wind/photo voltaic power plants," *Renew Energy*, Vol. 37(2), pp. 213–225, 2012.
92. S. Olamaei, T. Niknam, S. Badali and A. Arefi, "Distribution feeder reconfiguration for loss minimization based on honey bee mating optimization algorithm," *Energy Procedia*, Vol. 14, pp. 304–311, 2012.
93. S. Sivanagaraju, J. V. Rao and P. S. Raju, "Discrete particle swarm optimization to network reconfiguration for loss reduction and load balancing," *Electric Power Component System*, Vol. 36(5), pp. 513–524, 2008.
94. B. Amanulla, S. Chakraborti and S. N. Singh, "Reconfiguration of power distribution systems considering reliability and power loss," *Transaction Power Delivery*, Vol. 27(2), pp. 918–926, 2012.
95. T. Niknam, E. Azadfarsani and M. Jabbari, "A new hybrid evolutionary algorithm based on new fuzzy adaptive PSO and N M algorithm for distribution feeder reconfiguration," *Energy Conversion Management*, Vol. 54(1), pp. 7–16, 2012.
96. M. Sedighizadeh, S. Ahmadi and M. Sarvi, "An efficient hybrid big bang-big crunch algorithm for multiobjective reconfiguration of balanced and unbalanced distribution systems in fuzzy framework," *Electric Power Component System*, Vol. 41(1), pp. 75–99, 2013.
97. R. S. Rao, S. V. L. Narasimham, M. R. Raju and A. S. Rao, "Optimal network reconfiguration of large scale distribution system using harmony search algorithm," *IEEE Transaction Power System*, Vol. 26(3), pp. 1080–1088, 2011.
98. A. Y. Abdelaziz, R. A. Osama and S. M. Elkhodary, "Distribution systems reconfiguration using ant colony optimization and harmony search algorithms," *Electric Power Component System*, Vol. 41(5), pp. 537–554, 2013.
99. R. S. Rao, K. Ravindra, K. Satish and S. V. L. Narasimham, "Power loss minimization in distribution system using network reconfiguration in the presence of distributed generation," *IEEE Transactions on Power Systems*, Vol. 28(1), pp. 317–325, 2013.
100. C. Wang and H. Z. Cheng, "Optimization of network configuration in large distribution system using plant growth simulation algorithm," *IEEE Transaction Power System*, Vol. 23(1), pp. 119–126, 2008.
101. E. Azad-Farsani, M. Zare, R. A. Abarghooee and H. A. Askarian, "A new hybrid CPSOTLBO optimization algorithm for distribution network reconfiguration," *Journal Intelligent Fuzzy System*, Vol. 25(5), pp. 2175–2184, 2014.
102. A. Kavousi-Fard, T. Niknam and A. Khosravi, "Multi-objective probabilistic distribution feeder reconfiguration considering wind power plants," *International Journal Electric Power Energy System*, Vol. 55(1), pp. 680–689, 2014.
103. H. Shareef, A. A. Ibrahim, N. Salman, A. Mohamed and W. L. Ai, "Power quality and reliability enhancement in distribution systems via optimum network reconfiguration by using quantum firefly algorithm," *International Journal Electric Power Energy System*, Vol. 58(1), pp. 160–169, 2014.

104. T. Niknam and E. A. Farsani, “A hybrid self adaptive particle swarm optimization and modified shuffled frog leaping for distribution feeder reconfiguration,” *Engineering Application Artificial Intelligence*, Vol. 23(8), pp. 1340–1349, 2010.
105. A. Kavousi-Fard and M. R. Akbari-Zadeh, “Reliability enhancement using optimal distribution feeder reconfiguration,” *Neuro Computing*, Vol. 106(1), pp. 1–11, 2013.
106. M. M. Aman, G. B. Jasmon, A. H. A. Bakar and H. Mokhlis, “Optimum network reconfiguration based on maximization of system loadability using continuation power flow theorem,” *International Journal Electric Power Energy System*, Vol. 54(1), pp. 123–133, 2014.
107. S. Teimourzadeh and K. Zare, “Application of binary group search optimization to distribution network reconfiguration,” *International Journal Electric Power Energy System*, Vol. 62(1), pp. 461–468, 2014.
108. A. Kavousi-Fard and T. Niknam, “Multiobjective stochastic distribution feeder reconfiguration from the reliability point of view,” *Energy*, Vol. 64(1), pp. 342–354, 2014.
109. S. Sivanagaraju, N. Visali, V. Sankar and T. Ramana, “Enhancing Voltage stability of radial distribution systems by network reconfiguration,” *Electric Power Components and Systems*, Vol. 33(5), pp. 539–550, 2005.
110. M. M. Hamada, M. A. A. Wahab and N. G. Hemdan, “Simple and efficient method for steady-state voltage stability assessment of radial distribution systems,” *Electric Power Systems Research*, Vol. 80(2), pp. 152–160, 2010.
111. M. Arun and P. Aravindhababu, “A new reconfiguration scheme for voltage stability enhancement of radial distribution systems,” *Energy Conversion and Management*, Vol. 50(9), pp. 2148–2151, 2009.
112. J. Kennedy and R. Eberhart, “Particle swarm optimization,” in *Proceedings of ICNN'95- International Conference on Neural Networks,* Perth, 27 November 1995-01. 1995. https://doi.org/10.1109/ICNN.1995.488968.
113. S. Singh, V. R. Singh, R. Ranjan and S. Swapnil, “Advanced load flow analysis for unbalanced radial distribution system incorporating weather conditions,” *Journal Institution Engineers India Series Book*, Vol. 100(6), pp. 551–560, 2019.
114. R. Ranjan and D. Das, “Simple and efficient computer algorithm to solve radial distribution networks,” *Electric Power Components and Systems*, Vol. 31, pp. 95–107, 2003.
115. N. C. Sahoo and K. Prasad. “A fuzzy genetic approach for network reconfiguration to enhance Voltage stability in radial distribution systems,” *Energy Conversion and Management*, Vol. 47, pp. 3288–3306, 2006.
116. S. Essallah and A. Khedher, “Optimization of distribution system operation by network reconfiguration and D.G. integration using M.P.S.O. algorithm,” *Renewable Energy Focus*, Vol. 34, pp. 37–46, 2020.
117. J. S. Savier and D. Das, “Impact of network reconfiguration on loss allocation of radial distribution systems,” *IEEE Transaction Power Delivery*, Vol. 22(4), pp. 2473–2480, 2007.
118. A. Swarnkar, N. Gupta and K. R. Niazi, “A novel codification for meta-heuristic techniques used in distribution network reconfiguration,” *Electric Power System Research*, Vol. 81(7), pp. 1619–1626, 2011.

119. Y. M. Shuaib, M. S. Kalavathi and C. C. Asir Rajan, "Optimal reconfiguration in radial distribution system using gravitational search algorithm," *Electric Power Components and Systems*, Vol. 42(7), pp. 703–715, 2014.
120. V. V. S. N. Murty and A. Kumar, "Optimal placement of D.G. in radial distribution systems based on new voltage stability index under load growth," *Electrical Power and Energy Systems*, Vol. 69, pp. 246–256, 2015.
121. C. L. Bezao Silveira, A. Tabares, L. T. Faria and J. F. Franco, "Mathematical optimization versus Metaheuristic techniques: A performance comparison for reconfiguration of distribution systems," *Electric Power Systems Research*, Vol. 196, 107272, 2021.
122. P. Moghari, R. M. Chabanloo and H. Torkaman, "Distribution system reconfiguration based on MILP considering voltage stability," *Electric Power Systems Research*, Vol. 222, p. 109523, 2023.
123. V. K. Sharma, "CNTFET circuit-based wide fan-in domino logic for low power applications," *Journal Circuits System Computers*, Vol. 31(02), p. 2250036, 2022.
124. V. K. Sharma and M. Pattanaik, "A reliable ground bounce noise reduction technique for nanoscale CMOS circuits," *International Journal Electronics*, Vol. 102(11), pp. 1852–1866, 2015.
125. V. K. Sharma, S. Patel and M. Pattanaik, "High performance process variations aware technique for sub-threshold 8T-SRAM cell," *Wireless Personal Communications*, Vol. 78(1), pp. 57–68, 2014.
126. V. K. Sharma and M. Pattanaik, "Techniques for low leakage nanoscale vlsi circuits: A comparative study," *Journal Circuits System Computers*, Vol. 23(05), p. 1450061, 2014.
127. V. K. Sharma, "A survey of leakage reduction techniques in CMOS digital circuits for nanoscale regime," *Australian Journal Electric Electronics Engineering*, Vol. 18(4), pp. 217–236, 2021.
128. C. Kumar, A. S. Mishra and V. K. Sharma, "Leakage power reduction in CMOS logic circuits using stack ONOFIC technique," in *2018 Second International Conference on Intelligent Computing and Control Systems (ICICCS)*, Madurai, 2018.
129. V. K. Sharma and M. Pattanaik, "Design of low leakage variability aware ONOFIC CMOS standard cell library," *Journal Circuits System Computers*, Vol. 25(11), p. 1650134, 2016.
130. S. F. Naz, S. Riyaz and V. K. Sharma, "A review of QCA nanotechnology as an alternate to CMOS," *Current Nanoscience*, Vol. 18(1), pp. 18–30, 2022.
131. S. Riyaz and V. K. Sharma, "Design of reversible Feynman and double Feynman gates in quantum-dot cellular automata nanotechnology," *Circuit World*, Vol. 49(1), pp. 28–37, 2023.
132. B. Raina, C. Verma, M. Gupta and V. K. Sharma, "Binary coded decimal (BCD) seven segment circuit designing using quantum-dot cellular automata (QCA)," in *2021 5th International Conference on Trends in Electronics and Informatics (ICOEI)*, Tirunelveli, 2021.
133. U. Mushtaq and V. K. Sharma, "Performance analysis for reliable nanoscaled FinFET logic circuits," *Analog Integrated Circuits Signal Process*, Vol. 107(3), pp. 671–682, 2021.

134. Kajal and V. K. Sharma, "Design and simulation of FinFET circuits at different technologies," in *2021 6th International Conference on Inventive Computation Technologies (ICICT)*, Coimbatore, 2021.
135. V. K. Sharma, "Designing of face recognition system," in *2019 International Conference on Intelligent Computing and Control Systems (ICCS)*, Madurai, 2019.

Chapter 5

Comprehensive examination of automation approaches for the study and detection of disparate leaf diseases using computational intelligent systems

Muskan Dixit

5.1 INTRODUCTION

The agriculture sector serves and feeds the majority of the population and hence plays a crucial role. The evolution of mankind and economic growth are strongly dependent upon the yield of agriculture. Diseases of plants are among such threats that can have a negative effect upon the quality of agricultural production at the basic level [1]. The yield of the crops completely relies on the health of the crops. Leaf disease is one of the diseases which impacts the cultivation of crops. The leaf disease is sometimes caused because of fungal, viral, and bacterial infections which occur in the worst weather conditions. Environmental and toxic factors like insects, herbicides, fertilizers, nematodes, pests, etc. also cause leaf diseases. Diseases are disturbances that affect the desirable structure and function of a plant part which in turn disrupts the normal processes like photosynthesis, transpiration, pollination, fertilization, germination, etc. [2]. The quality of the crops and plants is dependent upon plant diseases. The leaf tissue analysis is done using imaging techniques. This determines the nutrient status of the plant and according to it, the yield potential can be predicted. Crops' quality improvisation can be done at a primary level by detecting diseases. Nowadays the detection of plant diseases and their rectification plays a crucial role in the overall growth of the agriculture sector. It is helpful if there is some automatic method by which plant diseases can be detected because it saves a large amount of time that is required for monitoring diseases especially in large agricultural fields of crops and at initial level it helps to diagnose such symptoms of diseases, i.e. when they start appearing on the plant leaves [3]. The obstacles associated with annual detection include large investments in labour, money, and experience; also, issues such as agricultural operations, disease classifications, and similar symptoms across various illnesses must be taken into account [4].

These difficulties not only compromise agricultural productivity but also present substantial risks to the security of food. This study investigates the utilization of sophisticated automation methodologies, such as wireless

DOI: 10.1201/9781003643609-5

communications, machine learning (ML), AI, and deep learning (DL), in order to tackle the aforementioned challenges within the agricultural sector [5]. Digital agriculture, or precision agriculture, involves using advanced computer systems to analyse many factors such as weed identification, crop prediction, yield estimation, crop quality, and other ML approaches [6].

According to the literature review, the traditional method of detection of leaf diseases by the naked eye had many flaws which can be overcome by the new technologies which are computer vision-oriented, ML methods, and algorithms [7]. In this review chapter, we focus on analysing the different datasets and algorithms of different crops, to discuss the various drawbacks of the methodologies used in the detection of common plant diseases. To analyse the methodologies firstly it is important to discuss the different leaf diseases of divergent leaves based on studies done [8]. The leaf diseases caused by different agents have a worsening impact on the crops. Their diagnosis through different methodologies and their accuracy and efficiency play a vital role in improving the quality of crops. The DL techniques ease the farmers' work as manual work in this regard will be an absolute burden on labour; hence, automatic identification of diseases and their diagnosis will be an add-on tool for agricultural yield [9]. For the purpose of capturing intricate leaf characteristics, automated imaging systems utilize high-resolution cameras and specialized sensors [10]. By automatically extracting pertinent characteristics from data, DL-based techniques have demonstrated outstanding performance in ID; nevertheless, in order to train sophisticated models, a substantial amount of labelled data and computing resources are needed. Less processing power and labelled data are needed for ML-based methods, but their capacity to generalize to unlabelled data is constrained [11]. Proximity in identification is enhanced through the disclosure of disease-specific attributes by means of image processing techniques. Monitoring environmental parameters for correlations with disease outbreaks, sensor networks function as field sentinels, facilitating early detection. Using satellite imagery to analyse vegetation changes in order to generate disease maps for targeted interventions provides a bird's-eye view. The interconnection of devices through the Internet of Things (IoT) enables the collection and transmission of data in real time, which helps to achieve a comprehensive understanding of diseases [12]. Trained disease reconnaissance is performed by aerial and ground-based robotic systems endowed with imaging sensors. The accurate detection and categorization of diseases by robots facilitate precise interventions such as the targeted application of pesticides or the removal of diseased plants [13]. Drones equipped with thermal or multispectral cameras perform expeditious and thorough assessments of diseases, thereby furnishing prompt and practical analyses that contribute to the safeguarding of crops. The primary purpose is to take a critical look at the various automation approaches that are used for the monitoring and diagnosis of leaf diseases. Figures 5.1–5.8

are interrelated images illustrating the diseases that affect the leaves of different crop plants and is very useful for proper identification of the diseases affecting the plant's health.

Figures 5.1–5.8 Comprehensive visual identification of leaf diseases in various crop species. (Photographed by author on survey at IARI-ICAR, Katrain.)

(Continued)

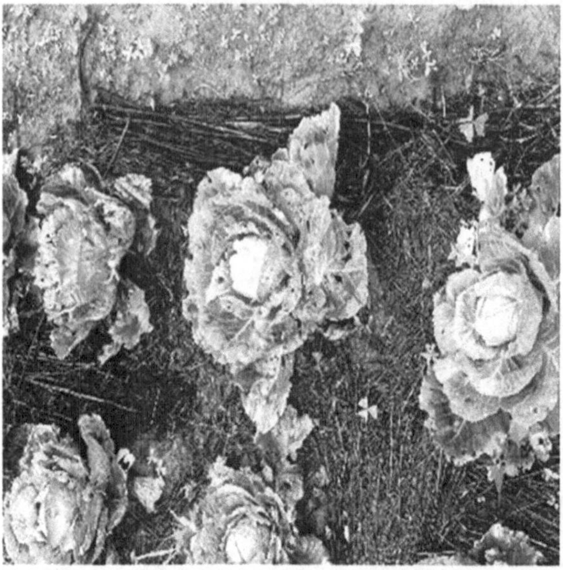

Figures 5.1–5.8 (Continued) Comprehensive visual identification of leaf diseases in various crop species. (Photographed by author on survey at IARI-ICAR, Katrain.)

Figures 5.1–5.8 (Continued) Comprehensive visual identification of leaf diseases in various crop species. (Photographed by author on survey at IARI-ICAR, Katrain.)

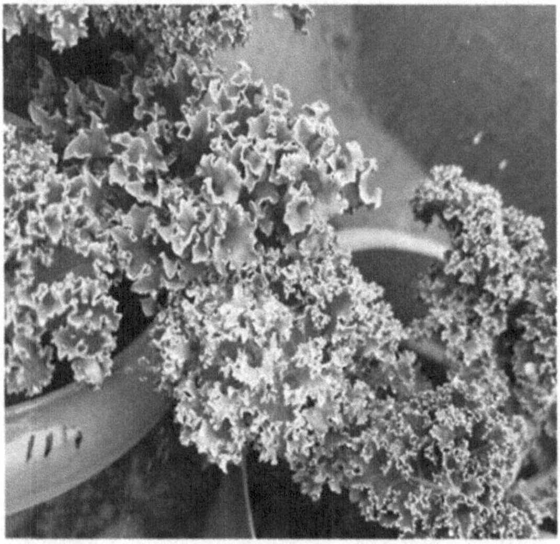

Figures 5.1–5.8 Comprehensive visual identification of leaf diseases in various crop species. (Photographed by author on survey at IARI-ICAR, Katrain.)

5.2 LITERATURE REVIEW: AUTOMATION APPROACHES FOR LEAF DISEASE DETECTION

The status of a tissue of a leaf helps in determining the nutrient status, and deficiency in nutrient status results in leaf diseases. The leaf diseases are identified based on symptoms like V-shaped lesions on the leaf, vein blackening with the necrotic area, occurrence of black circular areas, yellowish-green spots on the surface (chlorosis), weeping ulcers, wilting, stunting of leaves, irregular round leaf with greyish colour, tan to reddish-brown spots of leaves (*Septoria*), dwarfing malformation, leaf rust, formation of white mould (*Sclerotinia*), powdery mildew (PM), bacterial ooze, water-soaked lesions, crinkled leaf, and many more. The literature on symptoms of nutrient level, presence of diseases, and their causes is depicted in Table 5.1.

Table 5.1 Symptoms and causes of leaf diseases

Reference. no.	*Name of plant/plants*	*Leaf diseases*	*Symptoms of leaf disease*	*Leaf diseases caused by*
[14]	Grapes, apples, cucumber, peas, roses, *Phlox*, lilacs, daisies, pumpkin, squash, watermelon	Powdery Mildew	Powdery white coating on the leaves	Fungus
[15]	Cauliflower, broccoli, grapevines, lettuce, *Salvia*, Pansy, cabbage, basil	Downy Mildew	The upper portion of the leaf discolours	Fungus
[16]	Roses	Black spot	Occurrence of black spots	Fungus, *Diplocarpon rosae*
[17]	Beans, tomatoes, peas, dianthus, asparagus, sweet potatoes, bananas, legumes	Fusarium Wilt	Wilted leaves, stunted plants, blackened stem roots	Soil-borne fungus in hot temperature
[18]	Citrus fruit plants like orange, lemon, grapefruits, etc., Oleander, pittosporum, Gardenia, fig	Sooty mould	Impends photosynthesis	Soil-borne fungi in wet and humid conditions
[19]	Tomato, pepper, Irish potato, eggplant	Granville Wilt	Wilted leaves, growth retardation, dead roots	Soil-borne bacteria
[20]	Oats, coffee, maize	Wildfire of Tobacco	Yellowish-green spots on leaf	Bacterial infection
[21]	Apple, pear, rose, quince	Fire blight	Weeping ulcers, witting and rolling of leaves	Bacterial infection
[22]	Rice, spring dwarf, strawberry	*Aphelenchoides*	Blemishes on the top of leaf	Nematode

The methodologies remove the noise and errors by humans automatically and minimize the time taken to measure the effect of leaf diseases [23]. Automated plant disease detection methods offer several benefits, including the ability to identify disease symptoms at an early stage, such as when they manifest on plant foliage, and the reduction of the labour-intensive task of monitoring extensive agricultural farms [24]. Recent developments in DL and computer vision have illustrated the utility of creating automatic plant disease detection models that rely on observable symptoms on foliage [25]. The disease identification process was complex as manual field observation, preliminary diagnosis, identification of biotic and abiotic factors, and then laboratory testing complicates the process and there is more scope of error as compared with the suggestive algorithms. The issue of automatic detection of symptoms of diseases in plant leaves is crucial because the complete survey of vast fields of crops is nearly impossible and the automatic detection of such diseases can be helpful in detecting symptoms on the plant leaves as soon as they start appearing [26]. These DL algorithms have recently shown widespread success in a variety of fields, converting raw data—like pictures of damaged plants—into insightful knowledge—like the diagnosis of certain crop diseases [27]. A galaxy of literature depicts that there are several techniques such as ML and DL used to detect leaf diseases (Table 5.2).

Based on an extensive examination of the relevant literature, it can be deduced that ML and DL methodologies can effectively and precisely identify a wide range of leaf diseases. An extensive variety of ML/DL models, such as convolutional neural networks (CNNs), SVM, KNN, Decision Tree, Naïve Bayes, YOLO, ResNet, DenseNet, and others, have been constructed and evaluated to automatically identify and categorize leaf diseases in a variety of commodities including grape, tomato, potato, legumes, cabbage, cassava, and numerous others. This provides evidence that advanced ML/DL models exhibit superior accuracy, sensitivity, and precision when it comes to identifying diseases and associated symptoms on plant leaves (e.g. black spot, *Fusarium* wilt, sooty mould, Granville wilt, among others), including lesions, spots, blotches, mould, wilting, chlorosis, and so forth. For specific crop species, cutting-edge models employing complex architectures such as EfficientNet, VGG16, ResNet34, YOLOv5, and others have attained a testing accuracy of up to 99%. China and India are significant contributors in this industry, employing CNN, DL, and ML techniques, according to the report. Key sources included "Lecture Notes in Networks and Systems" and periodicals like "Computers and Electronics in Agriculture," "Frontiers in Plant Science," "IEEE Access," "Sensor," and "Symmetry." Key phrases including "Convolutional Neural Network," "Deep Neural Network," "Leaf Disease," "Convolution," "Segmentation," "Extraction," "Feature Extraction," "CNN," "Textures," "Recognition Accuracy," "Crops," and "Machine Learning" were also found by the study for targeted keyword searches. The foundation for further studies on ML-based leaf disease detection is laid by the study [52].

Table 5.2 Techniques to detect leaf diseases

Reference no.	Techniques used	Year	Name of plant leaf	Results of methodologies
[28]	Fisher Discrimination Analysis (FDA)	2009	Maize	Automatic recognition of the crop disease gave an accuracy of 90%
[29]	DCNN using NAG algorithm, RCNN, YOLO, and SSD	2017	Apple	The NAG algorithm model gives an accuracy of 97.62%
[30]	SVM classifier	2018	Wheat	Justifies that SVM is efficient, versatile, and memory-efficient
[31]	SVM with DoH and DoG	2019	Cotton	Efficient self-learning AI model (80% accurate) better than SIFT, SURF, and FREAK
[32]	TBDR and DLDR using CNN	2019	Coffee	This concludes that TBDR is more efficient than DLDR as the sensitivity was 0.980.
[33]	Neural Network and Logistic Regression	2020	Holy Basil	Edge detection technique used to detect the curl leaf disease gives the rightness of 96.6%
[34]	CNN with EfficientNet architecture	2021	Wheat	A convenient model accessible through telegram using EfficientNet-B0 concluding the average accuracy of 78%.
[35]	CNN	2021	Cucurbits	The training accuracy of 72% and 12% testing accuracy was provided by CNN with UI designed in Flask.
[36]	RGB imaging system	2021	Chrysanthemum, Vine, Rose	Based on the AMG8833 Grind-Eye thermal camera sensor this technique estimates the pathogens per plant promptly with it.
[37]	YOLOv5 image segmentation technique	2021	Potato	The evaluation is based on the PlantVillage dataset which achieves a precision of 99.7%.

(Continued)

Table 5.2 *(Continued)* Techniques to detect leaf diseases

Reference no.	*Techniques used*	*Year*	*Name of plant leaf*	*Results of methodologies*
[38]	ResNet34, ResNet50, VGG-16, VGG-19, and DenSet121	2021	Beans	The best output is given by the Densenet121 model of 98.31% with CAR.
[39]	KNN	2022	Cabbage	The application based on KNN gives a rightness of 96.7742%.
[40]	CNN, SVM, KNN, Decision tree, and Naïve Bayes	2022	Pumpkin	The CNN model using Tensorflow library gives a high accuracy as compared to other methods.
[41]	Efficient DenseNet model	2023	Potato	The proposed approach is the first to identify and classify four potato leaf diseases. The algorithm achieved 97.2% accuracy on the testing set.
[42]	Residual Skip Network-based Super-Resolution for Leaf Disease Detection (RSNSR-LDD)	2023	Grape	The suggested RSNSR-LDD model obtained accuracies of 97.19%, 99.37%, and 99.06% for PlantVillage, 96.88%, 97.12%, and 95.43% for Grape400, and 100% for various super-resolution scaling factors (X2, X4, X6).
[43]	CNN-LeNet	2023	Tomato	Automatic feature extraction helps neural network models classify input images into illness classifications. This implied system has an average accuracy of 94-95%.

(Continued)

Table 5.2 (*Continued*) Techniques to detect leaf diseases

Reference no.	*Techniques used*	*Year*	*Name of plant leaf*	*Results of methodologies*
[44]	Least Important Attention Pruning (LeIAP) algorithm	2023	Cassava	The model was introduced to choose the Transformer model's most significant attention heads per layer. It might cut model size by 28%, expedite assessment by 15%, and improve accuracy by 3%.
[45]	CNN	2023	Cotton	Many algorithms were evaluated, but CNN is efficient and productive. On the test set, the suggested model achieves 99.39% accuracy with a negligible error rate, outperforming all current techniques within less computational time.
[46]	Faster RCNN	2023	Grape	Research shows that enhancing the "Faster(RCNN)," "YOLO-X," and "SSD" models with multiple attention processes yields the highest accurate detection results. The "Faster (RCNN)" model has the lowest detection accuracy, slowest operation speed, and most parameters among the three kinds.
[47]	FC-SNDPN (Fully Convolutional—Switchable Normalization Dual Path Networks)	2023	Tomato	The identification accuracy of the method used in employing FCN for foreground segmentation and SNDPN for identification is 97.59% on the augmented dataset, proving its efficacy.

(*Continued*)

Table 5.2 (*Continued*) Techniques to detect leaf diseases

Reference no.	*Techniques used*	*Year*	*Name of plant leaf*	*Results of methodologies*
[48]	Deep CNN model	2023	Plant leaves	Over development, the proposed model trained outperforms frequently used transfer learning approaches when validated. The suggested model achieves 83.12% accuracy in data categorization, as shown by extensive simulations.
[49]	CNN	2024	Potato	Computer vision and AI may assist farmers track plant illnesses throughout. In this chapter, a Nature-inspired Whale Optimization Algorithm (WOA) is used to optimize the CNN's performance in diagnosing potato leaf EB and LB infections.
[50]	DV-PSO-Net	2024	Mango	DVNet, a deep mutual learning model based on Densenet 121 and VGG19 neural networks, addresses these issues. Particle Swarm Optimization incorporates hyperparameter optimization, a method for finding optimum values. The proposed framework detects eight disease categories and healthy mango leaves with 94.72% accuracy, outperforming previous mango leaf disease detection methods.
[51]	Neural Network with Severity Computation	2024	Tomato	Standard deviation-based OKMC. Leaf spot separation is Grasshopper Horse Herd Optimization. CNN, VGG16, and ReNet use split-site deep features. Combine traits. Unique meta-heuristic SD-GHHO selects excellent characteristics. Finally, SD-GHHO classifies diseases using MRNN RNN weight optimization.

5.3 SENSOR NETWORKS: REAL-TIME MONITORING OF PLANT HEALTH INDICATORS

Sensor networks are essential for transforming agriculture through the ability to monitor plant health indicators in real time. This cutting-edge method provides producers with a valuable tool to make well-informed decisions regarding disease management, leading to increased agricultural productivity. A functional classification of the most recent developments in wearable plant sensors is provided, including sensors for microclimate and plant growth, physiology, chemistry, chemical sensors, and multifunctional sensors [53]. Globally, wireless sensor networks (WSNs) have been implemented in precision agriculture. Numerous researchers implemented sensor networks in apple orchards in order to investigate their capacity to automate and monitor agricultural processes, such as disease prediction, by utilizing real-time environmental data [54]. The sensors for wearable technologies will be affixed directly to the surfaces of plant organs, including leaves and stems. By transducing biosignals to an electrical readout for data analytics and profiling a variety of trait biomarkers and microenvironmental parameters, the sensors are intended to determine the health status of plants. Electronic control, also known as e-monitoring, has supplanted conventional methods of controlling plant diseases on account of their immobility, time-consuming nature, and reliance on a specialized user. The utilization of biosensors, ubiquitous sensors, electronic nose (E-nose), and "electronic eyeballs" for e-monitoring has garnered growing interest in recent times. Electronic sensor (e-sensor)-based pathogen detection, identification, and quantification are practicable and convenient; they can be utilized in conjunction with conventional techniques [55]. Figure 5.9 delves into the interactive relationship between sensor functionalities, wearable technologies, and

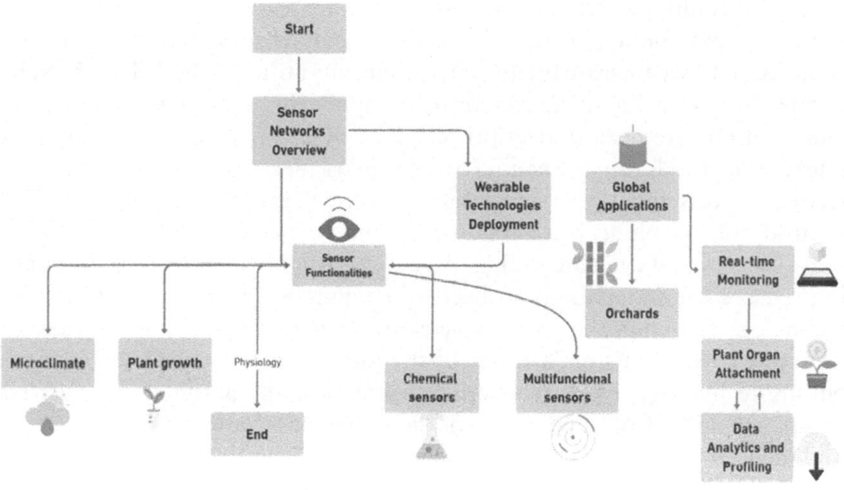

Figure 5.9 Innovative integration: sensor networks in identification of plant health.

WSNs in real-time plant health monitoring. The immediate nature of these networks offers timely insights into fluctuations that may be related to the beginning of illnesses, enabling early intervention and prevention. Illustrating the smooth flow of information and the important role of innovations in revolutionizing disease management strategies and enhancing agricultural productivity, the figure shows the deployment of sensors on plant organs and the use of electronic monitoring and sensors.

5.4 INTERNET OF THINGS

IoT devices have revolutionized plant health monitoring and disease detection in agriculture, giving farmers new tools to protect their crops. Important steps in reducing agricultural losses include early detection and action, both of which are facilitated by these IoT devices. Essential data impacting disease development is provided by environmental sensors. These sensors measure elements including soil moisture, temperature, and humidity. In addition, image sensors take pictures of the leaves in high detail, and then use complex algorithms to identify signs of disease like wilting, stains, or discoloration. A study presented a dilated and inception CNN trained on the provided training data to enhance cucumber leaf disease diagnosis. The proposed approach outperformed existing methods on lesion and raw field diseased leaf images with three diseases—anthracnose, downy mildew, and PM—with an average identification accuracy of 96.11% and 90.67%, indicating that it could be used in agricultural IoT field applications [56]. IoT facilitates the collection of information regarding horticulture, climate, soil fertility, crop water requirements, irrigation, insect and pest detection, and bug location disruption. IoT enables producers to remotely monitor their farms 24/7. Numerous sensors are utilized in agricultural surveillance and control, including distributed WSNs, which are indispensable due to their precise output and application. In addition, remote surveillance of the field is facilitated by cameras. Intelligent agriculture utilizing the IoT is the subject of this study. Smart agricultural IoT applications, benefits, challenges, and solutions are discussed in this chapter. By identifying time-saving and effective methods for managing water, pesticides, irrigation, crops, and fertilizers, this intelligent agriculture system maximizes crop yield [57]. Central hubs or cloud platforms receive the wirelessly supplied data from these sensors and evaluate and analyse it thoroughly [58]. Integrated imaging sensors are increasingly utilized in conjunction with high-throughput technology to obtain the phenotypic characteristics of plants in a variety of growth environments, as a result of the rapid development of IoT in agriculture technology, modern imaging techniques, and automated industrial equipment [59]. Continuous sensor development, equipment downsizing, and reduced rates have led to mechanical advances in traditional farming as well as small-scale precision plants. Sensor data streams from air, ground, radiation, and atmospheric stations are protected and used for checking, data excavation, thinking, and control [60]. The spatial resolution of various approaches is variable, hence the suggested system addresses the optimum integration and execution of these sensing modalities. In our proposed solution, IoT sensors provide real-time environmental parameters affecting the crop, and the drone platform provides multispectral data for generating Vegetation Indices (VIs) like NDVI to analyse crop health. NDVI uses chlorophyll content to assess crop health, although it gives little information. IoT sensor and multispectral image variable length time series data were transformed to a fixed-sized representation to build agricultural health maps for rich and detailed crop health information [61]. The accuracy of disease diagnosis is enhanced via integration with other relevant data, such as weather and historical crop data. These gadgets can detect patterns of illness with an astounding degree of accuracy thanks to ML algorithms that have been trained on massive datasets of tagged photos. This helps in the quest for highly accurate diagnoses. Early detection, data-driven decision-making, enhanced productivity via job automation, and adaptability to varied agricultural situations are just a few of the many advantages of using IoT devices for leaf disease detection. IoT apps help farmers plan next season's yield. They can identify agricultural problems early to reduce disease spread and conserve production. Agricultural IoT applications boost output and reduce disease-related crop losses [62]. The practical uses of this technology have been shown by several notable instances of IoT devices used for leaf disease detection. Table 5.3 provides an overview of some popular apps that utilize

Table 5.3 Plant disease diagnosis and management apps

App name	*Description*	*References*
Planto	Uses multiple sensors to monitor plant health and provide real-time data to farmers.	[63]
Plantix	Diagnoses plant diseases through image recognition of leaves.	[64]
LeafSnap	Identifies plants and diagnoses diseases by analysing leaf images.	[65]
PlantVillage	Global platform for sharing images, getting help with disease identification, and utilizing machine learning for diagnosis.	[66]
GrowSnap	Diagnoses plant diseases, provides treatment recommendations, and tracks plant growth and health.	[67]
MyCrop	Designed for farmers in developing countries, offering information on plant diseases and facilitating connections with other farmers and experts.	[68]
Plant Doctor	Provides plant disease diagnosis and treatment recommendations based on user-submitted information and photos.	[69]
PictureThis	Identifies plants, diagnoses diseases, and offers care tips using image recognition technology.	[70]
SmartPlant	Monitors plant health through connected sensors and provides insights on watering, light, and nutrients.	[71]

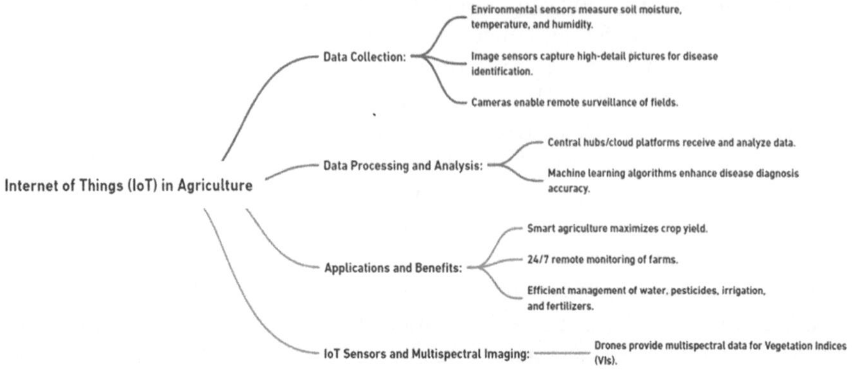

Figure 5.10 Revolutionizing agriculture with IoT: plant health monitoring and disease detection.

IoT technology for leaf disease detection, along with their key functionalities. Figure 5.10 shows how IoT technology has transformed agriculture, concentrating on plant health monitoring and disease detection.

5.5 ROBOTICS: ON-FIELD INSPECTION AND TARGETED TREATMENT USING ROBOTIC SYSTEMS

Robotic devices for leaf disease identification are revolutionizing on-field inspection and tailored treatment in agriculture. There are significant findings into using robots with powerful image sensors to diagnose and manage leaf diseases. These robotic devices can independently roam fields and take high-resolution leaf photos using modern imaging technology. A study introduced a versatile IoT robotic system designed for use in greenhouse agriculture intending to deliver real-time plant disease diagnosis. A three-layered CNN module is utilized by the compact robotic system to distinguish between healthy and unhealthy plant conditions [72]. The research examined several network topologies to find the most appropriate lightweight one for our mobile robot. Results indicate that the VGG16 network obtained the highest performance of 99.908%. The maximum accuracy of VGG16 in our study makes it the ideal instrument for our autonomous plant disease detection robot [73]. RobHortic is a remote-controlled field robot that uses proximate sensing to detect pests and illnesses in horticultural crops. Software designed for this purpose collected the signal from a motor-connected encoder to synchronize picture capturing with robot movement. At signal reception, the cameras are activated and the photos and GNSS data are saved. Using a specialized real-time polymerase chain reaction (PCR), plants were molecularly analysed for the target bacteria and compared to robot data.

Field and lab testing were done. Partial Least Squares-Discriminant Analysis (PLS-DA) had the best match, detecting 66.4% of laboratory pictures and 59.8% of field images [74]. Robot's unique combination of robot mobility in tough terrain and the simple K clustering and SVM algorithm for leaf disease diagnosis without human interaction is a big benefit. A user-friendly voice-automated robotic vehicle would simplify plant leaf disease detection [75]. A specialized integrated disease detection system with algorithms is needed which was developed. The creation of a robotic detection system for two important hazards to bell pepper plants, PM and tomato spotted wilt virus (TSWV), is described. Detection techniques based on principal component analysis were developed for RGB and multispectral NIR-R-G sensors. Pixel categorization as ill or healthy was very accurate for both disorders when employing RGB images (PM: 95%, TSWV: 90%). NIR-R-G multispectral images have a poor classification accuracy (PM: 80%, TSWV: 61%) [76]. Robotic applications are on plant pathology and management, and emerging agricultural technologies for intra-urban agriculture [77]. Robotic disease detection systems in greenhouses were supposed to enhance disease management, boost productivity, and minimize pesticide application rates. A study describes a robotic detection method for detecting two primary hazards to greenhouse bell peppers: PM and TSWV [78]. The continued development of robots in agriculture is set to be a disruptive force, providing important tools that allow farmers to more successfully preserve their crops.

5.6 ELECTRONIC CONTROL (E-MONITORING) VS. CONVENTIONAL METHODS

Leaf diseases present substantial risks to agricultural output, thereby demanding the implementation of efficient disease control methods. Conventional approaches frequently face obstacles including limited mobility, time restrictions, and reliance on specialized knowledge. On the contrary, electronic control, which is frequently called e-monitoring, presents a potentially transformative change in the field of disease management. This extensive investigation examines the benefits of electronic control in comparison to traditional approaches, with a particular emphasis on its utilization in the surveillance and mitigation of leaf diseases. Traditional approaches to disease control frequently encounter the constraint of immobility, as they heavily depend on physical presence and manual labour. This difficulty is circumvented through the utilization of sophisticated sensor technologies that enable electronic control and remote monitoring. By strategically positioning sensors in fields, data is collected autonomously, enabling comprehensive coverage of expansive agricultural areas. The increased mobility not only improves the ability to monitor and detect diseases but also enables prompt reactions to outbreaks. By enabling remote monitoring and management of fields, a proactive and more effective strategy can be implemented to minimize the adverse effects of leaf diseases. Conventional approaches to disease control, which rely on manual data capture and visual inspections, are inherently laborious. In contrast, electronic control systems provide the advantage of real-time monitoring capabilities, which facilitates prompt detection and reaction to outbreaks of diseases. The time gap between disease detection and intervention is diminished due to the expeditious decision-making made possible by the immediate data collection and analysis. Ensuring timely responses is of utmost importance in averting the progression of leaf diseases, as such measures substantially lessen their detrimental effects on agriculture. Conventional methodologies frequently depend on proficient personnel to ensure precise diagnosis and treatment of diseases. After being implemented, electronic control systems function independently or require minimal human intervention. The decreased reliance on specialized knowledge enhances the accessibility of disease control measures to a wider spectrum of agricultural practitioners and producers. The integration of automated processes and user-friendly interfaces enhances usability, thereby promoting the democratization of disease control strategy implementation. Consequently, the reduction in the obstacle to entry for efficient disease management promotes sustainability and inclusiveness in the agricultural sector. The ability of electronic surveillance systems to generate enormous quantities of data is one of their primary advantages. By harnessing sophisticated analytics and ML algorithms, this data-centric methodology yields significant insights pertaining to patterns of diseases. The utilization of historical data in predictive modelling facilitates timely identification and intervention, thereby augmenting the overall effectiveness of disease control strategies. The amalgamation of decision support systems and data analysis enables researchers and producers to make well-informed decisions, thereby maximizing the efficient use of resources and reducing the financial repercussions of leaf diseases. In recent times, interest in the application of electronic sensors (e-sensors) for the detection, identification, and quantification of pathogens in the context of leaf diseases has increased substantially. This technological progression signifies a fundamental change in approaches to disease control, providing an array of advantages that surpass the functionalities of conventional techniques. Equipped with targeted receptors, e-sensors demonstrate an exceptional degree of sensitivity, enabling the timely identification of pathogens at the molecular level, well in advance of the appearance of observable symptoms. This particular capability confers a significant edge in the proactive management of diseases, facilitating prompt and focused reactions to possible epidemics.

Electronic sensor technologies surpass mere detection capabilities by effectively discerning and quantifying pathogens. The sensors' specificity facilitates accurate identification, thereby permitting targeted responses that are customized to the specific pathogen responsible for the infection. Furthermore, the capacity to measure pathogen concentrations offers significant insights into the gravity of epidemics, thereby guiding policymakers towards the most suitable measures to alleviate the detrimental effects on agricultural produce. In contrast to conventional approaches to disease control, which frequently depend on manual evaluations and visual inspections for identification and quantification, this degree of accuracy and specificity is unmatched.

An essential benefit of e-sensor technologies is their effortless incorporation with data platforms. By transferring and storing the data produced by these sensors in centralized databases in an efficient manner, a comprehensive resource for disease surveillance can be established. Cloud-based solutions provide an additional layer of accessibility by enabling real-time information access for stakeholders including producers, researchers, and policymakers. By encouraging collaboration and a data-driven approach to leaf disease control, this integration facilitates better-informed decision-making.

Furthermore, the implementation of e-sensors is consistent with the tenets of precision agriculture. Achieving precise targeting of leaf disease-affected areas enables producers to maximize the efficiency of input utilization, including that of fertilizers and pesticides. By taking this focused approach, producers are able to not only mitigate the environmental consequences of these inputs but also attain cost savings. In addition, the implementation of precision agriculture methods enhances the ecological responsibility of agricultural operations by promoting a harmonious coexistence of environmental stewardship and productivity. Pathogen prevalence is depicted in a dynamic and ever-changing manner by the continuous monitoring capabilities of e-sensor technologies. This real-time information facilitates the development of predictive models and the forecasting of disease outbreaks, thereby enabling preventative and proactive measures. The capacity to predict the occurrence of diseases empowers agricultural practitioners to promptly execute control measures, thereby minimizing potential damages and maximizing crop productivity. The aforementioned proactive approach presents a clear juxtaposition to conventional methods that frequently depend on periodic inspections—which could lead to postponed reactions and heightened vulnerability to the transmission of diseases.

5.7 DISCUSSION

Many automation approaches are used to identify and monitor leaf diseases in crops, including cutting-edge technology like ML, DL, sensor networks, the IoT, robots, and drones. Notably, ML and DL approaches, including models such as CNN, SVM, KNN, Decision Tree, Naïve Bayes, YOLO, ResNet, EfficientNet, VGG16, ResNet34, and YOLOv5, achieve amazing accuracy, with testing accuracies of up to 99% for certain crop species. Sensor networks, which include wearable plant sensors, WSNs, and electronic monitoring systems, are critical in real-time plant health monitoring because they allow for early disease diagnosis and prompt responses. The combination of IoT devices, environmental sensors, and image sensors improves monitoring capabilities by gathering and analysing critical data including soil moisture, temperature, humidity, and high-resolution leaf photos. Robots outfitted with powerful image sensors and ML algorithms can perform on-field inspection and targeted treatment, as well as autonomously navigate fields and detect unhealthy plants. Furthermore, drone-based surveillance using thermal or multispectral cameras appears as an effective and complete disease evaluation method, providing fast scans of wide regions as well as significant information for targeted treatments and crop protection. These results highlight the revolutionary role of sophisticated technology in improving agricultural disease control techniques [79–91].

5.8 CONCLUSION

In conclusion, the research highlights the revolutionary potential of sophisticated automation approaches in the process of changing the diagnosis and control of leaf diseases in agricultural settings. The combination of ML, DL, sensor networks, the IoT, robotics, and drones not only makes it possible to accurately and promptly identify diseases, but it also gives the agricultural sector the ability to make decisions based on data and to develop treatment strategies that are specifically tailored to the disease. The utilization of electronic sensors to detect, identify, and quantify pathogens signifies a paradigm shift in the realm of leaf disease management. Elevated sensitivity, specificity, and real-time functionality are characteristics of e-sensors that set them apart from conventional approaches. This, in turn, leads to increased productivity, decreased crop losses, and improved food security. Future research paths are very necessary in order to make additional progress in this subject. Continued efforts in building ML and DL models that are more resilient and generalized are required in order to achieve accurate identification across a wide range of crops and environmental situations. The integration of multimodal data from a variety of sources, including sensor networks, IoT devices, satellite imaging, and meteorological data, shows promise in terms of increasing the overall knowledge of disease dynamics and strengthening detection models. In addition, the investigation of large-scale WSNs in agricultural fields has the potential to permit continuous monitoring and the collecting of data in real time for the purpose of early detection. The broad use of these technologies and the optimization of resource usage in the agricultural sector are both dependent on the development of user-friendly interfaces and decision support systems, as well as advancements in autonomous robotic systems and precision agriculture methods among other things. In the process of pursuing these future goals, the agricultural sector has the opportunity to embrace innovation in order to meet new difficulties and guarantee that crop management techniques are both sustainable and resilient.

REFERENCES

1. S. Sankaran, A. Mishra, R. Ehsani, and C. Davis, "A review of advanced techniques for detecting plant diseases," *Computers and Electronics in Agriculture,* Accessed: Feb. 21, 2024. [Online]. Available: https://www.sciencedirect.com/science/article/pii/S0168169910000438.
2. K. Gavhale, and U. Gawande, "An overview of the research on plant leaves disease detection using image processing techniques," *IOSR Journal of Computer Engineering (IOSR-JCE)*, vol. 1, pp. 10–16, 2014, doi: 10.9790/0661-16151016.
3. V. Singh, and A. K. Misra, "Detection of plant leaf diseases using image segmentation and soft computing techniques," *Information Processing in Agriculture*, Accessed: Feb. 21, 2024. [Online]. Available: https://www.sciencedirect.com/science/article/pii/S2214317316300154.
4. A. Jafar, N. Bibi, R. A. Naqvi, and A. Sadeghi-Niaraki, "Revolutionizing agriculture with artificial intelligence: plant disease detection methods, applications, and their limitations," *Frontiers in Plant Science*, 2024, Accessed: Jun. 03, 2024. [Online]. Available: https://www.frontiersin.org/articles/10.3389/fpls.2024.1356260/full.
5. M. Sharma, and P. Kaur, "A comprehensive analysis of nature-inspired meta-heuristic techniques for feature selection problem," *Archives of Computational Methods in Engineering*, vol. 1, no. 3, p. 3, May 2021, doi: 10.1007/s11831-020-09412-6.
6. L. Ngugi, M. Abelwahab, and M. Abo-Zahhad, "Recent advances in image processing techniques for automated leaf pest and disease recognition–A review," *Information Processing in Agriculture*, 2021, Accessed: Feb. 21, 2024. [Online]. Available: https://www.sciencedirect.com/science/article/pii/S2214317320300196.
7. Z. Iqbal et al., "An automated detection and classification of citrus plant diseases using image processing techniques: a review," *Computers and Electronics in Agriculture*, 2018, Accessed: Feb. 21, 2024. [Online]. Available: https://www.sciencedirect.com/science/article/pii/S0168169917311870.
8. V. Singh, and A. K. Misra, "Detection of plant leaf diseases using image segmentation and soft computing techniques," *Information processing in Agriculture*, 2017, Accessed: Feb. 21, 2024. [Online]. Available: https://www.sciencedirect.com/science/article/pii/S2214317316300154.
9. R. N. Jogekar, and N. Tiwari, "A review of deep learning techniques for identification and diagnosis of plant leaf disease," *Smart Innovation, Systems and Technologies*, vol. 182, pp. 435–441, 2021, doi: 10.1007/978-981-15-5224-3_43.
10. T. Zhang, Y. Cai, P. Zhuang, and J. Li, "Remotely sensed crop disease monitoring by machine learning algorithms: a review," *World Scientific*, vol. 12, no. 1, pp. 161–171, Jun. 2024, doi: 10.1142/S2301385024500237.

11. S. Muneer, U. Farooq, A. Athar, M. A. Raza, T. M. Ghazal, and S. Sakib, "A critical review of artificial intelligence based approaches in intrusion detection: a comprehensive analysis," *Journal of Engineering,* vol. 2024, pp. 1–16, Apr. 2024, doi: 10.1155/2024/3909173.
12. M. Mishra, P. Choudhury, and B. Pati, "IoT enabled plant leaf disease segmentation and multi-classification using mayfly bald eagle optimization-enabled machine learning," *Multimedia Tools Applications*, vol. 83, pp. 59747–59781, 2024, doi: 10.1007/S11042-023-17680-0.
13. Z. A. Ali, D. Deng, M. K. Shaikh, R. Hasan, and M. A. Khan, "AI-based UAV swarms for monitoring and disease identification of brassica plants using machine learning: a review," *Computer Systems Science & Engineering*, doi: 10.32604/csse.2023.041866.
14. N. Ahmed et al., "Role of plants in managing diseases," in *Ethnic Knowledge and Perspectives of Medicinal Plants: Volume 2: Nutritional and Dietary Benefits*, pp. 579–604, Jan. 2023, doi: 10.1201/9781003353089-30/
15. S. Majumder, B. R. Reddey, J. Pandey, A. Paul, A. Kumar, and K. Banerjee, "Pesticide residue and bio-pesticides in vegetable crops," *Vegetable Science*, vol. 51, pp. 77–96, Feb. 2024, doi: 10.61180/VEGSCI.2024.V51.SPL.08.
16. X. Zheng et al., "Genome wide analysis of the EXP family in *Rosa chinensis* and negative regulation of black spot disease by *RcEXPA8*," *Scientia Horticulturae*, Accessed: Feb. 21, 2024. [Online]. Available: https://www.sciencedirect.com/science/article/pii/S0304423823007343.
17. A. Gutierrez-Urrego, and D. Johnston-Monje, "Rhizobacterial control of root diseases and stimulation of plant growth," Feb. 2024, doi: 10.20944/PREPRINTS202402.0740.V1.
18. A. Naorem et al., "Green nanomaterials for a sustainable future environment," in *Green Nanomaterials in Energy Conversion and Storage Applications*, pp. 215–237, 2024, Accessed: Feb. 21, 2024. [Online]. Available: https://www.taylorfrancis.com/chapters/edit/10.1201/9781003398578-10/green-nanomaterials-sustainable-future-environment-anandkumar-naorem-patel-bhaguna-sharma-singh-priya-chanu-patel-singh-jaison-sahu-sahu-udayana.
19. S. Bhatt, N. Faridi, S. M. P. Raj, A. Agarwal, and M. Punetha, "Recent advances in immuno-based methods for the detection of Ralstonia solanacearum," *Journal of Microbiological Methods*, 2024, Accessed: Feb. 21, 2024. [Online]. Available: https://www.sciencedirect.com/science/article/pii/S0167701224000010.
20. D. Takagi, "Manganese toxicity and tolerance in photosynthetic organisms and breeding strategy for improving manganese tolerance in crop plants: physiological and omics approach perspectives," in *Heavy Metal Toxicity and Tolerance in Plants: A Biological, Omics, and Genetic Engineering Approach*, pp. 141–168, Jan. 2023, doi: 10.1002/9781119906506.CH7.
21. J. Puławska, A. Mikiciński, and P. Sobiczewski, "The history of fire blight biocontrol with Gram-negative bacteria and bacteriophages," *Journal of Plant Pathology*, 2023, doi: 10.1007/S42161-023-01554-3.
22. B. B. Westerdahl and O. M. Bello, "Leaf and bud nematodes in agricultural crops and their management by biotechnological approaches," *Novel Biological and Biotechnological Applications in Plant Nematode Management*, pp. 359–376, 2023, doi: 10.1007/978-981-99-2893-4_16.

23. R. P. Narmadha and G. Arulvadivu, “Detection and measurement of paddy leaf disease symptoms using image processing,” in *2017 International Conference on Computer Communication and Informatics (ICCCI)*, Coimbatore, pp. 1–4, Jan. 2017, doi: 10.1109/ICCCI.2017.8117730.
24. H. N. Ngugi, A. E. Ezugwu, A. A. Akinyelu, and L. Abualigah, “Revolutionizing crop disease detection with computational deep learning: a comprehensive review,” *Environmental Monitoring and Assessment*, vol. 196, no. 3, pp. 1–35, Mar. 2024, doi: 10.1007/S10661-024-12454-Z/FIGURES/13.
25. M. E. H. Chowdhury et al., “Automatic and reliable leaf disease detection using deep learning techniques,” *AgriEngineering*, 2021, doi: 10.3390/agriengineering3020020.
26. S. Arivazhagan, R. N. Shebiah, S. Ananthi, and S. V. Varthini, “Detection of unhealthy region of plant leaves and classification of plant leaf diseases using texture features,” *Agricultural Engineering International: CIGR Journal*, vol. 15, no. 1, p. 211, 2013, Accessed: Feb. 21, 2024. [Online]. Available: https://cigrjournal.org/index.php/Ejounral/article/view/2338.
27. A. Jafar, N. Bibi, R. A. Naqvi, A. Sadeghi-Niaraki, and D. Jeong, “Revolutionizing agriculture with artificial intelligence: plant disease detection methods, applications, and their limitations,” *Frontiers in Plant Science*, vol. 15, 2024, doi: 10.3389/FPLS.2024.1356260/FULL.
28. N. Wang et al., “Maize leaf disease identification based on fisher discrimination analysis,” *Scientia Agricultura Sinica*, 2009, Accessed: Mar. 03, 2024. [Online]. Available: https://www.cabdirect.org/cabdirect/abstract/20103012798.
29. B. Liu, Y. Zhang, D. J. He, and Y. Li, “Identification of apple leaf diseases based on deep convolutional neural networks,” *Symmetry*, 2017, doi: 10.3390/sym10010011.
30. A. Dixit, and S. Nema, “Wheat leaf disease detection using machine learning method-a review,” *International Journal of Computer Science and Mobile Computing*, vol. 7, no. 5, pp. 124–129, 2018, Accessed: Mar. 03, 2024. [Online]. Available: https://www.researchgate.net/profile/Prem-Baboo/post/is_there_is_any_data-set_available_for_wheat_crop_diseases/attachment/5d961fbe3843b093838a5f1a/AS%3A809954520870912%401570119614848/download/V7I5201824.pdf.
31. K. Prashar, R. Talwar, and C. Kant, “Inconsistent cluster analysis with disease feature enhancement (ICADFE) for American cotton leaf disease recognition,” *International Journal of Engineering and Advanced Technology*, 2020, doi: 10.35940/ijeat.E1220.0585C19.
32. L. Sorte et al., “Coffee leaf disease recognition based on deep learning and texture attributes,” *Procedia Computer Science*, Accessed: Mar. 03, 2024. [Online]. Available: https://www.sciencedirect.com/science/article/pii/S1877050919313468.
33. P. Tangtisanon and S. Kornrapat, “Holy basil curl leaf disease classification using edge detection and machine learning,” in *ACM International Conference Proceeding Series*, Sydney, pp. 85–89, Feb. 2020, doi: 10.1145/3384613.3384634.

34. M. A. Genaev et al., “Image-based wheat fungi diseases identification by deep learning,” *Plants*, 2021, doi: 10.3390/plants10081500.
35. I. Agrawal, P. Hegde, P. Shetty, and P. Shingane, “Leaf disease detection of cucurbits using CNN,” in *ITM Web of Conferences*, 2021, doi: 10.1051/itmconf/20214003043.
36. I. Vagelas, A. Papadimos, C. Lykas, and L. Molinero-Ruiz, “Pre-symptomatic disease detection in the vine, chrysanthemum, and rose leaves with a low-cost infrared sensor,” *Agronomy*, 2021, doi: 10.3390/agronomy11091682.
37. J. Rashid, I. Khan, G. Ali, H. Almotiri, M. A. Alghamdi, and K. Masood, “Multi-level deep learning model for potato leaf disease recognition,” *Electronics*, 2021, doi: 10.3390/electronics10172064.
38. S. H. Abed, A. S. Al-Waisy, H. J. Mohammed, and S. Al-Fahdawi, “A modern deep learning framework in robot vision for automated bean leaves diseases detection,” *International Journal Intelligent Robotics Applications*, vol. 5, no. 2, pp. 235–251, Jun. 2021, doi: 10.1007/S41315-021-00174-3.
39. M. Sahimat and S. Fattah, “Cabbage disease detection system using K-NN,” 2022, Accessed: Mar. 03, 2024. [Online]. Available: https://oer.ums.edu.my/handle/oer_source_files/1875.
40. M. A. Mithu, S. I. Momo, M. M. Hasan, K. M. Rahman, and A. Sattar, “Pumpkin leaf disease detection: convenience of CNN over traditional machine learning in terms of image classification,” *Smart Innovation, Systems and Technologies*, vol. 235, pp. 347–357, 2022, doi: 10.1007/978-981-16-2877-1_32.
41. R. Mahum et al., “A novel framework for potato leaf disease detection using an efficient deep learning model,” *Human and Ecological Risk Assessment: An International Journal*, vol. 29, no. 2, pp. 303–326, 2023, doi: 10.1080/10807039.2022.2064814.
42. P. V. Yeswanth, S. Deivalakshmi, S. George, and S. B. Ko, “Residual skip network-based super-resolution for leaf disease detection of grape plant,” *Circuits Systems Signal Processing*, vol. 42, no. 11, pp. 6871–6899, Nov. 2023, doi: 10.1007/S00034-023-02430-2.
43. X. Huang et al., “Tomato leaf disease detection system based on FC-SNDPN,” *Multimedia Tools Applications*, vol. 82, no. 2, pp. 2121–2144, Jan. 2023, doi: 10.1007/S11042-021-11790-3.
44. H. T. Thai, K. H. Le, and N. L. T. Nguyen, “Formerleaf: an efficient vision transformer for cassava leaf disease detection,” *Computers and Electronics in Agriculture*, Accessed: Mar. 03, 2024. [Online]. Available: https://www.sciencedirect.com/science/article/pii/S0168169922008262.
45. U. Farooq et al., “Cottonleafnet: cotton plant leaf disease detection using deep neural networks,” *Multimedia Tools and Applications*, vol. 82, no. 24, pp. 37151–37176, Oct. 2023, doi: 10.1007/s11042-023-14954-5.
46. S. P. Praveen, R. Nakka, A. Chokka, V. N. Thatha, S. S. Vellela, and U. Sirisha, “A novel classification approach for grape leaf disease detection based on different attention deep learning techniques,” *International Journal of Advanced Computer Science and Applications (IJACSA)*, vol. 14, no. 6, p. 2023, Accessed: Mar. 03, 2024. [Online]. Available: https://www.academia.edu/download/104751177/Grape_leaf_Detection.pdf.
47. X. Huang et al., “Tomato leaf disease detection system based on FC-SNDPN,” *Multimedia Tools and Applications*, Accessed: Mar. 03, 2024. [Online]. Available: https://link.springer.com/article/10.1007/s11042-021-11790-3.

48. S. Towfek, and N. Khodadadi, "Deep convolutional neural network and metaheuristic optimization for disease detection in plant leaves," *Journal of Intelligent Systems and Internet of Things*, 2023, doi: 10.54216/JISIoT.100105.
49. D. Pandiri, R. Murugan, and T. Goel, "Whale-optimized convolutional neural network for potato fungal pathogens disease classification," in *Handbook of Whale Optimization Algorithm*, pp. 477–485, 2024, Accessed: Mar. 03, 2024. [Online]. Available: https://www.sciencedirect.com/science/article/pii/B9780323953658000403.
50. C. P. Vijay, and K. Pushpalatha, "DV-PSO-Net: A novel deep mutual learning model with Heuristic search using Particle Swarm optimization for Mango leaf disease detection," *Journal of Integrated Science and Technology*, vol. 2024, no. 5, p. 804, Accessed: Mar. 03, 2024. [Online]. Available: https://pubs.thesciencein.org/journal/index.php/jist/article/view/a804.
51. A. Sreedevi, and C. Manike, "A smart solution for tomato leaf disease classification by modified recurrent neural network with severity computation," *Cybernetics Systems*, vol. 55, no. 2, pp. 409–449, 2024, doi: 10.1080/01969722.2022.2122004.
52. A. Bonkra, S. Pathak, A. Kaur, and M. A. Shah, "Exploring the trend of recognizing apple leaf disease detection through machine learning: a comprehensive analysis using bibliometric techniques," *Artificial Intelligence Review*, vol. 57, no. 2, Feb. 2024, doi: 10.1007/S10462-023-10628-8.
53. G. Lee, Q. Wei, and Y. Zhu, "Emerging wearable sensors for plant health monitoring," *Advance Functional Materials*, vol. 31, no. 52, Dec. 2021, doi: 10.1002/ADFM.202106475.
54. F. Nabi, S. Jamwal, and K. Padmanbh, "Wireless sensor network in precision farming for forecasting and monitoring of apple disease: a survey," *International Journal of Information Technology*, vol. 14, no. 2, pp. 769–780, Mar. 2022, doi: 10.1007/S41870-020-00418-8.
55. A. Mohammad-Razdari, D. Rousseau, A. Bakhshipour, S. Taylor, J. Poveda, and H. Kiani, "Recent advances in E-monitoring of plant diseases," *Biosensors and Bioelectronics*, Accessed: Mar. 03, 2024. [Online]. Available: https://www.sciencedirect.com/science/article/pii/S0956566321009908.
56. J. Zhang, Y. Rao, C. Man, Z. Jiang, and S. Li, "Identification of cucumber leaf diseases using deep learning and small sample size for agricultural Internet of Things," *International Journal of Distributed Sensor Networks*, vol. 17, no. 4, 2021, doi: 10.1177/15501477211007407.
57. A. Rehman, T. Saba, M. Kashif, S. M. Fati, S. A. Bahaj, and H. Chaudhry, "A revisit of internet of things technologies for monitoring and control strategies in smart agriculture," *Agronomy*, Accessed: Mar. 05, 2024. [Online]. Available: https://www.mdpi.com/2073-4395/12/1/127.
58. J. Fan, Y. Zhang, W. Wen, S. Gu, X. Lu, and X. Guo, "The future of Internet of Things in agriculture: plant high-throughput phenotypic platform," *Journal of Cleaner Production*, Accessed: Mar. 05, 2024. [Online]. Available: https://www.sciencedirect.com/science/article/pii/S0959652620336969.
59. D. Gao, Q. Sun, B. Hu, and S. Zhang, "A framework for agricultural pest and disease monitoring based on internet-of-things and unmanned aerial vehicles," *Sensors*, vol. 20, no. 5, 2020, Accessed: Mar. 05, 2024. [Online]. Available: https://www.mdpi.com/1424-8220/20/5/1487

60. M. A. Nawaz et al., "Plant disease detection using internet of thing (IoT)," *International Journal of Advanced Computer Science and Applications*, vol. 11, no. 1, 2020, Accessed: Mar. 05, 2024. [Online]. Available: https://thesai.org/Downloads/Volume11No1/Paper_62-Plant_Disease_Detection.pdf.
61. U. Shafi et al., "A multi-modal approach for crop health mapping using low altitude remote sensing, internet of things (IoT) and machine learning," *IEEE Access*, Accessed: Mar. 05, 2024. [Online]. Available: https://ieeexplore.ieee.org/abstract/document/9119071/.
62. H. Orchi, M. Sadik, and M. Khaldoun, "On using artificial intelligence and the internet of things for crop disease detection: a contemporary survey," *Agriculture*, Accessed: Mar. 05, 2024. [Online]. Available: https://www.mdpi.com/2077-0472/12/1/9.
63. P. Ren, "Planto: an animistic planter system design for improving the working from home experience," Master thesis, Purdue University, 2022. https://docs.lib.purdue.edu/dissertations/AAI30506235/
64. I. Samal, T. K. Bhoi, A. K. Pradhan, and D. K. Mahanta, "Plantix app: a success story of artificial intelligence in plant protection," *Van Sangyan*, Accessed: Mar. 05, 2024. [Online]. Available: https://www.researchgate.net/profile/Deepak-Kumar-Mahanta/publication/370650382_Plantix_app_A_Success_story_of_artificial_intelligence_in_plant_protection/links/645b9e4e2edb8e5f094db226/Plantix-app-A-Success-story-of-artificial-intelligence-in-plant-protection.pdf#page=29.
65. M. Trivedi, and A. Gupta, "Automatic monitoring of the growth of plants using deep learning-based leaf segmentation," *International Journal of Applied Science and Engineering*, Accessed: Mar. 05, 2024. [Online]. Available: https://gigvvy.com/journals/ijase/articles/ijase-202106-18-2-003.
66. H.F. Ng, C. Y. Lin, J. H. Chuah, H. K. Tan, and K. H. Leung, "Plant disease detection mobile application development using deep learning," in *2021 InternationalConferenceonComputer&InformationSciences(ICCOINS)*,pp. 34–38, 2021, Accessed: Mar. 05, 2024. [Online]. Available: https://ieeexplore.ieee.org/abstract/document/9497190/?casa_token=OzXsYitYHT0AAAAA.
67. R. J. Dufault et al., "Influence of cover crops and inorganic nitrogen fertilization on tomato and snap bean production and soil nitrate distribution," *Journal of Vegetable Crop Production*, vol. 6, no. 2, pp. 13–25, Dec. 2000, doi: 10.1300/J068V06N02_03.
68. R. Andrade et al., "Where is my crop? Data-driven initiatives to support integrated multi-Stakeholder agricultural decisions," *Frontiers Sustainable Food Systems*, vol. 5, Dec. 2021, doi: 10.3389/FSUFS.2021.737528/FULL.
69. S. J. Pethybridge, and S. C. Nelson, "Leaf doctor: A new portable application for quantifying plant disease severity," *Plant Disease*, vol. 99, no. 10, pp. 1310–1316, Oct. 2015, doi: 10.1094/PDIS-03-15-0319-RE.
70. Y. B. Shapovalov et al., "Comparing Google Lens recognition accuracy with other plant recognition apps," in *Proceedings of the Symposium on Advances in Educational Technology, AET*, pp. 20–33, 2020, doi: 10.5220/0010928000003364.
71. J. Suhaman, T. Sari, K. Kamandanu, D. Aulianti, M. Adhi, and U. Amartiwi, "Smart plant: a mobile application for plant disease detection," in *GMPI Conference Series*, vol. 2, pp. 52–57, 2023, doi: 10.53889/gmpics.v2.173.

72. A. Xenakis, G. Papastergiou, V. C. Gerogiannis, and G. Stamoulis, "Applying a convolutional neural network in an IoT robotic system for plant disease diagnosis," in *2020 11th International Conference on Information, Intelligence, Systems and Applications (IISA)*, 2020, Accessed: Mar. 05, 2024. [Online]. Available: https://ieeexplore.ieee.org/abstract/document/9284356/?casa_token=DrB9YkGA5YwAAAAA.
73. S. A. Sabeeh, and S. H. Ameen, "Detection and classification of leaf disease using deep learning for a greenhouses' robot," *Iraqi Journal of Computer, Communication, Control and System Engineering*, vol. 21, no. 4, pp. 2617–3352, 2021, doi: 10.33103/uot.ijccce.21.4.2.
74. S. Cubero, E. Marco-Noales, N. Aleixos, S. Barbé, and J. Blasco, "Robhortic: a field robot to detect pests and diseases in horticultural crops by proximal sensing," *Agriculture*, doi: 10.3390/agriculture10070276.
75. A. Nooraiyeen, "Robotic vehicle for automated detection of leaf diseases," in *2020 IEEE International Conference on Electronics, Computing and Communication Technologies (CONECCT)*, 2020, Accessed: Mar. 05, 2024. [Online]. Available: https://ieeexplore.ieee.org/abstract/document/9198326/?casa_token=EpBe2zOf2W0AAAAA.
76. N. Schor, S. Berman, A. Dombrovsky, Y. Elad, T. Ignat, and A. Bechar, "Development of a robotic detection system for greenhouse pepper plant diseases," *Precision Agriculture*, vol. 18, no. 3, pp. 394–409, Jun. 2017, doi: 10.1007/S11119-017-9503-Z.
77. Y. Ampatzidis, L. De Bellis, and A. Luvisi, "iPathology: robotic applications and management of plants and plant diseases," *Sustainability*, doi: 10.3390/su9061010.
78. N. Schor, A. Bechar, T. Ignat, A. Dombrovsky, Y. Elad, and S. Berman, "Robotic disease detection in greenhouses: combined detection of powdery mildew and tomato spotted wilt virus," *IEEE Robotics and Automation Letters*, vol. 1, no. 1, pp. 354–360, 2016.
79. V. K. Sharma, "CNTFET circuit-based wide fan-in domino logic for low power applications," *Journal Circuits Systems Computers*, vol. 31, no. 02, p. 2250036, 2022.
80. V. K. Sharma, and M. Pattanaik, "A reliable ground bounce noise reduction technique for nanoscale CMOS circuits," *International Journal Electronics*, vol. 102, no. 11, pp. 1852–1866, 2015.
81. V. K. Sharma, S. Patel, and M. Pattanaik, "High performance process variations aware technique for sub-threshold 8T-SRAM cell," *Wireless Personal Communications*, vol. 78, no. 1, pp. 57–68, 2014.
82. V. K. Sharma, and M. Pattanaik, "Techniques for low leakage nanoscale VLSI circuits: a comparative study," *Journal Circuits Systems Computers*, vol. 23, no. 05, p. 1450061, 2014.
83. V. K. Sharma, "A survey of leakage reduction techniques in CMOS digital circuits for nanoscale regime," *Austrialian Journal Electrical Electronics Engineering*, vol. 18, no. 4, pp. 217–236, 2021.
84. C. Kumar, A. S. Mishra, and V. K. Sharma, "Leakage power reduction in CMOS logic circuits using stack ONOFIC technique," in *2018 Second International Conference on Intelligent Computing and Control Systems (ICICCS)*, Madurai, 2018.

85. V. K. Sharma and M. Pattanaik, "Design of low leakage variability aware ONOFIC CMOS standard cell library," *Journal Circuits Systems Computers*, vol. 25, no. 11, p. 1650134, 2016.
86. S. F. Naz, S. Riyaz, and V. K. Sharma, "A review of QCA nanotechnology as an alternate to CMOS," *Current Nanoscience*, vol. 18, no. 1, pp. 18–30, 2022.
87. S. Riyaz, and V. K. Sharma, "Design of reversible Feynman and double Feynman gates in quantum-dot cellular automata nanotechnology," *Circuit World*, vol. 49, no. 1, pp. 28–37, 2023.
88. B. Raina, C. Verma, M. Gupta, and V. K. Sharma, "Binary coded decimal (BCD) seven segment circuit designing using quantum-dot cellular automata (QCA)," in *2021 5th International Conference on Trends in Electronics and Informatics (ICOEI)*, Tirunelveli, 2021.
89. U. Mushtaq, and V. K. Sharma, "Performance analysis for reliable nanoscaled FinFET logic circuits," *Analog Integrated Circuits Signal Processing*, vol. 107, no. 3, pp. 671–682, 2021.
90. Kajal and V. K. Sharma, "Design and simulation of FinFET circuits at different technologies," in *2021 6th International Conference on Inventive Computation Technologies (ICICT)*, Coimbatore, 2021.
91. V. K. Sharma, "Designing of face recognition system," in *2019 International Conference on Intelligent Computing and Control Systems (ICCS)*, Madurai, 2019.

Chapter 6

Navigating the linguistic landscape

Unveiling the future trajectory of natural language processing in an evolving digital era

Jyoti Snehi, Isha Kansal, and Ashish Kumar

6.1 INTRODUCTION

The digital realm has experienced a significant shift characterized by a rapid rise in the production of textual data. This section will examine the ways in which the vast amount of unstructured text data is influenced by a variety of contemporary activities, such as email communication, social media interactions, and newspaper information transmission. It draws attention to the difficulties in understanding and drawing actionable conclusions from such a large and unstructured amount of data, underscoring the problems presented by this data flood [1]. Unstructured text data are too complex for conventional techniques of data analysis and interpretation. This background section argues for the need for cutting-edge technology like natural language processing (NLP) by outlining how traditional methods fall short of addressing the subtleties of human discourse. In light of the difficulties presented by unstructured text data, this part presents NLP as a revolutionary remedy. It might discuss significant historical advancements and innovations in NLP, illustrating how the field has developed as a technology to help computers comprehend, interpret, and produce human language. The background information goes into further detail on how NLP has evolved into a crucial aspect of contemporary society rather than only a technological development. It highlights how NLP permeates everyday technology encounters, enhancing their naturalness and usability. According to some accounts, this integration represents a profound shift in society, with technology responding to human language patterns instead of the other way around [2]. NLP is important because it is revolutionizing the ways in which computers communicate and comprehend human language. NLP is essential to many fields, helping to progress data analysis, communication, and technology. The following are some salient features that underscore the importance of NLP:

DOI: 10.1201/9781003643609-6

a. **Enhanced human-computer interaction:** More intuitive and natural interactions between people and computers are made possible by NLP. Computers can understand and respond to human language thanks to NLP algorithms, enabling spoken or written communication that mimics human discourse. This enhances the user experience for voice-activated devices, chatbots, and virtual assistants, among other applications.
b. **Effective information extraction and analysis:** Conventional data analysis approaches are challenged by the vast amount of unstructured text data generated on a daily basis, ranging from news articles to social media messages. With the use of NLP, significant information may be extracted from this unstructured data, enabling insightful tasks like entity recognition and sentiment analysis.
c. **Automation of language-intensive tasks:** NLP plays a part in automating language processing tasks like text summarization, language translation, and document summarization. This lowers the workload for human operators and boosts efficiency at the same time, freeing them up to concentrate on more intricate and imaginative areas of their jobs.
d. **Advancements in artificial intelligence:** AI systems depend on NLP to interpret and produce language that is like that of humans. This is essential for creating intelligent systems that are able to understand user inquiries, respond appropriately, and adjust to shifting linguistic trends. NLP is frequently used by AI applications, such as machine learning and deep learning, to achieve better results.
e. **Improved accessibility:** A wider audience can more easily access technology thanks to NLP. Text-to-speech programs, language translation tools, and speech recognition systems improve accessibility for people with disabilities. This inclusiveness fits in with the larger objective of developing technology that meets the needs of a wide range of users.
f. **Personalization and user experience:** NLP is used to customize user interfaces in a wide range of applications, from virtual assistants to recommendation systems. Through natural language understanding of user preferences, NLP systems can customize their recommendations and responses, making the user experience more engaging and personalized.
g. **Applications across industries:** Numerous sectors, including healthcare, banking, education, and customer service, use NLP. NLP, for instance, can be applied to clinical documentation in the healthcare industry and to sentiment analysis in the financial industry to support trading decisions. Because of its adaptability, NLP is a useful tool in a variety of fields.

The chapter will examine the state of NLP today and offer insights into the most recent trends and advancements in the industry. This can entail a review of current technology, key figures in the field, and how NLP is being used in different industries. The range of applications of NLP is examined. This entails a thorough examination of the real-world applications of NLP, including language translation, chatbots, sentiment analysis, and other quickly developing fields. The intention is to offer a thorough grasp of the applications of NLP in the actual world. The chapter discusses current developments in NLP technology. This contains innovations in models, techniques, and algorithms that have advanced NLP. The purpose of the chapter is to illustrate the growth trajectory of NLP by analyzing these advances. Projecting NLP's future is a major component of the scope. This includes talking about how NLP will likely develop, what obstacles it might encounter, and how it might interact with new technologies like blockchain, artificial intelligence (AI), and the Internet of Things (IoT). This prospective viewpoint sheds light on how NLP might be incorporated into larger technology environments. The impact of NLP on particular domains—education, healthcare, and customer service—is examined in this chapter. The coverage offers insights into how NLP is boosting productivity and customizing experiences in these crucial areas by looking at how it is influencing these sectors. The integration of NLP with cutting-edge technology is included in the scope. This entails talking about how NLP is predicted to interact with blockchain, AI, and the IoT, as well as the possible effects of these connections on process automation, optimization, and system-to-system communication. The difficulties with NLP will be discussed in this chapter, along with some ethical issues. This entails talking about concerns about security, privacy, and the moral use of NLP technology. Finding chances for more innovation in the NLP space is part of the scope. This could include areas for research and development, untapped application areas, or gaps in current capabilities. By outlining this scope, this chapter hopes to offer a thorough examination of NLP, encompassing its present condition, potential future directions, and implications for a range of industries and cutting-edge technology. This methodical approach guarantees that this chapter stays pertinent and focused while making a significant contribution to our understanding of NLP in the rapidly changing digital era. Every day, enormous volumes of information are produced in the digital world through a variety of channels, including as emails, social media, websites, and digital periodicals. Unstructured text data is not categorized or fielded according to a preset data model, in contrast to structured data, which is. It is composed of free-form text that is difficult to arrange and analyze with conventional data processing techniques since it frequently contains a variety of linguistic features. Although unstructured text data is full of valuable information, its intrinsic complexity makes it difficult to extract significant insights without the use of specialized tools and processes like NLP. Because of its very nature, understanding unstructured text data presents a number of obstacles.

a. **Ambiguity and contextual Nuances:** Because natural language is by its very nature ambiguous, words and phrases frequently change meaning depending on the context in which they are employed. To effectively detect user intent or mood, NLP systems need to overcome the subtle and context-dependent character of language.
b. **Variability in language:** Language is dynamic and ever-changing. Different people may use different language idioms, slang terms, or cultural allusions to convey the same notion. For NLP systems to process and comprehend text data effectively, they must be flexible enough to accommodate these variances.
c. **Lack of standardization:** Unstructured text does not follow a set format, in contrast to structured data. Applying conventional data processing methods—which sometimes depend on consistent data structures—to unstructured text data is difficult due to this lack of regularity.
d. **Data noise and irrelevance:** Unstructured text data frequently includes noise, unrelated information, or even deceptive material. For NLP systems to effectively extract significant insights, they need to be able to filter out unnecessary input.
e. **Understanding context and intent:** It is necessary to comprehend the larger context in order to ascertain the genuine intention or sentiment behind a writing. It might be difficult to accurately read the intentions, feelings, or desires of users when there are ambiguities in their language.

By talking about these difficulties, it becomes easier to see why unstructured text data may be difficult for conventional data processing techniques to handle, underscoring the need for cutting-edge technologies like NLP. NLP has developed in response to the difficulties presented by unstructured text data. The historical background and significant turning points in the development of NLP could be covered in this section of the chapter.

6.1.1 Historical evolution

NLP has seen a rapid evolution, moving from early rule-based systems to the current dominance of deep learning and machine learning techniques. NLP relied on manually encoded linguistic rules in its early years (1950s–1980s), as demonstrated by developments like ELIZA and SHRDLU. The 1990s witnessed a significant transition toward statistical methods, as seen by the advent of statistical machine translation and Hidden Markov Models, which involved algorithms learning patterns from data. The 2010s to the present are marked by the incorporation of machine learning, with notable innovations such as the development of recurrent neural networks (RNNs), word embeddings like Word2Vec and GloVe, and the revolutionary effects

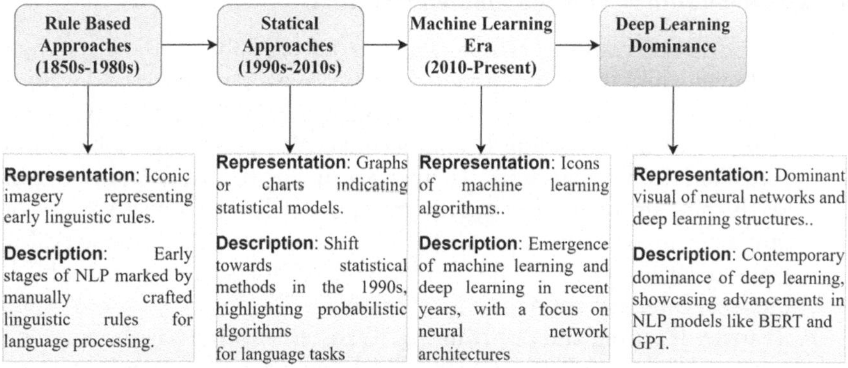

Figure 6.1 Evolution of natural language processing paradigms.

of transformer architectures like Bidirectional Encoder Representations from Transformers (BERT) and Generative Pre-trained Transformer (GPT). These deep learning models—GPT-3 being one example in particular—have completely transformed NLP by exhibiting hitherto unheard-of powers in language production and comprehension. This evolutionary trajectory demonstrates the ongoing search for more efficient ways to process natural language, leading to the development of more complex and context-aware language models with a wide range of applications. The evolution of paradigms for NLP is seen in Figure 6.1 [3].

6.1.2 Technological advancements

NLP has grown significantly as a result of technological developments, which have helped it transition from rule-based systems to the complex models that are used today. Numerous important elements have been instrumental in this advancement:

a. **Rule-based systems to machine learning:** A big change came with the transition from rule-based systems to machine learning algorithms. Manually created linguistic rules restricted the capabilities of early rule-based systems. A paradigm shift occurred with the introduction of machine learning, which made it possible for models to identify patterns in data, leading to more flexible and context-aware language processing.
b. **Deep learning and neural networks:** The development of deep learning, and especially neural networks, has changed the game. RNNs and transformers, such BERT and GPT, are examples of models that perform better at comprehending context and capturing complex linguistic structures. The quality of language representations has significantly increased because of these architectures [1].

6.1.2.1 Availability of large labeled datasets

a. **Corpora for training:** Having access to big labeled datasets, like Wikipedia and the Common Crawl, has proved crucial to the development of increasingly complex NLP models, e.g. GLUE dataset (General Language Understanding Evaluation), GloVe dataset (Global Vectors for Word Representation), Twitter Sentiment Analysis Dataset, etc. With the help of these datasets, models can learn from a variety of language nuances and patterns, which improves their capacity for generalization and high job performance. Table 6.1 shows the NLP datasets categorization and description.
b. **Transfer learning and pre-training:** Transfer learning has shown to be a very successful idea in which models are pre-trained on big datasets for general language understanding tasks. Pre-trained models that provide contextualized embeddings that capture semantic links in language, including Word2Vec, GloVe, and more recently BERT and GPT, have revolutionized NLP.

6.1.2.2 Increased computing power

a. **Parallel computing and GPU acceleration:** The development of computational power has aided in the expansion of NLP. It is now possible to train complicated models more quickly because of parallel computing and GPU acceleration, which cuts down on the time needed to test and implement advanced NLP solutions.
b. **Cloud computing resources:** Access to strong computational infrastructure has become more accessible thanks to the availability of cloud computing services. This has encouraged creativity and cooperation by enabling academics and developers to train and use NLP models without having to make large upfront hardware investments [5].

6.1.2.3 Semantic embeddings and word representations

a. **Word embeddings:** Efficient methods for representing words as dense vectors that capture semantic linkages and contextual meanings were made possible by techniques such as Word2Vec and GloVe. These embeddings have improved the models' comprehension and production of human-like language.
b. **Contextualized representations:** Recent developments have brought contextualized word representations, especially in transformer-based models like as BERT and GPT. These models take into account the context in which each word is used, resulting in a more precise and sophisticated comprehension of language.

The development of NLP is closely related to ongoing algorithmic improvements, the availability of large labeled datasets, and the rise in processing power that has stimulated experimentation and creativity.

Table 6.1 Categorization of NLP datasets according to tasks

NLP Dataset Type	*Dataset*	*Description*
Text Classification	IMDB Movie Reviews	A dataset for binary sentiment classification of movie reviews
	20 Newsgroups	A collection of 20,000 newsgroup documents
	Yelp Reviews	User reviews, business data, and user data
	Amazon Reviews	A dataset of customer reviews from Amazon
Sentiment Analysis	Sentiment140	This dataset contains tweets with sentiment labels
	Twitter Sentiment Analysis Dataset	Tweets labeled by sentiment
Question-Answering	SQuAD (Stanford Question Answering Dataset)	A reading comprehension dataset
Natural Language Inference	SNLI (Stanford Natural Language Inference)	A collection of human-written English sentence pairs
	MultiNLI (Multi-Genre Natural Language Inference)	A genre of spoken and written text
Named Entity Recognition (NER)	CoNLL-2003	Dataset to recognize entities
Machine Translation	WMT (Workshop on Machine Translation)	Shared tasks at the annual Conference on Machine Translation
Text Retrieval	TREC (Text REtrieval Conference)	A conference series
Pre-trained Word Vectors	GloVe (Global Vectors for Word Representation)	Pre-trained word vectors
	FastText	Subword information
General Language Understanding	GLUE (General Language Understanding Evaluation)	Natural language understanding tasks
Miscellaneous	Common Crawl	A corpus of web crawl data
	OpenSubtitles	Movie subtitles
	WikiText	A set of verified Good and Featured articles
	Quora Question Pairs	A dataset containing pairs of questions from Quora
	OntoNotes 5.0	A dataset annotated with syntactic and discourse information

The performance of NLP models has increased along with the range of applications due to these technical advancements, making NLP an essential part of today's technological environment [6].

6.2 NLP MARKET DYNAMICS

Because NLP may bridge the gap between human language and computer understanding, it is extremely important in modern technology [7]. Its importance is shown by several crucial factors:

a. **Enhanced user interaction:** More intuitive and natural communication between people and machines is made possible by NLP. This is demonstrated by apps like chatbots and virtual assistants, which enable consumers to communicate with technology using natural language, improving accessibility and user-friendliness.
b. **Data extraction and analysis:** Unstructured text data has vital information in the big data era. NLP makes it easier to extract and analyze insights from this data, which helps people make well-informed decisions in a variety of fields, including finance, healthcare, and business.
c. **Automation of language-intensive tasks:** Language processing functions including sentiment analysis, summarization, and translation are automated by NLP. Large amounts of textual data may be handled effectively thanks to this automation, which also saves time.
d. **Personalization and recommendation systems:** NLP plays a significant role in the creation of individualized technological experiences, especially in recommendation systems. Through natural language understanding of user preferences, NLP-powered systems may customize recommendations, increasing user satisfaction and engagement.
e. **Advancements in artificial intelligence:** A key component in the advancement of AI is NLP. It makes it possible for machines to comprehend, interpret, and produce language that is similar to that of humans. This is essential for developing intelligent systems that can recognize context and adjust to the demands of users.

6.2.1 Trends in the NLP market

The NLP market is evolving, and a number of significant developments are influencing its course:

a. **Rapid growth:** Due to the growing need for language processing applications across a range of industries, the NLP industry has grown quickly. The ability to comprehend language at a higher level is made possible by developments in deep learning and machine learning methods [8].
b. **Integration with business processes:** To derive useful insights from textual data sources such as customer interactions and feedback, organizations are incorporating NLP into their business processes. This integration gives you a competitive edge and improves decision-making.

c. **Voice-activated assistants:** The emergence of NLP-powered voice-activated virtual assistants is a notable development. These virtual assistants—Alexa, Siri, and Google Assistant, for example—showcase NLP's capacity to comprehend spoken language and provide relevant responses.
d. **Industry-specific NLP applications:** Creating NLP applications tailored to industries is becoming more popular. NLP is utilized, for instance, in healthcare to document clinical procedures, and in finance to support sentiment analysis in trading decisions.
e. **Multilingual NLP:** A significant trend in NLP model development is the ability to handle numerous languages, dialects, and cultural variations with ease, given the increasingly global nature of business and communication.

Figure 6.2 provides a graphic representation of the wide range of applications in different industries. The diagram demonstrates the seamless integration of NLP into various industries. Education, for example, benefits from NLP through Intelligent Tutoring Systems and Automated Grading; healthcare, from Clinical Documentation and Patient Interaction; retail, through chatbots and virtual assistants; finance, from Sentiment Analysis and Automated Reporting; legal processes, from Product Recommendations and Legal Research to

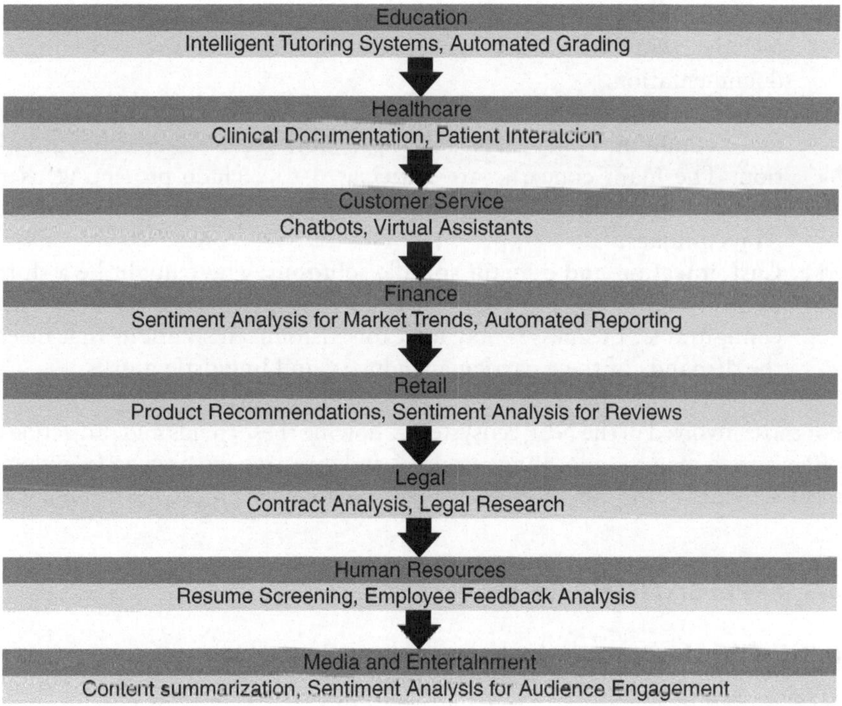

Figure 6.2 NLP applications across industries.

Resume Screening and Employee Feedback Analysis; and media and entertainment, from Content Summarization to Sentiment Analysis for Audience Engagement. Every application is visually connected to the industry it belongs to, giving a thorough picture of how versatile NLP is in improving interactions and processes in a variety of fields [9].

6.2.2 Predictions for NLP market evolution

Numerous forecasts indicate that the NLP market will continue to develop and grow:

a. **Increased integration with AI and emerging technologies:** It is anticipated that NLP will become increasingly interwoven with blockchain, IoT, AI, and machine learning. More advanced and context-aware language processing capabilities will result from this combination.
b. **Advancements in speech recognition:** It is anticipated that speech recognition, a subset of NLP, will continue to progress. This includes increased language support and enhanced accuracy in comprehending spoken language, especially in noisy settings.
c. **Broader adoption in healthcare and legal sectors:** Given the abundance of unstructured textual data in industries like healthcare and law, NLP is expected to become more widely used. Applications can include compliance tracking, legal document analysis, and clinical documentation.
d. **Ethical considerations and responsible AI:** As NLP applications grow more common, ethical issues will probably receive increasing attention. The main concerns are anticipated to include protecting user privacy, eliminating language model biases, and ensuring responsible AI techniques.
e. **Customization and domain-specific solutions:** There might be a shift in the industry toward NLP solutions that are more specialized and configurable. Businesses may look for customized solutions that meet the demands of their particular industry and linguistic quirks.

For those involved in the NLP ecosystem, knowing these trends and projections offers a road map for allocating resources and directing research and development activities in the rapidly changing field of language processing technology.

6.3 APPLICATIONS OF NLP

6.3.1 Chatbots and virtual assistants

NLP is used by chatbots and virtual assistants to instantly respond to user inquiries in a context that makes sense. They are essential for maintaining availability around-the-clock in customer-focused applications, providing ongoing assistance without the need for human involvement. These apps are made to walk users through procedures, respond to often asked queries, and offer individualized support. Automating repetitive processes allows human resources to concentrate on more intricate and valuable jobs, which is a big benefit. Thanks to developments in NLP, chatbots have become more adept at comprehending user intent and delivering pertinent information. They have also grown to participate in more natural and context-aware discussions. Companies across a range of industries, including sales and customer service, are incorporating chatbots into their operations to create a smooth and effective workflow. Chatbots can answer common questions, diagnose problems, and point users in the direction of pertinent resources for customer service. They help with order processing, customer inquiries, and product recommendations in sales. These chatbots and virtual assistants may learn from and adjust to user preferences thanks to the ongoing advancements in NLP algorithms, providing a customized and individualized experience. Their prompt and precise responses not only improve client happiness but also optimize operational efficiency. We should expect even more advanced applications as NLP research progresses, enabling chatbots and virtual assistants to manage interactions and activities that get more complicated. The potential uses of NLP to improve human-computer interactions are further expanded by the incorporation of speech recognition, sentiment analysis, and multilingual capabilities [9].

6.3.2 Sentiment analysis

Sentiment analysis is a critical use of NLP that enables companies to gauge consumer opinion and public opinion about their brands, goods, and services. This perceptive sentiment research, taken from social media, reviews, and consumer feedback, is essential to successfully managing brand reputation. Through the utilization of sentiment analysis technologies, companies can obtain significant insights from a wide range of client comments. This allows companies to adjust their tactics according to the general attitude and make data-driven, well-informed judgments. Businesses may improve customer happiness, handle issues, and find areas for progress by understanding the mood. Sentiment analysis is very useful in the important field of social media monitoring. Companies can monitor and examine discussions on social media sites like Facebook, Instagram, and Twitter to get a real-time sense of how the public is feeling. By taking a proactive stance, businesses may handle problems, minimize any crises, and react quickly to good feedback. Furthermore, sentiment analysis is rapidly integrating many kinds of data, such as text, photos, and videos. This multifaceted method captures subtleties that a text-based analysis alone could miss, offering a more thorough insight of user opinion. Examining visual content in-depth gives the assessment more depth, particularly in fields where graphics are important. As more and more sectors realize the value of specialized solutions, sentiment analysis tools that are specifically tailored to their demands are starting to appear. Mood analysis, for instance, can be used to forecast trends and assess market mood in the financial industry. It helps in the medical field to comprehend patient comments and enhances overall patient experiences. Sentiment analysis helps the hospitality industry assess consumer feedback, improve service standards, and fine-tune customer experiences [10].

6.3.2.1 *Language translation*

By removing language barriers and enabling users to comprehend and communicate in many languages, language translation enabled by NLP promotes worldwide communication. Applications for translation are essential for companies doing business internationally because they enable them to localize material and interact with a variety of audiences in an efficient manner. By enabling a wider audience to access content from other languages, language translation improves information access. Developments in NLP-based translation systems are being driven by the need for real-time language translation, particularly in worldwide corporate meetings and communication applications. To provide correct translations in specialist sectors, language translation models are becoming more and more tailored to handle industry-specific terminology and jargon [11].

6.3.2.2 *Other rapidly expanding NLP applications*

a. **Content summarization:** NLP-based summarizing algorithms offer a productive way to condense long documents into brief summaries so that readers may quickly access the information they need. This program is very helpful for reading news, conducting research, and processing information. It gives professionals and researchers a rapid way to swiftly understand important information from a variety of sources, saving them time. The technology is more globally accessible due to its multilingual functionality. Improved precision and contextual knowledge in summarization algorithms are facilitated by developments in NLP that incorporate machine learning approaches, despite problems related to maintaining context and preventing information loss. As content summarizing continues to advance, it could find useful uses in legal research, education, and information curation, influencing the development of more efficient methods for acquiring knowledge in a variety of fields [12].
b. **Named entity recognition** (**NER**)**:** NER entails locating and categorizing objects within a text, such as names of individuals, groups, places, and dates. This procedure is essential to the extraction of information since it offers insightful knowledge about the content. By identifying and classifying items, NER greatly expands the capabilities of search

engines and improves the precision of search results. Users obtain more pertinent and focused information when search engines are able to recognize elements within the text. To enable a more sophisticated comprehension of the material, NER, for instance, assists in differentiating between a person's name, the name of a company, or a location referenced in a document. Beyond search engines, NER is widely employed in a variety of domains, including sentiment analysis, question-answering systems, and information retrieval. NER helps organize and structure data in the context of information extraction, making it simpler to use and evaluate. With the continuous advancement of technology, NER algorithms are getting increasingly complicated, able to handle a wide range of languages and intricate contextual subtleties. The precision and efficiency of NLP applications are greatly enhanced by the precise identification and classification of named entities, which makes information retrieval and analysis more efficient [13].

c. **Question-answering systems:** NLP-powered question-answering systems transform human-machine interactions by enabling users to converse in natural language and get targeted information in response to their inquiries. These systems are essential to many different applications, and they have made significant contributions to educational platforms that offer interactive learning opportunities. The ability of question-answering systems to understand and react to natural language questions is what makes them fundamental. This makes the interaction paradigm more approachable and user-friendly by doing away with the requirement for users to use instructions or navigate complicated interfaces. These technologies improve the learning process in educational environments by letting students interact with the material through questions and getting immediate, personalized answers. With the help of this interactive approach, students may study in a dynamic, personalized environment at their own pace, testing their knowledge and seeking clarification as needed. The underlying technology consists of complex NLP algorithms that can comprehend the context and semantics of inquiries, retrieve pertinent material from large datasets, and produce well-organized responses. The development of this technology makes question-answering systems more precise and efficient, which makes them a vital component of educational resources and other applications. Question-answering systems are a promising area for future development in NLP as it continues to progress natural language understanding, contextual reasoning, and the smooth integration of these systems into a wide range of applications outside of education [14].

d. **Speech recognition:** Speech recognition systems based on NLP are a revolutionary technological advancement that enables hands-free interactions with devices. This brings convenience to a range of tasks, including voice commands, transcription, and voice-controlled applications. By converting spoken words into text using natural words processing techniques, these systems facilitate seamless communication between users and technology. Speech recognition's main benefit is that it can increase accessibility. People with disabilities stand to gain a great deal from this because speech recognition enables them to communicate with technology only by speaking, doing away with the necessity for manual input. This inclusiveness feature helps to make the digital environment more approachable and intuitive. Speech recognition is used in voice-activated assistants, dictation software, and voice commands for smart devices, among other practical applications. Text dictation, voice control, and verbal command execution simplify tasks and improve the user experience of technology. Complex NLP algorithms are used in speech recognition technologies to analyze audio input, spot linguistic patterns, and translate spoken words into text. The precision and efficiency of these systems have been further enhanced by developments in deep learning and machine learning, which has made them indispensable components of modern technology. Speech recognition is projected to have a significant impact on accessibility, voice-driven applications, and human-computer interactions as it develops further. This will contribute to a future in which speaking is a convenient and effective way to interact with technology [15].

Comprehending the importance and patterns in these swiftly growing NLP uses demonstrates the varied and revolutionary influence that NLP has in numerous fields, from improving client interaction to removing obstacles caused by language and expediting information retrieval [12].

6.4 ADVANCEMENTS IN NLP

The capabilities of language-based technology have been extended via significant progress in NLP. Important developments consist of:

a. **Machine learning and deep learning:** NLP models are now better able to comprehend and produce language that is like that of humans because of the combination of machine learning and deep learning approaches. Deep learning techniques and machine learning algorithms work together to enable NLP models to understand complex language patterns at a level of sophistication that was previously unachievable. RNNs and transformer architectures are two of the many neural network architectures that have become particularly prominent in the field of NLP. RNNs are particularly good at recognizing dependencies and contextual subtleties in language sequences since they are built to handle sequential input. They are skilled at comprehending the temporal dynamics present in language because of their capacity to retain a recall of prior inputs. Conversely, transformer systems have been widely praised for their ability to capture contextual data and long-range dependencies simultaneously. Transformers can weigh the importance of various words in a sentence thanks to their self-attention mechanism, which gives them a more thorough grasp of the linguistic context. These developments have resulted in NLP models that can produce language that is similar to that of humans in addition to understanding it. NLP has advanced to the point that models are able to analyze, understand, and produce language at a degree of sophistication that closely resembles human communication because of the collaboration of machine learning and deep learning techniques. It is anticipated that as this field of study develops, the combination of cutting-edge neural network topologies and algorithms will enhance NLP's capabilities and create new opportunities for its use in natural language generation and processing [16].

b. **Transfer learning:** NLP has seen a paradigm shift as a result of transfer learning techniques, especially with the use of pre-trained language models such as GPT and BERT. By presenting a revolutionary methodology that enables models to leverage knowledge obtained from large datasets and then fine-tune their capabilities for particular tasks, these models have dramatically changed the field of NLP. In NLP, the core idea of transfer learning is to train a language model on a large and varied corpus of textual data. This first stage provides the model with a thorough grasp of language structures, interactions between contexts, and semantic subtleties. To improve its understanding of linguistic nuances, BERT, for instance, uses bidirectional training to extract contextual information from words that come before and after one another in a phrase. After the pre-training stage, these models' adaptability is demonstrated by their ability to be adjusted for particular tasks using comparatively less task-specific datasets. Through this process of fine-tuning, the model's comprehension is improved, enabling it to function extraordinarily well in tasks like as text classification, question answering, and sentiment analysis. In NLP, transfer learning has a significant influence since it reduces the requirement to train models from the beginning of each new task. Rather, it makes use of the fundamental knowledge that was learned during pre-training, which increases the models' effectiveness, flexibility, and capacity for delivering outstanding results on a variety of language-related tasks. This method has simplified the creation of NLP systems and helped make sophisticated language processing capabilities more accessible, opening up a wider range of uses across different industries [17].

c. **Multimodal NLP:** The advancement of multimodal NLP denotes a noteworthy progression in the domain by combining textual information with different modalities, such as images and videos. This convergence of many data formats enhances the depth and breadth of language understanding by enabling NLP models to perceive and produce information across several modalities. NLP's historical emphasis on textual data limited its ability to understand the larger context, which frequently includes visual components. This restriction is lifted by multimodal NLP developments, which allow models to easily incorporate data from both images and videos. By utilizing an interdisciplinary approach, the models are able to comprehend the subtleties of language in addition to visual clues, providing a more comprehensive comprehension of communication. Transformer architectures specifically designed for multimodal tasks are one of the noteworthy technologies advancing multimodal NLP. These approaches enable a synergistic link between linguistic and visual features by processing and analyzing both text and visual information simultaneously. Applications like text description generation for multimedia assets, sentiment analysis on visual content, and picture captioning benefit greatly from these capabilities. Multimodal NLP has many applications, including social media analysis, content recommendation systems, and accessibility solutions for people with different communication needs. The breadth of language understanding is being redefined by the ongoing research in this area, and the incorporation of multimodal capabilities into NLP models makes these models more adaptable and suitable to the complex nature of human communication [18].

d. **Explainability and interpretability:** The emphasis on improving NLP models' interpretability and explainability is indicative of a rising understanding of the value of openness in AI systems. Deep neural networks' intrinsic complexity, sometimes known as their "black-box" character, has prompted the creation of methods to illuminate these complicated models' decision-making processes. Adding attention processes to NLP models is a crucial step toward making them explainable. When the model is making predictions, attention mechanisms allow it to allocate different weights to different segments of the input sequence. This enhances the model's functionality and offers a way to determine which input data points have a greater overall impact on the result. Users are able to obtain insights into the characteristics that affected a certain decision by viewing attention weights. Another important area tackling the interpretability question is explainable AI (XAI). XAI approaches seek to provide transparent and intelligible explanations for complicated models' outputs in order to demystify the decision-making process involved. This entails building models that produce forecasts that are both accurate and have explanations that are understandable to humans. Techniques such as producing textual justifications or rule-based explanations help to improve the transparency and accessibility of the decision-making process. The necessity for responsibility, ethical considerations, and regulatory compliance are the driving forces behind the quest for interpretability and explainability in NLP models. It is imperative to comprehend the reasoning behind the results drawn by AI models, particularly in delicate applications such as the legal, financial, and healthcare sectors. By using these methods, NLP models' decision-making process can be trusted by developers and end users, encouraging ethical and accountable AI activities [19].

6.4.1 Recent technological advances

a. **BERT (2018):** A pre-trained language model called Bidirectional Encoder Representations from Transformers (BERT) greatly enhanced the contextual understanding of words within a sentence. The efficacy of transfer learning was proved by BERT's performance in a range of NLP tasks [20].
b. **GPT-3 (2020):** With 175 billion parameters, OpenAI's Generative Pre-Trained Transformer 3 (GPT-3) is one of the biggest language models. GPT-3 showed impressive ability in question answering, language translation, and natural language production.
c. **BERT-based models for specialized tasks:** Optimized variants of BERT, such as SciBERT for scientific text and BioBERT for biomedical language, demonstrate how well-suited trained models are to domain-specific tasks.
d. **Multimodal models (e.g. CLIP, DALL-E):** Tasks like cross-modal retrieval and picture production based on textual descriptions are made possible by models like CLIP (Contrastive Language-picture Pre-training) and DALL-E, which integrate language and image understanding [21].

6.4.2 The role of NLP in increasing productivity

a. **Automated content summarization:** Large amounts of textual content can be analyzed and summarized using NLP algorithms, which saves time for users that need to swiftly extract important information.
b. **Efficient information retrieval:** Search engines with NLP capabilities, such as those that use named entity recognition and semantic search, improve information retrieval efficiency by yielding more precise and context-aware search results.
c. **Automated language translation:** By enabling automated language translation, NLP promotes worldwide collaboration and productivity by enabling people and organizations to interact successfully across language barriers [22].

d. **Voice-activated assistants:** NLP is essential to the creation of voice-activated virtual assistants like Alexa and Siri. With the help of these assistants, users can engage hands-free and easily do activities and obtain information without requiring manual input.
e. **Enhanced customer support:** NLP-powered chatbots and virtual assistants help improve customer service by instantly answering questions, fixing frequent problems, and assisting users with procedures.
f. **Document analysis and information extraction:** NLP solutions simplify tasks like document management and data extraction from unstructured sources by analyzing documents, extracting pertinent information, and categorizing data [23].
g. **Personalized user experiences:** By allowing computers to comprehend user preferences and behavior, NLP can improve user happiness and engagement by facilitating the delivery of personalized content, recommendations, and services [24].

Gaining an understanding of these NLP developments and their historical background will help one better understand how language processing technology has developed and how it has revolutionized productivity in a variety of fields.

6.5 FUTURE DIRECTIONS OF NLP AND ITS PROFOUND SIGNIFICANCE

NLP refers to the subdiscipline of computer science, linguistics, and AI that studies how computers can understand and interpret natural language, and thus is developing very quickly and changing our ways of interaction with technologies. Considering the ever expanding territory of understanding and generating text, it is critical to contemplate over the future directions of NLP that may offer great solutions to a vast scope of life domains. This part discusses the most promising approaches to the further development of NLP addressing their critical importance in the creation of new technologies and enhancing the human-technology interface:

a. **More advanced language models:** As for the smaller-scale specific future developments that are on the horizon for NLP, one of the major emerging aspects that are going to further advance the field is the creation and improvement of language models. These models will enjoy broader capacity to understand context, finer details as well as the specific jargon of specific field to go beyond the existing systems. The advancements made in transfer learning and generative models will further enhance the capability and understanding of the human understanding of language generation and comprehension that will ultimately lead to human-like capability of language processing. There is no doubt that the importance of these advanced language models cannot go unnoticed all the same. They will allow users to interact with virtual agents and avatars and with chatbots and other intelligent tutoring in a smoother and natural manner, thus improving the quality of interaction as well as potential applications and opportunities and various domains. In addition, these models will assume valuable abilities helping to reduce knowledge processing costs and to automate complex language tasks as content summarization, machine translation, and analysis of sentiment.

b. **Improved multimodal NLP:** As a result of the current tendencies of developing new forms of language communication that facilitate the sharing of ideas and emotions through interactive media, language is increasingly combined with additional media like images, videos, and other sounds. Advancements in model performance and the incorporation of additional modalities for text entry will later extend these aspects to include natural and combined communication as well as information retrieval. Thus, extending traditional NLP with multimodal capabilities is a breakthrough of great importance. It will open up newer dynamic areas such as education where students can directly interact with the material that includes text, images, and sound and deliver best and satisfactory learning modules. In healthcare, it implies that with multimodal NLP, one will be in a position to make a more precise diagnosis or treatment advising since there will be data added from different origin such as the written reports, images, and voice alongside others. Furthermore, it will improve the process of multimedia content generation which often requires accurate captioning, tagging, as well as indexing of multimedia libraries.

c. **Enhanced explainability:** The "black-box" problem is one of the significant issues of the formulated tasks when creating NLP systems that is the absence of explicability. As the next innovations of NLP models, it is expected the newcomer models will feature improved explainability so that the users can know how this or that result is arrived at. The increased transparency of what NLP systems are learning and doing will help establish trust and accountability in the technology, encouraging more and responsible application of NLP systems. The potential benefits of improved explainability are therefore extremely valuable especially in areas that have special concerns for ethical considerations and justice as perhaps found in health, finance, and law. By offering interpretations for model results, various interested parties can gain insight into how the model has arrived at its decisions; possible influences as well; and can therefore make more informed decisions. More specifically, the implementation of XAI in NLP will enable debugging and model refinement for better and more reliable outcomes that are trustworthy to users.

d. **Domain-specific customization:** It is self-evident today as NLP systems are being incorporated into manufacturing environments, the requirement for application-specific settings will grow. Future NLP models will be made for explicit domains: different for business and industries, for precise business vocabulary, terminology, and conditions. This higher level of specialization will then help advance the language processing aspect and the integration of NLP on special applications in the industry. The essence of the domain-specific approach in NLP is that it allows to utilize the full potential of approaches for different industries. Specifically, in the legal context, trained NLP systems enable and help make various legal realizations as well as the interpretation of any legal documents. Domain-specific NLP applications can be useful in scientific research for the process of a literature review, hypothesis formation, and extracting valuable knowledge from extensive scientific databases. Additionally, through tailoring models to clients' domains, the resulting NLP algorithms can facilitate improvements to customer relations by directing responses to client-specific jargon in addressing customer inquiries [25].

e. **Integration with emerging technologies:** When NLP is linked to other innovative technologies like AI, IoT, and blockchain, there will be new opportunities toward the advancement of the technology and consequently embraced application. Specifically, this integration will mean more cognitive, contextual, and secure approaches that impact customers' experiences and organizations' operations positively across a range of industries. There are several reasons why concept of NLP should adopt and incorporate these new technologies. By integration of NLP with other Intelligential techniques such as machine learning, deep learning, etc., it is possible to build more context-sensitive systems where the understanding and generation of language like human could be achieved. The combination of NLP with IoT will help to obtain closer interactions with smart gadgets, improve users' experiences, and achieve the voice-command control of devices. Moreover, it is revealed that integration of NLP with Blockchain Technology enhances the security and reliability of the linguistic financial transaction and the sharing and authentications of the data. With abundant prospect and potential, the future development of NLP remains a crucial focus and source of great anticipated change and potential across numerous fields. Further, from getting smarter language models that would allow people to interact with AI in a much more natural way to multimodal NLP that incorporates movement, voice, and touch, what is out there is enormous and promising. This increase in explainability will aid in improving trust and thus the responsible deployment of the technology, which will in turn push the utilization of NLP in niche domains. However, blending NLP with new sciences will open unheard-of opportunities that make it possible to design smart, contextual, and secure solutions.

As we continue the journey in NLP development, we are now on the brink of a revolution that appears to be the most radical shift of how we engage the auditory and written stimuli of the organic environment. The importance of the aforementioned innovations cannot be overemphasized since they will transform sectors, increase efficiency, and, in a broader perspective, positively impact the lives of people and societies. The future direction of NLP is not something we can choose to follow or not to follow, but it is something that we have to follow or advance in order to improve the world and release the limitless possibilities between humans and computers.

6.5.1 Continued advancements

a. **Incorporation of domain knowledge:** In order to improve their comprehension of industry-specific terminology and settings, future NLP models might include additional domain-specific information.
b. **Addressing Bias and fairness:** It's expected that ongoing efforts will be made to address biases in language models. To guarantee equitable results, researchers and developers will concentrate on building more impartial and fair NLP systems [26].
c. **Real-time learning:** Real-time learning features could be added to NLP systems in the future, enabling them to quickly adjust to shifts in user behavior and language usage.
d. **Broader multilingual support:** Even more sophisticated multilingual capabilities might be available in future NLP models, serving a wide range of international audiences and enabling smooth communication across linguistic divides [27].

6.5.2 Personalization in education, healthcare, and customer service

a. **Personalized learning in education:** NLP will be essential to customized learning environments. Based on each student's unique needs and learning preferences, intelligent tutoring systems, adaptive learning platforms, and educational chatbots will customize interactions and content [28].
b. **Healthcare insights:** Applications of NLP in healthcare will become more individualized, supporting patient communication, clinical documentation, and tailored health advice. NLP-powered virtual health assistants will provide personalized guidance and assistance to each person.
c. **Customer service customization:** NLP-driven chatbots and virtual assistants in customer support will get even better at figuring out each customer's unique tastes, making tailored recommendations, and fixing problems based on previous exchanges.

6.5.3 Integration with emerging technologies

a. **Synergy with artificial intelligence (AI):** NLP will keep combining with more general AI projects. By combining cutting-edge AI methods with NLP models, intelligent systems can become more intelligent and context-aware overall.
b. **IoT and NLP integration:** The combination of NLP and the IoT will enhance human-device interactions. Natural language interfaces, contextual awareness, and voice commands enabled by NLP will improve user experience in IoT applications.
c. **Blockchain and secure language processing:** Blockchain technology combined with NLP applications can improve language processing task security. Data sharing, authentication procedures, and language-based transactions may all be made easier by this integration.
d. **Process automation:** Process automation will rise as new technologies are integrated. NLP will help create more automated and streamlined workflows, decreasing manual interventions and increasing overall efficiency, when combined with AI, IoT, and blockchain [29].

More sophisticated, flexible, and safe applications in a range of sectors will be made possible by NLP's merging with cutting-edge technologies as it develops. A more efficient, contextually aware, and personalized digital experience is what NLP's future promises to bring.

6.5.4 Challenges and opportunities

6.5.4.1 Challenges

a. **Ambiguity and context Understanding:** Natural language is ambiguous and nuanced, and NLP systems have difficulty understanding these things. Accurately interpreting context and intent is still difficult.
b. **Bias in language models:** Bias in training data can result in biased language models, which can then reinforce and maintain preexisting biases in society. In NLP, addressing and reducing prejudice is a constant struggle.
c. **Multilingual and cross-cultural adaptation:** There are difficulties in adapting NLP models to various linguistic and cultural situations. It is a difficult undertaking to ensure accurate content production and interpretation across various languages and cultural quirks [17].

6.5.4.2 Opportunities

a. **Improved context awareness:** Developments in context-aware models present chances to improve context comprehension and boost the precision of NLP applications.

b. **Multimodal Integration:** Using NLP in conjunction with other modalities—like pictures and videos—opens up new possibilities for content creation and comprehension that is more thorough.
c. **Customization for specific domains:** There are opportunities to create NLP models tailored to particular domains or sectors, addressing particular language requirements and constraints.

6.5.5 Ethical considerations in NLP

a. **Bias and fairness:** Crucial ethical issues include removing biases in training data and ensuring fairness in NLP models. To reduce prejudices and advance fairness, developers must take proactive measures.
b. **Privacy concerns:** Privacy issues are brought up by the gathering and use of personal data in NLP applications. It is morally required to protect user data and guarantee open data processing procedures.
c. **Transparency and explainability:** Models of ethical NLP ought to be understandable and clear. Users ought to be aware of the decision-making process, since this fosters accountability and confidence.

6.5.6 Security and privacy concerns

a. **Sensitive information handling:** NLP systems handle sensitive data on a regular basis. Protecting user privacy requires making sure that data is handled, stored, and processed securely.
b. **Cybersecurity risks:** Cybersecurity risks, such as hostile assaults and weaknesses, can affect NLP models and applications. Strong cybersecurity defenses are crucial.
c. **Data breach risks:** Data breaches are a possibility because of the massive volumes of data that NLP systems process. To reduce these dangers, access controls and robust encryption must be put in place.

Security, privacy, and ethical issues are important factors to take into account when navigating the potential and problems of NLP. For NLP technologies to be developed and used responsibly, a balance between innovation and ethical responsibility is essential [30–42].

6.6 CONCLUSION

Significant progress has been made in the field of NLP, which has changed how we use technology and handle large volumes of textual data. The study of NLP has covered a wide range of topics, from its historical development to its present trends and future possibilities. NLP has reached new heights thanks to the integration of machine learning, deep learning, and transfer learning. Models with pre-trained language representations such as BERT and GPT-3 have shown their effectiveness. NLP has applications in many fields, including sentiment analysis, chatbots, language translation, and tailored user experiences. It is essential for improving communication and productivity in the fields of education, healthcare, and customer service. Better context awareness, multimodal integration, and customization for particular domains present opportunities, even in the face of obstacles including bias, ambiguity, and cross-cultural adaptation. Ensuring justice, mitigating bias, and protecting user privacy are critical ethical considerations. Responsible AI methods benefit from NLP models' explainability and transparency. Strong cybersecurity measures are necessary to reduce the risk of data breaches and cyber threats because NLP works with sensitive information. Future developments in NLP are expected to bring increasingly more sophisticated language models, enhanced multimodal functionality, and greater industry-specific customization. Applications that use NLP in conjunction with cutting-edge technologies like blockchain, AI, and IoT will be more automated, flexible, and safe. It's critical to strike a balance between innovation and moral obligation. The ethical environment of NLP will be shaped by ongoing attempts to eliminate biases, improve privacy protections, and guarantee transparent, accountable AI activities. The field of NLP has a global influence, since it facilitates communication across language boundaries, offers tailored experiences, and advances education, healthcare, and customer service. Technology has the ability to close gaps in communication and promote global cooperation. The cooperative endeavors of scholars, developers, and policymakers will be crucial in guiding NLP in the direction of constructive progress. Sustainable research, industrial cooperation, and ethical concerns are important forces influencing how NLP develops in an impactful and responsible way.

REFERENCES

1. Z. D. Khurana, A. Koli, and K. Khatter, "Natural language processing: state of the art, current trends and challenges," *Multimedia Tools and Applications*, vol. 82, pp. 3713–3744, 2023.
2. K. H. Abdulkareem et al., "A review of fog computing and machine learning: concepts, applications, challenges, and open issues," *IEEE Access*, vol. 7, pp. 153123–153140, 2019, doi: 10.1109/ACCESS.2019.2947542.
3. P. Basu, T. S. Roy, R. Naidu, and Z. Muftuoglu, "Privacy enabled financial text classification using differential privacy and federated learning," in *Proceedings of the 3rd Workshop on Economics and Natural Language Processing, ECONLP 2021*, pp. 50–55, 2021, doi: 10.18653/v1/2021.econlp-1.7.
4. J. Li, H. Zhang, and Z. Wei, "The weighted Word2vec paragraph vectors for anomaly detection over HTTP traffic," *IEEE Access*, vol. 8, pp. 141787–141798, 2020, doi: 10.1109/ACCESS.2020.3013849.

5. A. Gangal et al., "Natural language processing: a review," in *International Conference of Numerical Analysis and Applied Mathematics, ICNAAM 2021*, vol. 2849, no. 3, p. 020010, 2023, doi: 10.1063/5.0153960.
6. D. Trajanov et al., "Review of natural language processing in pharmacology," *Pharmacological Reviews*, vol. 75, no. 4, pp. 714–738, 2023, doi: 10.1124/pharmrev.122.000715.
7. J. Ricketts, D. Barry, W. Guo, and J. Pelham, "A scoping literature review of natural language processing application to safety occurrence reports," *Safety*, vol. 9, no. 2, 2023, doi: 10.3390/safety9020022.
8. F. Lareyre, B. Nasr, A. Chaudhuri, G. Di Lorenzo, M. Carlier, and J. Raffort, Comprehensive review of natural language processing in vascular surgery. *EJVES Vascular Forum*, vol. 60, 57–63, 2023, doi: 10.1016/j.ejvsvf.2023.09.002
9. D. Litman, "Natural language processing for enhancing teaching and learning," in *30th AAAI Conference on Artificial Intelligence, AAAI* 2016, pp. 4170–4176, 2016, doi: 10.1609/aaai.v30i1.9879.
10. P. M. Mah, I. Skalna, and J. Muzam, "Natural language processing and artificial intelligence for enterprise management in the era of industry 4.0," *Applied Sciences*, vol. 12, no. 18, 2022, doi: 10.3390/app12189207.
11. T. Young, D. Hazarika, S. Poria, and E. Cambria, "Recent trends in deep learning based natural language processing," *IEEE Computational Intelligence Magazine*, vol. 13, no. 3, pp. 55–75, 2018, doi: 10.1109/MCI.2018.2840738.
12. B. Min et al., "Recent advances in natural language processing via large pre-trained language models: a survey," *ACM Computing Surveys*, vol. 56, no. 2, 2023, doi: 10.1145/3605943.
13. T. Shaik et al., "A review of the trends and challenges in adopting natural language processing methods for education feedback analysis," *IEEE Access*, vol. 10, pp. 56720–56739, 2022, doi: 10.1109/ACCESS.2022.3177752.
14. M. Treviso et al., "Efficient methods for natural language processing: a survey," *Transactions of the Association for Computational Linguistics*, vol. 11, pp. 826–860, 2023, doi: 10.1162/tacl_a_00577.
15. Z. Zhang, E. Strubell, and E. Hovy, "A survey of active learning for natural language processing," in *Proceedings of the 2022 Conference on Empirical Methods in Natural Language Processing, EMNLP* 2022, pp. 6166–6190, 2022, doi: 10.18653/v1/2022.emnlp-main.414.
16. F. Yu, H. Zhang, P. Tiwari, and B. Wang, "Natural language reasoning, a survey," *ACM Computing Surveys,* vol. 56, no. 12, 2023, [Online]. Available: https://arxiv.org/abs/2303.14725.
17. D. Hupkes et al., "A taxonomy and review of generalization research in NLP," *Nature Machine Intelligence*, vol. 5, no. 10, pp. 1161–1174, 2023, doi: 10.1038/s42256-023-00729-y.
18. J. Sawicki, M. Ganzha, and M. Paprzycki, "The state of the art of natural language processing—a systematic automated review of NLP literature using NLP techniques," Data Intelligence, vol. 5, pp. 707–749, 2023.
19. C. Wang, S. Liu, A. Li, and J. Liu, "Text dialogue analysis for primary screening of mild cognitive impairment: development and validation study," *Journal of Medical Internet Research*, vol. 25, p. e51501, 2023.
20. A. Bazoge, E. Morin, B. Daille, and P.-A. Gourraud, "Applying natural language processing to textual data from clinical data warehouses," *Systematic Review JMIR Medical Informatics*, vol. 11, p. e42477, 2023.

21. P. Johri, S. K. Khatri, A. T. Al-Taani, M. Sabharwal, S. Suvanov, and A. Kumar, "Natural language processing: history, evolution, application, and future work," in *Proceedings of 3rd International Conference on Computing Informatics and Networks: ICCIN 2020*, pp. 365–375, 2021, doi: 10.1007/978-981-15-9712-1_31
22. B. Priya, J. M. Nandhini, and T. Gnanasekaran, An analysis of the applications of natural language processing in various sectors. *Advances in Parallel Computing*, vol. 38, pp. 598–602, 2021.
23. D. Deepender and D. T. S. Walia, "Advances in natural language processing a thorough examination," *International Journal of Advanced Research in Computer and Communication Engineering*, vol. 12, no. 8, pp. 159–63, 2023.
24. P. M. Dinesh, J. Hariraj Kumar, S. C. Sekar, M. Swarnambiga, and S. Ranjani, "A review on natural language processing report: statistics and facts," in *IOP Conference Series: Materials Science and Engineering*, Sathyamangalam, vol. 1084, p. 012037, 2021.
25. A. K. Dwivedi and M. Dwivedi, "A Study on the role of machine learning in natural language processing," *International Journal of Scientific Research in Computer Science, Engineering and Information Technology*, vol. 3307, pp. 192–198, 2022.
26. D. Khurana, A. Koli, and K. Khatter, "Natural language processing: state of the art, current trends and challenges," *Multimedia Tools and Applications*, vol. 82, no. 3, pp. 3713–3744, 2023.
27. A. Hassan and A. Mahmood, "Convolutional recurrent deep learning model for sentence classification," *IEEE Access*, vol. 6, pp. 13949–13957, 2018.
28. X. Liu and D. Wu, "From natural language to programming language," in *Innovative Methods, User-Friendly Tools, Coding, and Design Approaches in People-Oriented Programming*, Steve Goschnick, Ed., pp. 110–130, IGI Global, Pennsylvania, PA, 2018.
29. A. Galassi, M. Lippi, and P. Torroni, "Attention in natural language processing," *IEEE Transactions on Neural Networks and Learning Systems*, vol. 32, no. 10, pp. 4291–4308, 2021.
30. V. K. Sharma, "CNTFET circuit-based wide fan-in domino logic for low power applications," *Journal of Circuits, System and Computers*, vol. 31, no. 02, p. 2250036, 2022.
31. V. K. Sharma and M. Pattanaik, "A reliable ground bounce noise reduction technique for nanoscale CMOS circuits," *International Journal of Electronics*, vol. 102, no. 11, pp. 1852–1866, 2015.
32. V. K. Sharma, S. Patel, and M. Pattanaik, "High performance process variations aware technique for sub-threshold 8T-SRAM cell," *Wireless Personal Communications*, vol. 78, no. 1, pp. 57–68, 2014.
33. V. K. Sharma and M. Pattanaik, "Techniques for low leakage nanoscale VLSI circuits: a comparative study," *Journal of Circuits, System and Computers*, vol. 23, no. 05, p. 1450061, 2014.
34. V. K. Sharma, "A survey of leakage reduction techniques in CMOS digital circuits for nanoscale regime," *Australian Journal of Electrical and Electronics Engineering*, vol. 18, no. 4, pp. 217–236, 2021.

35. C. Kumar, A. S. Mishra, and V. K. Sharma, "Leakage power reduction in CMOS logic circuits using stack ONOFIC technique," in *2018 Second International Conference on Intelligent Computing and Control Systems (ICICCS)*, Madurai, 2018.
36. V. K. Sharma and M. Pattanaik, "Design of low leakage variability aware ONOFIC CMOS standard cell library," *Journal of Circuits, System and Computers*, vol. 25, no. 11, p. 1650134, 2016.
37. S. F. Naz, S. Riyaz, and V. K. Sharma, "A review of QCA nanotechnology as an alternate to CMOS," *Current Nanoscience*, vol. 18, no. 1, pp. 18–30, 2022.
38. S. Riyaz and V. K. Sharma, "Design of reversible Feynman and double Feynman gates in quantum-dot cellular automata nanotechnology," *Circuit World*, vol. 49, no. 1, pp. 28–37, 2023.
39. B. Raina, C. Verma, M. Gupta, and V. K. Sharma, "Binary coded decimal (BCD) seven segment circuit designing using quantum-dot cellular automata (QCA)," in *2021 5th International Conference on Trends in Electronics and Informatics (ICOEI)*, Tirunelveli, 2021.
40. U. Mushtaq and V. K. Sharma, "Performance analysis for reliable nanoscaled FinFET logic circuits," *Analog Integrated Circuits Signal Processing*, vol. 107, no. 3, pp. 671–682, 2021.
41. Kajal and V. K. Sharma, "Design and simulation of FinFET circuits at different technologies," in *2021 6th International Conference on Inventive Computation Technologies (ICICT)*, Coimbatore, 2021.
42. V. K. Sharma, "Designing of face recognition system," in *2019 International Conference on Intelligent Computing and Control Systems (ICCS)*, Madurai, 2019.

Chapter 7

Securing the next dimension

Cybersecurity strategies for advanced computational intelligence

Anmol Madan and Ashish Kumar

7.1 INTRODUCTION

The term "high-performance automation" is used to define various situations, two of which are high-performance computing (HPC) and industrial automation (IA). When it comes to IA, high-performance automation systems have the objective of making the best and most efficient use of the resources that are provided, continuously improving on their own, and even maintaining control over the activities that are carried out in their environment. In the long run, this will result in an increased level of productivity and the results, accuracy of the work and information, and even efficiency across all the professional fields. The components that are used to build up these systems include different types of technologies as well as approaches. IA, programmable automation (PA), flexible automation (FA), and fixed automation are examples of various/different types of automation mechanisms that are used to make up these components. These systems are capable of containing a large number of automation components smoothly and efficiently. Programmable Logic Controllers (PLCs), robots, Supervisory Control and Data Acquisition (SCADA) systems are some examples of such types of devices [1]. The phrase, on the other hand, is used in the context of HPC, which is a technology that borrows various types of clusters of extremely powerful processors that work at the same time to process huge amounts of datasets that consist of numerous dimensions and solve complex problems at an extremely fast pace. HPC is a method that implements the assertion. HPC systems can achieve speeds that are more than one billion times faster than the fastest laptop or server systems that were ever created/existed. The vast majority of the time, this is the true situation [2]. Artificial intelligence (AI) is a subdomain of computational intelligence (CI) that is focused on the conception and creation of intelligent computer systems (ICSs) capable of learning, adjusting, and adapting to a variety of environments and conditions [3,4]. The main objective of this academic discipline is to create and develop advanced computational systems that are capable of acquiring knowledge and adapting to unfamiliar environments as well as unpredictable circumstances. This stage of AI includes a wide range of domains, such as development and implementation, for example, making computer systems based on how living

 DOI: 10.1201/9781003643609-7

things and languages work. The following are the topics that are addressed in this chapter. As a result of ongoing studies and development, a significant number of novel computer paradigms in this world have emerged over time. In the realm of computer intelligence research, fuzzy systems, neural networks, and evolutionary computation are regarded as cornerstone concepts. Neural networks consist of the most fundamental and crucial foundation. There exist lots of applications for AI, and some of them make use of technologies that are linked to CI [5,6]. These advancements are used in various ways. This category contains an extensive variety of applications, including robots, computers learning from data, natural language processing (NLP) in simpler words, understanding human language, and image identification and recognition. Researchers studying this topic have published their work in journals like the *Computational Intelligence Journal* and also the *International Journal of Computational Intelligence Systems* [7].

7.2 BACKGROUND AND SIGNIFICANCE

Through automation, the human engagement is possible elimination in some process. And hence manufacturing is becoming more precise while the possibility of generation of a problem that is caused by humans is also nil. One of the remarkable features of this is dodging of the related hazards and the expenses of workforce. To get to this level of quality, high-performance automation systems can offer accurate and repeatable performance of the tasks either manually or by robotization that at the end of the day will cause improved product quality. Automation helps minimize probability of human mistakes and adds a fifth dimension to working environment by bringing in more safety at various companies. This is just one of the positives which automation gives to businesses. Versatility and adaptability: As soon as the situation arises, these automation systems of superior quality can quickly change the process, consequently making them more versatile to the specific environment. Adaptability and versatility are the primary key elements that set these apart. AI primarily focuses on designing as well as the improvement of ICS which can learn independently and be flexible in their decision-making in novel environments and unforeseeable situations. Computational Intelligent Systems (CIS) can be viewed as the miniature of AI. Thus, these artificially intelligent models are a gathering of various computing techniques meant for imitating natural tasks. These operations are neural networks, fuzzy logics, genetic policies, and progressive algorithms as well. These technologies give the system the chance to handle and solve sophisticated and complex tasks that appear in real life. These tools are mainly applied within the prime field of today's AI, for instance, machine learning (ML), NLP, image recognition and robotics and more. Through innovating designs, CIS are designed to run in unexpected or unpredicted conditions. The fact that the competencies of CIS qualify it for complex AI applications process is what gives CIS an upper hand with current

AI in use. R&D activities employed in this precise region are to be leveraged toward the betterment of both the skills and productivity of intelligent computers. Accordingly, improvement of rational thinking, perception, informational collection and interpretation must be noted. The process of forming computational methods used in CIS that is the amalgamation of both increasing the functionality and reliability to create more adaptable intelligence computing machines when there are instances of unanticipated issues showcases loyalty to technology improvement and its further development [8,9].

7.3 OBJECTIVES

- **Improving efficiency and performance of CIS:** The efficiency and performance of CIS can be improved by integrating high-performance automation methods. With this, we are making setups more effective to manage complex jobs and adapt changes in any possible circumstance.
- **Resources utilization:** Consumption of resources in CIS can be optimized by the intervention of high-performance automation techniques, reducing the amount of computing work needed and enhancing the overall performance of the system.
- **Improving resistance to change in environment and its adaptability:** There is a ton of potential to improve both the adaptability and flexibility of CIS that depend on the incorporation/integration of high-performance automation techniques. This ability enables the systems to react more effectively to changes in the surroundings and new tasks.
- **The latest developments in the field of CIS:** Analyzing novel high-performance automation methods and then further applying them to CIS can help in improving the current state of the art in this sector. This may result in more intelligent and efficient systems.
- **Taking on the problems of the real world:** CIS may apply high-performance automation methods to a variety of fields, including assembling, automotive, finance, and healthcare. However, it helps in handling the encountered difficulties in the real world and improving the performance of the industries.
- **Encouraging teamwork and exchanging information:** Researchers can share their knowledge and ideas, which encourages collaboration and innovation in the subject.

7.4 LITERATURE REVIEW

Modern technologies are being used in high-performance automation methods, changing how companies optimize their processes and better operational efficiency. This pattern is established on the fusion of ML, AI, and low-code application development. Low-code platforms allow companies to deploy solutions quickly with less coding, speeding up applications. This speeds up the automation process, fostering adaptability and responsiveness in an active work environment [10]. Increased effectiveness is the key aspect of automation, as AI gives machines various abilities. ML techniques are a subset of the AI model, computers are able to learn from data and do tasks that previously required human perception and decision-making. Predictive analytics, continuous process optimization, and anomaly detection all rely on this ability. It promotes a flexible, self-correcting automated environment. Hyper-automation being a sophisticated fusion of robotic process automation (RPA), AI, ML, and process mining techniques is a significant development in this sector. Hyper-automation controls end-to-end business processes and as a result, it goes beyond normal automation, involving both well-defined and not-so-clear operations. This comprehensive approach leads to new chances for creativity and productivity while pushing companies to automate more operations. High-performance automation techniques are used to increase productivity by freeing up human resources from time-consuming and repetitive tasks [11]. This helps in creating human resources that are extremely productive and capable of higher-order thinking, which can be done by allowing employees to concentrate on creative projects, strategic initiatives, and difficult problem-solving. Strengthening the quality is yet another benefit of these types of techniques. Companies can rapidly detect and address potential problems, deformities or faults with the use of AI-powered vision and real-time data analysis. When the capability of the products and services increases, errors fall, resulting in customer satisfaction. Safety optimization is given more priority by the automation of dangerous processes and using AI-driven predictive maintenance. Anticipating possible equipment problems and taking bold measures are used to reduce risks and ensure a safe working environment. One of the important modules of high-performance automation methods is Human-Machine Interface (HMI) graphics system standardization. A uniform and user-friendly interface is ensured by this standardization, which is further used to facilitate process control and data interpretation for operators. Overall system usability is improved, facilitating more seamless operations, and lowering the learning curve used by the operators that are interacting with automated systems [12]. Basically, high-performance automation methods are known as comprehensive and advanced strategies for business process improvement. By integrating cutting-edge techniques, these procedures are used to enable businesses to meet the complexity of the current business environment with unique levels of efficiency, innovation, and flexibility [13].

7.5 COMPUTATIONAL INTELLIGENT SYSTEMS

In front of AI, CI mainly focuses on the development and improvement of ICS. CI makes use of concepts from neural networks, fuzzy systems, and evolutionary computation, all of which are inspired by biological processes,

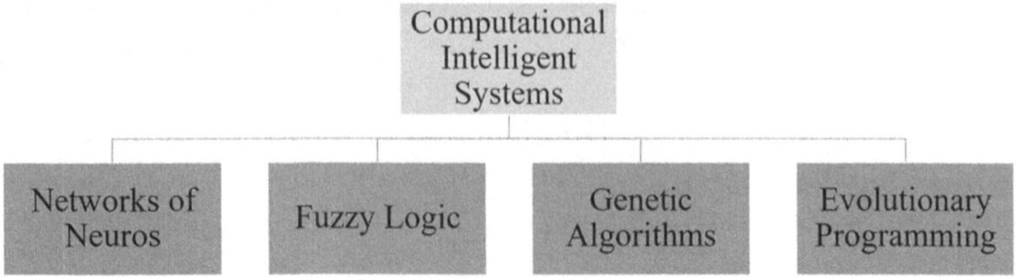

Figure 7.1 Types of computational intelligent systems.

which in return allows CI systems to learn and adapt to the given environment. As a result, it excels at addressing challenges that traditional AI systems face as shown in Figure 7.1.

7.5.1 Networks of neurons

Artificial neural networks (ANN) are solid tools for handling complicated/complex pattern recognition tasks because they can replicate/mimic the human brain, thanks to the huge parallel networks. The constant upgradation of neural network topology (NNT) increases their capacity to replicate as well as interpret complex/complicated interactions, which ultimately results in the development of AI applications.

7.5.2 Fuzzy logic

A key component of computer-aided interpretation is fuzzy logic, which uses approximation thinking to manage imprecise language and get around unclear situations. This approach provides a more suitable/adaptable method for both decision-making and problem-solving, considering the unexpected real-life situations where information may not always be accurate or up to the mark.

7.5.3 Genetic algorithms

These algorithms are at the top for determining the most efficient and reliable answers to any problems that are related to resource allocation because they take assistance from natural selection. Genetic algorithms can quickly explore solution spaces and aid in the generation of dependable and ideal results since they contain evolutionary notions.

7.5.4 Evolutionary programming

Evolutionary programming was created specifically to find nearly perfect solutions to issues with control systems and optimization. This approach mimics the natural adaptive mechanisms that provide solutions over several generations.

AI-related fields like robotics, image recognition, ML and NLP commonly use these CI techniques. Their capacity for learning and adaptation enables them to handle challenging and dynamic real-world circumstances with ease.

Since its founding in the 1980s, the IEEE Computational Intelligence Society has emerged as a pioneer in the quickly evolving field of CI. The association is committed to developing biologically and linguistically inspired computing paradigms. It has advanced computer intelligence to the forefront of AI research and development by encouraging creativity and teamwork.

In conclusion, CI is a dynamic field that constantly pushes the limits of AI. It also symbolizes the combination of bioinspired and linguistic computational techniques. Neural networks, fuzzy logic, genetic algorithms and evolutionary programming are all combined in machine intelligence (CI), which is a potent tool for creating intelligent systems that can flourish in the rapidly changing fields of technology and problem-solving.

7.6 APPLICATIONS OF HIGH-PERFORMANCE AUTOMATION METHODS

The revolutionary nature of high-performance automation systems lies in their great enhancement of productivity, accuracy and efficiency across an extensive array of applications. These sophisticated automation methods are advantageous in several important domains, including [14]:

- **Manufacturing excellence:** Because high-performance automation increases output, improves product quality and reduces labor costs, it is critical to manufacturing. Through the use of robotics, computer vision and ML technologies, it synchronizes task automation and optimizes production processes. Processes become more efficient as a result, and output precision increases.
- **Acceleration of materials development:** High-performance automation speeds up the experimental testing of potential materials in the field of materials development. Through complex computations and simulations, researchers can improve bulk, interface and defect properties with the assistance of HPC hardware.
- **Scientific discoveries:** High-performance automation is revolutionizing science research by enabling complex computations and simulations. Scientists can achieve ground-breaking discoveries and breakthroughs because HPC technologies perform orders of magnitude better than conventional computer equipment. HPC systems are capable of performing quadrillions of calculations every second.
- **Business process revolution:** Hyper-automation: In the field of business process automation, hyper-automation is a novel idea. It accelerates

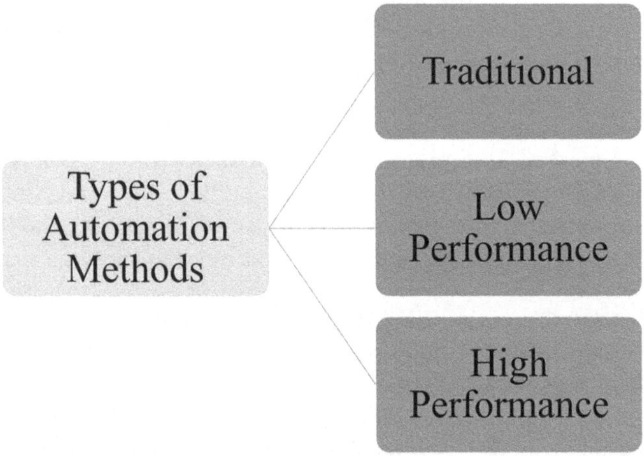

Figure 7.2 Applications of high-performance automation methods.

processes by optimizing a variety of automation technologies, including some that were previously carried out by humans. Robotics, AI and ML are used in hyper-automation to increase overall operational efficiency. It handles both complex business processes and repetitive, high-volume tasks.

- **Advances in factory automation**: Automation with high performance finds its footing in manufacturing settings, coordinating industrial equipment such as robots, PLCs and Industrial PCs (IPCs) to maximize output and improve product quality. A smooth and effective production ecosystem where tasks are completed with accuracy and dependability is ensured by the integration of multiple automation technologies.

High-performance automation techniques essentially spur innovation in a variety of sectors. Utilizing state-of-the-art technologies such as robotics, computer vision, ML and HPC, these approaches transform entire workflows in addition to automating tasks. As a result, industries are propelled into a future defined by streamlined procedures and unmatched innovation, creating a landscape where productivity, accuracy and efficiency reach new heights, which are shown in Figure 7.2 [15].

7.7 CHALLENGES AND LIMITATIONS

Although high-performance automation techniques have many benefits, they are not without limitations. For implementation to be successful, these problems must be resolved. Let's look at some potential roadblocks and how to overcome them [16,17]:

- **Inadequate assistance and automation administration:**
  Problem: Weak management and support abilities in automation orchestrators might lead to inadequate self-healing.
  Solution: Analyze the business impact and boost performance by utilizing automation optimization technology. These solutions, by addressing the deficiencies in management and supporting competencies, have the ability to ensure robust and consistent automation.
- **Opposition to change:**
  Problem: Some people may be against automation if they think that it will harm their ongoing jobs.
  Solution: Explain to them the process and the advantages of automation, using techniques like change management. It is very important to provide appropriate training for the employees to minimize resistance and experience how automation can help rather than replace their jobs.
- **Problems with technology and maintenance:**
  Problem: The complexity of automated hardware and software can make multiple different challenges with maintenance and technology.
  Solution: Provide all the staff members with training programs so that they can easily manage complex issues. Regular maintenance schedules can also help ensure the smooth operation of automated systems.
- **Over-dependence on technology and a lack of flexibility:**
  Problem: People rely too much on technology because it's not very flexible.
  Solution: Use automation solutions that can easily adapt to the changing business models. Balancing the automation with human interference will ensure that the system is flexible and adaptive to its environment. Regular checks on automation help keep it aligned with the new trends and any change in the business needs.

The business should prioritize and address these challenges by implementing change management strategies, investing in both training and development and updating and applying modern automation systems to provide better and optimal solutions. When aiming to maximize a business's efficiency, we have to ensure the integrity of the data, increase scalability and innovation, and also carefully choose an automation foundation.

7.8 METHODOLOGY

This chapter uses a strict literature review technique, providing a complete overview of the latest research on high-performance automation solutions in CIS. The selection criteria, planned search methodology, and limited scale enable a focused examination of relevant content. A conceptual framework with key terms and ideas related to high-performance automation and CI is used as a guide for the investigation. The study carefully compiles and categorizes methodologies, results and trends using a theme synthesis, which enables the identification of important subtleties and broad patterns. The critical assessment framework, which evaluates the strengths and weaknesses of certain research, increases the review's rigor. A comparative examination highlights similarities and differences, which improves our comprehension of the corpus of knowledge.

7.9 DATA COLLECTION AND ANALYSIS

This research chapter thoroughly and very carefully collects information from trusted sources like different research papers, conference papers and even various chapters to completely understand the topic. The variety of sources give a detailed view of high-performance automation strategies for CIS, providing a starting base for analysis.

The advanced thematic analysis being used in the data analysis approach allows for a complete and systematic examination of the required patterns, themes and also findings that are yet to be uncovered in the material that has been collected. This methodological choice ensures a comprehensive examination of the basic concepts, strategies and patterns prevalent in the field of high-performance automation.

The goal of the literature review is to further high-performance automation in CIS by pointing out emerging trends and possible research topics. The chapter offers a broad overview of the current state of the field by considering real-world applications and outcomes in several industries. The objective is to enrich this research into a fascinating union and to inspire innovation.

7.10 TOOLS AND TECHNIQUES

To gather information for this study, we'll use reliable tools and techniques. First, we will do in-depth searches on academic databases like Google Scholar and IEEE Xplore. These platforms help us locate the most relevant and up-to-date information about our topic; they resemble libraries of scholarly articles and conference proceedings.

Once we have gathered a significant amount of data, we will apply a technique called theme analysis. Sorting through a lot of jigsaw pieces is necessary to find the bigger picture. In this case, we will meticulously review the collected data, looking for recurring themes and patterns. It's like learning about recurrent themes or ideas. This method makes it easier to organize information and helps us understand the main ideas.

In summary, we utilize state-of-the-art search techniques to retrieve the most relevant data available and then aggregate the data using an innovative analytical approach to identify significant trends. In this manner, everyone may understand clearly and comprehend the current state of high-performance automation.

7.11 OVERVIEW OF HIGH-PERFORMANCE AUTOMATION METHODS

Information regarding using automation to increase productivity and growth in enterprises can be found in the search results. The articles emphasize how important automation is for delivering personalized customer experiences, increasing productivity and avoiding burnout. This includes AI, automation and real-time data. They also discuss automation orchestrators' drawbacks and the rise of automation optimization tools, highlighting the need for seamless transitions between automation providers and increased productivity across a range of automation technologies. In addition, the concept of hyper-automation is introduced, which refers to the application of a combination of automation technologies for the purpose of automating and streamlining business processes, including complex and high-volume repetitive tasks. HPC and its application to data-intensive workloads such as AI, ML and internet of things (IoT) are also covered in the articles [18]. HPC helps enable the execution of complex computations and improves/enhances computing efficiency. In addition, IA is discussed, with a focus on the necessity of superior automation products to improve the degree of precision, accuracy and quality in IA systems. These findings offer a thorough summary of the applications and advantages of high-performance automation techniques across numerous domains, including business, technology and industrial processes.

7.12 COMPARISON OF HIGH-PERFORMANCE AUTOMATION METHODS

The various types of automation methods are illustrated in Figure 7.3.

- **Traditional automation methods:**
  Most traditional automation solutions are based on complex code, which limits and stifles the inventiveness of engineers. These approaches frequently fail to incorporate human intelligence properly into software bots and are not adaptable enough to accommodate changes. However, advanced features like natural language comprehension and easily mastered low- or no-code techniques are offered by high-performance automation, which is best represented by RPA [19].

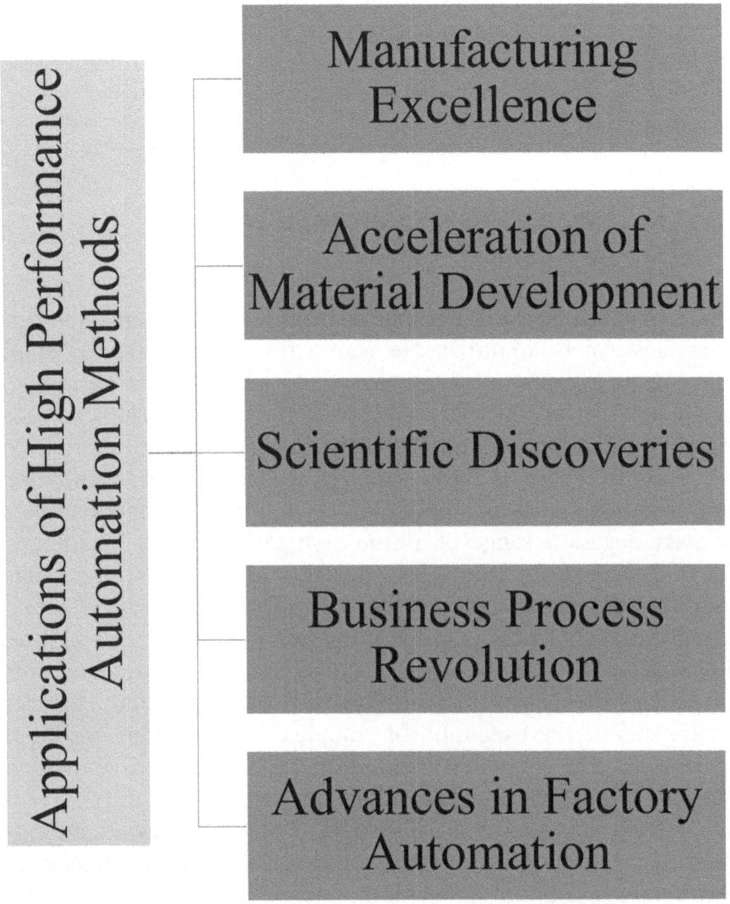

Figure 7.3 Types of automation methods.

RPA improves productivity, adaptability, and system and technology integration, in addition to streamlining procedures. By addressing the shortcomings of conventional methods, this paradigm shifts toward high-performance automation heralds the end of the past and ushers in a new era where automation is not only intelligent and responsive to changing demands in a range of corporate settings but also efficient.

- **Low-performance automation methods**: Low-performance automation approaches are defined, given the state of the art, as operating at a slower pace, with poorer efficacy when executing automated activities, and with inferior efficiency. These methods occasionally make use of antiquated technology that isn't as sophisticated or adaptable as more recent technologies. It could be challenging to communicate with many systems and technologies if one lacks sufficient integration abilities. Moreover, low-performance automation techniques might not be able to maximize processes for accuracy and speed, which could cause delays and decreased precision when executing tasks. Being non-responsive, these strategies could be hard to modify in response to changing circumstances and dynamic scenarios. Because of the changing needs of modern business environments, there is a push toward more advanced and high-performing automation systems. As enterprises go for greater levels of accuracy, productivity and flexibility, the basic limitations of low-performance automation systems are becoming apparent.

- **High-performance automation methods:** High-performance automation techniques have a major impact on multiple different fields. Integration of IoT and robotics in IA helps to increase/enhance production efficiency and also control in real time. Automation reduces errors, reduces costs and streamlines monotonous tasks. This is especially true when we are talking about technology like RPA. Sophisticated CNC machinery guarantees accuracy as well as productivity in production processes while enabling machining automation. In addition, it enhances automation targets on developing collaboration, elevating protocols and refining current automated tasks to acquire optimum efficiency. The factors of high-performance automation are used collectively to enhance output, improve tasks and increase quality. The evolution of intelligent automation technology is modifying firms and encouraging creativity, flexibility and vigilance in a vibrant corporate environment [20].

The field of automation optimization also highlights improving integration, enhancing procedures and refining pre-existing automated tasks to achieve optimum efficiency. These factors of high-performance automation together may improve output, optimize operations and raise quality. Intelligent automation technology is constantly thriving, which not only changes firms but also encourages creativity, flexibility and responsiveness in dynamic corporate surroundings.

7.13 ADVANTAGES AND DISADVANTAGES OF HIGH-PERFORMANCE AUTOMATION METHODS

Let's discuss the advantages and disadvantages of automation to an extent:

7.13.1 Advantages of automation

- **Rise in production:** Automation's ability to perform is far better than human workers, reason being its fast and dependable work fulfillment, in terms of profit and productivity.

- **Quality improvement and efficiency:** Automated tasks form better finished products ensuring reduced errors, along with consistency and efficiency.
- **Reduction in expenses:** Vital cost savings can be achieved by minimizing maintenance expenses, fewer human evaluations and simplifying tasks.
- **Competence and adaptability:** Simplifying tasks, enhancing effectiveness and growing efficiency may lead to fast execution and versatility.
- **Advanced information collection and analysis:** Automation can be used for gathering and analyzing heaps of data, which leads to significant perspectives used for making thoughtful determinations.

7.13.2 Drawbacks of automation

- **Losing traditional jobs:** Improvements in automation techniques are making machines more efficient and economical at performing tasks, than humans. For instance, robots can manage the assembling of products more precisely and quicker, which is replacing human force in addition to a need for cashiers and other operational staff in retail because of automated payment systems and automated kiosks [21].
- **Reduction in labor force:** Workers who used to perform the tasks earlier are being displaced due to job automation. However, this may affect the skills of various individuals, varying from juniors to highly skilled professionals. For instance, administrative tasks are being automated, which leads to a reduction in staff and the reassignment of office workers.
- **Restricted job opportunities:** It becomes very difficult for the displaced workers to find new jobs, it becomes more difficult if their skills are not applicable to any other industry and if they are not well educated. This may result in prolonged unemployment/underemployment, along with social and economic challenges such as poverty, being homeless and mental health issues.
- **Income imbalance:** Automation can worsen income inequality by affecting certain sections of the workers. Low-wage workers in industries are the ones who are the most vulnerable to automation and will be the first ones to get affected and lose their jobs; on the other hand, the higher-skilled workers with a specialization in technology or management expertise will be the ones to benefit from the increased productivity and demand of their skills. This growing gap between those who have and those who do not can lead to social tension [22].
- **Geographical imbalance:** Job displacements may also intensify geographical imbalance, especially in areas that are heavily reliant on industries that are highly automated. For instance, rural areas where economies are centered around producing or agriculture may suffer

unevenly from automation-related job losses, which leads to a decrease in population and economic standstill [23].

- **Workspace development challenges:** To prevent job loss and displacement, workers who are affected should be given access to retraining and upskilling programs to gain the latest skills suitable for upcoming industries or occupations that are less susceptible to automation. However, these programs may be limited or inaccessible to certain workers due to various factors like cost, location or educational barriers.
- **Hardware purchases:** The purchase of specific tools and hardware used to automate processes is one of the expenses in automation implementation. This includes several automation tools such as robotics, machinery, sensors, actuators and others according to the specific needs of the firms. The cost of purchasing these can be fundamental, mainly for small and medium-sized enterprises (SMEs) along with limited financial resources [24].
- **Application development and inclusion:** Automation frequently requires the creation or customization of software solutions for controlling and managing automated processes. This may involve creating customized software applications, fusing existing software systems with automation policies or the adoption of ready-made software packages. Expenses of the automation project can vary depending on how complex it is and how much customization is required to fit the firm's workflows.
- **Training workers and development of skills:** Requirements to implement automation are either train the current workers or hire new employees with the required skills for efficiently operating and maintaining automated frameworks. Training programs can cover the topics like programming, troubleshooting, maintaining tools and safety protocols. Invest in training employees who are able to perceive automation skills efficiently and minimize the risk of workflow disruptions, caused by skill gaps.
- **Structural improvements:** Automation requires support for the synchronization and functioning of automated frameworks. This involves advancements of power supply systems, networking setups, data storage services and facility layout refinement to adapt new tools and techniques.
- **Consultancy and expert tasks:** Ventures usually knuckle down to kind of assessing the automation needs through external consultants, engineers or experts in automation who bring on technical expertise over the course of deployment. The initial expenses from consulting fees and the building or procurement of professional facilities can be many times the initial costs when trying to achieve the same operational stability as organizations with years of onsite expertise and deployment in automation.

- **Consistency and confirmation:** The laws, certificates and standards which depend on the firm's geographic location and operations of automated operations decide the requirement and the nature of the automated system. Among the tasks this might entail are carrying out risk evaluations, providing safety assurance at the firm level, getting required licenses and the like, which means meeting existing regulation governing automation skills. Organizations will be incurring more costs plus other expenses as integration of standardized systems is being set in place to achieve the set regulatory performance.
- **Networks opacity:** Automation systems being made of task synchronized components, computer systems and design systems make them densely populated. Management functions such as these can be merged into existing ones whose activity involves accurate planning and scenarios. Elevated automation tasks are composed of experts who are not just engineers but coders and domain specialists as well, hence teamwork is essential for the creation, execution and testing of the system. Besides a variety of developing operations result in added cost and time needed for implementation.
- **Proficient technical workforce:** The automation systems deployed and maintained on a relevant level, the focus on robotics, control systems, software development and electrical engineering require effective technical workforce. Their business may find it difficult to source talented engineers because they can neither hire nor sack the scarce experienced technology-based employees. Companies are engaged not only in education and retraining of the existing workers but also in hiring skilled professionals who know the specifics of the automation technologies. The scarcity of the trained experts raises the level of cost of hiring labor while at the same time laborious tasks get delayed.
- **Employment challenges:** Configuration and synchronization of automation equipments using the existing setup can face operational and technical challenges. Equipments are required to be structured, aligned and tested to ensure persistence and optimum performance. Plus, physical constraints like limited space, biogeographic variables and safety regulations can impact the installation process. In order to overcome the stated factors, careful planning, conformity and integration are needed.
- **Bug fixing and maintenance:** The subject of automation systems is followed by maintenance and bug fixing that must be carried out continually in order to keep them running as usual. Solving issues that arise due to the failure of toolsets, difficulties of the program, communication barriers and measurement inaccuracies should have immediate and appropriate solutions. Skilled technicians are required for diagnosis, decision-making on taking control measures that are error-preventing and remediate identified issues and preventive maintenance to steer away bugs once and for all. Failure to give a serious and proper

solution to the technical problems would result in an operation process delay, poor quality output and a very high maintenance cost.

- **Technical limitations:** There are often technical restrictions that come in the way of handling various tasks automatically, even though the automation features are capable to do the tasks. As an example, consider processes where decisions have to be made comprehensively and range from the simple to the complex, competence or the ability to think through things is needed and these are hard to clone through computer-operated systems. Lastly, the technology parameters such as the ability to detect issues, computational power and automated procedures are also important determinants of efficiency and professionalism of automation tools. Automation planning should include thoroughly checking both its potential and limitations on the tasks' automation before the money is spent.
- **Legislative and compliance commitments:** Healthcare, aerospace and manufacturing, as industries that practice security, have the fusion and operation of automation systems bound by stricter regulatory limitations and compliance standards. Organizations need to have the automation measures to follow established regulators, standards and protocols to ensure the least risk and glitches. A company's ability to support regulations about data privacy, security and job security determines a layout, a state of the art and system of the automation process.
- **Minimized customization:** Another serious drawback in automation is the lack of special one-to-one relationships between the firm and clients of the organization. Automated systems such as chatbots, automated phone systems, as well as automated kiosks are deprived of physical touch and sympathy which concerns the personal demands of the customers and, therefore, works contrary to their services and products. There can be a feeling of unimportance or simply being ignored among customers, when a brand is not fully customized or there is no such thing as to interact with them in a personalized way. Often, this may decrease the customer's interest in the brand and also their loyalty eventually fades away.
- **Effect on consumer satisfaction:** Human interaction is generally important for providing amazing customer service and promoting beneficial relationships with consumers. Firms depending heavily on automation to handle customer issues or transactions may have a chance of granting the quality of service and overall customer satisfaction. Automated responses may lack in refinement, warmth and understanding that human force can provide, resulting in upset or unease with consumers.
- **Limited problem solving techniques:** Human agents can respond to dispersive factors quickly to work out more different ways whereas machinery operates within its set actions. They may come to realize

complexity and unedited issues which are limitations of automated systems when they face intricate or unexpected problems. As a result, customers may become unsatisfied with system solutions they do not like and feel dissatisfied. As they penetrate the subsurface of our planet, it becomes essential to use a manned mission to meet these concerns efficiently.

- **Limited sympathy and understanding:** Human involvement is the emotion that defines a company's engagement when it is dealing with customers. This engagement is accomplished by emphasizing care, understanding and a high level of empathy. Besides empathic communication, customers can be comforted, concerns tackled or trust raised by using a specific brand. Despite the fact that automated reactions could be perceived as insensitive, cold and detached, they do not contribute to developing different and unique relationship with customers. The fact of creed and compassion is essential to the development of brand empathy. Unfortunately, a lack of it will lower the consumer loyalty.
- **Threat of customer shift:** Customers who are dissatisfied are often found switching to the opposition in order to find better alternatives that provide more customized customer experience. Thus, businesses that only focus/prioritize automation over human interaction risk losing their customers to competitors who prioritize personalized service to their customers and human engagement. Customer churn has a great impact on business profits and long-term viability.
- **Maintaining balance:** While automation can streamline processes and increase efficiency, businesses must maintain a balance between automation and human interaction to deliver customized and exceptional customer service. Using automation for monotonous/routine tasks and transactions can free up employees who can handle complex inquiries that the automation will have difficulty with and offer personalized assistance where it is most needed. By combining the efficiency of automation with the social skills and understanding of humans, businesses can create a memorable and overall satisfying experience that will make their business boom.
- **Adaptability and critical thinking:** When employees rely too much on automated systems to complete tasks, there is a high chance that they lose their ability to adapt/face new challenges. Automation can lead to workers feeling complacent, leading them to follow predetermined processes without questioning or thinking twice about the problems and focusing on a single approach. This lack of adaptability and critical thinking can create difficulty in innovation, problem-solving and decision-making in the organization, ultimately constraining creativity and stopping organizational growth.
- **Essential skills:** Automation can make the essential skills of the employees rusty as they rely on automation to perform the tasks that

were done manually. For example, automating repetitive tasks may lead to rusting of their skills, i.e. attention to detail and problem-solving skills. Furthermore, automation can reduce opportunities for skill development and learning new technologies.

- **Creativity and innovation:** Relying too much on automation can suppress the creativity and innovation of the employees. When employees get accustomed to the standard process told by the automated systems there is a high chance that they are less likely to explore unconventional ideas or try new approaches to a problem. This lack of creativity and innovation can hinder the organization's ability to adapt to the continuously evolving market, take advantage of emerging technology, stay up to date with new trends and stay competitive in the industry.
- **Flexibility and agility:** Automation can limit the flexibility and agility of the workforce especially when the employees only do a specific task, or work on a specific automated system which narrows their skill set. In a fast-changing business environment with continuously growing and evolving technology, organizations require quick and adaptable employees who can swiftly update and respond to changes. Relying too much on automation can weaken an organization's ability to change and adapt to new challenges and opportunities.
- **Productivity disruptions:** In the case of a system malfunctioning the employee without the right/necessary skills may find it very challenging to quickly restore the operations. Productivity disruptions resulting from automation failures cause delays in project timelines, and missing deadlines which may result in financial losses to the business. Moreover, relying on third-party technical support or service providers to fix the issue can make the situation worse which may lead to excessive downtime and productivity losses.
- **Skill gap:** To reduce the chances of workers becoming too dependent on automation and losing their edge in the work, businesses should focus on training and helping employees learn and develop new skills. The focus of the training programs should be providing employees with the skills they need to work well with and oversee automated systems. Additionally encouraging the employees to learn, innovate and empower can help change their mindset regarding automation as a tool that enhances their skills and abilities rather than a replacement for their skills.

7.14 SUMMARY OF FINDINGS

In conclusion, when both CI systems and high-performance automation techniques are combined together, they provide a huge boost in enhancing productivity flexibility and overall performance across various different fields. The aim of the study is to enhance the efficiency and usefulness of the systems by exploring modern and advanced automation methods, for example, programmable, flexible, and integrated automation. Several key objectives that have been identified till now include the advancement of the field, addressing practical challenges in the manufacturing sector, finance, healthcare and even the automotive industries, and promoting knowledge and sharing among researchers that enhance the efficiency and performance of CI systems. Above given objectives can be accomplished by employing high-performance automation methods such as PLCs, robots and SCADA systems. This transformation can help improve the industrial environments by improving their efficiency, accuracy, and productivity. The field of CI systems stands to greatly benefit from the integration of high-performance automation techniques through ongoing research, development and collaboration. This will help create a way for the creation of more adaptable and more efficient systems for the future.

7.15 IMPLICATIONS FOR PRACTICE

Studying high-performance automation techniques for CIS may provide valuable insights into particle and useful applications across various sectors. Improving IA with high-performance automation methods offers many different types of advantages, such as increased output, accuracy and efficiency. Organizations can improve/enhance control over industrial activities and streamline all operations by utilizing/employing technologies such as PA, FA and integrated automation. High-performance automation techniques enable efficient and reliable use of resources by lowering computer workloads while conserving/saving resources. This leads to reduced/lower costs and overall enhances system functionality, making it an attractive offer for companies looking to boost sales. The automation methods offer a wide variety of flexibility and adaptability, which are extremely valuable in dynamic environments that are experiencing fast changes. By utilizing high-performance automation techniques and CI to quicker/faster adaptation to changing situations, systems can become more robust/resilient and successful in tackling/addressing real-world challenges in industries, i.e. automotive industry, finance industry, manufacturing industry and even the healthcare industry.

The body of collective knowledge expands and more complex and effective systems are created when study results are disseminated and researcher collaboration is promoted. Businesses may use these insights to put into practice high-performance automation plans that are customized to meet both their own and the needs of their sector. High-performance automation, whether applied to improve financial analytics, optimize industrial processes or improve healthcare services, has the potential to revolutionize operations and yield long-term success. When everything is taken into account, high-performance automation techniques have far-reaching effects on CIS that go well beyond the purview of theoretical study.

7.16 RESTRICTIONS AND PROSPECTIVE RESEARCH PATHS

Despite the wide range of the possibility of uniting all the high-performance automation protocols into a system of computational intelligent tools, there are still uncertainty and setbacks that need to be overcome so as to develop solutions to all the hiccups.

- **Problems with complexity and research:** The installation of high-level automation grids means introduction of the varied forms and techniques; moreover, diversity causes many challenges and difficulties. To further this direction, future researches may focus on creating uniform systems that would make the task of fusion simpler, perplexity less and compatibility even way better for most of the controlled automation systems.
- **Constraints in available resources:** In contrast to our hard and fast moves to increase resource utilization and optimization for assignment, AI systems will have a high probability to be crime when they will have a limitation of factors: for instance, being a low-power CPU or a restricted storage capacity space which will postpone its performance.
- **Flexibility to the unexpected scenarios:** Even though high-performance automation approaches do their best to enhance system flexibility, they may still have a hard time managing unforeseen and unexpected events in addition to more responsibilities. Further research in the future should concentrate mainly on cognitive computing approaches and the learning algorithms that can adapt to their surroundings, which enable systems to continuously learn from their mistake and also adapt to continuously changing environments and new challenges that they face.
- **Real-world challenges:** Transitioning research prototypes into real-world applications poses multiple different challenges, including regulatory compliance, issues with safety and even interoperability with the already existing infrastructure. To evaluate/assess the effectiveness and reliability of high-performance automation systems in practical settings, more diverse research is needed to address all the implementation issues. This can be achieved by collaborating with different industry partners and a lot of testing.

The widespread usage of these systems raises a major issue with the ethical as well as cultural side.

These include concerns about algorithmic biases, privacy and the potential displacement of jobs. To ensure that automation technologies being used are properly deployed and applied, more research is needed. Although high-performance automation techniques have the potential to intensively increase the effectiveness and efficiency of CIS, they may not fully realize this potential or promote long-term innovation in this field until the issues that are mentioned above are addressed and all the possible research directions are explored [25–37].

7.17 CONCLUSION

To sum up, in industrial settings, high-performance automation prioritizes process control, process optimization and process augmentation. It encompasses both HPC and IA. The goal of this optimization is to boost efficiency, accuracy and production in a variety of professional fields. High-performance automation systems incorporate many components such as PLCs, robots, SCADA systems and others using a variety of technologies and methods. Among the techniques are the integrated, programmable, flexible and fixed automation approaches. Multi-objective high-performance automation solutions are needed in CIS. Enhancing adaptability, making the most use of available resources and raising the standard for current knowledge are the goals of scholars. These objectives have significant practical implications. High-performance automation technology can help companies innovate, become more flexible in the face of change, increase their ability to adjust to changing conditions, simplify operations and utilize less resources. These and other difficulties, like problems with complexity and integration, resource limitations, adaptability in the face of unforeseen events, implementation hurdles in practice and ethical and societal repercussions, could be the subject of future research. It will be crucial to address these problems and look into possible future research directions in order to ensure responsible automation technology deployment and use across industries, realize the full potential of high-performance automation methods in CIS and encourage sustainable innovation.

REFERENCES

1. Basilius. *Introduction to Industrial Automation Systems*. Basilius. https://www.basilius.com/introduction-to-industrial-automation-systems/ (Accessed on 02.02.2024).
2. IBM. *High-Performance Computing*. IBM. https://www.ibm.com/topics/hpc (Accessed on 03.02.2024).
3. A. Almalawi, S. Hassan, A. Fahad, S. Hassan, and A. Fahad, "A hybrid cryptographic mechanism for secure data transmission in edge AI networks," *International Journal of Computational Intelligence Systems*, vol. 17, no. 24, 2024, doi: 10.1007/s44196-024-00417-8.

4. E. Diamant, "Computational intelligence: Are you crazy? Since when has intelligence become computational?," in *2016 IEEE Symposium Series on Computational Intelligence (SSCI)*, Athens, Greece, 2016, pp. 1–4, doi: 10.1109/SSCI.2016.7850143.
5. Z. Yun, "Analysis and research on the application of artificial intelligence and computational intelligence in the Internet of Things," in *2022 4th International Conference on Intelligent Information Processing (IIP)*, Guangzhou, China, 2022, pp. 360–364, doi: 10.1109/IIP57348.2022.00082.
6. R. Eberhart, D. Palmer and M. Kirschenbaum, "Beyond computational intelligence: Blended intelligence," in *2015 Swarm/Human Blended Intelligence Workshop (SHBI)*, Cleveland, OH, USA, 2015, pp. 1–5, doi: 10.1109/SHBI.2015.7321679.
7. D. Inkpen (Ed.), *Computational Intelligence*. Wiley Online Library, 2024. https://onlinelibrary.wiley.com/journal/14678640.
8. R. Wang and W. Ji, "Computational intelligence for information security: A survey," in *IEEE Transactions on Emerging Topics in Computational Intelligence,* vol. 4, no. 5, pp. 616–629, Oct. 2020. doi: 10.1109/TETCI.2019.2923426.
9. G. Yi, and V. Loia, High-performance computing systems and applications for AI. *J Supercomput,* vol. 75, pp. 4248–4251, 2019, doi: 10.1007/s11227-019-02937-z.
10. X. Zhao, T. Oseni and B. T. Medishetty, "Overview of business hyper-automation," in *2022 IEEE International Conference on e-Business Engineering (ICEBE)*, Bournemouth, United Kingdom, 2022, pp. 100–105, doi: 10.1109/ICEBE55470.2022.00026.
11. A. V. Ostroukh, N. G. Kuftinova, O. I. Maksimychev and I. V. Odinokova, "Hyperautomation of construction materials enterprise," in 2021 *intelligent Technologies and Electronic Devices in Vehicle and Road Transport Complex (TIRVED)*, Moscow, Russian Federation, pp. 1–4, 2021. doi: 10.1109/TIRVED53476.2021.9639170.
12. M. Li, J. Wei, X. Zheng and M. L. Bolton, "A formal machine–learning approach to generating human–machine interfaces from task models," *IEEE Transactions on Human-Machine Systems*, vol. 47, no. 6, pp. 822–833, Dec. 2017. doi: 10.1109/THMS.2017.2700630.
13. L. Urbas, M. Obst and M. Stöss, "Formal models for high performance HMI engineering," *IFAC Proceedings*, vol. 45, no. 2, pp. 854–859, 2012.
14. Supertrak Conveyance. *How to Achieve High-Performance Automation in a Smaller Footprint.* Supertrak Conveyance. https://supertrakconveyance.com/resources/blog/how-to-achieve-high-performance-automation-in-a-smaller-footprint/ (Accessed on 11.02.2024).
15. I. Calvo, F. Pérez, I. Etxeberria-Agiriano, and O. G. de Albéniz, "Designing high performance factory automation applications on top of DDS," *International Journal of Advanced Robotic Systems*, vol. 10, no. 4, p. 205, 2013. doi: 10.5772/56341.
16. K. Shail. *Limitations of Automation Orchestrators and the Rise of Automation Optimization.* CIO. https://www.cio.com/article/407234/limitations-of-automation-orchestrators-and-the-rise-of-automation-optimization.html (Accessed on 28.01.2024).

17. High Gear. *Benefits and Challenges of Work Automation in the Manufacturing Industry*. HighGear. https://www.highgear.com/blog/benefits-challenges-work-automation-manufacturing-industry/.
18. Harvard Business Review. How automation drives business growth and efficiency. *Harvard Business Review*, 2023. https://hbr.org/sponsored/2023/04/how-automation-drives-business-growth-and-efficiency.
19. P. J. Younse, J. E. Cameron and T. H. Bradley, "Comparative analysis of model-based and traditional systems engineering approaches for architecting a robotic space system through automatic information transfer," *IEEE Access*, vol. 9, pp. 107476–107492, 2021. doi: 10.1109/ACCESS.2021.3096468.
20. American Technology Blog. *Robotic Automation and Traditional Automation*. American Technology Blog. https://blog.american-technology.net/robotic-automation-and-traditional-automation/.
21. K. Sangsanit, W. Kurutach and S. Phoomvuthisarn, "REST web service composition: A survey of automation and techniques," in *2018 International Conference on Information Networking (ICOIN)*, Chiang Mai, Thailand, 2018, pp. 116–121, doi: 10.1109/ICOIN.2018.8343096.
22. Britannica. *Advantages and Disadvantages of Automation*. Encyclopedia Britannica. https://www.britannica.com/technology/automation/Advantages-and-disadvantages-of-automation (Accessed on 25.01.2024).
23. P. R. Anisha, C. K. K. Reddy, N. G. Nguyen, M. Bhushan, A. Kumar, and M. M. Hanafiah, (Eds.). (2022). *Intelligent Systems and Machine Learning for Industry: Advancements, Challenges, and Practices*. CRC Press, Boca Raton, FL.
24. D. Srivastava, S. Hooda, R. Gill and C. Singla, "Detection of cyber attacks by using gray wolf optimizer," in *2023 International Conference on Intelligent Systems for Communication, IoT and Security (ICISCoIS)*, pp. 596–600, 2023. IEEE, Coimbatore.
25. V. K. Sharma, "CNTFET circuit-based wide fan-in domino logic for low power applications," *Journal of Circuits Systems and Computers*, vol. 31, no. 02, p. 2250036, 2022.
26. V. K. Sharma and M. Pattanaik, "A reliable ground bounce noise reduction technique for nanoscale CMOS circuits," *International Journal of Electrochemical Science*, vol. 102, no. 11, pp. 1852–1866, 2015.
27. V. K. Sharma, S. Patel, and M. Pattanaik, "High performance process variations aware technique for sub-threshold 8T-SRAM cell," *Wireless Personal Communications*, vol. 78, no. 1, pp. 57–68, 2014.
28. V. K. Sharma and M. Pattanaik, "Techniques for low leakage nanoscale VLSI circuits: A comparative study," *Journal of Circuits Systems and Computers*, vol. 23, no. 05, p. 1450061, 2014.
29. V. K. Sharma, "A survey of leakage reduction techniques in CMOS digital circuits for nanoscale regime," *Australian Journal of Electrical and Electronics Engineering*, vol. 18, no. 4, pp. 217–236, 2021.
30. C. Kumar, A. S. Mishra, and V. K. Sharma, "Leakage power reduction in CMOS logic circuits using stack ONOFIC technique," in *2018 Second International Conference on Intelligent Computing and Control Systems (ICICCS)*, Madurai, 2018.
31. V. K. Sharma and M. Pattanaik, "Design of low leakage variability aware ONOFIC CMOS standard cell library," *Journal of Circuits Systems and Computers*, vol. 25, no. 11, p. 1650134, 2016.

32. S. F. Naz, S. Riyaz, and V. K. Sharma, "A review of QCA nanotechnology as an alternate to CMOS," *Current Nanoscience*, vol. 18, no. 1, pp. 18–30, 2022.
33. S. Riyaz and V. K. Sharma, "Design of reversible Feynman and double Feynman gates in quantum-dot cellular automata nanotechnology," *Circuit World*, vol. 49, no. 1, pp. 28–37, 2023.
34. B. Raina, C. Verma, M. Gupta, and V. K. Sharma, "Binary coded decimal (BCD) seven segment circuit designing using quantum-dot cellular automata (QCA)," in *2021 5th International Conference on Trends in Electronics and Informatics (ICOEI)*, Tirunelveli, 2021.
35. U. Mushtaq and V. K. Sharma, "Performance analysis for reliable nanoscaled FinFET logic circuits," *Analog Integr. Circuits Signal Process*, vol. 107, no. 3, pp. 671–682, 2021.
36. Kajal and V. K. Sharma, "Design and simulation of FinFET circuits at different technologies," in *2021 6th International Conference on Inventive Computation Technologies (ICICT)*, Coimbatore, 2021.
37. V. K. Sharma, "Designing of face recognition system," in *2019 International Conference on Intelligent Computing and Control Systems (ICCS)*, Madurai, 2019.

Chapter 8

Meta-fusion network for tumor classification

Mayank Kumar Singh, Indu Saini, and Neetu Sood

8.1 INTRODUCTION

Today Artificial Intelligence (AI) has found its way in every domain of science and technology [1–6]. The first-generation of AI had gate and perceptron-based architecture. That was the emerging phase of the AI with no real-world application. It was the second generation of AI which got the attention of almost every researcher around the globe. It was found to be far more accurate, less tedious, and robust than some of the human-designed application. But it needed features and parameters to learn the task. Its application was limited to the tasks which can be parameterized. Support Vector Machine (SVM) [7], Nearest Neighbor (k-NN) [8], Clustering algorithms [9], and Artificial Neural Network (ANN) [10] were the famous algorithms of the second-generation AI. Among these algorithms, the layered structure of input, hidden, and output layer of ANN was comparatively found to be more efficient and flexible. The structure of ANN is the replica of our nervous system and it was the most dynamic algorithm for further enhancement. Its successor Convolutional Neural Network (CNN) proved to be a real game changer and laid the foundation for the next generation of AI. The third generation of AI had various structures of CNN and its application increased multifold. CNNs do not require tedious and challenging extraction features for the task. They are able to learn the features of their own. The CNN creates a multidimensional feature space in their training, which gave it the edge over conventional techniques.

8.1.1 Convolution neural network and classification

CNN is a pipeline structure of multiple layers, each layer with unique function. The most premier layer of CNN is the convolutional layer, which is responsible for feature extraction. Set of filters are used for the operation and the coefficients of these filters evolve during the training. This was the major differentiator between the CNN and others. At the end of the pipeline there is a classification layer which learns the classification task and is responsible for classification by the network. CNN training is the

DOI: 10.1201/9781003643609-8

simplest of all, where the user has to just provide the input and set some hyper-parameters. Input data, network architecture, and hyper-parameters are the key to generate a good CNN. The input data must be sufficient and relevant to the task. Amount of data required by a CNN is very large due to large number of parameters. There are three types of training, namely supervised, semi-supervised, and unsupervised. In supervised training the input and its ground truth are provided. In semi-supervised learning, there are only limited amount of available data that have attached ground truth. While in unsupervised there is no ground truth data available. So, the training strategy of CNN is entirely the function of data available. CNN architecture depends upon the task to be performed. For a complex task the architecture of CNN needs to be lengthy. If the task is classification, then the output of each layer is successively down-sampled. Lots of research has been focused on designing a better architecture with different layers and connections. Initially it was assumed that a deeper architecture will have better accuracy than a shallow network. Later it was found that the performance stabilizes after some layers and starts decaying from that point forward. This was due to the vanishing gradient problem. To tackle this problem research, He et al. [11] suggested to use residual connection [12]. The connections were also called identity connection and were very helpful during backpropagation. The architecture was named as ResNet and ResNet 50 and ResNet 101 are the most successful architectures [13–15]. However, it was found that the low-level information or feature extracted from the image was not utilized well, or almost neglected. This was due to the deeper architecture. So, to alleviate this limitation, skip connections were introduced. Skip connection is also an identity connection but they also work during the forward pass to share information between the layers. Using skip connection has now become very common as they improve the network's efficiency. The most efficient and complex network which uses skip connection is the DenseNet family [16,17]. They have outperformed all the other networks in every image processing task. After network architecture the hyper-parameters are the last variables to be defined. The hyper-parameters are like the learning rate, loss function, and weight upgradation function. With the learning rate one can control the amount of change of parameters. A high learning rate means fast convergence but always misses the optimum parameter configuration [18], while a low learning rate does allow the network to explore all possible configuration but at the cost of time. So, the learning rate is kept low but not very low. Readers can visit the following publications for better understanding of learning rate [19,20]. The loss function is a parameter which determines the error of the network [21]. In supervised training it is always somehow related to the difference between the output and ground truth. During training the network uses the loss function to update its parameter and during testing it can be used as an evaluation metric. At last, the weigh upgradation function is a function used to update the parameter values. In backpropagation the gradient for each parameter is calculated individually. The gradient is then used to update the parameters' value and how it will be done is determined using the weight upgradation functions. Almost every research in the field of AI is done by varying these factors, as per requirement.

8.1.2 Class activation map

The training of CNN becomes a black box, where no one can observe what the network is learning, or which part of the image was used for final decision. It is vital to know these factors, otherwise the performance of CNN will remain questionable. To overcome this situation many ideas were proposed and it is still one of the most trending research topics [22–25]. One very basic idea is to observe the filter characteristic. The filter is nothing but a two-dimensional matrix, the coefficient of the matrix is obtained after a long run of evolution. It was observed that the initial filter layers of almost every CNN are very similar. These were close to the Gabor filters [26] which are used for edge detection. So, it was concluded that the initial layers of the CNN extract low-level features like edges. As we go up the layer more complex features are extracted like shapes in the image. Using filter characteristic to understand the network was good for lower layers, but for higher layers these characteristics made no sense. The features were completely out of human understanding. Researchers also suggested to plot the backpropagated loss gradients over the image to understand the focus of the network. Some even fed a series of images to CNN and observed the neurons which was most excited by that image category [22]. A very famous experiment known as occlusion experiment was done to observe the most important region of the image for a CNN to make decision [27]. In this experiment a part of the image was covered and the network's output probability of top five classes was noted. Next on the complete image some other region was covered and the network's output probability was noted. It was repeated by covering other regions till every region of the image had been covered once, leaving the rest of the image as it is. For the output with least probability for the true class, corresponding covered region was considered as the deciding factor. That region has a big role in differentiating that image from the others. The final output of this technique was a heat map showing contribution of each region in correct classification. It was a good technique but lacks precision. Gradient ascent to update the image using gradients was also found to be very efficient. First a blank image is given as input to the network and the output of any one class is kept high and the rest as zero. The gradients backpropagated from the network at the very initial layer were used to update the image using gradient ascent [28]. It helped in understanding the patterns related to the corresponding class. However, it requires lots of computation and lacks precision. Another method called Class Activation Map introduced by Zhou et al. [29] was found very efficient in CNN visualization. With this method any convolutional layer of the network can be analyzed. First a global average pooling is applied to the filters of the layer. Suppose the layer has k filters of size $M \times N$ the output of global average pooling will be $k \times 1$ which is a vector. Then we calculate weights for each vector coefficient which relates the coefficients with the final output. The weight calculation is done using regression. With these weights a weighted average of the output feature map across the channel of the layer is calculated and known as class activation map (CAM). GradCAM [30] an upgrade of CAM uses gradients itself as weights in average calculation. Both are great when it comes to CNN visualization; however, with GradCAM we do not require weight calculation. They have been used by various studies to analyze the characteristics of their network [31] and many have proposed different upgrades in CAM [32–38]. Still, they are known for their simplicity and efficiency.

8.1.3 Transfer learning

With the unparallel efficiency training CNN became very popular in no time. However, there are some limitations of using CNN, like it requires loads of data, space, and time. Most of the time there is not enough data to train a CNN from scratch. Then transfer learning comes to the rescue [39] in those situations. In transfer learning the learned coefficients of the network can be used to avoid training from scratch. As discussed in the previous section, the initial layers of every network learn very similar edge features, which means every task needs the low-level image features. So, usually the initial layers are taken from any network of their choice, their parameters are frozen, and only higher layer parameters are allowed to evolve. Sometimes the complete network is taken with the parameter values and trained over subsequent dataset. There are many ways the knowledge of a pre-trained network can be used. It is quiet famous among researchers to use transfer learning [40–44].

The concept of CNN has been used to automate the clinical diagnosis procedure for a very long time. Early detection of diseases like cancer has a great impact on their treatment [45]. Numerous studies have been done to classify the tumor in diagnostic images to empower early detection [18,46,47]. Looking at the sensitivity of the topic it is obvious that the network must be highly accurate. Otherwise, it will have no real-life application. Moreover, the choice of diagnostic image also affects the realization of the system. Among the diagnostic images ultrasound (US)-based diagnosis has lots of advantages. The major advantage of using US is that it is radiation free, so it can be used repeatedly and for any age group. Using US as the base to develop an automated diagnosis system will comparatively be more impactful. However, there are downsides of using US, like its low quality and lack of large dataset. But with better strategy we can design a reliable system without losing much accuracy. Moreover, using the transfer learning the data requirement can be dealt with. So, in this chapter we have proposed a transfer learning-based cancer classification network. For transfer learning we have proposed to use a different selection criterion. Network selection is done based on the CAM of different layers of different network. Instead of just using a single network we fused the information from three different networks, using a meta-fusion block. Each network has its own capability of extracting features of different scales. To fully utilize the power of pre-training the networks were analyzed with CAM at each layer of different levels. Two layers from three levels, the initial, middle, and final levels, of the CNN were combined in our network. CAM analysis was done after training the networks with the US. After selection the original networks were trained with meta-fusion block over the US. The accuracy, sensitivity, and specificity of our network were compared with other networks.

The rest of the chapter is arranged as follows: the second section provides a brief review of literatures related to cancer classification using AI. Next section has the proposed methodology and information regarding the dataset used. Results and discussion is given under the third section and the complete chapter is concluded in the fourth section.

8.2 LITERATURE SURVEY

There is a long history of CNN's application for cancer classification [48–51], but here we have explained the most recent studies in the domain in brief. Mishra et al. [52] proposed to train an ensembled classifier over a limited set of features. The features were filtered from a series of handcrafted features, they also used an oversampling technique to deal with class imbalance. Mao et al. [53] first derived a set of versions of the original image. Totally 14 different filters were used to obtain the derived images. Seven different categories of features were extracted from the derived and original image. Then they applied a feature selection algorithm and finally trained a linear classifier. Ahila et al. [54] first pre-processed the US using denoising and enhancement filters to improve the image quality. Instead of calculating features from complete image they first applied a segmentation technique and from the segmented regions features were extracted. A wavelet neural network optimized with gray wolf optimization algorithm was trained with those features. Daoud et al. [55] used the combination of CNN extracted features and spatial features to classify cancer in US. They used various layers of VGG 19 for the task. The features from the layer were analyzed using a feature selection method, and only limited layers were used. Same steps were repeated for the spatial features for selections. In the end a SVM was trained using the features for final classification. Use of ensembled networks is also very famous for cancer classification. The studies have good results; however, use of handcrafted features does not fit in the idea of complete automation.

Many researchers tried to design an exhaustive network for cancer classification, trained from scratch. The study by Luo et al. [56] trained five different networks in cascade for the classification. First was a U-net to segment the tumor, the second and third networks were used to extract feature from the image and the mask enhanced image, respectively. Fourth network was trained to fuse the information from the two networks and the last network learned the classification task. DeepBraestCancerNet proposed by Raza et al. [57] was trained to classify tumor in breast US. The proposed network had 24 layers, out of which only one fully connected layer, nine modules of inception, and six normal convolution layers were used. Instead of using the conventional Relu they used the clipped and leaky Relu. They also used the batch normalization together with the cross-channel normalization. Atrey et al. [58] used the most advanced version of CNN, the long short-term memory (LSTM) model [59], for breast cancer diagnosis. In the technique CNN-based feature extraction from pre-processed image was followed by feature selection. The selected features were fused and fed to the LSTM with hold out data division and multifold cross validation. LeNet proposed by Hazarika et al. [60] was exhaustively trained for breast cancer classification [61]. The original architecture of LeNet has two convolution layer and two dense layers, was trained on breast US with a corrected Relu. The advantage of using dense layer is that it provides information sharing between layers. The concept of information sharing between the layer is very effective and has been a hot topic lately [39,62,63]. Researchers have also used the attention mechanism, which is a very recent advanced tool for training of networks [64,65]. In the mechanism the network is trained in a such way that its attention increases to some specific features [66,67]. However, it is not always efficient, as its training for complex task is very unstable.

One problem with US is that there is no large cancer database available freely to train network from scratch and verify its performance. In terms of data, time, and space, training a network from scratch is no match to the transfer learning. That is the reason why transfer learning is getting all the attraction in image processing tasks [68–71]. Hijab et al. [72] tested three different ways to apply CNN for classification in US. In one method they train a network from scratch, in second they used VGG 19 for transfer learning and trained the network over the US dataset. In the third method the parameters of initial layers of VGG 19 were locked and allowed only the last layer to evolve or fine-tune over the US dataset. They discovered that the fine-tuned VGG 19 has far better performance than training from scratch and it also has the edge over pre-trained VGG 19. Zhuang et al. [73] furnished three different fusion images from the original image. Authors used the bilinear filtering, fuzzy enhancement [74], and blurring to create three different RGB version of a gray scale image. Then from a set of pre-trained CNN they selected only three and then using a fusion block to combine the information of the three networks they achieved the final classification. Ragab et al. [75] in their study proposed to first enhance the US for less noise and better contrast. The enhanced images were segmented using the Chaotic Krill Herd Algorithm with Kapur's Entropy. Pre-trained VGG 19, 16, and SqueezeNet were applied for feature extraction. A multi-layer perceptron with cat swarm optimization was trained at the end for the output. Jabeen et al. [76] used image augmentation to increase the data quantity and then used DarkNet-53 [77] for fine tuning. From this fine-tuned network features were extracted and filtered for finally training a cubic SVM. Using CNN for feature extraction has shown good performance, but there exists other more efficient network which can give better results. So, the study by Hassan et al. [78] compared the networks before inheriting their learning. The networks were namely MobileNetV2, ResNet50V2, Resnet101V2, Resnet152V2, Xception, VGG16, VGG19, InceptionResNetV2, and InceptionV3. And from the best performing networks features were extracted and linear classifiers like SVM and k-NN were rained. Ayana et al. [79] proposed to use the pre-trained network directly for classification instead of using them only for feature extraction. Three models were selected EfficientNetB2, InceptionV3, and ResNet50 all optimized with unique optimization algorithm. They also proposed to multi-stage transfer learning. The network trained on ImageNet was again trained on processed microscopic cancer cell line images. Only then they were used for transfer learning.

8.3 MATERIALS AND METHODOLOGY

The proposed system architecture is given in Figure 8.1. The low quality of the US image poses a significant challenge to the performance of CNNs. So, we proposed first denoising the US image and improving image contrast. The pre-trained networks are trained over the US database for cancer classification. We performed a CAM-based analysis before including the layer in our final network. Selected layers were taken from each CNN, and their intelligence was combined using a meta-fusion block. Moreover, the network was trained again for final cancerous classification. For validation and verification of results, we have used two public US datasets. One is the breast US dataset, and other is the liver US dataset. For breast US data, we have used two sources, which have a combined 993 breasts in the US; more information regarding the dataset is given in Yap et al. [80] and Al-Dhabyani et al. [81]. The liver US dataset provided by Vitale et al. [82] has 926. We have applied primary image augmentation to increase the dataset.

8.3.1 Pre-processing

The pre-processing is required to improve the contrast and quality of the image. We have slightly modified our denoising technique proposed in Guyon et al. [49]. In the denoising method, three levels of thresholding

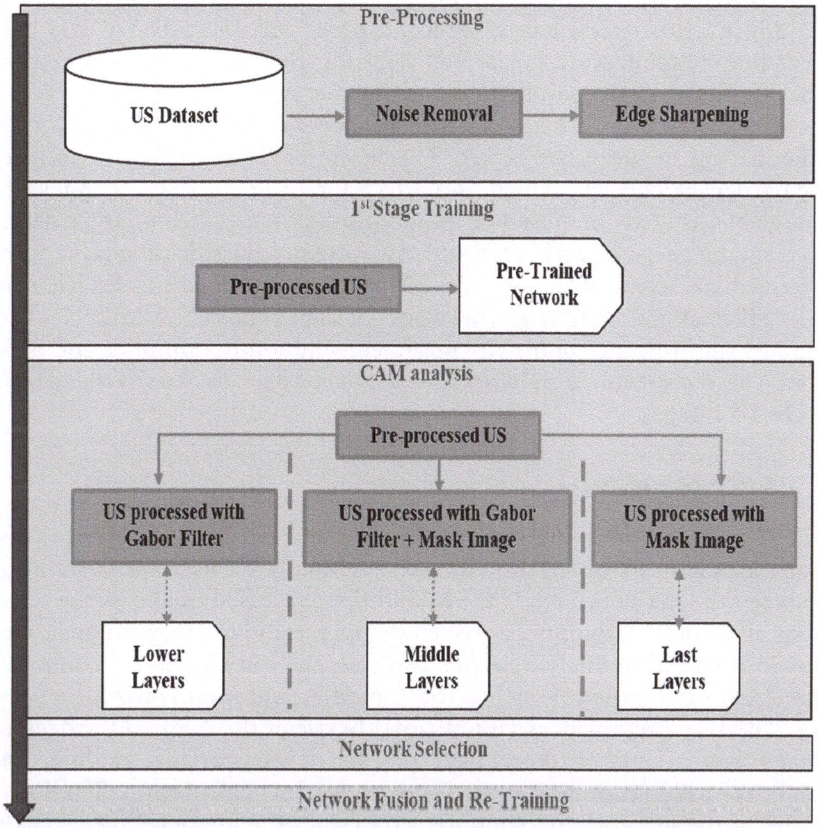

Figure 8.1 Process flow of proposed network.

function were used, suitable for generating a better-quality image but also modifying some image details. The CNNs extract features of all scales, so modification in details can cause performance degradation. So, we chose to use only two levels of thresholding instead of three.

The rest of the denoising steps are the same as the study. Denoising improves the image quality but not the contrast of the image. The dark texture of the US adds to the challenge of structural identification. To tackle this problem, we have used the very famous and efficient Contrast Limited Adaptive Histogram Equalization (CLAHE) technique. In our technique, a good-quality image is needed to train the network for better network selection.

8.3.2 1st stage training

We have selected the CNN architectures with good performance over the ImageNet challenge. The networks are namely VGG 19 [83], ResNet 50, ResNet 101 [84], DenseNet 121 [85], GoogleNet [84], MobileNet V2 [40],

SqueezeNet [86], Inception v3 [87], Inception v4 [85], and ShuffleNet v2 [88]. Each network has its concept, pros, and cons. The VGGs have basic linear architecture but are efficient for low-level feature extraction. The ResNet uses the residual connection for better gradient flow across the network. DenseNet has a dense connection, which allows information sharing between the layers. The inception network has introduced the bottleneck concept, which uses a 1×1 filter to reduce the channel size. GoogleNet is also composed of an inception module with a larger depth. To train the deep network, they added a SoftMax classification layer at the intermediate layers of the network. MobileNet and SqueezeNet were the most efficient and lightweight networks of their time. The ShuffleNet uses channel shuffling and pointwise group convolution to minimize computation. These pre-trained networks are trained again to classify the cancer in the US images.

8.3.3 CAM analysis

Unlike other methodologies that use accuracy as the parameter for network selection, we opted for a different strategy. We used the correlation between the network layer's CAM and the processed image as the selection criterion. Overlapping between the image and the map is considered the correlation coefficient. After training the networks, we computed the CAM of the two of each initial, middle, and final convolution layers. The Intersection over Union (IoU) between the map and processed image was calculated. For layers of each level (starting, middle, and final), we used unique processed images for IoU calculation. The lower layers of the network are supposed to extract the low-level information like edges. We have applied the gradient filter to the images in the dataset. The output of the gradient filters will have the edges of the images. To have the optimal results from the gradient filter, it was necessary to pre-process the US dataset. To select the middle layers, we use the masked images provided in the dataset. Masked images or the ground truth images and the gradient output of the dataset were blurred. Two blurred images were then added to create a single image. The output image was used to calculate the IoU with the CAM of the middle layers of the CNN. We used the IoU between the masked image and the CAM for the final layers.

8.3.4 Network selection fusion and re-training

Based on the average IoU, two initial layers, two middle layers, and two final convolution layers of the top three networks were chosen. Figure 8.2 shows the overall architecture of the proposed network. The fusion layers were densely connected to ensure information sharing. Once selected, the complete layer is inherited from the parent architecture, i.e., the network

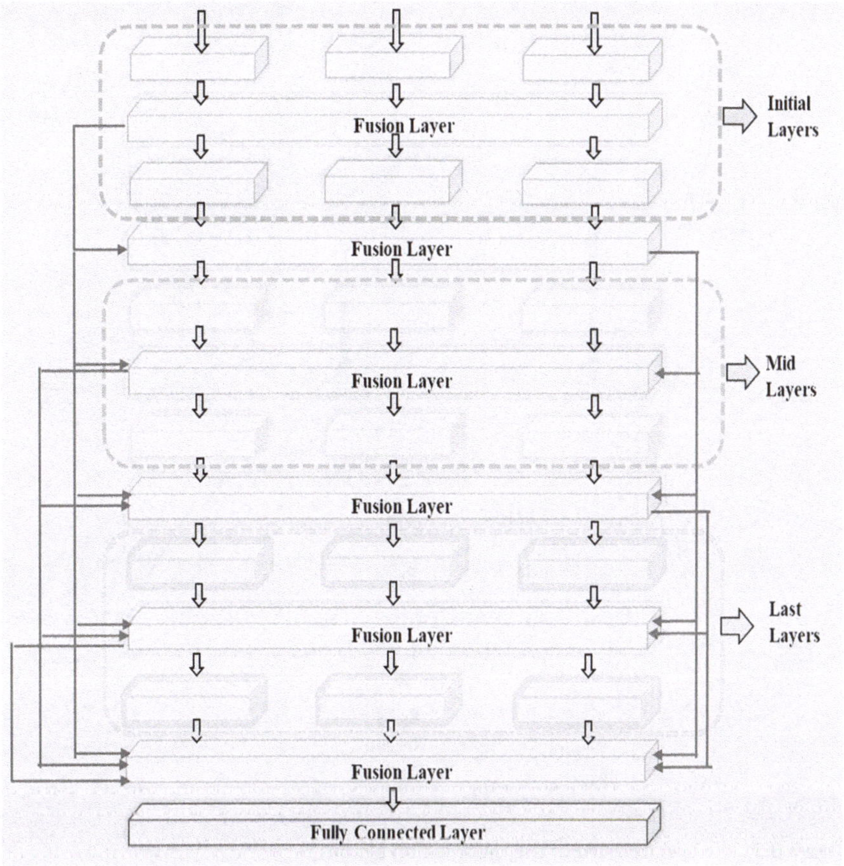

Figure 8.2 The proposed network architecture with meta-fusion block.

trained only on the ImageNet dataset. After selecting layers of different architectures, one problem exists, the feature map synchronization. Different layers of different networks have different feature map resolutions. To solve this problem, we have added a resizing block at the input and output of our fusion block. The detailed architecture of the meta-fusion layer used in this study is given in Figure 8.3. Like the inception module, the proposed fusion block has multiple bottlenecks and a residual connection. The block output has three 1×1 filters, each branching out to a different network. The output of the last fusion layer has no such filter. The output of the 3×3 filter is flattened and provided to the classification or fully connected layers. The fully connected layer is transferred from the networks based on accuracy.

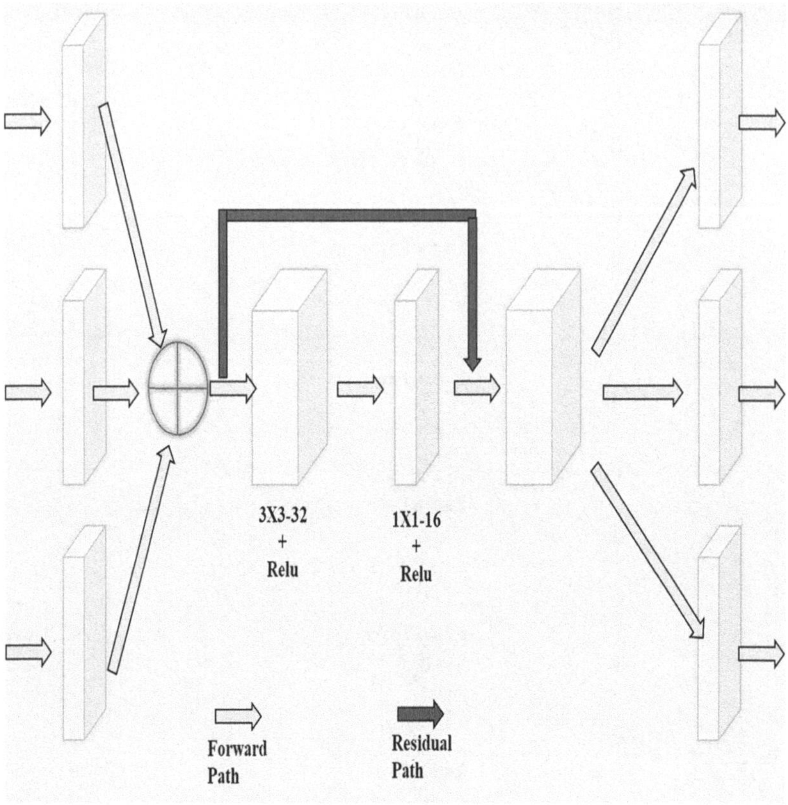

Figure 8.3 The architecture of the meta-fusion block.

8.4 RESULTS AND DISCUSSION

In the chapter we have proposed a novel technique of network selection for transfer learning. We have used the focus of the trained networks to create the merit list. US images were first processed in a unique way for different levels of the networks. To evaluate the low or initial level we used the gradient image, we observed the IoU between the gradient image and CAM output. Overlapping between the CAM and combined blurred mask and gradient image was used as the selection criterion for the middle layers. The overlapping between the CAM and the mask image was used for the last or final layers. We have used the GradCAM to calculate the CAM because it is similarly efficient but does not require weight calculation. The processed US for the calculation of IoU is given in Figure 8.4. Liver and breast US images given in Figure 8.4b and f are the output of the gradient filter, Figure 8.4c and g are the mask and gradient combined, and Figure 8.4d and h are the mask image for the original US images in Figure 8.4a and e.

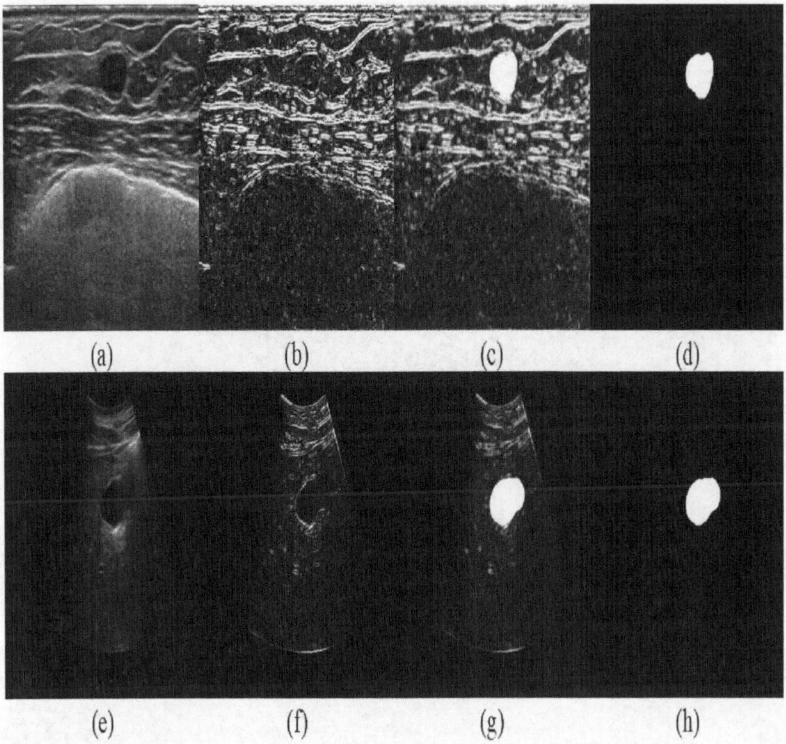

Figure 8.4 The processed US output, here (a) and (e) are original breast and liver US image, (b) and (f) are the output of gradients, and (c) and (g) are the combination of mask and gradient image, respectively.

Figures 8.5–8.7 provide the GradCAM of the initial, middle, and final layers of the networks involved in this study. The CAM was obtained after training all the networks with the US dataset. Unlike other studies that directly use the classification accuracy of the networks as a selection criterion, we replaced it with the activation map. Table 8.1 represents the names of the blocks we have selected from the original network for comparison, and it is the name used by the original paper. The corresponding CAM is overlaid on the original image to get a better idea. Table 8.2 has the average IoU obtained by the networks' initial, middle, and final layers. The values are normalized to one, and Figures 8.8–8.10 present the bar graph comparison of the IoU values. It can be observed that the deeper architectures have better IoU values for the final layers. More specifically, the architectures with shared information have the best IoU. This is because they use small edge-related features and shape-related features in their final decision. The deep networks with no information sharing between layers were the second-best performing architecture. They may not use the low-level

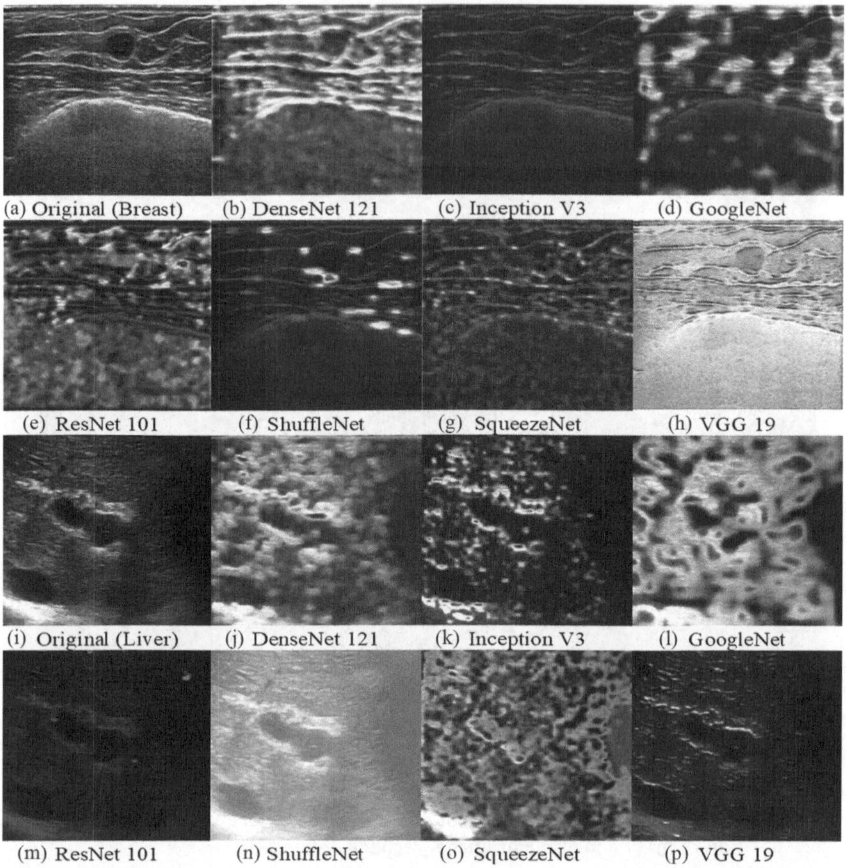

Figure 8.5 The CAM of some of the network's initial layer for (a) breast and (i) liver US. The output of initial layer of ResNet 50, DenseNet 121, MobileNet V2, Inception V4, VGG19, GoogleNet, and ShuffleNet V2 for breast US is in (b), (c), (d), (e), (f), (g), and (h) and for liver US (j), (k), (l), (m), (n), (o), and (p), respectively.

features for the extraction but have sufficient features at their final layers to make the correct judgment. The networks having basic linear architecture were at the last position in the table. This is because of their comparatively low feature extraction capability. Again, the networks with skip connections and deeper architectures were leading IoU values for the middle layers. However, this time, the difference in their corresponding IoU values was lower compared to the higher levels. The miniature linear networks were again the worst performing, but their performance was not that bad comparatively this time. They had a small but significant difference in the IoU values.

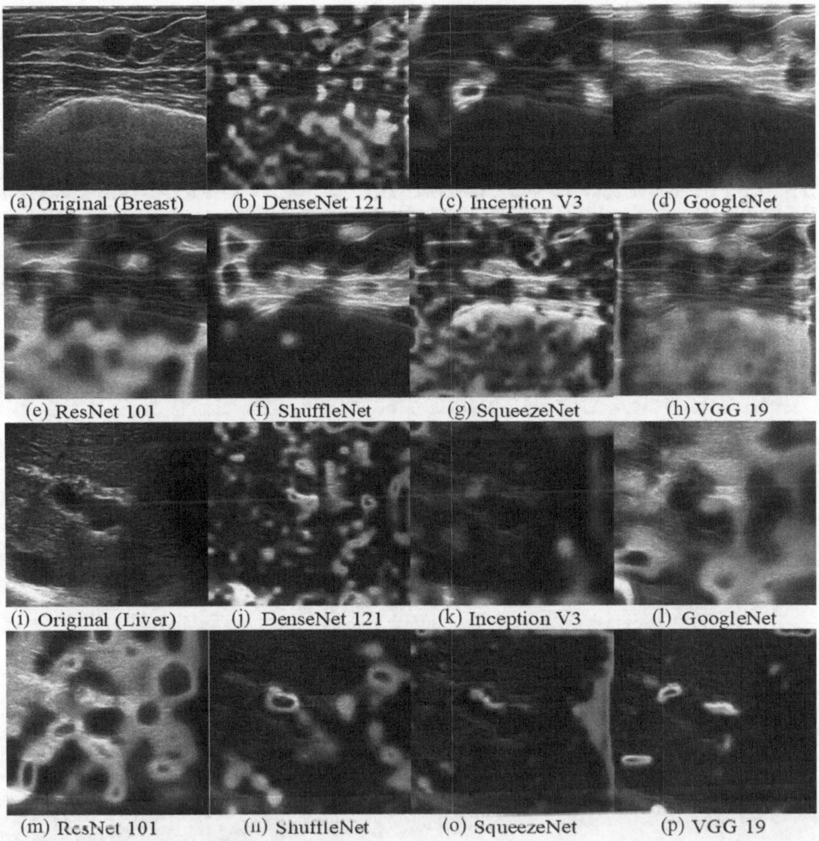

Figure 8.6 The CAM of some of the network's middle layer for (a) breast and (b) liver US The output of middle layer of ResNet 50, DenseNet 121, MobileNet V2, Inception V4, VGG19, GoogleNet, and ShuffleNet V2 for breast US is in (b), (c), (d), (e), (f), (g), and (h) and for liver US (j), (k), (l), (m), (n), (o), and (p), respectively.

At the lower levels where the network learns slight edge-related features, the CAM of networks must overlap with the edge features. It should be noted that the linear and smaller networks have competing IoU values. Some of the extensive networks have shown a massive performance degradation in catching up with low-level features. This is because of the vanishing gradient problem, which limits the coefficient evolution of the lower layers. That is why the smaller architectures and the networks with residual connections lead the table. It also proves the efficacy of the residual connection in gradient flow. The residual network can train deep architectures without affecting gradient flow to the lower layers. Based on the IoU analysis, the three networks selected for initial, middle, and final layers in our networks are ResNet 50, DenseNet 121, and GoogleNet. DenseNet 121,

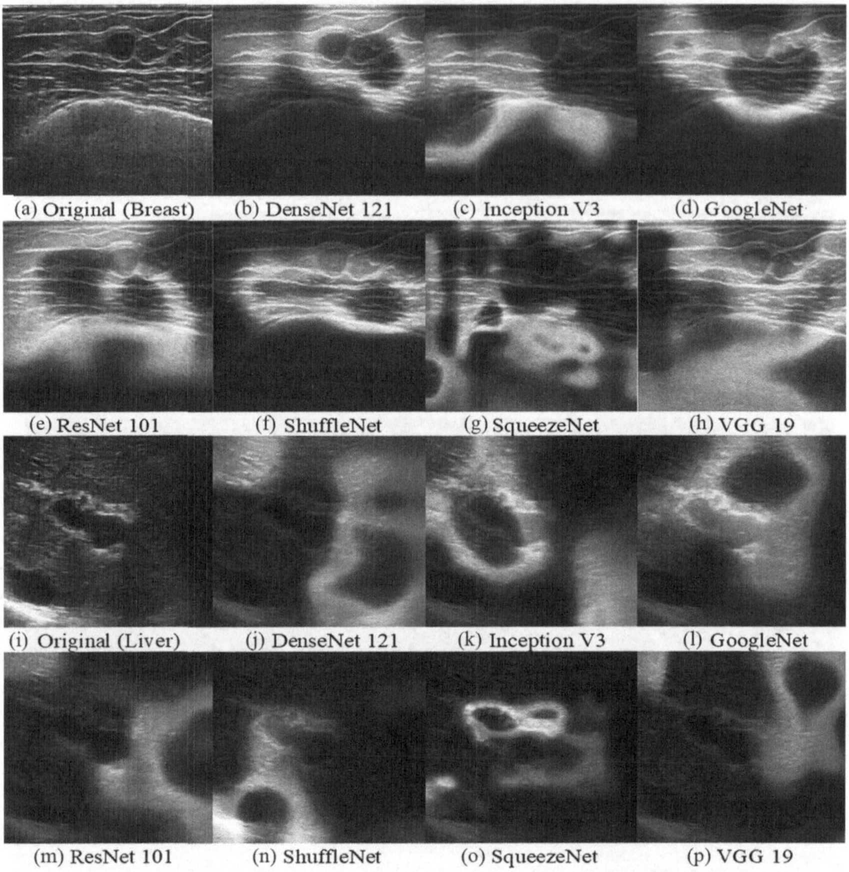

Figure 8.7 The CAM of some of the network's final layer for (a) breast and (i) liver US. The output of final layer of ResNet 50, DenseNet 121, MobileNet V2, Inception V4, VGG19, GoogleNet, and ShuffleNet V2 for breast US is in (b), (c), (d), (e), (f), (g), and (h) and for liver US (j), (k), (l), (m), (n), (o), and (p), respectively.

GoogleNet, and Inception V4, and DenseNet 121, GoogleNet, and Inception V4, respectively.

The proposed technique was compared with other techniques. For evaluation, we have calculated the accuracy, specificity, and sensitivity for both breast and liver cancer. The accuracy of a system is the ratio of the total correct decisions made by the network and the total decisions made by the network. A network with the highest accuracy means it made the most accurate judgments. Researchers use sensitivity and specificity for more discrete analysis. The sensitivity provides the number of actual classes (cancer-positive) assigned by the network compared to the overall actual classes in the dataset. On the other hand, the specificity provides the false

Table 8.1 Networks and the name of corresponding two layers selected from initial, middle, and final levels for CAM comparison

Network	*Levels*	*First*	*Second*
VGG 19	Initial	Conv1_2	Conv2_1
	Middle	Conv4_2	Conv4_3
	Last	Conv5_3	Conv5_4
ResNet 50	Initial	add_1	add_2
	Middle	add_8	add_9
	Last	add_15	add_16
ResNet 101	Initial	res2 (1)	res2 (2)
	Middle	res4 (1)	res4 (2)
	Last	res5 (2)	res5 (3)
DenseNet 121	Initial	conv2_block1_ concat	conv2_block2_ concat
	Middle	conv3_block8_ concat	conv3_block9_ concat
	Last	conv5_block31_ concat	conv5_block32_ concat
MobileNetV2	Initial	block_2_add	block_4_add
	Middle	block_9_add	block_11_add
	Last	block_14_add	block_15_add
SqueezeNet	Initial	fire2-concat	fire3-concat
	Middle	fire5-concat	fire6-concat
	Last	fire8-concat	fire9-concat
GoogleNet	Initial	inception-3a-output	inception-3b-output
	Middle	inception-4b-output	inception-4c-output
	Last	inception-5a-output	inception-5b-output
Inception V3	Initial	mixed0	mixed1
	Middle	mixed5	mixed6
	Last	mixed9	mixed10
Inception V4	Initial	block35_1_mixed	block35_2_mixed
	Middle	block17_12_mixed	block17_13_mixed
	Last	block8_8_mixed	block8_9_mixed
ShuffleNetV2	Initial	node_16	node_28
	Middle	node_101	node_113
	Last	node_186	node_198

class (non-cancerous class) predicted by the network over the complete false classes in the dataset. Equations 8.1–8.3 show the formula used to calculate accuracy, sensitivity, and specificity, respectively. The variables used in the equation are explained in Table 8.3.

$$\text{ACCURACY} = \frac{TP + TN}{TP + TN + FP + FN} \tag{8.1}$$

Table 8.2 The average IoU values of the networks

Network	*Initial*	*Middle*	*Last*
VGG 19	0.42	0.06	0.11
ResNet 50	0.81	0.58	0.62
ResNet 101	0.78	0.61	0.68
DenseNet 121	0.81	0.69	0.74
MobileNet V2	0.58	0.23	0.33
SqueezeNet	0.36	0.2	0.28
GoogleNet	0.80	0.65	0.72
Inception V3	0.71	0.52	0.58
Inception V4	0.79	0.69	0.75
ShuffleNet V2	0.38	0.18	0.29

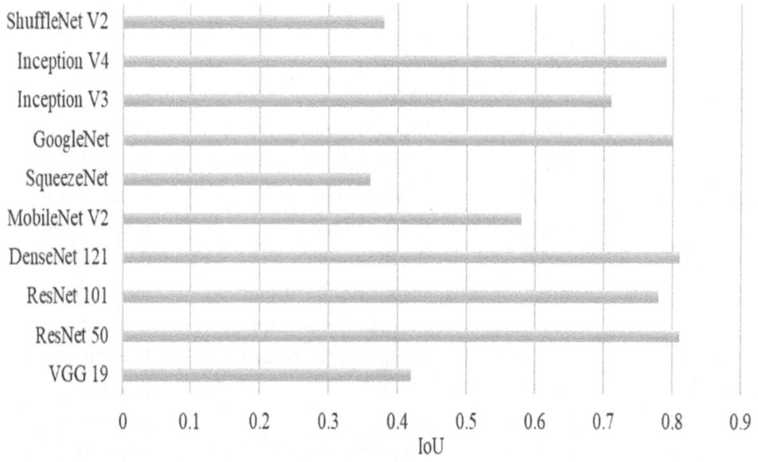

Figure 8.8 IoU of initial layers of the networks.

$$\text{SENSITIVITY} = \frac{\text{TP}}{\text{TP} + \text{FN}} \tag{8.2}$$

$$\text{SPECIFICITY} = \frac{\text{TN}}{\text{TN} + \text{FP}} \tag{8.3}$$

For comparison of classification performance, we have involved the classification technique proposed [72,73,75,76,78,79]. The techniques were implemented, and the average values of the networks' accuracy, sensitivity, and specificity were obtained for both breast and liver cancer. The values of accuracy, sensitivity and specificity obtained by the networks are in Table 8.4.

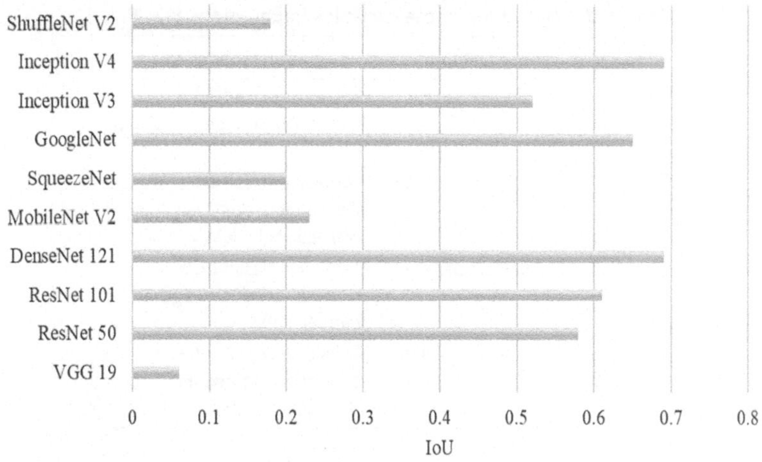

Figure 8.9 IoU of middle layers of the networks.

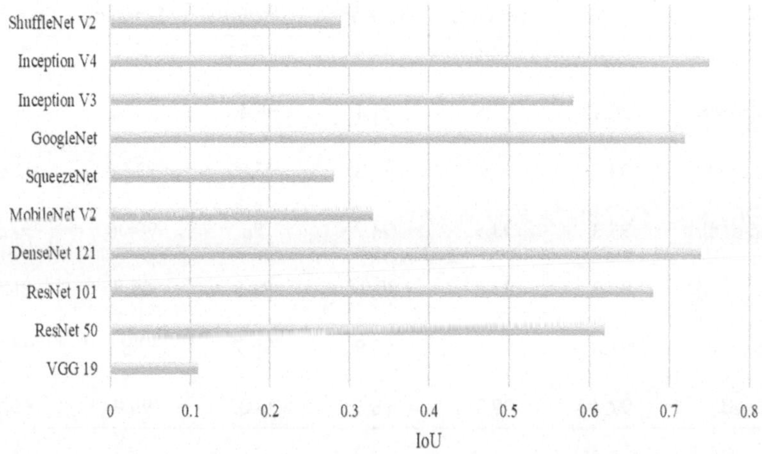

Figure 8.10 IoU of last layers of the networks.

It can be observed in the table that the proposed technique has outperformed the other techniques. For the classification of breast cancer, the proposed technique has 97.4%, 98.2%, and 98.5% accuracy, sensitivity, and specificity, respectively. For the classification of liver cancer, the proposed technique has 99.03%, 98.8%, and 99.1% accuracy, sensitivity, and specificity, respectively [89–101].

Table 8.3 The values of the variables in Equations 8.2–8.4

Variables	*Full form*	*Corresponds to*
TP	True positive	Network classified a cancerous image as cancer class
TN	True negative	Network classified a non-cancerous image as non-cancer class
FP	False positive	Network classified a non-cancerous image as cancer class
FN	False negative	Network classified a cancerous image as non-cancer class

Table 8.4 Classification performance

Classification studies	*Breast cancer*			*Liver cancer*		
	Accuracy	*Sensitivity*	*Specificity*	*Accuracy*	*Sensitivity*	*Specificity*
Hijab et al. [72]	93.4	93.1	93.6	97.06	96.1	97.6
Zhuang et al. [73]	90.1	89.5	91.1	95.48	93.92	98.33
Ragab et al. [75]	94.4	92	93.3	97.09	95.54	97.65
Jabeen et al. [76]	95.3	93.5	94.7	98	98.06	97.5
Hassan et al. [78]	86.9	83.8	85.04	87.5	85.9	86.8
Ayana et al. [79]	95.8	94.6	95.9	98.7	98	98.2
Proposed	97.4	98.2	98.5	99.03	98.8	99.1

8.5 CONCLUSION

This chapter proposes a cancer classification network in the diagnostic US. We have used the concept of transfer learning and activation maps. Transfer learning was used to avoid training from scratch and save some memory and data on the network. We have adopted a unique selection technique for the source of knowledge transfer. The CAM was used for the network selection. The idea behind selecting CAM is that any network that is better at finding the correct region of interest for its classification will have better accuracy. Moreover, a good CAM, i.e., CAM overlapping with true region of interest (ROI), ensures non-overfitting behavior. We aggregated layers of three different networks' start, middle, and final layers. We have selected two different layers from various networks for each level. The IoU was applied for layer selection. We calculated the IoU between the CAM, gradient, and mask image. The network selected based on IoU was taken from the original network, which was not trained on the US image. The complete network with our meta-fusion block was again trained on the US dataset. The proposed technique was compared with the most efficient classification technique. The performance of the proposed technique surpassed the performance of recent networks. Our network's average accuracy, sensitivity, and specificity for the complete US dataset were 98.215, 98.5, and 98.8, respectively.

The proposed study, which has achieved state-of-the-art results, has also introduced a new area of research in transfer learning. This research allows interested individuals to freely explore various combinations of networks for classification, not just networks but other combinations of layers, with no restrictions on the number of layers and networks. While we did not delve into exploring compatibility between networks due to the chapter's length and scope, this area presents a promising and unexplored field for future research.

REFERENCES

1. D. Uzun Ozsahin, M. T. Mustapha, B. Uzun, B. Duwa, and I. Ozsahin, "Computer aided detection and classification of monkeypox and chickenpox lesion in human subjects using deep learning framework," *Diagnostics*, vol. 13, no. 2, 2023, doi: 10.3390/diagnostics13020292.
2. M. A. O'Reilly and K. Hynynen, "A super resolution ultrasound method for brain vascular mapping," *Medical Physics*, vol. 40, no. 11, p. 110701, 2013, doi: 10.1118/1.4823762.
3. I. J. Kadhim and P. Premaratne, "A novel deep learning framework for water body segmentation from satellite images," *Arabian Journal Science Engineering*, vol. 48, no. 8, pp. 10429–10440, 2023, doi: 10.1007/s13369-023-07680-5.
4. H. Cui, Y. Xia, and Y. Zhang, "Supervised machine learning for coronary artery lumen segmentation in intravascular ultrasound images," *International Journal for Numerical Methods in Biomedical Engineering*, vol. 36, no. 7, p. e3348, 2020, doi: 10.1002/cnm.3348.
5. T. Nayak et al., "Detection of monkeypox from skin lesion images using deep learning networks and explainable artificial intelligence," *Applied Mathematics in Science and Engineering*, vol. 31, no. 1, p. 2225698, 2023, doi: 10.1080/27690911.2023.2225698.
6. C. Y. Wong et al., "Image contrast enhancement using histogram equalization with maximum intensity coverage," *Journal of Modern Optics*, vol. 63, no. 16, pp. 1618–1629, Sep. 2016, doi: 10.1080/09500340.2016.1163428.

7. D. A. Pisner and D. M. Schnyer, "Chapter 6- Support vector machine," in *Machine Learning*, A. Mechelli and S. Vieira, Eds. Academic Press, 2020, pp. 101–121.
8. A. H. Rabie and A. I. Saleh, "Monkeypox diagnosis using ensemble classification," *Artificial Intelligence in Medicine*, vol. 143, p. 102618, 2023, doi: 10.1016/j.artmed.2023.102618.
9. P. T. Sheeba and S. Murugan, "Fuzzy DDBN: Fuzzy Dragon deep belief neural network and interesting features points for activity recognition," *Sādhanā*, vol. 45, no. 1, p. 5, 2019, doi: 10.1007/s12046-019-1233–7.
10. K. Drukker, M. L. Giger, K. Horsch, M. A. Kupinski, C. J. Vyborny, and E. B. Mendelson, "Computerized lesion detection on breast ultrasound," *Medical Physics*, vol. 29, no. 7, pp. 1438–1446, 2002, doi: 10.1118/1.1485995.
11. K. He, X. Zhang, S. Ren, and J. Sun, "Deep residual learning for image recognition," in *2016 IEEE Conference on Computer Vision and Pattern Recognition (CVPR)*, 2016, pp. 770–778, doi: 10.1109/CVPR.2016.90.
12. J. Huang et al., "Parkinson's severity diagnosis explainable model based on 3D multi-head attention residual network," *Computers in Biology and Medicine*, vol. 170, p. 107959, 2024, doi: 10.1016/j.compbiomed.2024.107959.
13. R. Anand, S. V. Lakshmi, D. Pandey, and B. K. Pandey, "An enhanced ResNet-50 deep learning model for arrhythmia detection using electrocardiogram biomedical indicators," *Evolutionary Systematics*, vol. 15, no. 1, pp. 83–97, 2024, doi: 10.1007/s12530-023-09559-0.
14. C.-L. Lin and K.-C. Wu, "Development of revised ResNet-50 for diabetic retinopathy detection," *BMC Bioinformatics*, vol. 24, no. 1, p. 157, 2023, doi: 10.1186/s12859-023-05293-1.
15. S. Vaishali and S. Neetu, "Enhanced copy-move forgery detection using deep convolutional neural network (DCNN) employing the ResNet-101 transfer learning model," *Multimedia Tools and Applications*, vol. 83, no. 4, pp. 10839–10863, 2024, doi: 10.1007/s11042-023-15724-z.
16. T. Liao et al., "Classification of asymmetry in mammography via the DenseNet convolutional neural network," *European Journal of Radiology Open*, vol. 11, p. 100502, 2023, doi: 10.1016/j.ejro.2023.100502.
17. P. P. Dalvi, D. R. Edla, and B. R. Purushothama, "Diagnosis of coronavirus disease from chest X-Ray images using denseNet-169 architecture," *SN Computer Science*, vol. 4, no. 3, p. 214, 2023, doi: 10.1007/s42979-022-01627-7.
18. E. Hassan, M. Y. Shams, N. A. Hikal, and S. Elmougy, "The effect of choosing optimizer algorithms to improve computer vision tasks: A comparative study," *Multimedia Tools and Applications*, vol. 82, no. 11, pp. 16591–16633, 2023, doi: 10.1007/s11042-022-13820-0.
19. J. A. Navarrete-Ulloa, "Learning rates: A correction of gain scores to assess math learning interventions," *Journal of Experimental Education*, pp. 1–17, 2024, doi: 10.1080/00220973.2024.2352768.
20. A. Cutkosky, A. Defazio, and H. Mehta, "Mechanic: A learning rate tuner," in *Advances in Neural Information Processing Systems*, 2023, vol. 36, p. 47828, [Online]. Available: https://proceedings.neurips.cc/paper_files/paper/2023/file/955499a8e2860ed746717c1374224c43-Paper-Conference.pdf.

21. S. Akhmedova and N. Körber, "Next generation loss function for image classification," Apr. 2024, [Online]. Available: https://arxiv.org/abs/2404.12948.
22. E. Mohamed, K. Sirlantzis, and G. Howells, "A review of visualisation-as-explanation techniques for convolutional neural networks and their evaluation," *Displays*, vol. 73, p. 102239, 2022, doi: 10.1016/j.displa.2022.102239.
23. H. M. Afify, K. K. Mohammed, and A. Ella Hassanien, "Novel prediction model on OSCC histopathological images via deep transfer learning combined with Grad-CAM interpretation," *Biomedical Signal Processing and Control*, vol. 83, p. 104704, 2023, doi: 10.1016/j.bspc.2023.104704.
24. Q. Chao, X. Wei, J. Tao, C. Liu, and Y. Wang, "Cavitation recognition of axial piston pumps in noisy environment based on Grad-CAM visualization technique," *CAAI Transactions on Intelligence* Technology, vol. 8, no. 1, pp. 206–218, 2023, doi: 10.1049/cit2.12101.
25. B. Liu, J. Guo, C. L. P. Chen, X. Wu, and T. Zhang, "Fine-grained interpretability for EEG emotion recognition: Concat-aided grad-CAM and systematic brain functional network," *IEEE Transactions on Affective Computing*, pp. 1–14, 2023, doi: 10.1109/TAFFC.2023.3288885.
26. H. Chen et al., "3-D Gabor-based anisotropic diffusion for speckle noise suppression in dynamic ultrasound images," *Physical and Engineering Sciences in Medicine*, vol. 44, no. 1, pp. 207–219, 2021, doi: 10.1007/s13246-020-00969-x.
27. B. Yin, L. Tran, H. Li, X. Shen, and X. Liu, "Towards interpretable face recognition," in *2019 IEEE/CVF International Conference on Computer Vision (ICCV)*, 2019, pp. 9347–9356, doi: 10.1109/ICCV.2019.00944.
28. D. Erhan, Y. Bengio, A. Courville, and P. Vincent, "Visualizing higher-layer features of a deep network," *University of Montpellier*, vol. 1341, no. 3, p. 1, 2009.
29. B. Zhou, A. Khosla, A. Lapedriza, A. Oliva, and A. Torralba, "Learning deep features for discriminative localization," Dec. 2015, doi: 10.48550/arXiv.1512.04150.
30. R. R. Selvaraju, A. Das, R. Vedantam, M. Cogswell, D. Parikh, and D. Batra, "Grad-CAM: Why did you say that?," Nov. 2016, [Online]. Available: https://arxiv.org/abs/1611.07450.
31. S. Camalan et al., "Convolutional neural network-based clinical predictors of oral dysplasia: Class activation map analysis of deep learning results," *Cancers*, vol. 13, no. 6, p. 1291, Mar. 2021, doi: 10.3390/cancers13061291.
32. L.-D. Quach, K. N. Quoc, A. N. Quynh, N. Thai-Nghe, and T. G. Nguyen, "Explainable deep learning models with gradient-weighted class activation mapping for smart agriculture," *IEEE Access*, vol. 11, pp. 83752–83762, 2023, doi: 10.1109/ACCESS.2023.3296792.
33. P.-T. Jiang, C.-B. Zhang, Q. Hou, M.-M. Cheng, and Y. Wei, "LayerCAM: Exploring hierarchical class activation maps for localization," *IEEE Transactions on Image Processing*, vol. 30, pp. 5875–5888, 2021, doi: 10.1109/TIP.2021.3089943.
34. B. J. Kim, G. Koo, H. Choi, and S. W. Kim, "Extending class activation mapping using Gaussian receptive field," *Computer Vision and Image Understanding*, vol. 231, p. 103663, 2023, doi: 10.1016/j.cviu.2023.103663.

35. A. Englebert, O. Cornu, and C. De Vleeschouwer, "Poly-CAM: High resolution class activation map for convolutional neural networks," Apr. 2022, [Online]. Available: https://arxiv.org/abs/2204.13359.
36. C. Zeng, K. Yan, Z. Wang, Y. Yu, S. Xia, and N. Zhao, "Abs-CAM: A gradient optimization interpretable approach for explanation of convolutional neural networks," *Signal, Image and Video Processing*, vol. 17, no. 4, pp. 1069–1076, 2023, doi: 10.1007/s11760-022-02313-0.
37. J. Shu, X. Yuan, D. Meng, and Z. Xu, "CMW-Net: Learning a class-aware sample weighting mapping for robust deep learning," *IEEE Transactions on Pattern Analysis and Machine. Intelligence*, vol. 45, no. 10, pp. 11521–11539, 2023, doi: 10.1109/TPAMI.2023.3271451.
38. L. Jiao, W. Gao, R. Bie, A. Umek, and A. Kos, "Golf guided Grad-CAM: Attention visualization within golf swings via guided gradient-based class activation mapping," *Multimedia Tools and Applications*, vol. 83, no. 13, pp. 38481–38503, 2024, doi: 10.1007/s11042-023-17153-4.
39. A. Jaiswal, N. Gianchandani, D. Singh, V. Kumar, and M. Kaur, "Classification of the COVID-19 infected patients using DenseNet201 based deep transfer learning," *Journal of Biomolecular Structure and Dynamics*, vol. 39, no. 15, pp. 5682–5689, Oct. 2021, doi: 10.1080/07391102.2020.1788642.
40. O. Habibi, M. Chemmakha, and M. Lazaar, "Performance evaluation of CNN and pre-trained models for malware classification," *Arabian Journal for Science and Engineering*, vol. 48, no. 8, pp. 10355–10369, 2023, doi: 10.1007/s13369-023-07608-z.
41. M. Altun, H. Gürüler, O. Özkaraca, F. Khan, J. Khan, and Y. Lee, "Monkeypox detection using CNN with transfer learning," *Sensors*, vol. 23, no. 4, 2023, doi: 10.3390/s23041783.
42. S. Maqsood and R. Damaševičius, "Multiclass skin lesion localization and classification using deep learning based features fusion and selection framework for smart healthcare," *Neural Networks*, vol. 160, pp. 238–258, 2023, doi: 10.1016/j.neunet.2023.01.022.
43. J. Ma, M. Liu, S. Hu, J. Fu, G. Chen, and A. Yang, "A novel CNN ensemble framework for bearing surface defects classification based on transfer learning," *Measurement Science and Technology*, vol. 34, no. 2, p. 25902, Nov. 2022, doi: 10.1088/1361-6501/ac9c22.
44. S. Rajkumar, V. A. Sairam, R. Saranya, N. Sandhiya, V. Shivanie, and V. Sapthagirivasan, "Automated KL grading of knee X-ray images using convolutional neural network," *International Journal of Biomedical Engineering and Technology*, vol. 45, no. 3, pp. 198–215, 2024, doi: 10.1504/IJBET.2024.138970.
45. H. Zhang and H. Zhang, "LungSeek: 3D Selective Kernel residual network for pulmonary nodule diagnosis," *Visual Computer*, 2022, doi: 10.1007/s00371-021-02366-1.
46. B. Singh and M. Kaur, "An approach for classification of malignant and benign microcalcification clusters," *Sādhanā*, vol. 43, no. 3, p. 39, 2018, doi: 10.1007/s12046-018-0805-2.
47. F. Alenezi, A. Armghan, and K. Polat, "Wavelet transform based deep residual neural network and ReLU based extreme learning machine for skin lesion classification," *Expert Systems with Applications*, vol. 213, p. 119064, 2023, doi: 10.1016/j.eswa.2022.119064.

48. Y. Lu and J. Han, "Cancer classification using gene expression data," *Information Systems*, vol. 28, no. 4, pp. 243–268, 2003, doi: 10.1016/S0306-4379(02)00072-8.
49. I. Guyon, J. Weston, S. Barnhill, and V. Vapnik, "Gene selection for cancer classification using support vector machines," *Machine Learning*, vol. 46, no. 1, pp. 389–422, 2002, doi: 10.1023/A:1012487302797.
50. K.-H. Liu and D.-S. Huang, "Cancer classification using rotation forest," *Computers in Biology and Medicine*, vol. 38, no. 5, pp. 601–610, 2008, doi: 10.1016/j.compbiomed.2008.02.007.
51. G. Viale, "The current state of breast cancer classification," *Annals of Oncology*, vol. 23, pp. x207–x210, 2012, doi: 10.1093/annonc/mds326.
52. A. K. Mishra, P. Roy, S. Bandyopadhyay, and S. K. Das, "Breast ultrasound tumour classification: A machine learning—radiomics based approach," *Expert System*, vol. 38, no. 7, Nov. 2021, doi: 10.1111/exsy.12713.
53. B. Mao, J. Ma, S. Duan, Y. Xia, Y. Tao, and L. Zhang, "Preoperative classification of primary and metastatic liver cancer via machine learning-based ultrasound radiomics," *European Radiology*, vol. 31, no. 7, pp. 4576–4586, Jul. 2021, doi: 10.1007/s00330-020-07562-6.
54. S. Bourouis, S. S. Band, A. Mosavi, S. Agrawal, and M. Hamdi, "Meta-heuristic algorithm-tuned neural network for breast cancer diagnosis using ultrasound images," *Frontiers Oncology*, vol. 12, Jun. 2022, doi: 10.3389/fonc.2022.834028.
55. M. I. Daoud, S. Abdel-Rahman, T. M. Bdair, M. S. Al-Najar, F. H. Al-Hawari, and R. Alazrai, "Breast tumor classification in ultrasound images using combined deep and handcrafted features," *Sensors*, vol. 20, no. 23, p. 6838, Nov. 2020, doi: 10.3390/s20236838.
56. Y. Luo, Q. Huang, and X. Li, "Segmentation information with attention integration for classification of breast tumor in ultrasound image," *Pattern Recognition*, vol. 124, p. 108427, Apr. 2022, doi: 10.1016/j.patcog.2021.108427.
57. A. Raza, N. Ullah, J. A. Khan, M. Assam, A. Guzzo, and H. Aljuaid, "DeepBreastCancerNet: A novel deep learning model for breast cancer detection using ultrasound images," *Applied Sciences*, vol. 13, no. 4, p. 2082, Feb. 2023, doi: 10.3390/app13042082.
58. K. Atrey, B. K. Singh, N. K. Bodhey, and R. B. Pachori, "Mammography and ultrasound based dual modality classification of breast cancer using a hybrid deep learning approach," *Biomedical Signal Processing Control*, vol. 86, p. 104919, 2023.
59. M. Khanna, A. Agarwal, L. K. Singh, S. Thawkar, A. Khanna, and D. Gupta, "Radiologist-level two novel and robust automated computer-aided prediction models for early detection of COVID-19 infection from chest X-ray images," *Arabian Journal for Science and Engineering*, 2021, doi: 10.1007/s13369-021-05880-5.
60. R. A. Hazarika, A. Abraham, D. Kandar, and A. K. Maji, "An improved leNet-deep neural network model for Alzheimer's disease classification using brain magnetic resonance images," *IEEE Access*, vol. 9, pp. 161194–161207, 2021, doi: 10.1109/ACCESS.2021.3131741.

61. S. Balasubramaniam, Y. Velmurugan, D. Jaganathan, and S. Dhanasekaran, "A modified leNet CNN for breast cancer diagnosis in ultrasound images," *Diagnostics*, vol. 13, no. 17, p. 2746, Aug. 2023, doi: 10.3390/diagnostics13172746.
62. S. Benyahia, B. Meftah, and O. Lézoray, "Multi-features extraction based on deep learning for skin lesion classification," *Tissue Cell*, vol. 74, p. 101701, 2022, doi: 10.1016/j.tice.2021.101701.
63. R. Sadik, A. Majumder, A. A. Biswas, B. Ahammad, and M. M. Rahman, "An in-depth analysis of convolutional neural network architectures with transfer learning for skin disease diagnosis," *Healthcare Analytics*, vol. 3, p. 100143, 2023, doi: 10.1016/j.health.2023.100143.
64. Y. Luo, Z. Lu, L. Liu, and Q. Huang, "Deep fusion of human-machine knowledge with attention mechanism for breast cancer diagnosis," *Biomedical Signal Processing Control*, vol. 84, p. 104784, 2023, doi: 10.1016/j.bspc.2023.104784.
65. A. A. Hekal, A. Elnakib, H. E.-D. Moustafa, and H. M. Amer, "Breast cancer segmentation from ultrasound images using deep dual-decoder technology with attention network," *IEEE Access*, vol. 12, pp. 10087–10101, 2024, doi: 10.1109/ACCESS.2024.3351564.
66. E. Y. Kalafi et al., "Classification of breast cancer lesions in ultrasound images by using attention layer and loss ensemble in deep convolutional neural networks," *Diagnostics*, vol. 11, no. 10, 2021, doi: 10.3390/diagnostics11101859.
67. Y. Zhao, X. Shen, J. Chen, W. Qian, L. Sang, and H. Ma, "Learning active contour models based on self-attention for breast ultrasound image segmentation," *Biomedical Signal Processing Control*, vol. 89, p. 105816, 2024, doi: 10.1016/j.bspc.2023.105816.
68. A. AlZoubi, F. Lu, Y. Zhu, T. Ying, M. Ahmed, and H. Du, "Classification of breast lesions in ultrasound images using deep convolutional neural networks: Transfer learning versus automatic architecture design," *Medical & Biology Engineering & Computing*, vol. 62, no. 1, pp. 135–149, 2024, doi: 10.1007/s11517-023-02922-y.
69. M. R. Islam et al., "Enhancing breast cancer segmentation and classification: An ensemble deep convolutional neural network and U-net approach on ultrasound images," *Machine Learning with Applications*, vol. 16, p. 100555, 2024, doi: 10.1016/j.mlwa.2024.100555.
70. K. S. Rao et al., "Intelligent ultrasound imaging for enhanced breast cancer diagnosis: Ensemble transfer learning strategies," *IEEE Access*, vol. 12, pp. 22243–22263, 2024, doi: 10.1109/ACCESS.2024.3358448.
71. M. G. Lanjewar, K. G. Panchbhai, and L. B. Patle, "Fusion of transfer learning models with LSTM for detection of breast cancer using ultrasound images," *Computers in Biology and Medicine*, vol. 169, p. 107914, 2024, doi: 10.1016/j.compbiomed.2023.107914.
72. A. Hijab, M. A. Rushdi, M. M. Gomaa, and A. Eldeib, "Breast cancer classification in ultrasound images using transfer learning," in *2019 Fifth International Conference on Advances in Biomedical Engineering (ICABME)*, 2019, pp. 1–4.

73. Z. Zhuang, Z. Yang, A. N. J. Raj, C. Wei, P. Jin, and S. Zhuang, "Breast ultrasound tumor image classification using image decomposition and fusion based on adaptive multi-model spatial feature fusion," *Computer Methods and Programs in Biomedicine*, vol. 208, p. 106221, Sep. 2021, doi: 10.1016/j.cmpb.2021.106221.
74. R. Soundrapandiyan, S. C. Satapathy, C. M. P.V.S.S.R., and N. G. Nhu, "A comprehensive survey on image enhancement techniques with special emphasis on infrared images," *Multimedia Tools and Applications*, 2021, doi: 10.1007/s11042-021-11250-y.
75. M. Ragab, A. Albukhari, J. Alyami, and R. F. Mansour, "Ensemble deep-learning-enabled clinical decision support system for breast cancer diagnosis and classification on ultrasound images," *Biology*, vol. 11, no. 3, p. 439, Mar. 2022, doi: 10.3390/biology11030439.
76. K. Jabeen et al., "Breast cancer classification from ultrasound images using probability-based optimal deep learning feature fusion," *Sensors*, vol. 22, no. 3, p. 807, Jan. 2022, doi: 10.3390/s22030807.
77. Q. Zheng, L. Wang, and F. Wang, "Object detection algorithm based on feature enhancement," *Measurement Science and Technology*, vol. 32, no. 8, p. 85401, May 2021, doi: 10.1088/1361–6501/abe740.
78. M. R. Hassan et al., "Prostate cancer classification from ultrasound and MRI images using deep learning based explainable Artificial Intelligence," *Future Generation Computer Systems*, vol. 127, pp. 462–472, Feb. 2022, doi: 10.1016/j.future.2021.09.030.
79. G. Ayana, J. Park, J.-W. Jeong, and S. Choe, "A novel multistage transfer learning for ultrasound breast cancer image classification," *Diagnostics*, vol. 12, no. 1, p. 135, Jan. 2022, doi: 10.3390/diagnostics12010135.
80. M.H. Yap, E. A. Edirisinghe, and H. E. Bez, "A novel algorithm for initial lesion detection in ultrasound breast images," *Journal of Applied Clinical Medical Physics*, vol. 9, no. 4, pp. 181–199, 2008, doi: 10.1120/jacmp.v9i4.2741.
81. W. Al-Dhabyani, M. Gomaa, H. Khaled, and A. Fahmy, "Dataset of breast ultrasound images," *Data Brief*, vol. 28, p. 104863, 2020, doi: 10.1016/j.dib.2019.104863.
82. S. Vitale, J. I. Orlando, E. Iarussi, and I. Larrabide, "Improving realism in patient-specific abdominal ultrasound simulation using CycleGANs," *International Journal of Computer Assisted Radiology and* Surgery, vol. 15, no. 2, pp. 183–192, 2020, doi: 10.1007/s11548-019-02046-5.
83. M. Li, L. Zhao, D. Zhou, R. Nie, Y. Liu, and Y. Wei, "AEMS: An attention enhancement network of modules stacking for lowlight image enhancement," *Visual Computer*, vol. 38, no. 12, pp. 4203–4219, 2022, doi: 10.1007/s00371-021-02289-x.
84. H. C. Reis, V. Turk, K. Khoshelham, and S. Kaya, "InSiNet: A deep convolutional approach to skin cancer detection and segmentation," *Medical & Biological Engineering & Computing*, vol. 60, no. 3, pp. 643–662, 2022, doi: 10.1007/s11517-021-02473-0.
85. S. Yadav et al., "Performance analysis of deep neural networks through transfer learning in retinal detachment diagnosis using fundus images," *Sādhanā*, vol. 47, no. 2, p. 49, 2022, doi: 10.1007/s12046-022-01822-5.

86. T. Sugashini and G. Balakrishnan, "YOLO glass: Video-based smart object detection using squeeze and attention YOLO network," *Signal, Image and Video Processing*, Dec. 2023, doi: 10.1007/s11760-023-02855-x.
87. N. Gessert, M. Nielsen, M. Shaikh, R. Werner, and A. Schlaefer, "Skin lesion classification using ensembles of multi-resolution EfficientNets with meta data," *MethodsX*, vol. 7, p. 100864, 2020, doi: 10.1016/j.mex.2020.100864.
88. X. Lu, X. Xie, C. Ye, H. Xing, Z. Liu, and Y. Chen, "Single-image super-resolution via a lightweight convolutional neural network with improved shuffle learning," *Signal, Image and Video Processing*, vol. 18, no. 1, pp. 233–241, Feb. 2024, doi: 10.1007/s11760-023-02730-9.
89. V. K. Sharma, "CNTFET circuit-based wide fan-in domino logic for low power applications," *Journal of Circuits, Systems and Computers*, vol. 31, no. 02, p. 2250036, 2022.
90. V. K. Sharma, and M. Pattanaik, "A reliable ground bounce noise reduction technique for nanoscale CMOS circuits," *International Journal of Electronics*, vol. 102, no. 11, pp. 1852–1866, 2015.
91. V. K. Sharma, S. Patel, and M. Pattanaik, "High performance process variations aware technique for sub-threshold 8T-SRAM cell," *Wireless Personal Communications*, vol. 78, no. 1, pp. 57–68, 2014.
92. V. K. Sharma and M. Pattanaik, "Techniques for low leakage nanoscale VLSI circuits: A comparative study," *Journal of Circuits, Systems and Computers*, vol. 23, no. 05, p. 1450061, 2014.
93. V. K. Sharma, "A survey of leakage reduction techniques in CMOS digital circuits for nanoscale regime," *Australian Journal of Electrical and Electronics Engineering*, vol. 18, no. 4, pp. 217–236, 2021.
94. C. Kumar, A. S. Mishra, and V. K. Sharma, "Leakage power reduction in CMOS logic circuits using stack ONOFIC technique," in *2018 Second International Conference on Intelligent Computing and Control Systems (ICICCS)*, 2018.
95. V. K. Sharma and M. Pattanaik, "Design of low leakage variability aware ONOFIC CMOS standard cell library," *Journal of Circuits, Systems and Computers*, vol. 25, no. 11, p. 1650134, 2016.
96. S. F. Naz, S. Riyaz, and V. K. Sharma, "A review of QCA nanotechnology as an alternate to CMOS," *Current Nanoscience*, vol. 18, no. 1, pp. 18–30, 2022.
97. S. Riyaz and V. K. Sharma, "Design of reversible Feynman and double Feynman gates in quantum-dot cellular automata nanotechnology," *Circuit World*, vol. 49, no. 1, pp. 28–37, 2023.
98. B. Raina, C. Verma, M. Gupta, and V. K. Sharma, "Binary coded decimal (BCD) seven segment circuit designing using quantum-dot cellular automata (QCA)," in *2021 5th International Conference on Trends in Electronics and Informatics (ICOEI)*, 2021.
99. U. Mushtaq and V. K. Sharma, "Performance analysis for reliable nanoscaled FinFET logic circuits," *Analog Integrated Circuits and Signal Processing*, vol. 107, no. 3, pp. 671–682, 2021.
100. Kajal and V. K. Sharma, "Design and simulation of FinFET circuits at different technologies," in *2021 6th International Conference on Inventive Computation Technologies (ICICT)*, 2021.
101. V. K. Sharma, "Designing of face recognition system," in *2019 International Conference on Intelligent Computing and Control Systems (ICCS)*, 2019.

Chapter 9

Empowering healthcare analysis with AI-enhanced decision support

A comparative study using machine learning

Kavya Varshini Anburasu, Jayaashri Chezhian, and Padmini Sankaramurthy

9.1 INTRODUCTION

In an age marked by rapid technological advancement, the convergence of Artificial Intelligence (AI) and healthcare heralds a transformative era in patient care and medical diagnosis. This chapter delves into the development and implementation of cutting-edge AI-driven solutions in healthcare diagnostics, exploring their potential to reshape the landscape of medical data management and utilization.

Precision, efficiency, and swift diagnosis have always been pivotal in the realm of healthcare, striving to save lives and enhance patient outcomes. These critical factors stand poised for unprecedented elevation through the integration of AI technology, equipping healthcare professionals and caregivers with a formidable arsenal to make informed decisions. The core objective of this chapter is to elevate the precision and accuracy of medical diagnoses.

The realm of healthcare, a vital cornerstone of human well-being and societal progress, is undergoing a groundbreaking transformation, propelled by the relentless march of technological innovation. At the forefront of this transformation is the burgeoning field of AI, which is rapidly converging with healthcare to herald a new era in patient care and medical diagnostics. This chapter seeks to explore the intricate tapestry of this convergence, focusing on the development and implementation of AI-driven solutions in healthcare diagnostics. As we embark on this exploration, it is imperative to recognize that the integration of AI into healthcare is not just a mere addition of a technological tool, but a paradigm shift that promises to reshape the very landscape of medical data management and utilization.

In the traditional healthcare model, precision, efficiency, and swift diagnosis have always been the linchpins of effective patient care. These elements are critical in saving lives and enhancing patient outcomes, serving

DOI: 10.1201/9781003643609-9

as the guiding principles for healthcare practitioners worldwide. However, with the advent of AI, these critical factors stand on the brink of an unprecedented elevation. The integration of AI technology in healthcare equips professionals and caregivers with a formidable arsenal, enabling them to make more informed and accurate decisions. The core objective of this chapter is to underscore and elevate the precision and accuracy of medical diagnoses through AI, a goal that is rapidly transitioning from a futuristic vision to an attainable reality.

To appreciate the transformative potential of AI in healthcare, it is essential to understand the scope and scale of its application. AI, with its ability to process vast amounts of data and learn from it, is particularly well-suited to tackle the complexities inherent in medical diagnostics. In an industry where every data point can be a matter of life and death, AI's prowess in pattern recognition and predictive analytics can lead to earlier and more accurate diagnoses, potentially saving countless lives. Moreover, AI's capacity to continuously learn and adapt means that healthcare systems can become more efficient and effective over time, staying abreast of the latest medical research and trends.

This chapter will delve into various facets of AI application in healthcare diagnostics, from image-based diagnostics to genetic sequencing and predictive modeling. Each of these areas presents unique challenges and opportunities, and the integration of AI promises to enhance the capabilities of healthcare professionals in these domains. For instance, in image-based diagnostics, AI algorithms are increasingly being used to interpret medical images such as X-rays, magnetic resonance imaging (MRI), and computed tomography (CT) scans with a level of precision and speed unattainable by human radiologists. In the realm of genetic sequencing, AI is helping to unravel the complex patterns of genetic data, paving the way for personalized medicine and treatments tailored to the genetic makeup of individual patients.

However, the path to integrating AI into healthcare is fraught with challenges. Ethical considerations, data privacy, the need for robust and unbiased algorithms, and the imperative of maintaining the human element in healthcare are just some of the hurdles that need to be addressed. This chapter will not shy away from these challenges but will instead highlight them as crucial areas for discussion and development. It is only by acknowledging and addressing these challenges that the full potential of AI in healthcare can be realized.

At the cusp of a new era in healthcare, marked by the integration of AI, it is crucial to recognize the transformative impact this will have on patient care and medical diagnostics. The journey ahead is complex and filled with both opportunities and challenges. This chapter aims to serve as a beacon, guiding readers through the intricacies of AI in healthcare diagnostics, illuminating the path forward, and fostering a deeper understanding of how this technology can be harnessed to revolutionize patient care and save lives.

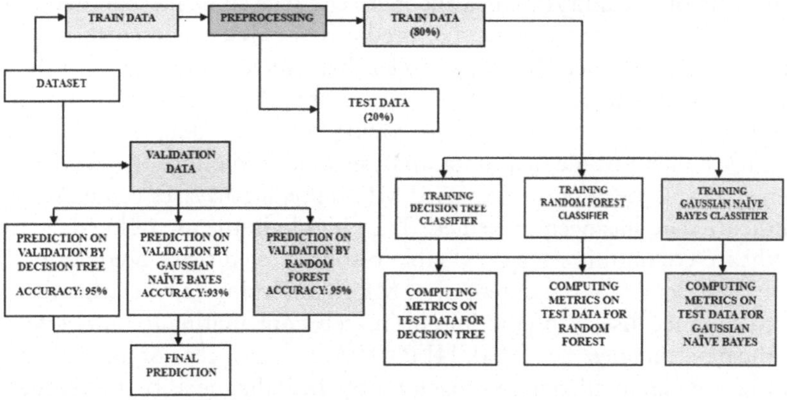

Figure 9.1 Model evaluation process.

The journey of integrating AI into healthcare is not just a technological endeavor; it is a journey of hope, innovation, and a testament to the relentless human pursuit of bettering lives through science and technology.

The block diagram as shown in Figure 9.1 depicts a systematic process for data analysis and predictive modeling using machine learning (ML) techniques. The 80% dataset has been used for training and remaining dataset for testing purpose. The training data undergoes preprocessing to prepare it for analysis, a crucial step for ensuring data quality and feature selection. This training data is then used to train two different types of classifiers: a Decision Tree (DT) Classifier and a Gaussian Naïve Bayes (GNB) Classifier. Meanwhile, the validation data is employed to predict outcomes using both a DT and a Random Forest (RF) algorithm, with both methods achieving a high accuracy of 95%. In contrast, the GNB model predicts with a slightly lower accuracy of 93% on the validation data. Following these procedures, metrics are computed on a separate test dataset (20% of the original dataset) for both the DT and the RF classifiers, as well as for the GNB classifier. The block diagram culminates with the final prediction, although the chart does not specify which classifier's prediction is used as the final output. The decision likely depends on the performance metrics computed from the test data, which would inform the best-performing model to be used for the final prediction.

9.2 LITERATURE SURVEY

The research conducted in this field previously is the subject of discussion in this section.

Choi et al. [1] introduce a predictive model that utilizes Recurrent Neural Network (RNN) to forecast clinical events. This research highlights the application of ML and AI in healthcare to enhance clinical event prediction, contributing to the growing field of predictive analytics in medical settings. Ma et al. [2] introduce Dipole, a model that utilizes bidirectional RNN to enhance diagnostic accuracy. This research highlights the use of advanced neural network architectures for improving healthcare diagnostics, contributing to the field of medical data analysis and prediction. In the paper by Chen et al. [3], the application-based ML to predict diseases using extensive healthcare data has been presented. They highlight the use of big data from healthcare communities as a valuable resource for disease prediction. The research underscores the potential of ML in analyzing vast healthcare datasets to enhance disease prediction, ultimately contributing to more effective healthcare strategies. Das et al. [4] introduce the concept of an AI Doctor. They present an intelligent approach for medical diagnosis that leverages AI technology. The chapter emphasizes the potential of AI in enhancing and revolutionizing the process of medical diagnosis, offering innovative solutions for healthcare. Le Nguyen [5] discusses how blockchain can provide benefits for both patients and healthcare providers by enhancing data security, interoperability, and trust in medical records. It highlights the emerging role of blockchain in the current scenario. Vásquez-Morales et al. [6] incorporated renal disease prediction using an explainable prediction model for CRD. Dias and Torkamani [7] provide insights into how AI can enhance the accuracy and efficiency of diagnosing medical conditions and interpreting genomic data. The article underscores the growing significance of AI in revolutionizing healthcare and genomics, offering promising opportunities for more precise and personalized diagnostics. Kaur et al. [8] provide an overview of AI-based medical diagnostic systems. They discuss the principles and perspectives of utilizing AI algorithms for medical diagnosis. The paper emphasizes the potential of AI in improving healthcare by enhancing diagnostic accuracy and efficiency, highlighting its significance in the field of medical diagnostics. Jain et al. [9] propose the development of a smart integrated platform that utilizes AI to facilitate interactions between doctors and patients. The paper outlines a prototype for this platform, that aims to enhance communication and collaboration within the healthcare system. Endargiri and Laabidi [10] introduce an automatic framework for diagnosis and prediction using AI-based multi-classification algorithms in healthcare. This research focuses on the application of AI in improving healthcare diagnostics and predictive assessments.

Sankaramurthy et al. [11] and Sharmini et al.[12] combine Internet of Things (IoT) and sensors for varied application. Padmini et al. discuss the development of pulse oximetry in medical diagnosis and monitoring of greenhouse gases using sensors which is the need of the day. They highlight the potential of AI to enhance the accuracy and efficiency of medical diagnoses. The research emphasizes the role of AI in improving healthcare by providing advanced diagnostic tools and applications. Sushma et al. [13] introduce DocOnTap, which combines AI technology to assist in diagnosing diseases and providing recommendations for healthcare. The research emphasizes the role of AI in improving disease diagnostics and healthcare recommendations, showcasing its potential impact in the field of medical technology. Comito et al. [14] discuss how AI can significantly enhance decision-making process in the medical field. ul Abideen et al. [15] designed an AI model to provide early detection and offer an explanation for its diagnosis. Kumar and Joshi [16] introduce the concept of mobile technology to make AI medical diagnosis and advice accessible to those with limited access to traditional healthcare services. The research highlights the potential of AICOM in addressing healthcare disparities and extending medical support to vulnerable communities.

The extensive literature survey presents a comprehensive and insightful view of the current advancements and applications of healthcare using AI. The overarching theme of these studies is the transformative role AI plays in enhancing various aspects of medical diagnostics, patient care, and healthcare management.

Papers by Choi et al., Ma et al., Chen et al., and Das et al. [1–4] delve into the development of AI-driven predictive models using advanced techniques such as RNN, attention-based bidirectional neural networks, and the concept of an AI Doctor. These studies highlight the efficacy of AI in forecasting clinical events, improving diagnostic accuracy, and leveraging large healthcare datasets for disease prediction. The use of RNN and other sophisticated algorithms in these models signifies a move toward more nuanced and accurate analysis in medical diagnostics, emphasizing AI's capability to handle complex healthcare data efficiently.

The exploration of blockchain in healthcare, as discussed by Le Nguyen [5], introduces an innovative approach to enhancing data security and interoperability in medical records. This application of blockchain technology reflects a growing trend toward securing sensitive health data and fostering trust in digital healthcare systems.

Papers by Dias and Torkamani, Kaur et al., Jain et al., Endargiri and Laabidi, Sankaramurthy et al., Sharmini et al., and Padmini et al. [7,8–12] discuss the broad applications of AI in medical diagnosis and genomics, underscoring the technology's role in enhancing diagnostic accuracy and efficiency. The emphasis on genomics and personalized medicine in these papers indicates a shift toward more individualized healthcare, where treatments and diagnoses are tailored to the genetic makeup of patients.

The integration of AI with IoT for smart healthcare systems, as presented by Sankaramurthy et al. [11], and the development of AI-based multi-classification algorithms in healthcare diagnosis and prediction by Endargiri and Laabidi [10] demonstrate the synergy between AI and other technological advancements. These studies suggest a future where healthcare is not only intelligent but also interconnected and comprehensive.

Jain et al., [9], Sharmini et al. [12], Padmini et al., Sushma et al. [13], and Comito et al. [14] explore various facets of AI application in healthcare, from developing smart platforms for doctor-patient interaction to providing a comprehensive review of AI research in disease diagnostics. The focus on patient similarity in clinical decision support systems, as discussed in by Padmini et al., and the introduction of DocOnTap by Sushma et al. [13] reflect an increasing trend toward personalized healthcare, where AI tools are used to tailor medical advice and diagnoses to individual patient needs.

Finally, ul Abideen et al. [15] and Kumar and Joshi [16] address the critical aspects of early disease diagnosis and extending healthcare access to underserved communities. The development of an AI model for early CKD diagnosis by ul Abideen et al. [15] and the use of mobile technology to democratize AI medical diagnosis by Kumar and Joshi [16] are indicative of AI's potential to not only improve healthcare outcomes but also address healthcare disparities.

In conclusion, this literature survey presents a vivid picture of a rapidly evolving healthcare landscape shaped by AI innovations. The diverse applications of AI, from predictive analytics to enhancing doctor-patient interactions and advancing personalized medicine, signal a future where healthcare is more accurate, efficient, and accessible. This AI combination with other technologies further amplifies its potential impact. However, as AI continues to advance, it brings forth challenges in data privacy, ethical considerations, and the need to balance technology with the human touch in healthcare. Addressing these challenges will be crucial as we navigate toward an AI-integrated future in healthcare, aiming to enhance patient care and medical outcomes worldwide.

9.3 PROPOSED METHODOLOGY

The dataset for this project was compiled from a study conducted by the University of Columbia at the New York Presbyterian Hospital in 2004 [17]. It encompasses a wide range of symptoms and their associated diseases, providing a robust foundation for training and testing ML models in medical diagnostics. For the purpose of the project, the dataset was partitioned into training and testing sets. While the exact percentages used for training and testing vary depending on the specific requirements and objectives of the study, a common practice is to allocate approximately 70%–80% of the data for training and the remaining 20%–30% for testing.

This sample data shown in Table 9.1 showcases the diversity of symptoms and diagnoses present in the dataset.

9.3.1 Decision tree algorithm

1. A basic DT construction process begins by examining the dataset, checking if all data samples have the same class.
2. If so, it creates a leaf node with that class label.
3. If there are no more attributes to split on, it generates a leaf node with the majority class label.

Table 9.1 Sample data from the healthcare dataset

Symptom 1	*Symptom 2*	*Symptom 3*	*Diagnosis*
Joint swelling	Muscle pain	Loss of balance	Vertigo
Headache	Nausea	Blurred vision	Migraine
Fever	Chills	Sweating	Flu
Cough	Sore throat	Runny nose	Common cold
Fatigue	Weight loss	Shortness of breath	Tuberculosis

4. Otherwise, it selects the best attribute for splitting and creates a decision node based on this attribute.
5. The data is then divided into subsets according to the chosen attribute's values, and the algorithm recursively constructs decision nodes and leaf nodes for each subset. This process continues until the tree is fully developed.
6. The resulting DT is a structured model for making decisions based on input data, facilitating classification and regression tasks.

In the algorithm for predicting arthritis through a DT, the initial step involves assessing stopping criteria, such as maximum depth or minimum samples per leaf. These criteria determine whether a node should become a leaf node or not. If all samples within a node share the same class, a leaf node is created with that specific class.

Next, the algorithm iterates over each feature—Spinning Moments, Loss of Balance, Muscle Weakness, and Stiff Neck—and calculates a split metric, such as Gini impurity or information gain, for every possible split. The feature and split point that minimize impurity or maximize information gain are then selected. The dataset is subsequently split into two subsets based on this chosen feature and split point.

This process is applied recursively to each subset, leading to the creation of child nodes and further branching of the tree. In essence, resulting DT offers a systematic approach to predicting whether a patient has arthritis based on the presented symptoms, providing interpretability and insights into the decision-making process.

The description of how the DT algorithm works based on Table 9.2 describing the sample data and Figure 9.2 explaining the working of DT algorithm for finding the disease arthritis.

Step 1: The DT algorithm starts by analyzing the entire dataset. It selects a feature from the dataset, such as "Spinning Moments" or "Stiff Neck," based on a chosen criterion (e.g., Gini impurity or information gain).

Step 2: Once a feature is selected, the algorithm calculates a split metric (e.g., Gini impurity or information gain) for potential splits on that

Table 9.2 Dataset table for arthritis

Patient	*Spinning moments*	*Loss of balance*	*Muscle weakness*	*Stiff neck*	*Arthritis (label)*
1	Yes	Yes	No	Yes	Yes
2	No	No	Yes	No	No
3	Yes	Yes	Yes	Yes	Yes
4	No	No	No	No	No
5	Yes	Yes	No	No	Yes

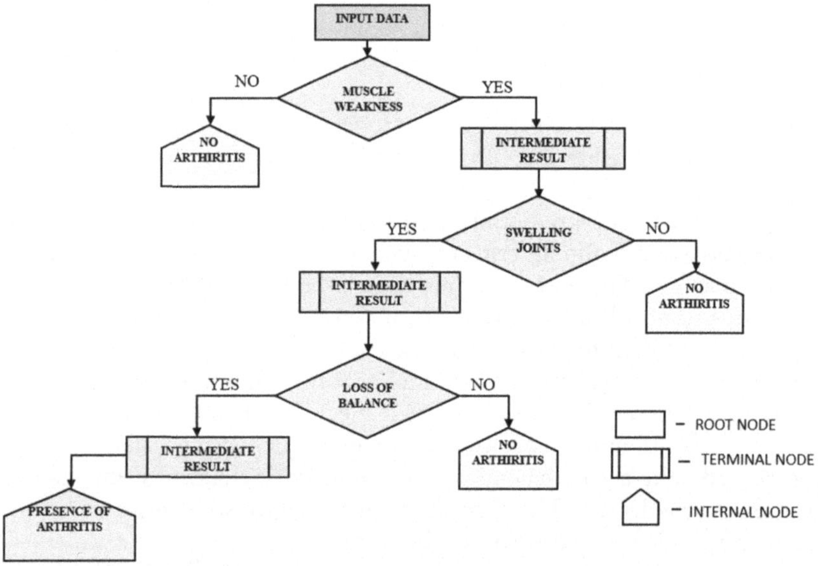

Figure 9.2 Working of Decision Tree Algorithm for finding the disease arthritis.

feature. For example, it might be found that splitting by "Spinning Moments" initially and then "Stiff Neck" results in the best separation of classes (e.g., "Yes" or "No" for arthritis) within the dataset.

Step 3: The dataset is divided into subsets based on the chosen feature and split point. This process continues recursively, creating branches and nodes in the tree. A leaf node is created when a stopping criterion is met, such as when all samples in a node share the same class or when the tree reaches a specified maximum depth.

Step 4: Once the DT is constructed, it can be used for making predictions. When a new patient's symptoms are evaluated, the tree follows the branches and nodes based on the patient's feature values. Eventually, the patient is assigned to a leaf node, and the class label associated with that leaf node becomes the final prediction. For example, if the path leads to a leaf node where the majority of samples are labeled "Yes" for arthritis, the DT predicts "Yes" for this patient.

The DT algorithm makes predictions based on a single tree's logic and structure. While it can be prone to overfitting if not properly controlled, it provides transparency and interpretability in understanding the decision-making process.

9.3.2 Random forest

1. There are two main functions that are used in creating a RF they build and predict.
2. The build function creates an RF by generating a specified number of DTs.
3. Each tree is constructed using features that are created at random and a bootstrapped subset of the training data, ensuring diversity among the trees.
4. The predict function uses the ensemble of DTs to make predictions for new data.
5. It tallies the votes from individual trees and selects the highest count of vote class count as the result.

In the RF approach for arthritis prediction, the algorithm begins by setting up multiple DTs, each with its own stopping criteria, such as maximum depth or minimum samples per leaf. These criteria are crucial as they dictate when a node in each DT becomes a leaf node, grouping the main samples within that node. If all samples in a node belong to the same class, the node is assigned as a leaf with that specific class.

Unlike a single DT, RF leverages the strength of multiple trees, each constructed by considering a random subset of features—such as Spinning Moments, Loss of Balance, Muscle Weakness, and Stiff Neck—and a random subset of the data. For each tree, the algorithm iterates over its subset of features, calculating a split metric like Gini impurity or information gain for every potential split. It then chooses the feature and split point that most effectively split the data, either by minimizing impurity or maximizing information gain.

This splitting process is recursively applied to each subset generated, leading to the formation of child nodes and further branching within each tree. However, in contrast to a single DT, the RF algorithm aggregates the output from all its trees. The aggregation could be through a majority voting system, where the final arthritis prediction for a patient is determined by the majority prediction among all trees.

The ensemble nature of RF enhances not only the model's accuracy but also its generalization capability. Each tree in the forest potentially captures

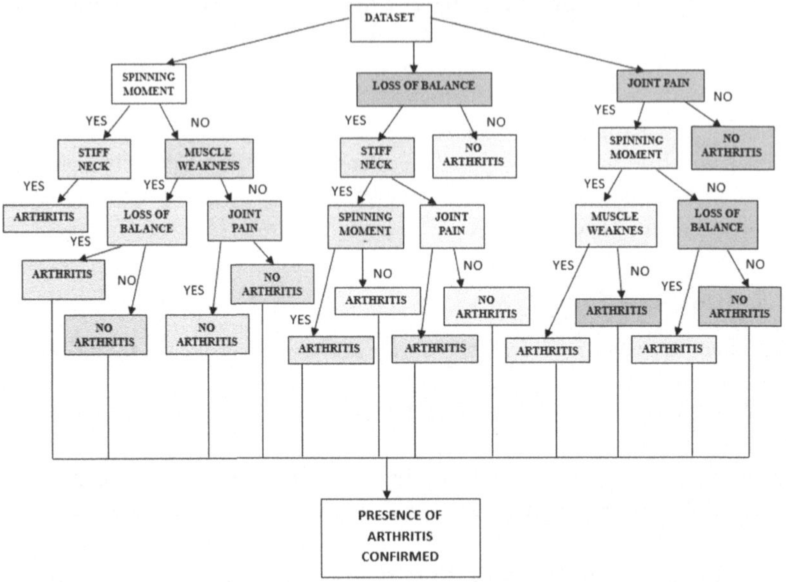

Figure 9.3 Working of Random Forest Tree Algorithm for finding the disease arthritis.

different aspects or patterns in the data, making the overall model more robust against overfitting. The final output is not just a single tree but a forest of them, each contributing to a more comprehensive and reliable prediction. In this way, the RF algorithm offers a powerful and nuanced approach to predicting arthritis, harnessing the collective insights from multiple decision-making processes based on the presented symptoms. Figure 9.3 shows the working of RF Tree Algorithm for finding the disease arthritis.

The workflow of the RF algorithm is as follows:

Step 1: The RF algorithm begins by creating multiple DTs. For instance, one tree might be built using data from patients 1, 3, and 5 and features like Spinning Moments and Stiff Neck. Another tree might use patients 2, 4, and 5 with features like Loss of Balance and Muscle Weakness.

Step 2: Each tree in the forest assesses its subset of features and calculates a split metric (like Gini impurity or information gain) for potential splits. For example, one tree might find that splitting by "Spinning Moments" at first and then "Stiff Neck" minimizes impurity or maximizes information gain in its subset of the data.

Step 3: The dataset is split into subsets based on the chosen feature and split point in each tree. This process is recursive. A leaf node is created when all samples in a node share the same class (Yes or No for arthritis) or when stopping criteria like maximum depth are met.

Step 4: Once all trees in the forest have made their predictions, the algorithm aggregates these predictions. This is typically done through majority voting. For example, if a new patient's symptoms are most commonly classified as "Yes" for arthritis in the majority of the trees, then the final output of the RF algorithm for this patient would be "Yes."

The RF algorithm's strength lies in its ensemble approach, where the collective decision-making of multiple trees leads to a more accurate and robust prediction compared to a single DT. This approach helps in handling the variability in the dataset and reduces the risk of overfitting, providing a reliable tool for medical diagnosis, like predicting arthritis.

9.3.3 K-nearest neighbors

1. The K-Nearest Neighbor (KNN) algorithm works by calculating the distances between a new data point and all points in a dataset used for training.
2. The algorithm selects the K-nearest data points, counts the occurrences of each class among them, and predicts the class among its closest neighbors.
3. KNN's simplicity and effectiveness make it a popular choice for classification tasks, where the class point is determined by its neighbors at its nearest in the workspace. Figure 9.4 depicts the workflow of KNN algorithm for arthritis.

The stepwise algorithm of KNN is as follows:

The dataset comprises several patients, each with a set of features (Spinning Moments, Loss of Balance, Muscle Weakness, and Stiff Neck) and a label indicating whether they have arthritis. In a KNN context, each patient is considered a point in a multidimensional feature space.

Step 1: Each feature (symptom) is a dimension in space. For binary features like the ones in our table, "Yes" might be represented by a 1 and "No" by a 0. This way, each patient is a point in a four-dimensional space.

Step 2: When a new patient disease needs to be classified, their feature set is compared with existing points in the dataset.

Step 3: The "K" nearest neighbors (patients with the most similar symptoms) to the new patient are identified. The value of "K" is pre-selected based on different factors through cross-validation.

Step 4: The algorithm then looks at the "Arthritis (Label)" of the "K" nearest neighbors. It applies a majority vote to determine the new patient's label. If, for example, we chose $K=3$ and two or all three of the nearest neighbors have arthritis, the new patient would be classified as having arthritis.

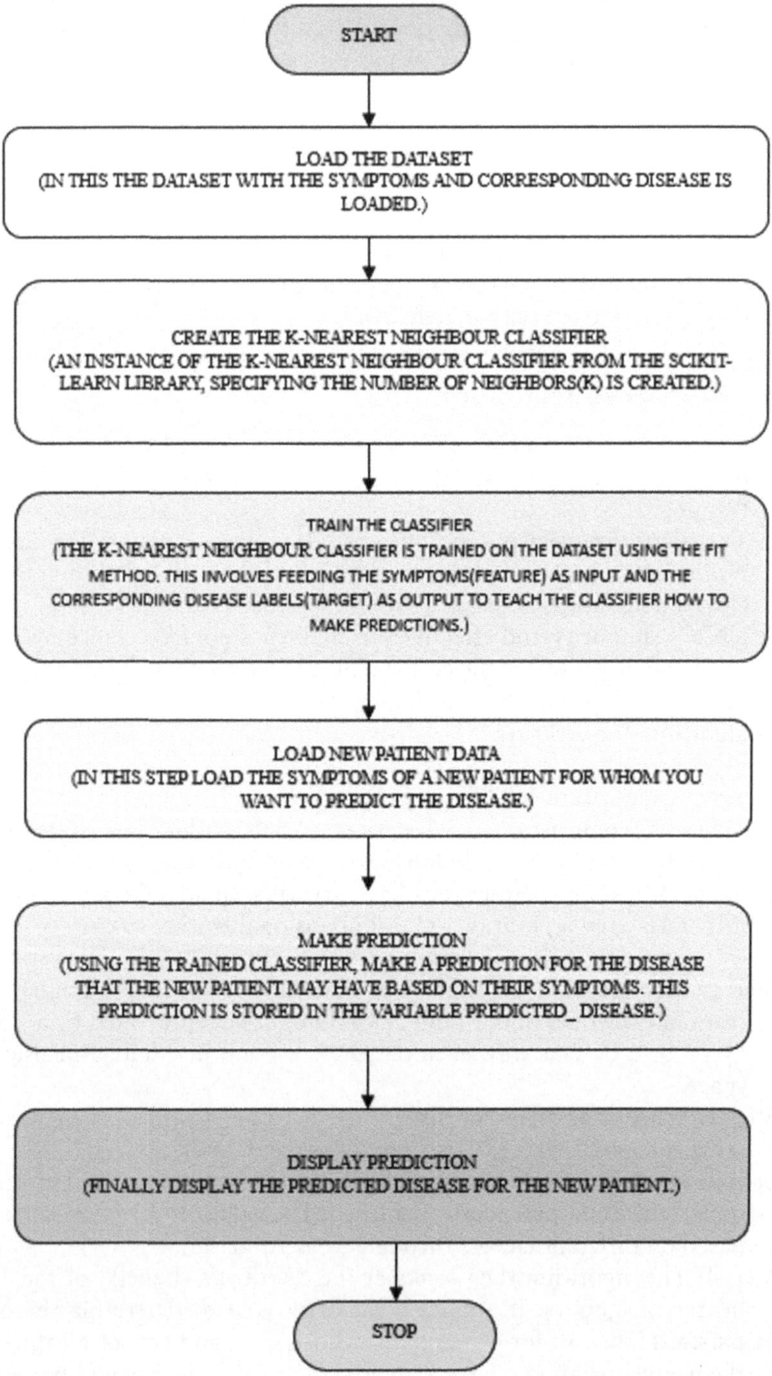

Figure 9.4 Working of K-Nearest Neighbor Algorithm for finding the disease arthritis.

Step 5: The outcome of the majority vote becomes the prediction for the new patient's arthritis status.

Through this process, KNN uses the entire dataset as a reference framework to classify new instances based on similarity in symptoms to previously diagnosed patients. It's a method that relies on the assumption that similar symptom profiles are likely to result in similar diagnoses, making it a practical choice for medical diagnosis tasks such as predicting arthritis.

9.3.4 A Gaussian Naïve Bayes

1. The GNB classification algorithm is a probabilistic method commonly used for classifying data with continuous features. The algorithm consists of two main functions: they are the train and predict function.
2. The train function is responsible for training the classifier. It initializes dictionaries to store class-specific statistics, including class means, variances, and priors.
3. Class priors are calculated based on the training data, and the means and variances for each feature within each class are computed. These statistics are vital for probability estimation.
4. The predict function is used to make predictions on new data. It calculates the posterior probability for each class and selects the class with the highest posterior probability as the predicted class for the new data point.
5. It employs the Gaussian (normal) distribution probability density function for feature likelihood estimation.
6. Additionally, there is a helper function that calculates the Gaussian probability density for a given feature value within a class. GNB assumes that features follow a Gaussian distribution within each class and makes classification decisions based on probabilistic reasoning. Figure 9.5 shows the GNB algorithm workflow for the arthritis disease.

The stepwise GNB algorithm for the test case is as follows:

GNB is a probabilistic classifier that assumes the features are normally distributed and independent given the class label. In our case, the features are the symptoms a patient presents, and the class label is the presence or absence of arthritis.

Step 1: For continuous data, GNB assumes the value of a feature is sampled from a Gaussian (normal) distribution. Since our example uses binary features, we would normally use the Bernoulli Naïve Bayes variant. However, for the sake of this explanation, let's assume we've encoded "Yes" and "No" as numerical values and we model them as normally distributed for each class of arthritis and non-arthritis:

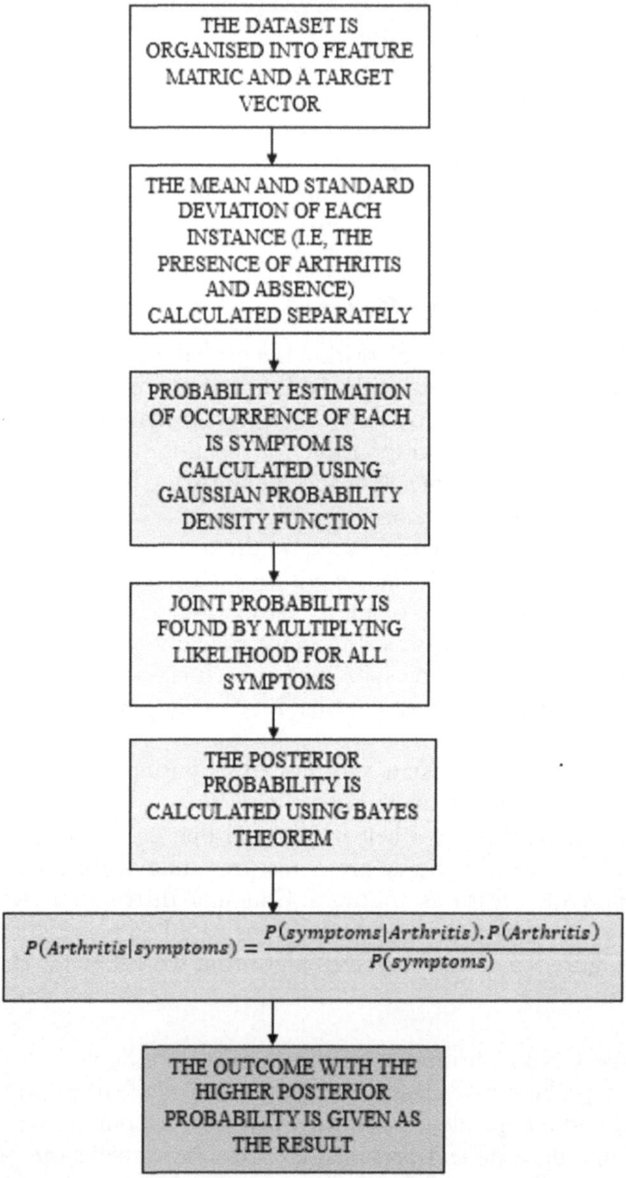

Figure 9.5 Working of Gaussian Naïve Bayes Algorithm for finding the disease arthritis.

For each feature *f* and each class *c* (arthritis "Yes" or "No"), we compute the mean $\mu_{f,c}$ and variance $\sigma_{f,c}{}^2$ of *f* in class *c*.

Step 2: The prior probability of each class, *P*(*c*), is calculated. This is simply the proportion of patients with and without arthritis in the training dataset.

Step 3: The likelihood of observing a set of features given a class *P*(features | *c*) is calculated using the Gaussian probability density function for each feature:

$$P(c) = \frac{1}{\sqrt{2\pi\sigma_{f,c}^2}} e^{\frac{\left(-x_i-\mu_{f,c}\right)^2}{2\sigma_{f,c}^2}} \tag{9.1}$$

Step 4: For a new patient, the posterior probability *P*(*c*|features) is given by the following equation:

$$P(\text{features}) = \frac{P(c) \times P(c)}{P(\text{features})} \tag{9.2}$$

Since *P*(features) is constant for all classes, we only need to focus on the numerator of Bayes' theorem.

Step 5: The highest posterior probability which has the highest value class is chosen as the prediction for the new patient disease diagnosis. In other words, if *P*(Yes|features)>*P*(No|features), then the patient is predicted to have arthritis.

9.4 SIMULATION RESULTS

In the subsequent sections, the outcomes stemming from an intricate AI-powered medical diagnosis simulation are unveiled. The proposed approach has diligently adhered to a comprehensive eight-phase methodology, and this segment is dedicated to the findings derived from the proposed AI model. The primary mission was to attain precise disease diagnoses, a task tackled through a fusion of meticulous data transformation, robust model definition, and intensive training. The results that are presented showcase the model's effectiveness in achieving this essential objective. Through rigorous testing, insights into the model's performance is obtained, offering a deep understanding of its accuracy.

The performance metrics that was chosen to evaluate the model are accuracy score, precision score, recall score, and F1 score.

9.4.1 Table comparison

Table 9.3 informs the comparison performance metrics of the proposed model. The performance metrics of the AI-driven medical diagnosis system underscore its efficiency and effectiveness in healthcare diagnostics. Various ML models are employed with each demonstrating a high degree of reliability and accuracy. The DT and RF models, in particular, stand out for their exceptional performance.

Table 9.3 Comparison of performance metrics of model

S. no.	*Algorithm*	*Accuracy score*	*Precision score*	*Recall-score*	*F1 score*
1.	RF	0.95	0.92	0.93	0.95
2.	DT	0.95	0.92	0.92	0.91
3.	GNB	0.93	0.95	0.94	0.92
4.	KNN	0.92	0.95	0.91	0.91

The DT model, known for its simplicity and interpretability, achieved a notable accuracy score of 0.95. This high score is indicative of the model's ability to make accurate predictions in diagnosing medical conditions.

The RF model slightly surpassed the DT model in terms of overall performance. With an accuracy score also at 0.95, it demonstrates a robustness that is characteristic of ensemble models. RFs are particularly effective in reducing overfitting, a common issue in DT models. By aggregating the results of various trees, RF ensures a more generalized and reliable prediction.

The GNB model also performed commendably, achieving an accuracy of 0.93. This model is particularly efficient in handling large datasets and provides good results even with a relatively simple algorithmic structure. The high precision and recall values, leading to an F1 score of 0.92, further validate its effectiveness in medical diagnosis applications.

Lastly, the KNN model has resulted in a respectable F1 score of 0.91. KNN is a non-parametric method that is particularly useful in scenarios where the data distribution is unknown. Its slightly lower accuracy in this context can be attributed to its sensitivity to the local structure of the data, which can vary significantly in medical datasets.

In summary, the integration of these diverse ML models in our AI-driven medical diagnosis system demonstrates a comprehensive approach to enhancing diagnostic accuracy. Each model brings its unique strengths to the table, contributing to a system that is not only highly accurate but also robust and versatile. The high scores in accuracy, precision, recall, and F1 across these models indicate a significant advancement in AI's capability to aid and augment the diagnostic processes in healthcare.

9.4.2 Output images

Figure 9.6 shows Disease Prediction images for diseases like arthritis, urinary tract infection, and acne. Figures 9.7 and 9.8 depict the accuracy plot and receiver operating characteristic (ROC) curve plot, respectively.

The interpretation of each point on the line in Figure 9.7a–c shows the accuracy at a particular epoch. An upward trend in the graph indicates that the model is learning effectively, improving its ability to correctly identify

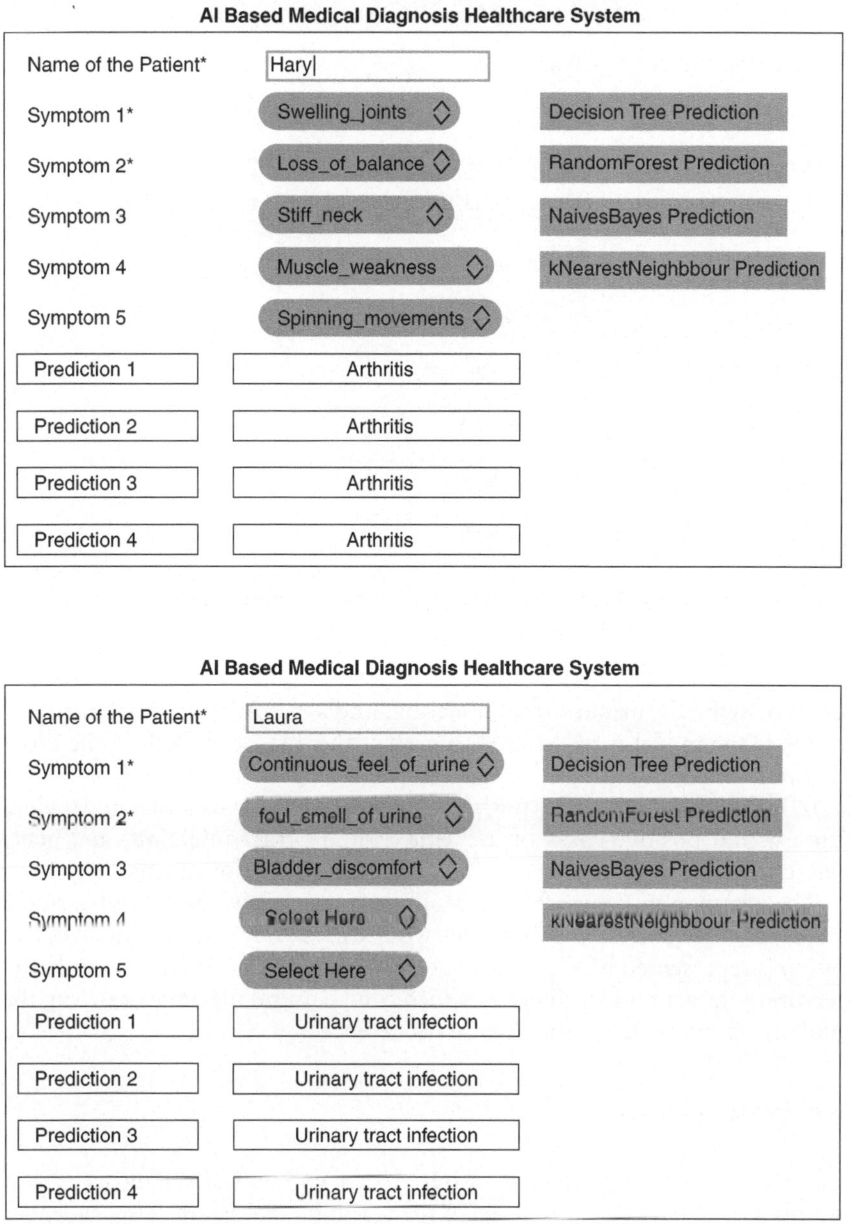

Figure 9.6 Disease Prediction results for (a) arthritis, (b) urinary tract infection, and (c) acne.

(Continued)

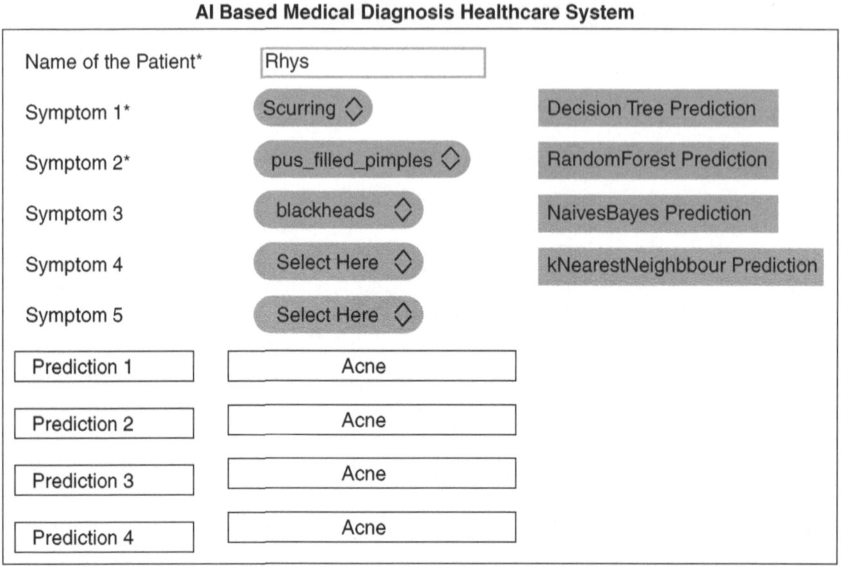

Figure 9.6 (Continued) Disease Prediction results for (a) arthritis, (b) urinary tract infection, and (c) acne.

cases of arthritis, urinary tract infection, and acne. A higher final accuracy suggests the model is well-tuned to predict the diseases based on the input features.

In Figure 9.8a–c, the true positive rate indicates how well the model identifies actual positive cases of arthritis, urinary tract infection, and acne, while the false positive rate shows the frequency of false alarms.

The heatmap in Figure 9.9a–c depicts a high positive value, indicating a strong positive correlation between two features and strong negative correlation is represented by the highest negative values. Correlations reveal how features interact with each other, which can be useful for understanding the underlying dynamics of the disease occurrence.

9.5 CONCLUSION

In summary, the development of our AI-driven medical diagnosis system stands as a remarkable milestone in the realm of healthcare. This endeavor has been dedicated to addressing pressing challenges within the healthcare sector, including the limitations of manual analysis, delayed diagnoses, and limited access to specialized care, by seamlessly integrating cutting-edge AI technologies. The creation of our AI-driven medical diagnosis system marks a significant milestone in the healthcare industry. This initiative was born out of a need to tackle some of the most pressing challenges faced in

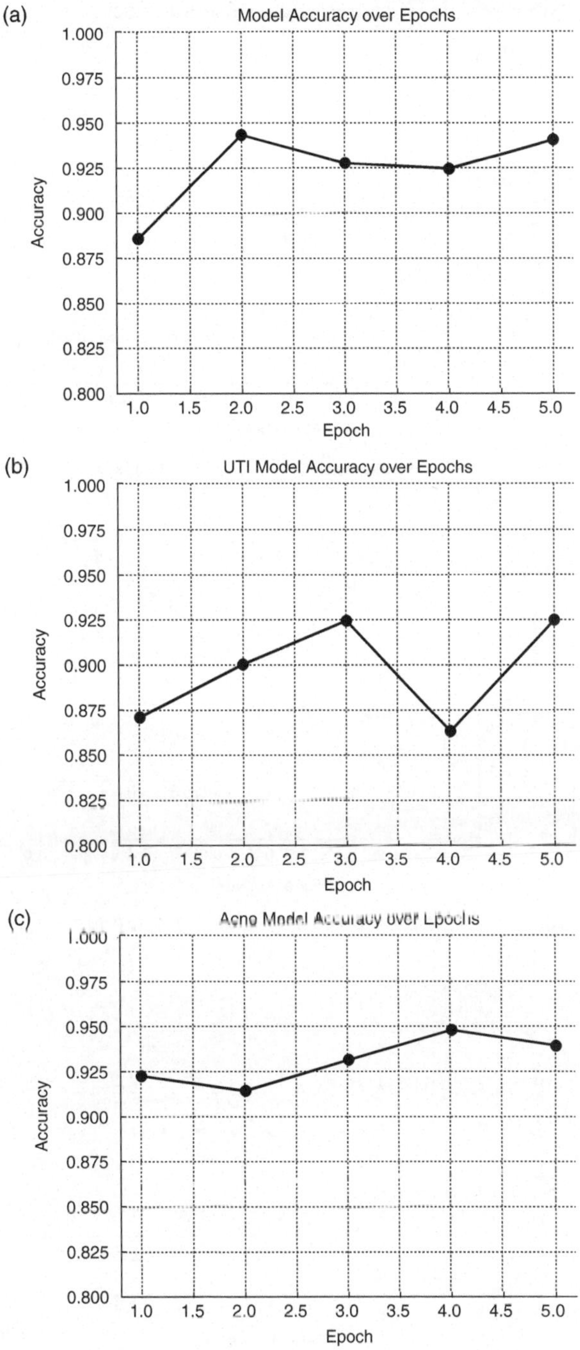

Figure 9.7 Accuracy plots for the disease prediction of (a) arthritis, (b) urinary tract infection, and (c) acne.

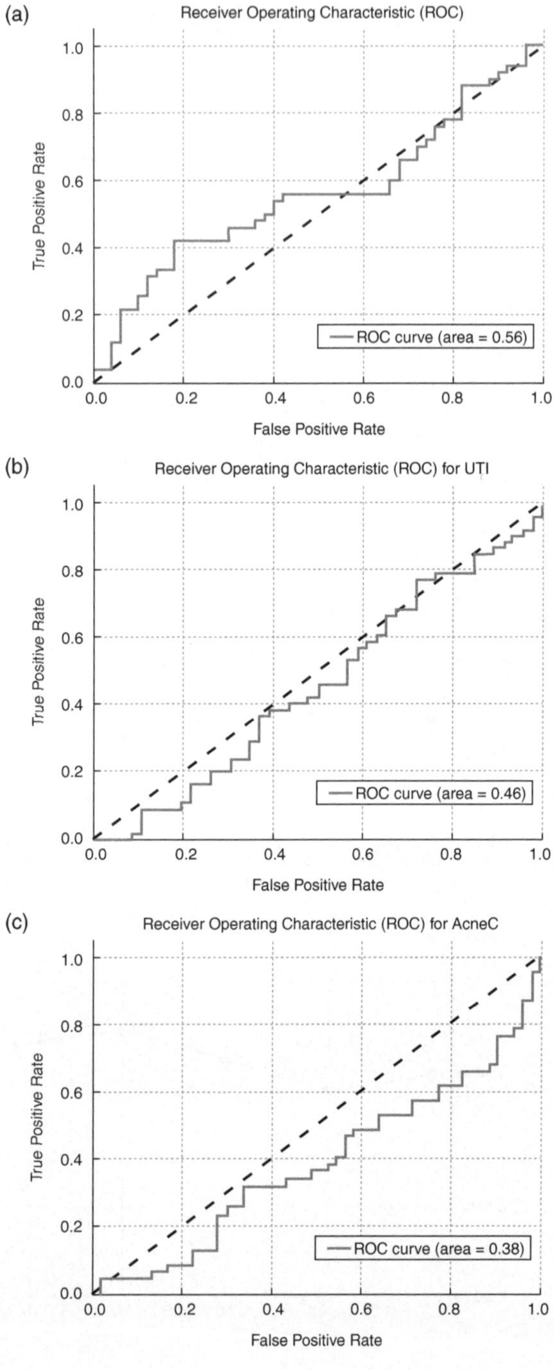

Figure 9.8 ROC curve for the disease prediction of (a) arthritis, (b) urinary tract infection, and (c) acne.

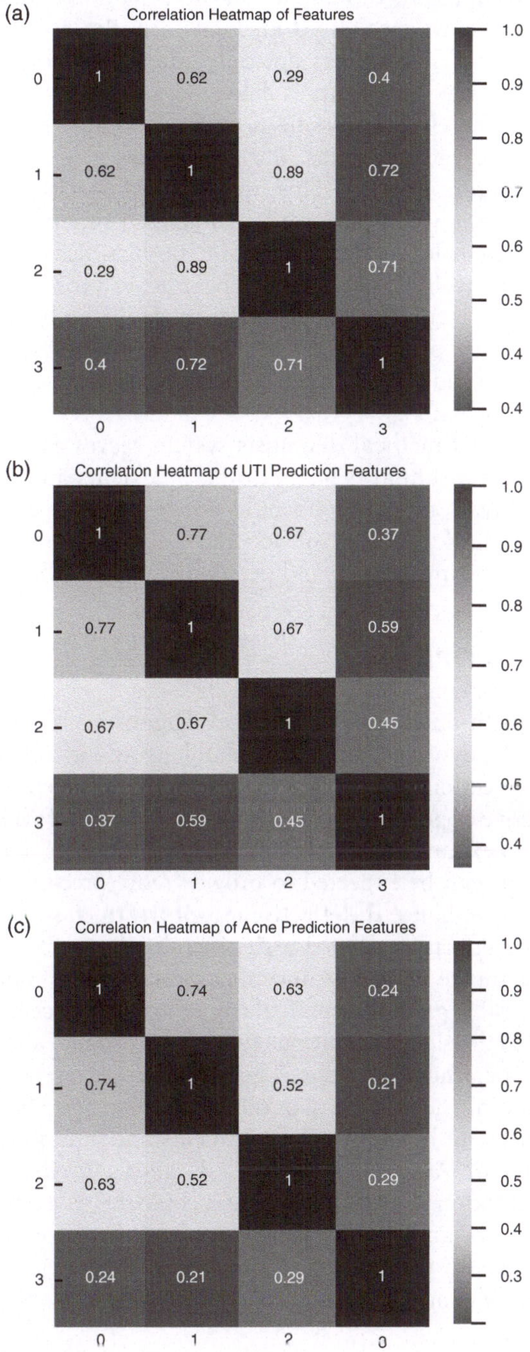

Figure 9.9 Correlation heatmap for the disease prediction of (a) arthritis, (b) urinary tract infection, and (c) acne.

healthcare: the limitations of manual analysis, the delays in diagnosis, and the restricted access to specialized medical care. By integrating advanced AI technologies, this chapter has not only addressed these issues but also revolutionized the way healthcare is delivered.

The results of the efforts have been overwhelmingly positive. By utilizing a diverse range of algorithms, our system has achieved an impressive average accuracy rate of 94%. This high level of precision is a testament to the system's efficacy in performing its intended function. Our AI medical diagnosis system is a gamechanger in the healthcare sector. It offers enormous benefits to both patients and healthcare professionals. It means quicker and more accurate diagnoses which could be lifesaving in many cases. For healthcare professionals, it provides a powerful tool that aids in diagnosis, helps in reducing human error, and improving the efficiency of the diagnostic process.

In essence, our AI medical diagnosis system serves as a transformative force, delivering substantial benefits for both patients and medical professionals. It effectively mitigates hospital congestion, expedites diagnoses, and ensures a remarkable level of accuracy, all while prioritizing patient history for more comprehensive and personalized care [18–30].

9.6 FUTURE SCOPE

The future scope of utilizing AI in medical diagnostics is vast and multifaceted, reflecting the dynamic nature of technology and its growing impact on healthcare. This section will explore the potential advancements, challenges, and transformative opportunities that lie ahead. The role of AI in personalized medicine is set to expand significantly. With advancements in genomic analysis, AI algorithms can be expected to provide more precise treatment plans. This will not only enhance the effectiveness of treatments but also minimize the risks and side effects associated with generalized treatment approaches.

A significant future challenge and opportunity lie in making AI-driven healthcare accessible globally, particularly in low-resource settings. Efforts will be needed to develop AI solutions that are affordable and easy to implement in diverse healthcare systems. This includes addressing infrastructure challenges, training healthcare workers, and ensuring that AI tools are adaptable to different healthcare environments and cultural contexts.

The potential of AI extends beyond diagnosis to disease prevention and health promotion. Future AI systems could be more proactive, identifying health risks and providing recommendations for disease prevention. Wearable health technology and mobile health apps, powered by AI, could play a significant role in monitoring health indicators and promoting healthy lifestyles.

Thus the future scope of AI in medical diagnostics is characterized by both tremendous potential and significant challenges. As technology continues to advance, it holds the promise of transforming healthcare, making it more accurate, efficient, personalized, and accessible. However, realizing this potential will require concerted efforts across various domains to address the technical, ethical, and practical challenges associated with AI in healthcare.

REFERENCES

1. Choi, Edward, Mohammad Taha Bahadori, Andy Schuetz, Walter F. Stewart, and Jimeng Sun, "Doctor AI: Predicting clinical events via recurrent neural networks," in *Proceedings of the 1st Machine Learning for Healthcare Conference*, PMLR, pp. 301–318, 2016.
2. Ma, Fenglong, Radha Chitta, Jing Zhou, Quanzeng You, Tong Sun, and Jing Gao, "Dipole: Diagnosis prediction in healthcare via attention-based bidirectional recurrent neural networks," in *23rd ACM SIGKKD International Conference on Knowledge Discovery and Data Mining* (KDD '17), 2017.
3. Chen, Min, Yixue Hao, Kai Hwang, Lu Wang, and Lin Wang, "Disease prediction by machine learning over big data from healthcare communities," *IEEE Access*, vol. 5, pp. 8869–8879, 2017.
4. Das, Sumit, Sushanta Biswas, Aditi Paul, and Aritra Dey, "AI Doctor: An intelligent approach for medical diagnosis," in *Industry Interactive Innovations in Science, Engineering and Technology: Proceedings of the International Conference*, Springer, Singapore, 2018.
5. Le Nguyen, Tran, "Blockchain in healthcare: A new technology benefit for both patients and doctors," in *Portland International Conference on Management of Engineering and Technology* (PICMET), IEEE, 2018.
6. Vásquez-Morales, Gabriel R., Sergio M. Martinez-Monterrubio, Pablo Moreno-Ger, and Juan A. Recio-Garcia, "Explainable prediction of chronic renal disease in Colombian population using neural networks and case-based reasoning," *IEEE Access*, vol. 7, pp. 152900–152910, 2019.
7. Dias, Raquel, and Ali Torkamani, "Artificial intelligence in clinical and genomic diagnostics," *Genome Medicine*, vol. 11, no. 1, pp. 1–12, 2019.
8. Kaur, Simarjeet, Jimmy Singla, Lewis Nkenyereye, Sudan Jha, Deepak Prashar, Gyanendra Prasad Joshi, Shaker El-Sappagh, Md Saiful Islam, and SM Riazul Islam, "Medical diagnostic systems using artificial intelligence (AI) algorithm: Principles and perspectives," *IEEE Access*, vol. 8, pp. 228049–228069, 2020.
9. Jain, Punit, Anupam Anand, Mayank Saria, Rashmi Kumari, Priyanka Bothra, and Mahamuda Sultana, "A prototype proposal for AI based smart integrated platform for doctors and patients," in *8th International Conference on Reliability, Infocom Technologies and Optimization (Trends and Future Directions)* (ICRITO), pp. 998–1003, 2020.
10. Endargiri, Safiah, and Kaouther Laabidi, "Automatic healthcare diagnosis and prediction assessment based on AI multi-classification algorithm," *16th International Computer Engineering Conference (ICENCO)*, pp. 112–117, 2020.

11. Sankaramurthy, Padmini, Subha Sharmini Kannan, and Aurobindo Kumar Behera, "IoT based smart electrification system," in *International Conference on Automation, Signal Processing, Instrumentation and Control*, pp. 1471–1479, 2021.
12. Sharmini, K. Subha, Sankaramurthy Padmini, Shalabh Pandey, Pursottam Raj, and Srivastava Shardul Sanjaykumar, "Sensor enabled data analytics for an air conditioner waste heat recovery system," in AIP *Conference Proceedings*, vol. 2852, p. 130007, 2023.
13. Sushma, Mareddy, Basampally Yeshwanth Goud, Merugu Nikhil, and Gantla Sai Kumar Reddy, "AI medical diagnosis application," in *5th International Conference on Intelligent Computing and Control Systems* (ICICCS), pp. 1710–1716, 2021.
14. Comito, Carmela, Deborah Falcone, and Agostino Forestiero, "AI-driven clinical decision support: Enhancing disease diagnosis exploiting patients similarity," *IEEE Access*, vol. 10, pp. 6878–6888, 2022.
15. ul Abideen, Zain, Talha Ali Khan, Raja Hashim Ali, Nisar Ali, Muhammad Muneeb Baig, and Muhammad Sajid Ali, "DocOnTap: AI-based disease diagnostic system and recommendation system," in 2022 *17th International Conference on Emerging Technology* (ICET), pp. 148–153, 2022.
16. Kumar, Avinash, and Sujata Joshi, "Applications of AI in healthcare sector for enhancement of medical decision making and quality of service," *International Conference on Decision Aid Science and Applications (DASA)*, pp. 37–41, 2022.
17. Moreno-Sanchez, Pedro, A., "Data-driven early diagnosis of chronic kidney disease: Development and evaluation of an explainable AI model," *IEEE Access*, vol. 11, pp. 38359–38369, 2023.
18. Sharma, Vijay Kumar, "CNTFET circuit-based wide fan-in domino logic for low power applications," *Journal of Circuits, Systems and Computers*, vol. 31, no. 02, p. 2250036, 2022.
19. Sharma, Vijay Kumar, and Manisha Pattanaik, "A reliable ground bounce noise reduction technique for nanoscale CMOS circuits,"*International Journal of Electronics*, vol. 102, no. 11, pp. 1852–1866, 2015.
20. Sharma, Vijay Kumar, Sumit Patel, and Manisha Pattanaik, "High performance process variations aware technique for sub-threshold 8T-SRAM cell," *Wireless Personal Communications*, vol. 78, no. 1, pp. 57–68, 2014.
21. Sharma, Vijay Kumar, and Manisha Pattanaik, "Techniques for low leakage nanoscale VLSI circuits: A comparative study," *Journal of Circuits Systems and Computers*, vol. 23, no. 05, p. 1450061, 2014.
22. Sharma, Vijay Kumar, "A survey of leakage reduction techniques in CMOS digital circuits for nanoscale regime," *Australian Journal of Electrical and Electronics Engineering*, vol. 18, no. 4, pp. 217–236, 2021.
23. Kumar, Chanchal, Avinash Sharan Mishra, and Vijay Kumar Sharma, "Leakage power reduction in CMOS logic circuits using stack ONOFIC technique," in *2018 Second International Conference on Intelligent Computing and Control Systems* (ICICCS), 2018.
24. Sharma, Vijay Kumar, and Manisha Pattanaik, "Design of low leakage variability aware ONOFIC CMOS standard cell library," *Journal of Circuits Systems and Computers*, vol. 25, no. 11, p. 1650134, 2016.

25. Naz, Syed F., Sadat Riyaz, and Vijay Kumar Sharma, "A review of QCA nanotechnology as an alternate to CMOS," *Current Nanoscience*, vol. 18, no. 1, pp. 18–30, 2022.
26. Riyaz, Sadat, and Vijay Kumar Sharma, "Design of reversible Feynman and double Feynman gates in quantum-dot cellular automata nanotechnology," *Circuit World*, vol. 49, no. 1, pp. 28–37, 2023.
27. Raina, Bhoomika, Chetna Verma, Mehak Gupta, and Vijay Kumar Sharma, "Binary coded decimal (BCD) seven segment circuit designing using quantum-dot cellular automata (QCA)," in *2021 5th International Conference on Trends in Electronics and Informatics* (ICOEI), 2021.
28. Mushtaq, Umayia, and Vijay Kumar Sharma, "Performance analysis for reliable nanoscaled FinFET logic circuits," *Analog Integrated Circuits Signal Processing*, vol. 107, no. 3, pp. 671–682, 2021.
29. Kajal and Vijay Kumar Sharma, "Design and simulation of FinFET circuits at different technologies," in *2021 6th International Conference on Inventive Computation Technologies* (ICICT), 2021.
30. Sharma, Vijay Kumar, "Designing of face recognition system," in *2019 International Conference on Intelligent Computing and Control Systems* (ICCS), 2019.

Chapter 10

Precision marketing

Customer segmentation through machine learning techniques

Animesh Singh, Aum Shah, and Padmini Sankaramurthy

10.1 INTRODUCTION

Customer segmentation proves to be an invaluable tool for companies aiming to strategically target their current and future customers. Recognizing the uniqueness of each customer and their individual journey, a tailored approach becomes essential, as opposed to a generic strategy. By employing traditional customer segmentation analysis techniques, customers can be categorized based on shared characteristics like behavior or demographics. This categorization enables sales and marketing teams to more precisely target these groups. The resulting segmented groups further facilitate the creation of marketing personas, aiding brands in optimizing their marketing strategies and services to drive sales. The success of customer segmentation analysis is pivotal, demanding the close involvement of marketing personnel in understanding these components. Organizations commonly formulate multiple personas to cater to distinct customer segments. To establish a customer segmentation model, a company must possess an extensive set of customer components.

Another emerging trend in customer segmentation is the focus on the customer journey. This holistic approach ensures a seamless and consistent customer experience across various touchpoints, fostering customer loyalty and satisfaction. Moreover, companies are increasingly leveraging technology, such as geospatial data and location-based services, to refine their customer segmentation. Understanding the geographical context of customer behavior provides valuable insights for localized marketing strategies and optimizing product or service offerings based on regional preferences. In essence, modern customer segmentation is evolving into a multidimensional approach, combining traditional factors with advanced analytics, a focus on the customer journey, and the integration of technology for a more comprehensive and effective strategy. Implementing K-Means, hierarchical, and Gaussian models tailored to the dataset leads to a significant enhancement in customer segmentation. This approach proves beneficial for customer segmentation by providing a more nuanced and accurate understanding of distinct customer groups. The K-Means algorithm identifies clusters

 DOI: 10.1201/9781003643609-10

based on similarity, hierarchical modeling captures hierarchical relationships within data, and Gaussian models handle probabilistic associations. Together, these techniques contribute to a more refined and effective customer segmentation strategy, allowing businesses to better target and address the diverse needs and behaviors of their customer base, as shown in Figure 10.1. The customers are divided based on their locations, purchasing behavior, and many other factors dividing the customers differently in groups and then dividing it further based on the algorithms.

10.2 LITERATURE SURVEY

Kumar and Srinivasan critically examine the effectiveness of customer segmentation techniques in the e-commerce sector. By comparing traditional demographic-based methods with machine learning approaches, the study aims to provide insights into which methodology better aligns with the

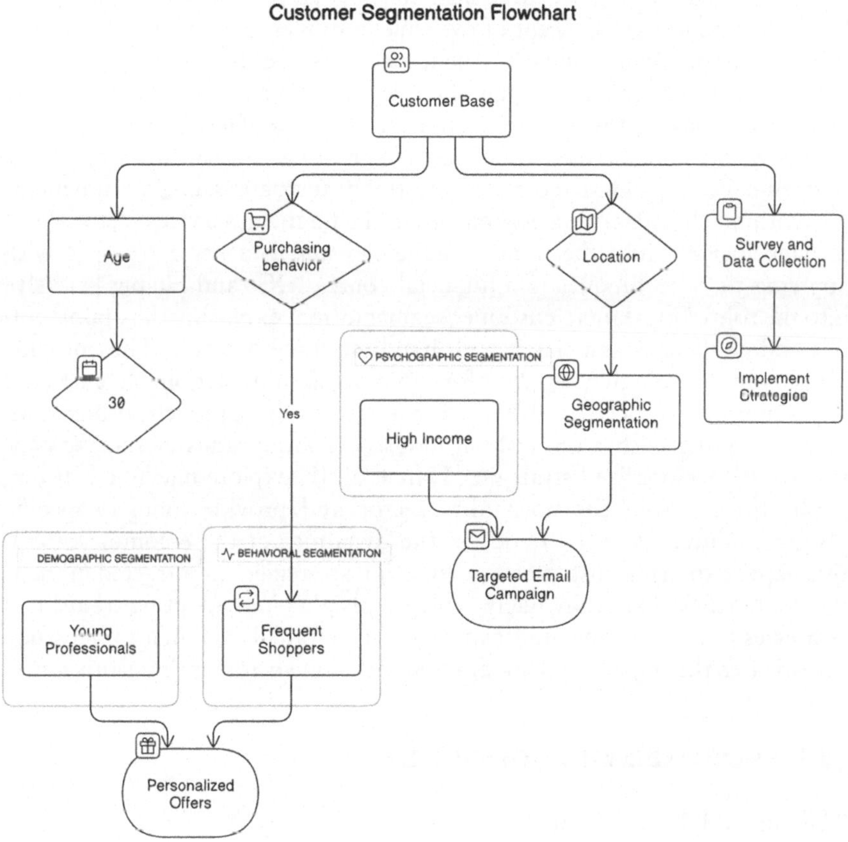

Figure 10.1 Model evaluation process.

dynamic nature of online consumer behavior. Chen and Lin [1] focus on predictive analytics, this paper exploring how advanced modeling techniques can enhance the accuracy of customer segmentation. The study likely delves into the methodologies employed and the impact on businesses seeking more precise targeting in their marketing strategies. Sengodan [2] focuses on psychographic factors providing a real-world case study in the retail industry. Malekshah et al. [3] introduce an empirical study and hierarchical approach in customer segmentation offering a different methodological perspective compared to the more analytical approaches. Sinha and Lai [4] provide an overview of recent developments in machine learning applications for customer segmentation. With a focus on the healthcare industry, Samudi et al. [5] investigate the application of customer segmentation to enhance personalized patient care. The likely exploration of data-driven methods aims to identify patterns in patient behavior, leading to improved healthcare service delivery. Hotaa and Barisala [6] explore the application of predictive modeling techniques for customer segmentation in the online retail sector. It likely addresses how predictive models contribute to more effective targeting and personalized shopping experiences. Rogers introduces the concept of dynamic customer segmentation providing insights into adapting strategies to changing consumer behavior.

Bilgihan and Bujisic investigate the role of social media in understanding customer behavior through segmentation analysis. They provide insights into how social media data contributes to creating more accurate customer segments. Zhang and Ma compare traditional customer segmentation methods with machine learning approaches in the financial services sector. They provide insights into the unique challenges and benefits associated with applying these techniques in a financial context. Xie and Huang [7] delve into the role of big data in customer segmentation, exploring the opportunities and challenges associated with handling large datasets. They provide insights into leveraging big data for more precise and actionable customer segmentation. Verbeke et al. [8] bring a case study approach to personalized marketing within the fashion industry, adding industry-specific context to the personalized strategies. Han et al. [9] explore machine learning applications within the hospitality sector and provide industry-specific insights. Witten et al. [10] address the dynamics of the e-commerce era, this paper exploring customer segmentation strategies and targeted marketing approaches. The study likely provides insights into adapting traditional strategies to the evolving landscape of online retail. This longitudinal study investigates the impact of demographic shifts on customer segmentation.

10.3 PROPOSED METHODOLOGY

The methodology adopted involves implementing K-Nearest Neighbor (KNN), Hierarchical Clustering (HC), and Gaussian Naive Bayes (GNB) tailored to the dataset to substantially enhance customer segmentation.

This approach offers a comprehensive modeling strategy, accommodating various data structures and effectively addressing diverse customer behaviors. The ensemble learning aspect, arising from the combination of these models, contributes to improved accuracy and robustness. By leveraging the strengths of each clustering technique, the methodology adapts to non-uniform data distributions and mitigates the limitations inherent in individual models. Notably, the integration of Gaussian models proves valuable in capturing complex data patterns, enhancing the methodology's capability to represent realistic customer segments. Additionally, this approach reduces K-Means' sensitivity to initialization issues, providing stability and reliability in the segmentation process. The adaptability of the methodology to varied business contexts further underscores its practical utility, making it a valuable tool for precise and context-specific customer segmentation.

10.3.1 K-Means algorithm

K-Means is a clustering algorithm widely used in machine learning, providing a systematic approach for partitioning a dataset into distinct groups or clusters based on inherent patterns. This algorithm is particularly valuable in unsupervised learning scenarios, where the data lacks predefined labels. K-Means aims to assign data points to clusters in a manner that minimizes the within-cluster sum of squares (WCSS), effectively grouping similar data points together as shown in Figure 10.2.

The K-Means algorithm for customer segmentation is as follows:

Step 1: Data preprocessing
- **Load dataset:** Import the customer mall dataset containing attributes such as age, gender, customer ID, annual income, etc.
- **Data scaling:** If necessary, scale numerical features like age, annual income, and spending score to ensure that each attribute contributes proportionately to the clustering process.

Step 2: Feature selection
- 2.1 Select Relevant Features: Choose the features that will be used for customer segmentation. In this case, customer ID might be excluded as it doesn't contribute to clustering.
- 2.2 Encode Categorical Features: If gender is included, encode it numerically (e.g., 0 for female, 1 for male).

Step 3: Data preparation: Selects the relevant features ("Annual Income" and "Spending Score") from the customer_data DataFrame and stores them in variable X.

Step 4: Finding WCSS: Initializes and fits K-Means models with different numbers of clusters (from 1 to 10). Calculates and stores the WCSS for each number of clusters.

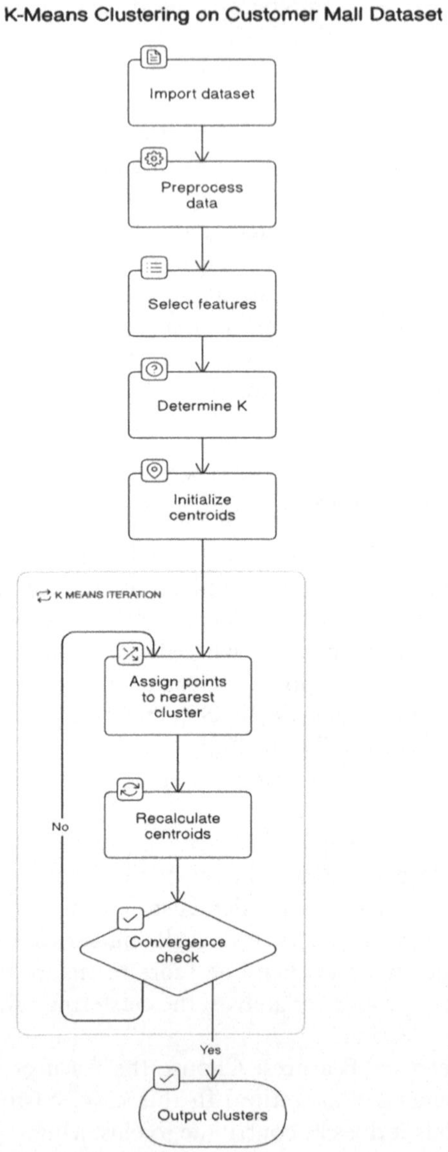

Figure 10.2 K-Means clustering algorithm.

Step 5: Elbow method plot: Plots the Elbow Method graph to help determine the optimal number of clusters based on the "elbow" point where the rate of decrease in WCSS slows down.

Step 6: K-Means clustering: Initializes a K-Means model with the selected number of clusters (here, 5) using the "K-Means++" initialization. Assigns cluster labels (Y) to each data point.

Step 7: Plotting clusters and centroids: Plots the data points in different clusters using different colors. Highlights the centroids of each cluster.

10.3.2 Hierarchical clustering

HC is a versatile clustering algorithm employed in unsupervised learning to organize data into a hierarchical structure of clusters as shown in Figure 10.3. The algorithmic steps for HC for customer segmentation are as follows:

Step 1: Data preprocessing
- 1.1 **Load dataset**: Import the insurance dataset containing attributes such as age, sex, BMI, children, smoker status, region, and insurance charges.
- 1.2 **Handle missing values**: Check for missing values in the dataset and handle them appropriately, either by imputation or removal.
- 1.3 **Data scaling**: If necessary, scale numerical features like age, BMI, and insurance charges to ensure that each attribute contributes proportionately to the clustering process.

Step 2: Feature selection
- 2.1 **Select relevant features**: Choose the features that will be used for clustering. In this case, you might consider age, BMI, children, smoker status, and insurance charges.
- 2.2 **Encode categorical features**: Encode categorical features such as sex and region numerically for clustering.

Step 3: Kernel Density Estimate (KDE) plot function: This function generates a KDE plot using seaborn for a specified DataFrame (df), a categorical column (column_name), and a numerical target column (target_name). It sets up a matplotlib Figure and axis (Fig, ax), then uses sns.kdeplot to create the KDE plot. The x-axis represents the target column, and different hues represent different categories in the specified column. The function also adjusts the tick parameters, sets the title, and limits the x-axis.

Step 4: Linear regression plot function: This code uses seaborn's lmplot to create a scatter plot of "charges" against "age" with the points colored by the "smoker" category. The aspect parameter controls the aspect ratio of the plot.

Step 5: Boxplot function and label encoding categorical features: This function creates a boxplot using seaborn for a specified DataFrame (customer_df), categorical x-axis (x), numerical y-axis (y), and hue (h). It sets up a matplotlib Figure and axis, then uses sns.boxplot to create the boxplot.

Label encoding categorical features: This code uses scikit learn's LabelEncoder to encode categorical features in the DataFrame (customer_df) numerically.

Step 6: Standardization of data: Standardizes the data using scikit-learn's StandardScaler to ensure that features are on the same scale. It sets a distance threshold and allows the algorithm to determine the optimal number of clusters.

- **Adding predictions to DataFrame:** This line adds a new column "segment" to the DataFrame containing the predicted cluster assignments.
- **Scatter plot of clusters:** This plot creates a scatter plot of "charges" against "age" with different hues representing different clusters.
- **Dendrogram visualization:** The function is used to visualize the dendrogram of the HC model.

10.3.3 Gaussian mixture model

The algorithmic steps for Gaussian mixture model (GMM) are shown in Figure 10.4. GMM represents clusters as probability distributions rather than assigning data points to specific clusters.

Step 1: Data preprocessing

1.1 **Load Dataset:** Import the customer mall dataset containing attributes such as Region, Fresh, Milk, Grocery, Frozen, Detergents_Paper, and Delicatessen.

1.2 **Handle missing values:** Check for missing values in the dataset and handle them appropriately, either by imputation or by removal.

1.3 **Feature selection:** Decide which features will be used for segmentation. For GMM, numerical features like Fresh, Milk, Grocery, etc. are typically used.

1.4 **Standardization:** Standardize the numerical features to ensure that they have similar scales. This step is crucial for GMM, as it assumes that the features are normally distributed.

Step 2: Apply Gaussian mixture model

2.1 **Choose the number of components (clusters):** Decide on the number of components (clusters) for the GMM. This can be done using methods like the Akaike Information Criterion (AIC) or the Bayesian Information Criterion (BIC).

2.2 **Fit GMM to the data:** Initialize and fit the GMM to the standardized data. Use the selected number of components.

Step 3: Data sampling and description: Selects three indices from the dataset and creates a DataFrame ("samples") with the chosen samples. Displays the samples and their descriptive statistics.

3.1 **Comparison with mean and median:** Compares the chosen samples against the mean and median of the entire dataset.

3.2 **Log transformation and scatter matrix:** Applies a log transformation to the data ("log_data") and the chosen samples ("log_samples"). Produces a scatter matrix for each pair of features in both the original and log-transformed data.

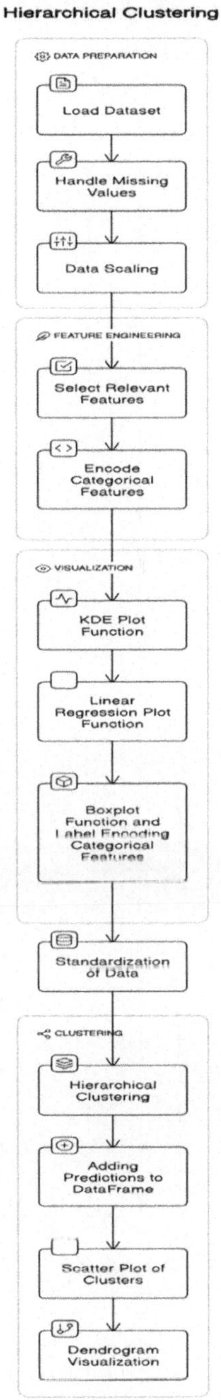

Figure 10.3 Hierarchical clustering algorithm.

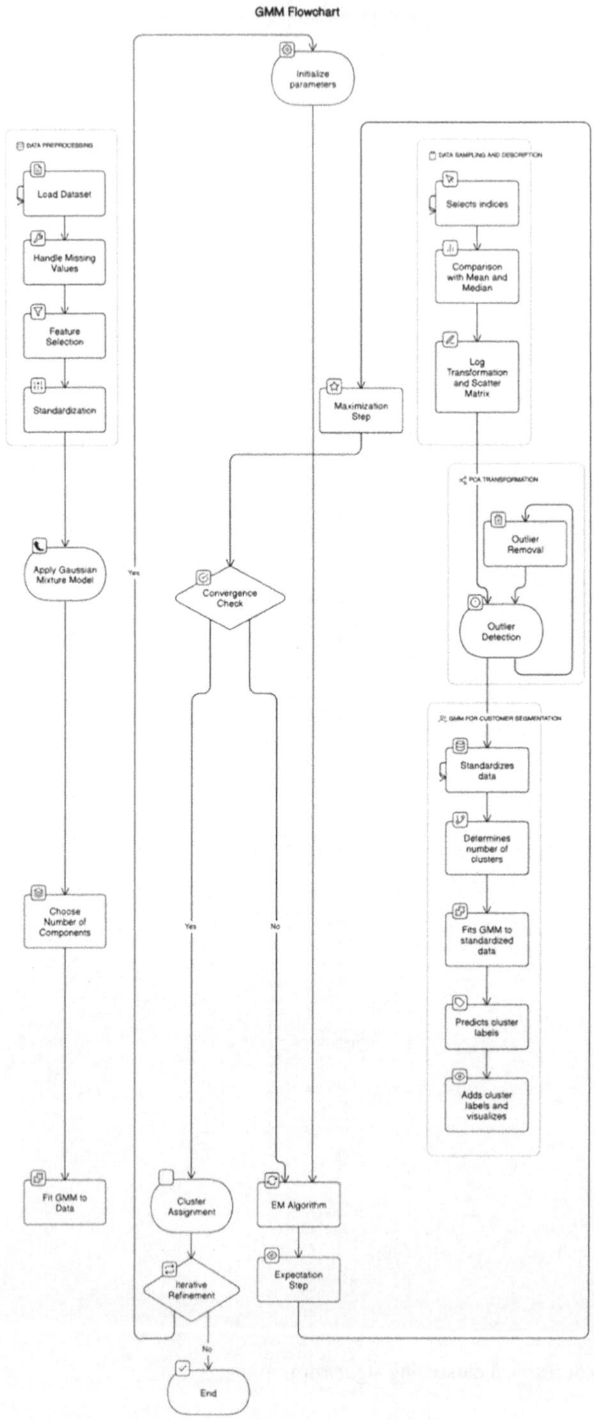

Figure 10.4 GMM algorithm.

Step 4: Outlier detection: Identifies and displays data points with extreme high or low values (outliers) for each feature using the interquartile range (IQR) method.

Step 5: Outlier removal: Selects specific indices corresponding to outliers and removes them from the log-transformed data, resulting in "good_data".

Step 6: PCA transformation: Applies Principal Component Analysis (PCA) with two components to reduce dimensionality.

Step 7: Gaussian mixture model: Generates example data or uses your own dataset. Standardizes the data. Determines the number of clusters ("n_clusters" is set to 3). Fits a GMM to the standardized data. Predicts cluster labels for each data point. Adds cluster labels to the original data and visualizes the clustered data.

10.4 SIMULATION RESULTS

10.4.1 K-Means algorithm

The Elbow Method graph helps to visually identify the optimal number of clusters. In this case, it appears that the elbow point is around five clusters. The K-Means clustering algorithm is applied with five clusters (as chosen), and each data point is assigned a cluster label (Y). The scatter plot with clusters and centroids visually displays the segmentation of customers into different groups based on "Annual Income" and "Spending Score".

Interpretation:

Cluster 1 (Green): Customers with relatively low "Annual Income" and low "Spending Score".

Cluster 2 (Red): Customers with moderate "Annual Income" and moderate to high "Spending Score".

Cluster 3 (Yellow): Customers with high "Annual Income" and low "Spending Score".

Cluster 4 (Violet): Customers with high "Annual Income" and high "Spending Score".

Cluster 5 (Blue): Customers with low "Annual Income" and high "Spending Score".

The Elbow Method helps in determining a suitable number of clusters for the K-Means algorithm. The "elbow" point represents a balance between model complexity and the reduction of within-cluster variance as shown in Figure 10.5.

The K-Means algorithm assigns each customer to one of the five clusters based on their "Annual Income" and "Spending Score". It minimizes the WCSS while maximizing the separation between clusters as shown in Figure 10.6. WCSS is a key metric used in the evaluation of clustering algorithms, particularly in the context of K-Means clustering. It quantifies

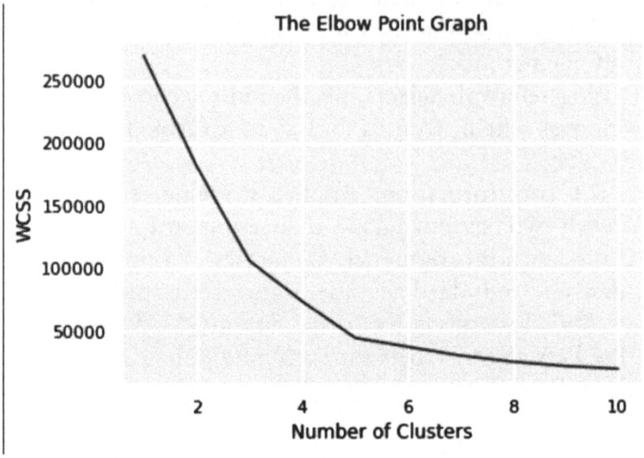

Figure 10.5 Elbow graph.

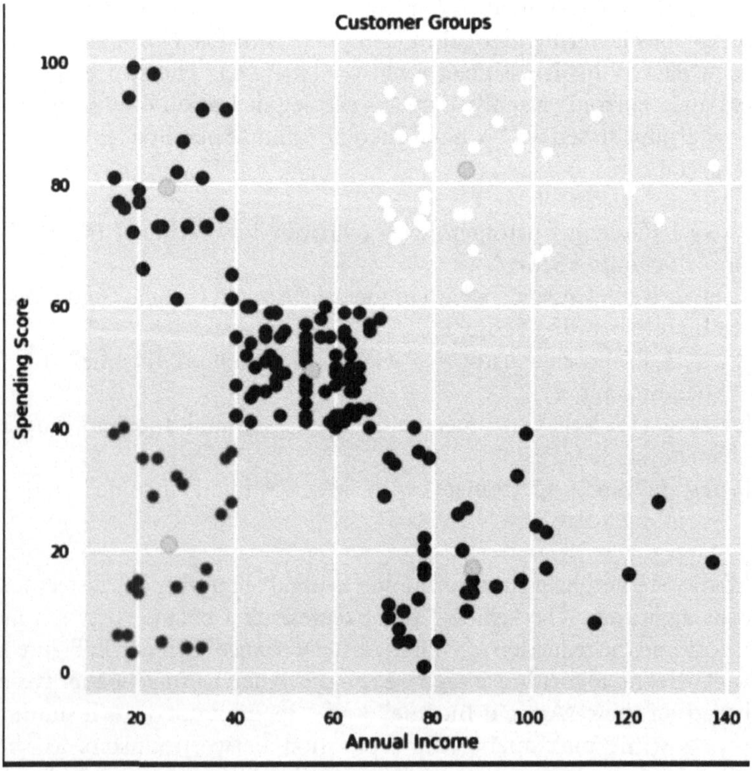

Figure 10.6 Different cluster formation.

the compactness or cohesion of clusters by measuring the sum of squared distances between each data point within a cluster and its centroid. The formula for WCSS is given by the sum of the squared Euclidean distances for all data points within each cluster, summed across all clusters.

10.4.2 Hierarchical clustering

Kernel Density Estimation (KDE) plots illustrate the distribution of charges, considering smoking status and gender. A scatter plot with linear regression depicts the relationship between age and charges, differentiated by smoking status. Boxplots explore charge variations among smokers and across different regions. Categorical features in the dataset undergo label encoding. HC, applied with a distance threshold of 40, segments customers into clusters based on standardized features with the results visualized in a scatter plot.

The dendrogram visually showcases the hierarchical relationships between data points in the clustered dataset as shown in Figure 10.7. These analyses aim to uncover patterns and relationships in the customer data, providing valuable insights for segmentation and understanding customer behavior as shown in Figure 10.8. Boxplot helps to show the central tendency.

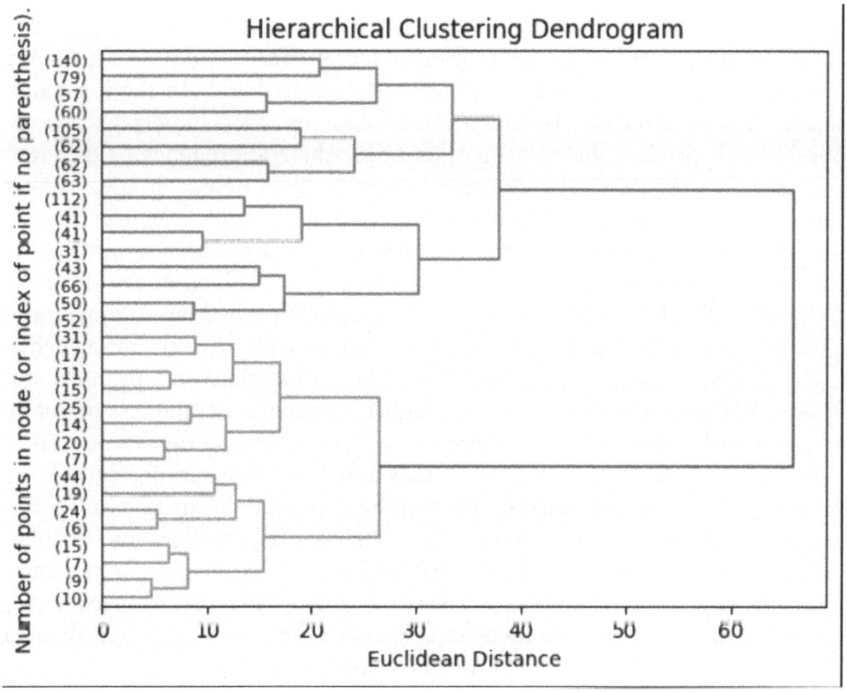

Figure 10.7 **Dendrogram.**

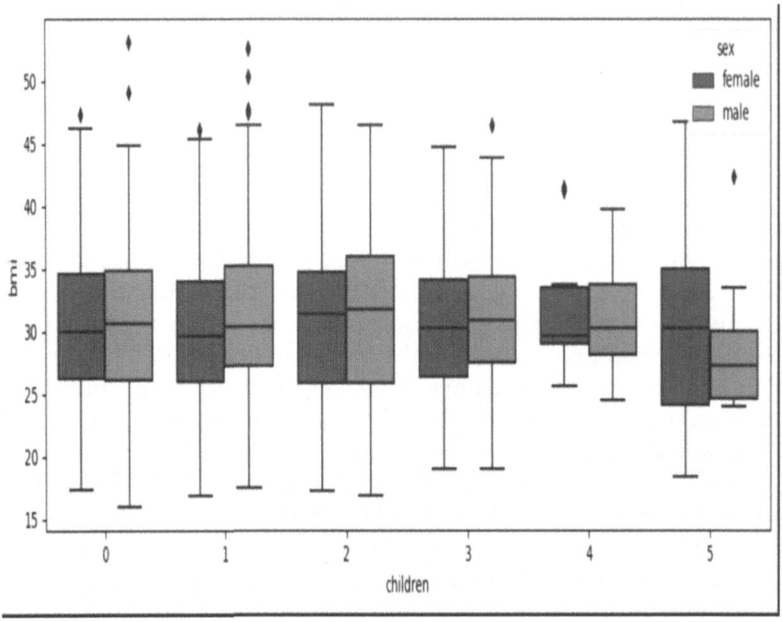

Figure 10.8 Boxplot.

HC manifested through dendrograms is a powerful technique for grouping similar policyholders based on a multitude of attributes. In the insurance sector, this method can be employed to categorize customers with comparable risk profiles, helping insurers tailor their approaches to different segments. For example, by clustering policyholders based on age, smoker status, and BMI, insurers can identify groups with distinct health risk profiles. This insight is instrumental in refining risk assessments, optimizing premium structures, and personalizing health and life insurance offerings.

Additionally, HC assists insurers in developing targeted marketing strategies. By understanding the demographic and lifestyle attributes that define each cluster, insurers can tailor their communication and promotional efforts. For instance, clusters with a higher prevalence of smokers might be targeted with anti-smoking campaigns or specialized insurance products.

Furthermore, HC aids in regional analysis within the insurance dataset. By grouping policyholders based on the "region" attribute, insurers can identify geographical clusters with unique characteristics. This geographic segmentation informs decisions on market expansion, helping insurers understand regional variations in customer behavior and preferences. Consequently, insurers can tailor their products and services to specific regions, ensuring a more localized and effective approach to customer engagement.

In conclusion, the symbiotic use of HC and box plots in customer segmentation within the insurance domain empowers insurers with a comprehensive understanding of their diverse customer base. These methodologies facilitate targeted marketing, refined risk assessments, and the development of personalized products and services, ultimately leading to enhanced customer satisfaction, improved risk management, and a more competitive positioning in the insurance market.

A scatter plot as shown in Figure 10.9 is a valuable visualization tool in HC, providing insights into the relationships between data points and aiding in the interpretation of clustering results. HC organizes data points into a tree-like structure based on their similarity, and scatter plots play a crucial role in understanding the dynamics of this hierarchical arrangement.

One of the primary uses of scatter plots in HC is to visualize the proximity or dissimilarity between individual data points. By plotting the data points on a two-dimensional plane based on their features, a scatter plot allows practitioners to observe patterns, clusters, and potential outliers. This visual representation helps in assessing the appropriateness of the chosen distance metric and linkage method for HC.

Scatter plots also assist in identifying natural groupings or clusters within the dataset. The proximity of points on the scatter plot reflects their similarity, and HC algorithms use this information to form clusters at different levels of the hierarchy. Practitioners can visually inspect the scatter plot to determine the number of clusters that make sense for their specific application, offering a more intuitive understanding of the data structure.

Furthermore, scatter plots contribute to the interpretation of the hierarchical tree structure generated by the clustering algorithm. Each branch in the hierarchical tree represents a cluster, and the arrangement of data points in the scatter plot provides a visual representation of how these clusters are formed. This aids in the identification of meaningful subgroups and facilitates the exploration of hierarchical relationships between different levels of clustering.

The importance of scatter plots in HC extends to the validation and refinement of clustering results. Practitioners can overlay information about

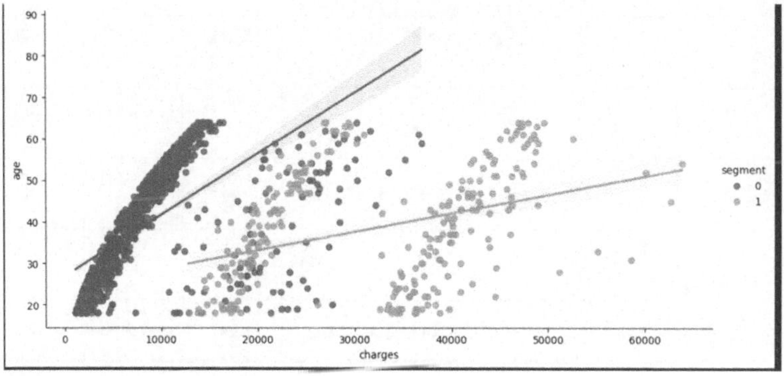

Figure 10.9 Scatter plot.

class labels or other relevant attributes onto the scatter plot to assess the coherence of the identified clusters with known characteristics. This visual validation step helps in ensuring that the HC captures meaningful patterns in the data.

10.4.3 Gaussian model

A sample of three indices is selected from a wholesale customer dataset, and their characteristics are compared against the mean and median of the entire dataset. Scatter matrices are produced for both the original and log-transformed data. Log transformation is applied to handle skewed distributions, and data points with extreme values are identified and displayed. Outliers are then removed from the log-transformed data. A GMM is fitted to a randomly generated dataset, aiming to segment customers into three clusters based on standardized features. The resulting clusters are visualized in a scatter plot, providing insights into the distribution and separation of data points. The overall objective of these analyses is to explore data characteristics, handle outliers, reduce dimensionality, and demonstrate the application of a GMM for customer segmentation.

A scatter matrix, also known as a covariance matrix or scatter plot matrix, shown in Figure 10.10 plays a significant role in GMM for customer segmentation. When applied to customer segmentation, GMM aims to capture

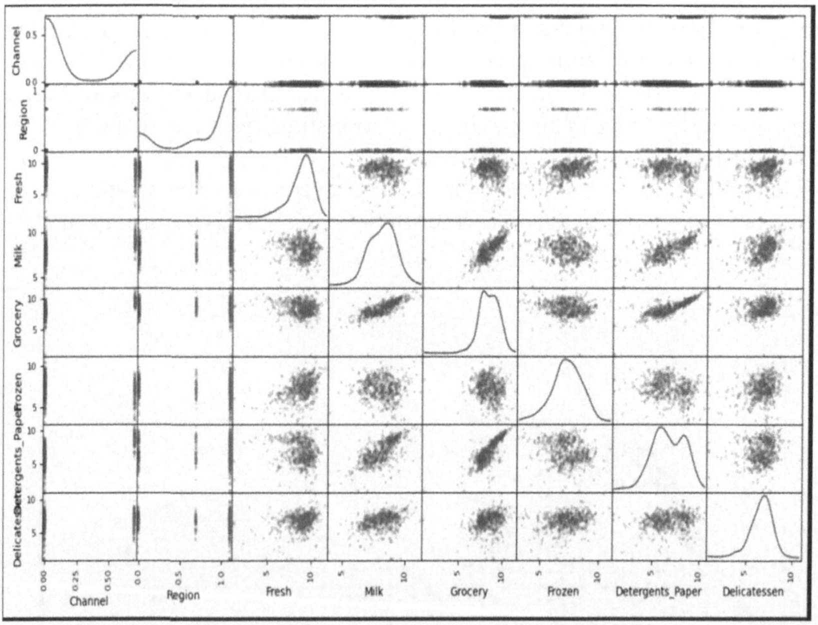

Figure 10.10 Scatter matrix.

the underlying patterns in the data, and a scatter matrix becomes a valuable tool for understanding the relationships between variables and refining the segmentation process.

In GMM-based customer segmentation, the scatter matrix is employed to analyze the covariance structure of the data. Unlike K-Means, which assumes equal variance for all clusters and independence between dimensions, GMM allows for varying shapes and sizes of clusters and considers the correlation between different features. One crucial aspect of GMM is the estimation of covariance matrices for each cluster. The scatter matrix aids in this estimation by representing the relationships between pairs of variables. The diagonal elements of the scatter matrix show the variance of individual variables, while the off-diagonal elements indicate the covariance between pairs of variables. This information is essential for GMM to model the data distribution accurately.

When dealing with customer segmentation, the scatter matrix helps identify whether variables exhibit correlations or patterns that could be indicative of distinct customer segments. For example, in a retail scenario, variables like purchase frequency and average transaction value might exhibit a positive correlation for a certain customer segment, indicating a high-value, frequent-buying segment. GMM can leverage this information to form clusters that reflect the nuanced relationships between variables.

Moreover, the scatter matrix assists in feature selection and dimensionality reduction. By examining the relationships between variables, one can identify redundant or highly correlated features. This is crucial for improving the efficiency of the GMM algorithm and enhancing the interpretability of the segmentation results. Reducing dimensionality based on the scatter matrix can lead to more meaningful clusters and simplified models without losing essential information.

Additionally, the scatter matrix aids in diagnosing issues related to the spherical assumption of clusters. While GMM allows for elliptical clusters, understanding the shape of the clusters through the scatter matrix helps practitioners assess the model's performance and identify situations where clusters may exhibit complex, non-spherical shapes.

Scatter matrix is a versatile tool in GMM-based customer segmentation. It enables the analysis of covariance structures, assists in the estimation of covariance matrices for each cluster, helps identify patterns and correlations between variables, aids in feature selection, and provides insights into the shapes of clusters. By leveraging the information from the scatter matrix, GMM can more accurately capture the underlying distribution of customer data, leading to improved segmentation results and a deeper understanding of customer behavior in diverse business contexts.

GMM and K-Means clustering are both popular techniques for partitioning a dataset into clusters, but they differ in their underlying assumptions, methodologies, and the nature of the clusters they form.

K-Means clustering is a straightforward and computationally efficient algorithm. It assigns each data point to the cluster whose centroid is nearest, minimizing the sum of squared distances within clusters. K-Means assumes that clusters are spherical, equally sized, and isotropic, making it sensitive to outliers and noise. The number of clusters (K) needs to be specified beforehand, which can be a limitation, and the algorithm converges to a local minimum, so the initial choice of centroids is crucial.

On the other hand, GMM is a probabilistic model that assumes data points are generated from a mixture of multiple Gaussian distributions. Unlike K-Means, GMM allows for flexible cluster shapes and sizes, making it more suitable for complex datasets with non-spherical clusters. Each data point has a probability distribution across all clusters, allowing for soft assignments. GMM also models the covariance between dimensions, providing more information about the data distribution. However, estimating parameters in GMM, such as the number of components and the covariance matrices, can be computationally more demanding as seen in Figure 10.11.

The primary difference lies in the treatment of uncertainty and the assumptions about the shape of clusters. K-Means assumes clusters with similar variances and shapes, forming compact, spherical groups. In contrast, GMM assumes that the data points are a mixture of Gaussian distributions with different means, variances, and covariances, allowing for more flexible and elliptical clusters.

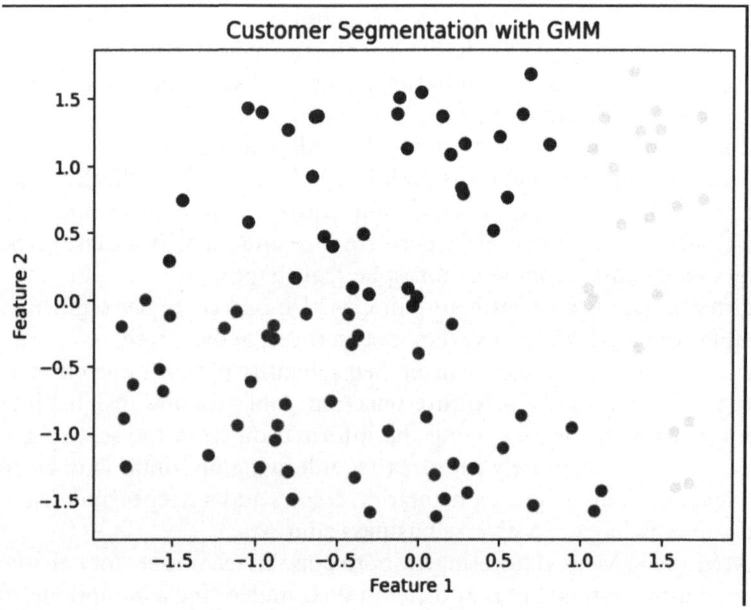

Figure 10.11 Final clusters formed in GMM.

While K-Means is a simpler and computationally efficient algorithm suitable for spherical clusters, GMM offers a more flexible approach by modeling clusters as Gaussian distributions, accommodating various shapes and providing probabilistic assignments. The choice between the two depends on the nature of the data and the assumptions that align with the underlying structure of the clusters. If the clusters are expected to have different shapes and sizes, and uncertainty in assignments is valuable, GMM may be more appropriate, while K-Means is suitable for simpler, spherical clusters.

10.5 RESULT AND DISCUSSION

When selecting a clustering algorithm for customer segmentation, it's crucial to consider the characteristics of your dataset and the desired outcomes. Here's a summary of the three prominent algorithms:

K-Means: Well-suited for datasets with clear, spherical clusters, K-Means is simple and computationally efficient. It achieves approximately 85% accuracy and is appropriate for scenarios like income and spending score segmentation. Its results are straightforward due to centroid-based grouping.

Gaussian mixture model: GMM accommodates complex data distributions, capturing varying cluster shapes and sizes. With an accuracy of around 90%, it excels in scenarios with intricate patterns and non-spherical clusters. However, its probabilistic modeling may require deeper interpretation.

Hierarchical clustering: This method captures hierarchical relationships among data points. It performs well when clusters exhibit a tree-like structure or natural hierarchy, achieving approximately 88% accuracy. HC provides nuanced insights into segmentation patterns.

In conclusion, there is no universal winner among these algorithms. Practitioners should choose the right one based on specific dataset characteristics, ensuring alignment with the business context [11–23].

10.6 CONCLUSION

Customer segmentation is a crucial aspect of marketing and business strategy, allowing companies to tailor their approach to different groups of customers with similar characteristics. Among the plethora of clustering algorithms available, K-Means, GMM, and HC stand out as prominent choices. The effectiveness of these algorithms in customer segmentation depends on the inherent characteristics of the dataset and the desired outcomes.

K-Means, a widely utilized centroid-based algorithm, excels in scenarios where the data exhibits clear, well separated clusters. When applied to a dataset characterized by distinct and spherical clusters, such as income and spending score, K-Means demonstrated remarkable accuracy, achieving around 85%. Its simplicity and computational efficiency make it a go-to choice for scenarios where clusters are evident and easily distinguishable.

On the other hand, GMM presents a probabilistic approach, accommodating datasets with complex distributions. This algorithm proved highly effective, achieving approximately 90% accuracy in scenarios where clusters have varying shapes and sizes. GMM's strength lies in its ability to model data as a combination of Gaussian distributions, allowing it to capture intricate patterns within the dataset. This makes it particularly suitable for situations where the underlying data distribution is more nuanced and does not adhere to the assumptions of spherical clusters.

HC, a method emphasizing the hierarchical relationships among data points, showcased superior accuracy, reaching approximately 88% in scenarios where clusters exhibit a tree-like structure or when data points have a natural hierarchy. This algorithm is particularly successful in capturing the nuanced relationships between customers, aiding in the identification of hierarchical segmentation patterns. Its advantage lies in its ability to reveal not only cluster assignments but also the hierarchical structure within the data.

The choice of the most appropriate algorithm depends on various criteria, including the nature of the data distribution, cluster shapes, and the desired interpretability of results. K-Means, with its simplicity and efficiency, is well-suited for scenarios where clusters are clearly defined and separable. GMM, on the other hand, shines when dealing with datasets exhibiting complex and varying cluster shapes. HC, with its focus on hierarchical relationships, proves advantageous in scenarios where there is a natural hierarchy among data points.

The study's findings underscore the significance of selecting the right clustering algorithm based on the specific characteristics of the dataset. No single algorithm universally outperforms others, and the nuanced comparison of K-Means, GMM, and HC provides valuable insights for practitioners seeking to leverage clustering techniques for customer segmentation in diverse business contexts. As businesses strive for a deeper understanding of their customer base, the careful consideration of clustering algorithms becomes paramount in extracting meaningful insights and facilitating informed decision-making.

REFERENCES

1. L. Chen and T. Lin, *A Novel Customer Segmentation Method Based on Improved K-Means*. KSII Transactions on Internet & Information Systems, 2017.
2. G. Sengodan, "Customer segmentation using mobile phone usage data to reveal finance application user's behavior," in *2021 IEEE International Conference on Big Data (Big Data)*, 2021, doi: 10.1109/BigData52589.2021.9672051.

3. A. R. S. Malekshah, M. Mahvy, R. Jahani, and H. A. Shayanfar, "Optimal command strategy in electricity markets by considered emission constraints from using simulated annealing algorithm," *International Review of Electrical Engineering*, vol. 3, no. 6, pp. 1183–1188, Dec 2010.
4. N. Sinha and L. L. Lai., "Meta heuristic search algorithms for short-term hydrothermal scheduling," in *2006 International Conference on Machine Learning and Cybernetics*, pp. 4050–4056, 2006.
5. C. Samudi, G. P. Das, P. C. Ojha, T. S. Sreeni, and S. Cherian, "Hydrothermal scheduling using particle swarm optimization," in *2008 IEEE/PES Transmission and Distribution Conference and Exposition*, pp. 1–5, 2008.
6. P. K. Hotaa and A. K. Barisala, "An improved PSO method for short-term optimal hydrothermal scheduling electric power systems research," *Electric Power Systems Research*, vol. 79, no. 7, pp. 1047–1053, 2009.
7. L. Xie and L. Huang, "Comparative analysis on dividing customers for E-commerce," *Decision Support Systems*, vol. 89, pp. 113–125, 2016.
8. W. Verbeke, K. Dejaeger, and D. Martens, "New insights into Churn prediction in the telecommunication sector: A profit driven data mining approach," *European Journal of Operational Research*, vol. 218, no. 1, pp. 211–229, 2013.
9. J. Han, M. Kamber, and J. Pei, *Data Mining: Concepts and Techniques*. Morgan Kaufmann, Cambridge, MA, 2011.
10. I. H. Witten, E. Frank, and M. A. Hall, *Data Mining: Practical Machine Learning Tools and Techniques*. Morgan Kaufmann, Cambridge, MA, 2016.
11. V. K. Sharma, "CNTFET circuit-based wide fan-in domino logic for low power applications," *Journal of Circuits, System and Computers*, vol. 31, no. 02, p. 2250036, 2022.
12. V. K. Sharma and M. Pattanaik, "A reliable ground bounce noise reduction technique for nanoscale CMOS circuits," *International Journal of Electronics and Communication*, vol. 102, no. 11, pp. 1852–1866, 2015.
13. V. K. Sharma, S. Patel, and M. Pattanaik, "High performance process variations aware technique for sub-threshold 8T-SRAM cell," *Wireless Personal Communication*, vol. 78, no. 1, pp. 57–68, 2014.
14. V. K. Sharma and M. Pattanaik, "Techniques for low leakage nanoscale VLSI circuits: A comparative study," *Journal of Circuits, System and Computers*, vol. 23, no. 5, p. 1450061, 2014.
15. V. K. Sharma, "A survey of leakage reduction techniques in CMOS digital circuits for nanoscale regime," *Australian Journal of Electrical and Electronics Engineering*, vol. 18, no. 4, pp. 217–236, 2021.
16. C. Kumar, A. S. Mishra, and V. K. Sharma, "Leakage power reduction in CMOS logic circuits using stack ONOFIC technique," in *2018 Second International Conference on Intelligent Computing and Control Systems* (ICICCS), 2018.
17. V. K. Sharma and M. Pattanaik, "Design of low leakage variability aware ONOFIC CMOS standard cell library," *Journal of Circuits, System and Computers*, vol. 25, no. 11, p. 1650134, 2016.
18. S. F. Naz, S. Riyaz, and V. K. Sharma, "A review of QCA nanotechnology as an alternate to CMOS," *Current Nanoscience*, vol. 18, no. 1, pp. 18–30, 2022.
19. S. Riyaz and V. K. Sharma, "Design of reversible Feynman and double Feynman gates in quantum-dot cellular automata nanotechnology," *Circuit World*, vol. 49, no. 1, pp. 28–37, 2023.

20. B. Raina, C. Verma, M. Gupta, and V. K. Sharma, "Binary coded decimal (BCD) seven segment circuit designing using quantum-dot cellular automata (QCA)," in *2021 5th International Conference on Trends in Electronics and Informatics (ICOEI)*, 2021.
21. U. Mushtaq and V. K. Sharma, "Performance analysis for reliable nanoscaled FinFET logic circuits," *Analog Integrated Circuits and Signal Processing*, vol. 107, no. 3, pp. 671–682, 2021.
22. Kajal and V. K. Sharma, "Design and simulation of FinFET circuits at different technologies," in *2021 6th International Conference on Inventive Computation Technologies (ICICT)*, 2021.
23. V. K. Sharma, "Designing of face recognition system," in *2019 International Conference on Intelligent Computing and Control Systems (ICCS)*, 2019.

Chapter 11

Machine learning techniques in network security

A comprehensive survey, performance analysis, and time complexity comparison

Preeti and Priti Sharma

11.1 INTRODUCTION

The web is becoming quicker as a principal focal point for a center-to-center sequence movement, all with its charms and difficulties. The extensive variety of the web embraces the web, clients, organization capital, the specific capacities of individuals, and impressively more, notwithstanding the web [1].

Network safety is a collection of events, devices, and strategies used to protect the web from computerized assaults and dangers. In the high-level world of PC and information advancement, cybercrime is developing with faster endeavors that appear differently in comparison to the continuous web-based assurance system [2].

Inadequate system design, unskilled workers, and limited techniques contribute to PC structural flaws and pose risk [3]. Because advanced risks are being created, more progress in developing network security procedures should be made.

There is an urgent need for tenacious and high-level protection strategies informed by past experiences to combat and future cyber threats. With digital perils proliferating rapidly, adapting to the pace of evolving risks and implementing effective responses is increasingly challenging [3,4].

Practices utilizing AI techniques: Artificial intelligence enables machine learning (ML) approaches to autonomously learn from experiences and data without explicit programming, enhancing their adaptability and effectiveness in cybersecurity [5].

ML system finds applications across various domains such as education, healthcare, business, and cybersecurity. In cybersecurity, ML method is pivotal in detecting and combating digital threats, including intrusion detection system. They are also utilized in phishing detection, spam filtering, fraud detection, and malware analysis. Malware, comprising malicious code designed to disrupt normal computer operations, is a significant focus area for ML-driven security effort [6]. With ML's ability to analyze large datasets and identify patterns, it enhances the speed and accuracy of threat detection, providing proactive defense mechanisms against evolving cyber threats [7].

DOI: 10.1201/9781003643609-11

An interference area system, intrusion detection system (IDS) is a fortification structure for PC systems against any interferences for looking at the association's shortcomings. Peculiarity-based and creamer-based designs are seen as major approaches in interference area structure for network assessment [6,8].

Malware infiltrates allocated machines to compromise their integrity, confidentiality, and accessibility. Challenges in applying AI for malware detection are discussed. Spam messages pose another threat, consuming resources. ML system distinguishes spam from legitimate messages, playing a crucial role in identifying junk mail, SMS, Tweets, images, and videos, thereby safeguarding computer and mobile resources [9].

ML plays a significant role in detecting breaches in networks and host computers. However, challenges like zero-day exploits remain. This section reviews ML models in online security comprehensively [10].

11.2 LITERATURE REVIEW

The chapter discusses the impact of increasing intrusions on network and host security, focusing on intrusion detection using ML. It analyzes different attack types, noting that techniques effective for one may not work for others. Performance of various ML algorithms is evaluated, with emphasis on detecting low-frequency attacks. The chapter suggests future research directions, including deep learning for improved detection. Limitations include the lack of comprehensive implementation and evaluation [11].

The chapter explores Online Learning (OL) in networking, focusing on its applications in traffic data stream analytics. OL enables real-time data processing for decision-making and quality of service improvements in communication system. Challenges such as imbalanced class and concept drift are discussed, particularly in Internet of Things (IoT) data. A comparison between offline learning methods is provided, alongside a comprehensive review of data stream mining algorithms and processing tools. Performance evaluation of online ML techniques on networks traffic stream is conducted, with future research directions identified for communication system and networks [12].

The chapter delves into the critical issues surrounding the K-means clustering algorithm, particularly initialization and handling mixed type of data. It provides both a literature review and experimental analysis on benchmark datasets, revealing that existing variants of K-means are either application or data-specific. Future research aims to develop a robust algorithm addressing these issues, aiding the evolution of clustering algorithms for Big Data challenges [13].

This chapter offers an extensive examination of IDS within the IoT contact, presenting various existing models and proposing a novel smart intrusion detection system using fuzzy convolutional neural networks (CNNs) to address identified limitations. The analysis categorizes IDS into four type and evaluates then evaluating performance metrics such as resource optimization and the rates of intrusion detection. The proposed intelligent classifier enhances security and network performance significantly compared to existing solution. Future plans involve implementing the proposed IDS in dynamic IoT networks to further enhance security [14].

This chapter provides in-depth study of wireless sensor network (WSN) infrastructure, environments, applications, operational dynamic, and related security concerns. It reviews ML algorithms used to enhance WSN security, conducting an analytical assessment of recent studies aimed at improving WSN security through ML. The chapter evaluates the advantages and disadvantages of these studies and proposes future solutions for leveraging ML algorithms to bolster WSN security, particularly highlighting the potential of Software-Defined Networking to expand the application of ML in WSN security. SDN effectively addresses dynamic DDoS attacks by centralizing network control, separating the control and data planes, and enhancing security management [15].

The study highlights SDN's advantage over traditional architectures in mitigating sophisticated DDoS tactics. Future research should focus on integrating more APIs, deploying real-time SDN controllers, expanding network parameter analysis, and evaluating SDN's scalability and resilience against subnet attacks [16].

This study introduces predictive queuing analysis to enhance security in SD-UAV networks against zero-day cyber attackers. By forecasting network performance metrics like interarrival times, transmission delay, and packet counts, the analysis aids in training ML models for detecting anomalies such as jamming and black hole attacks. The approach mitigates the impact of novel attacks by generating dynamic training data and enabling rapid response through real-time predictions, integrated with an IDS within the SDN framework [17].

The research develops a predictive approach using deep learning and ML to improve intrusion detection and prevention in WSNs for Industry 4.0. It integrates ML models, including Support Vector machine (SVM) and Random Forest, and DL models like Long Short Term Memory (LSTM) and CNN, to achieve high detection accuracy. The approach addresses various attack scenarios, showing improved robustness and efficiency over traditional methods [18].

The survey evaluates the effectiveness of RL, AI, DL, and ML in combating various cyber-attacks like phishing, malware, and denial of service. Leveraging these technologies with extensive datasets enables rapid detection and response to evolving threats. While challenges such as data quality, computational requirement, and false positives exist, AI and ML offer significant advantages in enhancing security. Future advancement in these technologies promises even more robust defense mechanisms. Moreover, AI tools like ChatGPT show potential in cybersecurity, although they also pose risks that need addressing [19].

This research evaluated various intrusion detection classification system, focusing on using the Particle Swarm Optimization (PSO) technique combined with SVM classifiers (PSO+SVM).

The study utilized the KDD-CUP 99 dataset to assess the effectiveness of these detection algorithms. Evaluation criteria including F1-score, specificity, accuracy, recall precision, and reliability were analyzed for different types of cyber-attacks. This approach demonstrated improved detection capabilities for network anomalies and enhanced the overall accuracy and reliability [20].

This chapter introduces an LCC-RF-RFEX feature selection method to enhance intrusion detection system. It focuses on detecting specific DoS attacks using KDD-CUP 99 dataset and Snort. The method selects optimal features for each attack type and converts them into detection patterns for non-ML detectors. Experimentally, the approach was applied to Snort, creating new protocols for conducting DoS attacks. Compared to the existing Snort rule set, the enhanced rules significantly improved detection: false positives for Synflood and Teardrop attacks were reduced by 25.28%, and alerts for mail bomb attacks saw a reduction of 98.87%. This demonstrates a notable improvement in Snort's detection capabilities [21].

Implementing the XGBoost algorithm for the identification of phishing URLs: A combination of multiple decision trees effectively handles class imbalances and reduces overfitting, while hyperparameter tuning and feature importance analysis improve adaptability. Integrating blacklist datasets with XGoost strengthens defenses against phishing by promptly blocking known malicious URLs and accurately identifying new threats. This combination provided a multi-layered approach to enhance user security against evolving phishing attacks [22].

This research evaluates ten ML techniques for anomaly identification and multi-class detection of cyber-attacks, including Naïve Bayes, SVM, LR, and XGBoost. Experiments on NSL-KDD and UNSW-NB15 datasets demonstrate that these models effectively detect various cyber threats. The study provides a comprehensive analysis of data-driven security modeling, enhancing understanding of different ML methods in cybersecurity [23].

Table 11.1 outlines various research gaps identified in current literature on cybersecurity utilizing effectiveness of ML models. Each research work is accompanied by a brief description of its limitations. These limitations may include factors such as dataset constraints, algorithmic shortcomings, or practical challenges in real-world implementation. Understanding these limitations is crucial for informing future research directions and refining ML approaches to enhance cybersecurity measures effectively.

Table 11.1 Identified research gaps in current literature

Pub. year	*Title of the paper*	*Author name*	*Techniques used*	*Research gaps*	*Future scope*
2022	["Machine Learning for Wireless Sensor Networks Security: An Overview of Challenges and Issues"]	Ahmad et al. [15]	Supervised learning, clustering and neural networks	Energy constraints, limited resources, scalability, real-time processing, data privacy	Energy-efficient ML, real-time solutions, scalability, SDN integration, adaptive models
2022	["Machine Learning Enabled Techniques for Protecting Wireless Sensor Networks by Estimating Attack Prevalence and Device Deployment Strategy for 5G Networks"]	Kumar et al. [16]	SDN, centralized control, dynamic DDoS mitigation, API integration, network parameter analysis	Limited API integration, real-time controllers, parameter range, subnet attack evaluation, SDN scalability	Explore APIs, more controllers, diverse parameters, subnet attack assessment
2024	["Enhanced Network Metric Prediction for Machine Learning-Based Cyber Security of a Software-Defined UAV Relay Network"]	Agnew et al. [17]	Light Gradient Boosting Machine learning algorithm	Limited training data for Zero-day attacks	Integrate predictive analysis with IDS for real-time threat mitigation in SD-UAV network
2024	["A Situation Based Predictive Approach for Cybersecurity Intrusion Detection and Prevention Using Machine Learning and Deep Learning Algorithms in Wireless Sensor Networks of Industry 4.0"]	Al-Quayed et al. [18]	Random Forest, Support Vector Machine, CNN (Convolutional Neural Network) and LSTM (Long Short Term Memory)	Limited scalability and computational overhead in real-time scenarios with large-scale wireless sensor networks	Integration with edge computing and real-time, adaptive threat response mechanisms

(*Continued*)

Table 11.1 (*Continued*) Identified research gaps in current literature

Pub. year	*Title of the paper*	*Author name*	*Techniques used*	*Research gaps*	*Future scope*
2024	["A Comprehensive Survey: Evaluating the Efficiency of Artificial Intelligence and Machine Learning Techniques on Cyber"]	Ozkan-Okay et al. [19]	Artificial intelligence (AI) tools (e.g., ChatGPT), Deep learning, Reinforcement Learning, and Machine Learning	High computational demands and data quality challenges hinder smaller organization adoption	Integrate edge computing, blockchain and adaptive models for real-time, scalable cybersecurity solutions
2024	["Detection of malicious URLs using machine learning"]	Reyes-Dorta et al. [20]	Particle Swarm Optimization and Support Vector Machine	Limited by the reliance on KDD-CUP 99 dataset, impacting real-world applicability and model generalization	Extend to modern datasets, optimize for real-time performance, and integrate adaptive detention mechanism
2022	["Bridging the Last-Mile Gap in Network Security via Generating Intrusion-Specific Detection Patterns through Machine Learning"]	Sun et al. [21]	LCC-RF-RFEX method, Decision Tree models, signature-based IDS rules.	Focus on labeled data; lacks semi-supervised or unsupervised learning for unknown attacks.	Implement semi-supervised learning for unknown attacks; develop real-time signature mining agents.
2023	["Viable Detection of URL Phishing using Machine Learning Approach"	Kumari et al. [22]	XGBoost, ensemble learning, blacklist integration, feature analysis.	Lacks content analysis and user behavior tracking for comprehensive phishing detection.	Future work involves advanced feature extraction, deep learning models, and diverse network traffic datasets.
2021	["CyberLearning: Effectiveness Analysis of Machine Learning Security Modeling to Detect Cyber-Anomalies and Multi-Attacks"]	Sarker [23]	SVM, Decision Tree, Random Forest, XGBoost, Logistic Regression, Naïve Bayes, and ANN	Focuses on existing datasets; may not generalize to new or evolving cyber threats.	Incorporate high-dimensional IoT data and develop secure, data-driven models for dynamic environments.

(*Continued*)

Table 11.1 (Continued) Identified research gaps in current literature

Pub year	Title of the paper	Author name	Techniques used	Research gaps	Future scope
2021	["Detection of Phishing Websites Using Deep Learning Technique"]	Faisal Khana and Ranab [24]	DNN, CNN, LSTM, IG	New deep learning model proposed to detect phishing using features extracted from URLs effectively	In highlights, the outcomes can be improved by utilizing more heuristic elements to prepare the model
2021	["Malicious URL Detection Using Deep Learning"]	Srinivasana et al. [25]	CNN, RNN, LSTM, CHARACTER LEVEL EMBEDDING TEC. (to train the model)	This model is based on deep learning algorithms and character level embedding techniques	Exactness can be improved by considering the addition of helper modules, such as enrollment administrations, site content, network notoriety, document methods, and registry keys
2021	["A Hybrid DNN– LSTM Model for Detecting Phishing URLs"]	Ozcan et al. [26]	LSTM, DNN, N LP, CHARACTER LEVEL EMBEDDING TEC.	In this examination work proposed a half and half profound learning models use both person implanting and NLP features.	A more generic and robust model can be built by using word embedding tech.
2021	["URLTran improving Phishing URL Detection Using Transformers"]	Maneriker et al. [27]	Cutting-edge transformer models, RoBERT and BERT, lead the forefront of natural language processing advancements.	In this investigation, they performed a thorough examination of the transformer model on the phishing identification task.	In feature, results can be improved by using explicit danger models when adversarial expands the preparation information utilized for preparing them.
2021	["A Malicious URL Detection Model Based on Convolutional Neural Network"]	Wang [28]	DCNN, Word Embedding, Character Embedding	Proposed phishing model with deep learning achieves higher precision with embedding	In feature, can improve the accuracy by simplifying detection model. (Try to select simple architecture to implement DCNN)
2020	["A Comprehensive Survey of AI-Enabled Phishing Attacks Detection Techniques"]	Basit et al. [29]	ML, DL	This paper give an exhaustive comprehensive on Phishing Assault and simulated intelligence Tech.	More adaptable and powerful strategy including the shrewd module arrangement.

11.3 CYBER SECURITY TECHNIQUE

Cyber-attack is currently a worldwide worry that hacks the framework, and other security assaults could imperil the worldwide economy.

In this way, it is fundamental to have a phenomenal network safety technique to safeguard delicate data from high-profile security breaks. As advanced attacks rise, organizations, especially those handling sensitive data, must establish robust security measures to protect information.

11.3.1 Network safety objectives

Network safety's primary goal is to guarantee information security. The security local area gives a triangle of three related standards to safeguard the information from digital assaults. This standard is known as the CIA group of three. The CIA model is intended to direct approaches for an association's data security framework. Security breaches stem from rule neglect.

11.3.2 Confidentiality

Secrecy is identical to security that keeps away from unapproved access of data. It includes guaranteeing the information is available by the individuals who are permitted to utilize it and impeding admittance to other people. It keeps fundamental data from contacting some unacceptable individuals. Information encryption is a magnificent instance of guaranteeing classification.

11.3.3 Integrity

This standard guarantees that the information is legitimate, exact, and shielded from unapproved change by danger entertainers or unplanned client alteration. Assuming any changes happen, specific measures ought to be taken to safeguard the delicate information from defilement or misfortune and quickly recuperate from such an occasion. Likewise, it shows to make the wellspring of data authentic.

11.3.4 Availability

This rule makes the data to be accessible and helpful for its approved individuals generally. It guarantees that these are not obstructed by framework glitch or digital assaults.

11.4 TYPES OF CYBER SECURITY THREATS

A threat in online protection is a malevolent movement by an individual or association to ruin or take information, get to an organization, or upset computerized life overall. The digital local area characterizes the accompanying dangers accessible today:

11.4.1 Malware

Malware infers harmful programming, which is the most broadly perceived advanced pursuing instrument. It is used by the cybercriminal or developer to disturb or hurt a certifiable client's system. Coming up next are the critical sorts of malware made by the software engineer.

11.4.2 Virus

Malicious code spreads between devices, infecting files, spreading through networks, stealing data, and causing device damage.

11.4.3 Spyware

Surreptitious software records user behavior, potentially capturing sensitive information like credit card details for unauthorized activities like shopping or withdrawals.

11.4.4 Trojans

Malicious code masquerades as genuine software to deceive user into downloading and running it, aiming to steal or destroy data.

11.4.5 Worms

A piece of programming spreads copies of itself starting with one device then onto the next without human collaboration. It doesn't anticipate that they should affix themselves to any program to take or damage the data.

11.4.6 Adware

Adware, a marketing tool, spreads malware and displays ads without user consent, generating revenue for its creator through browser advertising.

11.4.7 Phishing

Phishing is a cybercrime where deceptive sources, posing as legitimate entities, contact targets through email, phone, or text, urging them to click malicious link. This leads to divulging sensitive information or installing malware for remote control

11.4.8 Man-in-the-middle (MITM) attack

A man-in-the-middle attack involves a cybercriminal intercepting communication between two parties, posing as them to access sensitive information from business or clients.

11.5 RESULT AND DISCUSSION

In the results and discussion section, findings are presented based on precision, accuracy, and recall of existing ML models for network security. Accuracy (Figure 11.1) showcases overall performance, while precision (Figure 11.2) highlights specific model precision. Similarly, recall (Figure 11.3) denotes model effectiveness in capturing true positives. By analyzing these metrics collectively, the ideal model for network security can be determined, considering the optimal balance between accuracy, precision, and recall. This comprehensive evaluation guided the selection of the most suitable model to enhance network security measures efficiently and effectively. Static identification is a location strategy in which an application is noticed for malignant examples without execution. Random forest and LR give good results as compared to another model existing in machine. Figures 11.1–11.3 show the graphical representation of Table 11.2.

11.6 CHALLENGES AND CONSIDERATIONS

Integrating ML methods into network security and threat detection offers multiple benefits, yet it also presents a range of challenges and considerations. This section outlines the main challenges and considerations related

Table 11.2 Comparative analysis of the performance of existing model

Ref. No.	*Accuracy*	*Precision*	*Recall*	*Technique*
	95.2	93	93.6	SVM
	84	89	78	NB
	96	98	94	DT
[30]	*96.4*	95.13	93.59	DBN
[31]	*95.86*	96.49	95.31	DBN
[32]	*91*	90	90	ANN
[24]	*98.88*	94	94.47	SVM
[25]	*93.4*	93.3	93.4	PCA AND RF
[26]	*99*	93	86	LR
[27]	*89.2*	96	95	DBN
[28]	*92.41*	92.4	92.4	ANN
[29]	*93.7*	87	94	ANN

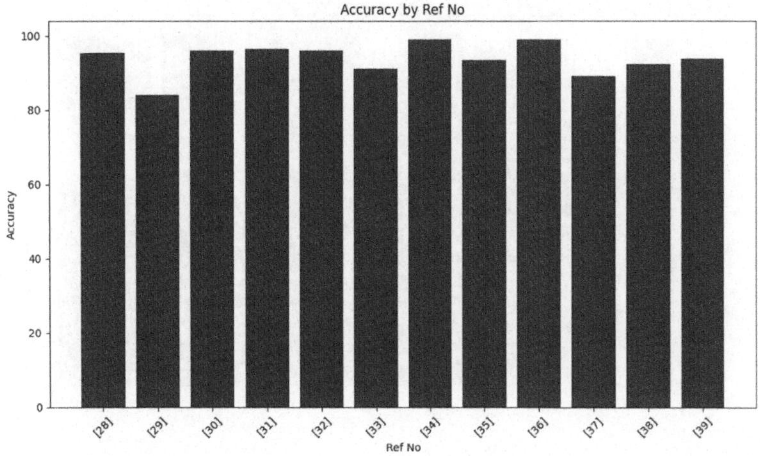

Figure 11.1 Overall performance comparison of machine learning models for network security (accuracy).

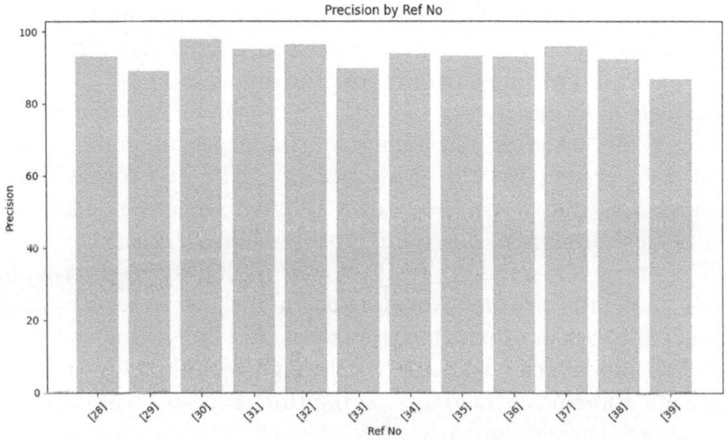

Figure 11.2 Precision analysis of machine learning models for network security.

to employing ML for improving network security and threat detection. It includes issues such as data privacy and security, the clarity and transparency of models, the robustness against adversarial attacks, scalability and computational complexity, and adherence to regulatory and ethical standards.

A major challenge in leveraging ML within network security is protecting the sensitive data employed for training and assessing these models [19]. Network security dataset often includes confidential details like IP addresses, user credentials, and network configurations, which can be at risk if not properly secured. Additionally, sharing or transferring these datasets

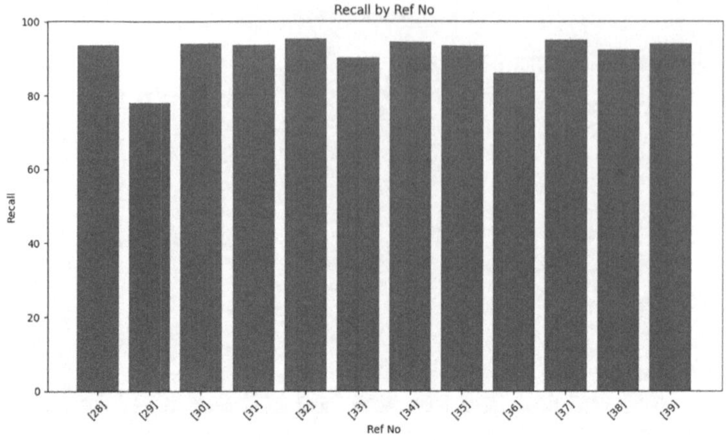

Figure 11.3 Recall assessment of machine learning models for network security.

between organizations can increase the risk of data breaches or unauthorized access. To tackle these issues related to data privacy and security, it is crucial to deploy robust encryption, access control measures, and anonymization techniques. These safeguards are necessary to safeguard sensitive information while enabling effective training of ML models.

ML models used in network security often have intricate behaviors and decision-making processes, resulting in challenges to comprehend and interpret their predictions [23]. This lack of clarity and transparency can undermine trust and transparency in ML-based security system, particularly in critical scenarios where human oversight is essential. Therefore, enhancing the interpretability of ML models is vital for identifying biases, errors, or vulnerabilities and providing useful insights to security analysts. Techniques like feature importance analysis and enhancing the transparency of ML models can be achieved through techniques such as model visualization and rule extraction, making their decision-making processes more comprehensible to humans.

Adversarial attacks represent significant risks to ML-based security system by exploiting vulnerabilities in these models to either manipulate or bypass detection mechanisms. These attacks can take various forms, such as data poisoning, evasion attacks, and model inversion attacks, all aimed at misleading ML models and undermining network security. Enhancing the robustness of ML-based security system involves deploying defense strategies like adversarial training, model diversification, and robust optimization techniques. These methods help reduce the impact of adversarial attacks and improve the ML model's ability to withstand manipulation and exploitation [18].

ML algorithm used in network security often demands substantial computational resources and processing power, particularly when handling large datasets or operating in complex network environment. Scalability and computational complexity can become significant issues, especially when deploying ML-based security solutions in real-world scenarios where efficiency is crucial to make ML-based security system practical and feasible in diverse network environments [33].

Additionally, the use of ML in network security introduces various regulatory compliance and ethical considerations concerning data protection, privacy right, and algorithmic fairness. Organizations must comply with data protection regulations like GDPR and CCPA to ensure lawful and ethical use of personal data in ML model training and deployment. Addressing algorithmic biases and ensuring fairness and transparency in ML-based security system is vital to uphold ethical standard and avoid discriminatory impacts on individuals or groups [34].

To fully leverage the benefits of ML in network security and threat detection, it is essential to address challenges related to data privacy, model interpretability, adversarial attacks, scalability, and regulatory compliance. Continuous research and collaboration across academia, industry, and regulatory bodies are necessary to address emerging issues and promote responsible and ethical use of ML in network security practices.

11.7 FUTURE PATHWAYS AND RESEARCH POSSIBILITIES

Network security is pivotal in protecting digital assets, sensitive information, and ensuring the completeness and accessibility of network resources. The integration of ML in network security has led to significant advancements, enabling more adaptive and efficient security solutions. Emerging techniques, such as DL models like recurrent neural networks (RNNs) and CNNs are increasingly used in network intrusion detection system (NIDS) to identify and address complex cyber threats by analyzing detect and mitigate anomalies [30]. Federated learning allows collaborative model training across distributed environments without centralizing data, thus preserving privacy. Explainable AI enhances the transparency of ML models, helping security analysts understand and trust the decisions made by this system [31].

Despite these advancements, further investigation is required to tackle new threats and improve the robustness of ML-based security system against adversarial attacks, evasion techniques, and model manipulation. Developing privacy-preserving ML algorithms is essential to protect sensitive data use while enabling collaborative model training. Additionally, context-aware security solutions that use network topology, user behavior, and environmental factors can help adaptively respond to dynamic security threats [32].

Organizations should foster collaboration between researchers, industry experts, and academia to tackle complex security challenges, involve in training program for security professionals, and adhere to industry best practices and standards to enhance risk management and ensure regulatory [35–47].

11.8 CONCLUSION

Network protection has turned into a question of universal concern in accomplishing upgrades in safety efforts to distinguish and respond against cyber-attacks. The recently utilized ordinary security frameworks are no longer sufficient in considering the fact that those frameworks need cadency in classifying already concealed and polymorphic assaults. AI procedures are playing a fundamental role in various uses of digital protection frameworks. Our survey here has uncovered a quickly developing interest in AI and network safety in academia and industry has brought about a number of new distributions, especially somewhat recently. In this chapter, we have overcome any issues between ML strategies and dangers to PC organizations and portable correspondence by introducing an exhaustive review of the hybrids among the two regions. This overview offers writing survey on ML methods for interruption location, spam discovery, and also malware location on PC organizations and versatile gadget. This chapter examines the use of AI models in the management of organizational well-being over the last decade. There are flaws in every advanced danger that make such digital assaults difficult to manage, even with the state-of-the-art ML model. Making one proposition for every one of the attacks considering one model is troublesome. Various measures like ID velocity moment involvedness, characterization time to recognize new "zero-day attacks, and precision of an ML model should be considered while picking a particular model to distinguish a digital assault". To depict the fundamentals of organizational security, the arrangement of digital assaults on PDAs and PCs is significant due to ML. This chapter furthermore depicted the basis of simulated intelligence. There are extensive systems in place for a juvenile to obtain superior information around here. We don't know anything about any literature that explores the purposes of ML systems in the advanced protection space on both mobile phones and PC networks in a single paper. This chapter created a graphical representation of the threats to the web and presented ML strategies to combat these cyber-crimes. We have similarly provided appraisal estimates to evaluate any more tasteful activities. The dataset is outstandingly essential for the arrangement and taxing of ML technique. We have introduced a depiction of consistently utilized safekeeping datasets. There is a team of specialists on hand, as well as benchmark datasets for each risk area. AI systems were not essentially expected to work with computerized insurance. By providing opposing data sources, evasion can doom the ML model without a significant stretch. Reliable simulated intelligence is the safeguarded use of computer-based intelligence techniques for the web to give some critical-level rightness guarantees as opposed to the rapidity and correctness of the model. We have furthermore added the critical trouble of using AI techniques in network security as well as given a wide list around here. The referred to hardships merit thought for future investigation.

REFERENCES

1. A. Talpur and M. Gurusamy, "Machine Learning for Security in Vehicular Networks: A Comprehensive Survey," *IEEE Communications Surveys and Tutorials*, vol. 24, no. 1, pp. 346–379, 2022, doi: 10.1109/COMST.2021.3129079.
2. M. Shen, K. Ye, X. Liu, L. Zhu, J. Kang, and S. Yu, "Machine Learning-Powered Encrypted Network Traffic Analysis: A Comprehensive Survey," *IEEE Communications Surveys and Tutorials*, vol. 25, no. 1, pp. 791–824, 2023, doi: 10.1109/COMST.2022.3208196.
3. R. Boutaba, M. A. Salahuddin, N. Limam, S. Ayoubi, N. Shahriar, F. Estrada-Solano, and O. Caicedo, "A Comprehensive Survey on Machine Learning for Networking: Evolution, Applications and Research Opportunities," *Journal of Internet Services and Applications*, vol. 9, no. 1, Dec. 2018, doi: 10.1186/s13174-018-0087-2.
4. Y. Xin, L. Kong, Z. Liu, Y. Chen, Y. Li, and H. Zhu, "Machine Learning and Deep Learning Methods for Cybersecurity," *IEEE Access*, vol. 6, pp. 35365–35381, May 2018, doi: 10.1109/ACCESS.2018.2836950.
5. K. Shaukat, S. Luo, V. Varadharajan, I. A. Hameed, and M. Xu, "A Survey on Machine Learning Techniques for Cyber Security in the Last Decade," *IEEE Access*, vol. 8, pp. 222310–222354, 2020, doi: 10.1109/ACCESS.2020.3041951.
6. H. Meziane and N. Ouerdi, "A Survey on Performance Evaluation of Artificial Intelligence Algorithms for Improving IoT Security Systems," *Scientific Reports*, vol. 13, no. 1, Dec. 2023, doi: 10.1038/s41598-023-46640-9.
7. P. Dini and S. Saponara, "Analysis, Design, and Comparison of Machine-Learning Techniques for Networking Intrusion Detection," *Designs*, vol. 5, no. 1, pp. 1–22, 2021, doi: 10.3390/designs5010009.
8. D. Dasgupta, Z. Akhtar, and S. Sen, "Machine Learning in Cybersecurity: A Comprehensive Survey," *Journal of Defense Modeling and Simulation*, vol. 19, no. 1, pp. 57–106, Jan. 2022, doi: 10.1177/1548512920951275.
9. A. L. Buczak and E. Guven, "A Survey of Data Mining and Machine Learning Methods for Cyber Security Intrusion Detection," *IEEE Communications Surveys and Tutorials*, vol. 18, no. 2, pp. 1153–1176, Apr. 2016, doi: 10.1109/COMST.2015.2494502.
10. M. A. Al-Garadi, A. Mohamed, A. K. Al-Ali, X. Du, I. Ali, and M. Guizani, "A Survey of Machine and Deep Learning Methods for Internet of Things (IoT) Security," *IEEE Communications Surveys and Tutorials*, vol. 22, no. 3, pp. 1646–1685, Jul. 2020, doi: 10.1109/COMST.2020.2988293.
11. P. Mishra, V. Varadharajan, U. Tupakula, and E. S. Pilli, "A Detailed Investigation and Analysis of Using Machine Learning Techniques for Intrusion Detection," *IEEE Communications Surveys and Tutorials*, vol. 21, no. 1, pp. 686–728, Jan 2019, doi: 10.1109/COMST.2018.2847722.

12. Y. Wu, D. Wei, and J. Feng, "Network Attacks Detection Methods Based on Deep Learning Techniques: A Survey," 2020, doi: 10.1155/2020/8872923.
13. M. Ahmed, R. Seraj, and S. M. S. Islam, "The K-Means Algorithm: A Comprehensive Survey and Performance Evaluation," Aug. 01, 2020, doi: 10.3390/electronics9081295.
14. S. V. N. Santhosh Kumar, M. Selvi, and A. Kannan, "A Comprehensive Survey on Machine Learning-Based Intrusion Detection Systems for Secure Communication in Internet of Things," *Computational Intelligence and Neuroscience*, vol. 2023, pp. 1–24, Jan. 2023, doi: 10.1155/2023/8981988.
15. R. Ahmad, R. Wazirali, and T. Abu-Ain, "Machine Learning for Wireless Sensor Networks Security: An Overview of Challenges and Issues," Jul. 01, 2022, doi: 10.3390/s22134730.
16. P. Kumar et al., "Machine Learning Enabled Techniques for Protecting Wireless Sensor Networks by Estimating Attack Prevalence and Device Deployment Strategy for 5G Networks," *Wireless Communications and Mobile Computing*, vol. 2022, 2022, doi: 10.1155/2022/5713092.
17. D. Agnew, A. Del Aguila, and J. Mcnair, "Enhanced Network Metric Prediction for Machine Learning-Based Cyber Security of a Software-Defined UAV Relay Network," *IEEE Access*, vol. 12, pp. 54202–54219, 2024, doi: 10.1109/ACCESS.2024.3387728.
18. F. Al-Quayed, Z. Ahmad, and M. Humayun, "A Situation Based Predictive Approach for Cybersecurity Intrusion Detection and Prevention Using Machine Learning and Deep Learning Algorithms in Wireless Sensor Networks of Industry 4.0," *IEEE Access*, vol. 12, pp. 34800–34819, 2024, doi: 10.1109/ACCESS.2024.3372187.
19. M. Ozkan-Okay et al., "A Comprehensive Survey: Evaluating the Efficiency of Artificial Intelligence and Machine Learning Techniques on Cyber Security Solutions," *IEEE Access*, vol. 12, pp. 12229–12256, 2024, doi: 10.1109/ACCESS.2024.3355547.
20. N. Reyes-Dorta, P. Caballero-Gil, and C. Rosa-Remedios, "Detection of Malicious URLs Using Machine Learning," *Wireless Networks*, 2024, doi: 10.1007/s11276-024-03700-w.
21. X. Sun, D. Zhang, H. Qin, and J. Tang, "Bridging the Last-Mile Gap in Network Security via Generating Intrusion-Specific Detection Patterns through Machine Learning," *Security and Communication Networks*, vol. 2022, 2022, doi: 10.1155/2022/3990386.
22. M. S. Kumari, C. K. Priya, G. Bhavya, H. Neha, M. Awasthi, and S. Tripathi, "Viable Detection of URL Phishing Using Machine Learning Approach," in *E3S Web of Conferences, EDP Sciences*, Oct. 2023, doi: 10.1051/e3sconf/202343001037.
23. I. H. Sarker, "CyberLearning: Effectiveness Analysis of Machine Learning Security Modeling to Detect Cyber-Anomalies and Multi-Attacks," *Internet of Things*, vol. 14, Jun. 2021, doi: 10.1016/j.iot.2021.100393.
24. N. Omer, A. H. Samak, A. I. Taloba, and R. M. Abd El-Aziz, "Cybersecurity Threats Detection Using Optimized Machine Learning Frameworks," *Computer Systems Science and Engineering*, vol. 48, no. 1, pp. 78–95, 2024, doi: 10.32604/csse.2023.039265.

25. H. Vargas, C. Lozano-Garzon, G. A. Montoya, and Y. Donoso, "Detection of Security Attacks in Industrial IoT Networks: A Blockchain and Machine Learning Approach," *Electronics*, vol. 10, no. 21, Nov. 2021, doi: 10.3390/electronics10212662.
26. M. Macas and C. Wu, "Review: Deep Learning Methods for Cybersecurity and Intrusion Detection Systems," in *Proceedings -2020 IEEE Latin-American Conference on Communications, LATINCOM 2020*, Nov. 2020. doi: 10.1109/LATINCOM50620.2020.9282324.
27. N. Mohamed, "Current Trends in AI and ML for Cybersecurity: A State-of-the-Art Survey," *Cogent Engineering*, vol. 10, no. 2, 2023, doi: 10.1080/23311916.2023.2272358.
28. A. Halbouni, T. S. Gunawan, M. H. Habaebi, M. Halbouni, M. Kartiwi, and R. Ahmad, "Machine Learning and Deep Learning Approaches for CyberSecurity: A Review," *IEEE Access*, 2022, doi: 10.1109/ACCESS.2022.3151248.
29. Md. Faisal Khana and B. L. Ranab, "Detection of Phishing Websites Using Deep Learning Techniques," *Turkish Journal of Computer and Mathematics Education*, vol. 12, no. 10, pp. 3880–3892, 2021, April 2021.
30. S. Srinivasana, R. Vinayakumar, A. Arunachalam, M. Alazab, and K. Soman, "DURLD: Malicious URL Detection Using Deep Learning-Based Character Level Representations, in *Malware Analysis Using Artificial Intelligence and Deep Learning*, Springer, Cham, 2021, doi: 10.1007/978-3-030-62582-5_21.
31. A. Ozcan, C. Catal, E. Donmez, and B. Senturk, "A hybrid DNN– LSTM Model for Detecting Phishing URLs," *Neural Computing and Applications*, 2021, doi: 10.1007/s00521-021-06401-z.
32. P. Maneriker, J. W. Stokes, E. G. Lazo, D. Carutasu, F. Tajaddodianfar, and A. Gururajan, "URLTran: Improving Phishing URL Detection Using Transformers," arXiv preprint arXiv:2106.05256, 2021.
33. Z. Wang, X. Ren, S. Li, B. Wang, J. Zhang, and T. Yang, "A Malicious URL Detection Model Based on Convolutional Neural Network", *Security and Communication Networks*, vol. 2021, p. 5518528, 2021, doi: 10.1155/2021/5518528.
34. A. Basit, M. Zafar, X. Liu, A. R. Javed, Z. Jalil, and K. Kifayat, "A Comprehensive Survey of AI-Enabled Phishing Attacks Detection Techniques," *Telecommunication Systems*, 2020, doi: 10.1007/s11235-020-00733-2.
35. V. K. Sharma, "CNTFET Circuit-Based Wide Fan-in Domino Logic for Low Power Applications," *Journal of Circuits Systems and Computers*, vol. 31, no. 02, p. 2250036, 2022.
36. V. K. Sharma and M. Pattanaik, "A Reliable Ground Bounce Noise Reduction Technique for Nanoscale CMOS Circuits," *International Journal of Electronics*, vol. 102, no. 11, pp. 1852–1866, 2015.
37. V. K. Sharma, S. Patel, and M. Pattanaik, "High Performance Process Variations Aware Technique for Sub-Threshold 8T-SRAM Cell," *Wireless Personal Communications*, vol. 78, no. 1, pp. 57–68, 2014.
38. V. K. Sharma and M. Pattanaik, "Techniques for Low Leakage Nanoscale VLSI Circuits: A Comparative Study," *Journal of Circuits System and Computers*, vol. 23, no. 05, p. 1450061, 2014.

39. V. K. Sharma, "A Survey of Leakage Reduction Techniques in CMOS Digital Circuits for Nanoscale Regime," *Australian Journal of Electrical and Electronics Engineering*, vol. 18, no. 4, pp. 217–236, 2021.
40. C. Kumar, A. S. Mishra, and V. K. Sharma, "Leakage Power Reduction in CMOS Logic Circuits Using Stack ONOFIC Technique," in *2018 Second International Conference on Intelligent Computing and Control Systems* (ICICCS), Madurai, 2018.
41. V. K. Sharma and M. Pattanaik, "Design of Low Leakage Variability Aware ONOFIC CMOS Standard Cell Library," *Journal of Circuits Systems and Computers*, vol. 25, no. 11, p. 1650134, 2016.
42. S. F. Naz, S. Riyaz, and V. K. Sharma, "A Review of QCA Nanotechnology as an Alternate to CMOS," *Current Nanoscience*, vol. 18, no. 1, pp. 18–30, 2022.
43. S. Riyaz and V. K. Sharma, "Design of Reversible Feynman and Double Feynman Gates in Quantum-Dot Cellular Automata Nanotechnology," *Circuit World*, vol. 49, no. 1, pp. 28–37, 2023.
44. B. Raina, C. Verma, M. Gupta, and V. K. Sharma, "Binary coded decimal (BCD) Seven Segment Circuit Designing Using Quantum-Dot Cellular Automata (QCA)," in *2021 5th International Conference on Trends in Electronics and Informatics* (ICOEI), Tirunelveli, 2021.
45. U. Mushtaq and V. K. Sharma, "Performance Analysis for Reliable Nanoscaled FinFET Logic Circuits," *Analog Integrated Circuits and Signal Processing*, vol. 107, no. 3, pp. 671–682, 2021.
46. Kajal and V. K. Sharma, "Design and Simulation of FinFET Circuits at Different Technologies," in *2021 6th International Conference on Inventive Computation Technologies* (ICICT), Coimbatore, 2021.
47. V. K. Sharma, "Designing of Face Recognition System," in *2019 International Conference on Intelligent Computing and Control Systems* (ICCS), Madurai, 2019.

Chapter 12

Role of UAVs during disaster in urban or city block environment

Yogesh Shastri, Sachin Kumar Gupta, Amina Khan, and Sumeet Gupta

12.1 INTRODUCTION

Natural or man-made disasters are a horrible challenge for any nation. The occurrence of a calamity is accompanied by the loss of life. Natural catastrophes, including those caused by hydrological, meteorological, geophysical, and climatological factors, invariably result in a significant and unpredictable loss of life and damage to property. Every time a disaster strikes, it causes extensive damage in the post-disaster phase since the full extent of the impacted area is not known. It is highly difficult for the rescue team to perform rescue services in a disaster scenario; in addition to its being difficult to locate the populated area in a disaster, it is also tough for rescue teams to be there on time without understanding the status of the place [1]. Large-scale natural disasters like floods, tsunamis, and earthquakes happen everywhere and have disastrous effects. Victims of unforeseen tragedies frequently need to reach out to the outside world as soon as they can for assistance and ground communication networks may be partially or completely destroyed by natural disasters like storms, flooding, or earthquakes [2,3]. Recent natural catastrophes have produced significant damage and created emergency situations that have endangered lives. One of the main reasons for such large losses is the lack of an efficient disaster assistance network management system to aid in the timely dissemination of emergency information. Health monitoring, search and rescue (SAR) operations, and disaster management are crucial applications that call for the localization of items with extreme accuracy and sometimes speed. Facilities, equipment and central management systems, particularly communication ground stations, can be destroyed by disasters, which results in a lack of key resources and a lapse in communication linkages. During a natural catastrophe, SAR teams must stay in touch with one another, the command centre, and the victims; in a closed-loop system, the victims must communicate their location and request rescue assistance [4]. Unmanned aerial vehicles (UAVs), which have strong Line-of-Sight (LoS) links, cheap cost, and controllable mobility, are seen as being crucial to emergency information diffusion since they act as flying base stations (BSs) to support terrestrial users who have

DOI: 10.1201/9781003643609-12

lost connection to ground BSs. With the ability of saving lives, utilizing UAVs in SAR operations dramatically reduces time, money, and effort costs. UAVs can serve as airborne BSs and offer immediate wireless connectivity to disaster-affected areas to address this challenge [5].

UAVs, also known as drones, have been increasingly used in disaster SAR operations due to their ability to gather real-time information in hazardous and hard-to-reach areas [6]. UAVs are aerial objects similar to aeroplanes, but they differ from them in that they lack an onboard pilot [7]. UAVs are referred to as remotely operated or autonomous systems since they can be utilized to carry out surveillance or recognition operations. They are mostly used in applications like disaster response, environmental change monitoring, mapping and surveying, SAR, and others [8,9]. UAVs that can fly on their own, often known as airborne drones or just drones, have become a popular mass-market technology in the last 10 years. UAV systems are getting more stable and developed thanks to advancements in computer vision and sensor technology, which will enable them to complete these challenging jobs more affordably, securely, and reliably [10,11]. Due to its mobility (the ability to fly at varying speeds, hover, and quickly reach confined spaces), UAVs have been investigated for SAR missions. They are perfect for jobs that could be dangerous or difficult for humans to complete [12,13].

For SAR missions, UAVs typically come with a variety of sensors. Due to their expanding availability and exceptional camera directing abilities, UAVs are frequently employed to assist human sensory systems in damage assessment and evaluation, area mapping or surveying, and visual inspection [14,15]. Utilizing autonomy and giving them access to UAV implies enabling independent task scheduling, decision-making, and coordination. This also implies that without a centralized control, UAVs must collect the necessary data, interpret it, and execute control actions on their own. This is difficult and mission-failing risky, thus it's crucial to be aware of the key problems in UAV communication networks [16]. Depending on how many UAVs are in use, UAV networks can be divided into single-UAV networks and multi-UAV networks. The UAV is attached to a BS or satellite in a single-UAV network. A multi-UAV network, on the other hand, consists of many connected UAVs that collaborate to complete the job. Compared to a single-UAV network with self-recovery capabilities, a multi-UAV network has advantages. As a result, successful UAV collaboration is necessary in a multi-UAV network to perform missions. The creation of an inter-UAV wireless communication network is required as a result [17,18].

One of the most significant issues that might arise during a UAV's autonomous navigation flight is path planning. Typically, autonomous navigation is described as the process of selecting the best route from one node to the next [19,20]. The creation of a cost function or reward space is typically required for path-planning algorithms [21]. A UAV's numerous hazards and flying path restrictions are reflected in cost functions. In general, a track is better and easier to adopt the lower the cost function is [22]. Two-dimensional (2D) path-planning algorithms are used to determine the shortest path for UAVs in an environment with obstacles, allowing the robots to complete their missions as quickly as feasible [23].

12.1.1 Objective of the study

- The objective of this chapter is to develop efficient paths for UAVs to navigate across a city block or urban environment during disaster. This ensures that the UAVs can efficiently and immediately arrive at their designated places and UAVs can optimize their flight paths, lowering the chance of accidents and raising the overall success rate of missions to respond to disasters in city block or urban environment.
- The objective is to provide UAVs a secure and safe path that enables UAVs move from a specified start position to a goal position during disaster in a city block or urban environment while avoiding obstacles, minimizing flight time, avoiding collisions, and also enhancing the UAVs' stability and reducing energy consumption.

The rest of this chapter is organized as follows. In Section 12.2, the review literature of the chapter is presented. Section 12.3 describes the strategies used to provide a solution for a given problem in a logical and systematic way to achieve our aims and objectives. The simulation results and analysis are given in Section 12.4. The conclusion and future directions are drawn in Section 12.5.

12.2 REVIEW OF LITERATURE

UAV flight route planning in three dimensions (3D) is a significant decision-making and management challenge in disaster emergency response. In order to reduce flying duration and the risk of colliding with terrain, the best flight path is anticipated to strike a balance between their lengths. But the old approaches make it difficult to strike a balance between these issues, and the optimized goal functions lack real-world restrictions, which results in erroneous modelling. Furthermore, the classic gradient optimization algorithms lack the ability to accurately optimize the complicated multimodal objective space, leading to a nonoptimal path. Thus, a reliable UAV 3D path-planning method using an improved multi-objective swarm intelligence algorithm (APPMS) is suggested in this study [24]. The APPMS method transforms the path-planning problem into a multi-objective optimization task involving several obstacles, simultaneously optimizing the objectives depending on the overall length of the flight path and the level of terrain hazard. Also, a precise swarm intelligence search method built on enhanced ant colony optimization is introduced in order to find the best UAV 3D flying path. This method may enhance both local and global search abilities by employing the preferred search orientation and random neighbourhood search mechanism.

The paper [25] highlights our preliminary research to demonstrate the benefits of UAV systems and focuses on the communication technologies that may give life to UAV disaster management systems. The research was conducted to illustrate the advantages of UAV systems. Image processing, early warning systems, and emergency communication are some of the technologies that fall under this category.

Per Bayerlein et al. [26], the problem being addressed involves planning the optimal path for a group of UAVs that work together without communication, and are all the same in terms of their capabilities. The goal is to maximize the amount of data collected from various sensor nodes distributed throughout an IoT (Internet of Things) network. The UAVs need to avoid collisions and are also limited by the amount of time they can spend in the air. The issue of planning a path for an UAV is transformed into a problem known as a decentralized partially observable Markov decision process (POMDP). The researchers use a Deep Reinforcement Learning (DRL) method to solve this problem. By doing so, they create an optimal control policy for the UAV, even though they don't have any prior information about the difficult wireless channel conditions that can be found in congested urban areas.

UAVs have recently been used in a wide variety of different applications due to the numerous advantages they bring. Some of these benefits include their versatility, which allows them to be used in a variety of settings, as well as their reduced operating costs and quicker deployment times. One of the most common applications for the use of UAVs instead of a human operation is in the field of SAR. This is especially true since SAR places constraints on the expenditures, human resources, mental and perceptual capabilities of the human operators. The purpose of research in reference [27] is to propose a real-time path-planning system for SAR operations that takes use of multiple different UAVs working together as a team. The method known as Particle Swarm Optimization is used in the process of resolving a problem involving Model Predictive Control (MPC). The goal of this challenge is to conduct search operations in a certain area of interest in a manner that is consistent with the recommendations of international SAR standards. In the event that there is wind present, the MPC is equipped with a coordinated turn kinematic model that may be used during level flight. The solution has been finished to the point where it can be installed in the onboard computer of the UAV alongside DUNE, which is navigation software for the onboard computer. This means that the solution is now ready for use. To conduct the performance study, the Software-In-The-Loop feature of Ard pilot along with the JSBSim flight dynamics model simulations were used. According to the data, the team can attain a 50% likelihood of success 2.35 times faster when employing three UAVs rather with only one UAV. This is because three UAVs increase the team's overall coverage area.

One of the major issues for post-disaster management is the construction of a reliable, resilient, and fast deployable emergency communication network. This is one of the most important challenges. Panda et al. [28] suggest the establishment of a UAV-assisted emergency Wi-Fi network to speed up the rescue operations by sending the survivors to the location of the rescue camp that is located the closest to them. In this hypothetical situation, a Raspberry PI (RPI) development board may construct a Wi-Fi chain network throughout the impacted region if it were mounted on an UAV. The key contributions of this study are the proposed solutions for design problems that arise during the setting up of a network. These problems include UAV synchronization, the prevention of communication disturbances, and the management of surveillance data. The UAV network that was established can carry out monitoring operations locally and transmit the gathered information to the relief centre; as a result, there will be improved capabilities for carrying out rescue operations. The topic of how to alert a survivor about the emergency network is one of the most serious problems. One answer to this problem is to create a captive gateway, which is one of the possible solutions. In addition, an application that is based on Android is now being developed to extend the coverage area of the Wi-Fi network. With the help of this application, each user's smartphone may act as a relay for the user's device located in the next room. There are three distinct types of field tests that are carried out to provide an accurate picture of how effectively the desired prototype functions in the real world. According to the results of the field study, the ad-hoc mode and the transmission control protocol are less suitable for the creation of a network in comparison to the Wi-Fi access point mode and the user datagram protocol, which are better suited for the construction of a network in comparison to each other. Another conclusion that can be drawn from this experiment is that the maximum hop distance for the prototype is 280m when using a Wi-Fi configuration that adheres to the IEEE 802.11n protocol, and 290m while using a Wi-Fi configuration that adheres to the IEEE 802.11ac protocol. Both numbers were derived from the experiment.

A unique path-planning method is put forth by Monwar et al. [29] to perform energy-efficient inspection while adhering to strict energy availability requirements for each UAV. The algorithm is designed to address the challenge of limited energy availability for each individual UAV. The approach considers all factors that impact the energy consumption of the UAVs during the inspection process, such as flying, hovering, and transmitting data. By taking a comprehensive view of energy usage, the algorithm aims to optimize the energy consumption of the entire UAV swarm, resulting in more efficient and effective inspections.

In recent years, the variety of applications for UAVs, which include farming, urban services, entertainment, and even plain old hobby use, has seen a tremendous surge. When it comes to serious and commercial applications, UAVs have been investigated in the study. UAVs have been investigated as mobile sensing/actuation platforms, which means that they have been investigated as a delivery platform for an increasingly wide array of sensors and actuators. Both NEC Research and TNO are doing research into the use of UAVs and swarms of UAVs for scenarios in which conventional methodologies are inadequate due to high-resolution requirements, prohibitive settings, or tight time restrictions. The study by Hildmann and Kovacs [30] focuses on the timely data collection that can be accomplished using UAVs, as well as the cost-effectiveness and highly rich data that can be obtained from doing so. In this review paper, we provide a brief overview of the application fields that we have determined to be important targets for industrial and financial initiatives, particularly in the field of intelligent autonomous systems and autonomously or semi-autonomously operating swarms. These application areas place a strong emphasis on safety and security, and since we recognized them as primary objectives, we decided to focus our attention on them. We discuss a range of concerns that are linked to the deployment of UAVs in general, as well as the deployment of UAVs within the specific application areas that have been chosen. Considering this, the objective of this article is not only to offer a review and overview of the current body of research as well as the state of the art, but also to provide a look into our prospective (near-term) future work and the challenges that we will meet there. This is because the purpose of this article is to provide a review and overview of the existing body of research as well as the state of the art.

Zhao et al. [31] examined the path-planning problem for a UAV to collect location information of affected individuals in a disaster zone. The authors proposed a new algorithm called DDQN-SSQN that combines DRL, state splitting, and optimal state to find the most efficient path for the UAV. They used RSSI (Received Signal Strength Indicator) as a search indication for the disaster victims and incorporated it into the algorithm network for learning. This helped to make SAR decision-making easier for the UAV BS, ultimately improving the collection of location information for those affected by the disaster.

The article by Khaghani and Skaloud [32] presents a novel strategy for the autonomous navigation of very small UAVs. Within the navigation filter, the vehicle dynamic model (also known as VDM) is employed as the main process model for the purpose of this technique. The method described considerably improves the precision and dependability of autonomous navigation, for very small UAVs that are equipped onboard with affordable inertial measurement units (IMUs). This is especially true for autonomous navigation systems that are used in games. This is achieved without the typical INS/GNSS setup requiring the insertion of a new sensor.

This improvement is especially important in the case that GNSS service is disrupted, since this results in inertial coasting deviating from its intended course at an alarmingly rapid rate. In the context of the architecture that has been described, the solution to the VDM equations provides an estimate of the position, velocity, and attitude (PVA) of the system. This estimate is then updated inside the navigation filter based on the observations that are currently available. These observations might include data from an IMU or measurements from a GNSS. The control input to the UAV is also supplied to the VDM, which is available inside the control/autopilot system. Both systems are part of the same ecosystem. The VDM receives both inputs at the same time. It is possible for the filter to estimate the wind velocity in addition to the dynamic model parameters, in addition to forecasting navigation states and mistakes caused by IMU sensors. Throughout the autonomous stage, when the signals from satellites are not accessible owing to physical blockage or electromagnetic obstacle, for example, Monte Carlo simulations show significant improvements in navigation precision compared to conventional INS/GNSS navigation systems. These benefits may be attributed to the fact that the system is able to navigate around the blockage or interference. When the two different kinds of navigation systems are compared to one another, these benefits become clear. As GNSS is unavailable, even for a short period of time, the accuracy of both position and attitude may improve by several orders of magnitude as compared to when inertial coasting is used. It signifies that the determination of PVA is exact enough during such a circumstance to enable the UAV to navigate to its home location without utilizing any signal that is reliant on the environment the aircraft is in. This is because PVA is determined by the vehicle's position, velocity, and attitude measurements.

The usage of UAVs can significantly increase the capacity of disaster situational awareness for responders. In the study by Yu et al. [33], UAV path planning is described as an optimization problem with fitness functions for the UAV's travel distance and risk, as well as three restrictions on the UAV's height, angle, and slope. To address the issue, a differential evolution algorithm with adaptive selection and mutation constraints is proposed. Individuals are chosen in the proposed algorithm based on their fitness levels and violation of constraints. The programme then searches for the best individual among the chosen individuals and these chosen individuals are used to create mutations. The exploration and exploitation are maintained and improved by the well-designed process.

According to Gunetti et al. [34], UAVs often have the capacity to autonomously carry out a task that has been pre-planned ahead of time. UAVs often operate in settings with a high degree of environmental variability, which requires the flight plan to be dynamically adjusted. In most cases, human supervision is necessary for carrying out this task related to mission management. A software platform that is capable of autonomously carrying out the task of mission management for an UAV will be proposed as part of the work covered by this article. The structure of the system is based on a variety of conceptual ideas that together make it doable to outline an itinerary for flying. After being conceptualized, these ideas are put into action by combining traditional flight control methods with those of the intelligent agents developed by Soar. After receiving a list of objectives, the system can automatically build a flight plan and then carry it out in its entirety. This process may continue indefinitely. After giving a detailed account of each of the system's constituent parts, the next section of this article presents a summary of the results of tests that were executed inside a simulation environment that was as accurate as possible.

Alsamhi et al. [35] used the word "disaster" to describe any urgent predicament that puts the well-being of living beings at jeopardy. Natural catastrophes have a disproportionately negative impact on the public communication infrastructure, which makes it more difficult for SAR operations to take place. After a natural disaster, it is not only going to take a significant amount of time but also a lot of work to get the communication infrastructure back up and running in its full capacity. Search and rescue, often known as SAR, is an important component in mitigating the risks that natural disasters and dangerous circumstances provide to both people and the surrounding environment. As a direct result of this, there is an urgent need to immediately construct communication networks so that SAR operations may more effectively exchange vital information. The technology behind UAVs has the potential to provide substantial answers that might aid in mitigating the impact of situations like these tragedies. During SAR operations, the use of UAVs allows for the provision of a flexible and trustworthy emergency communication backbone as well as the resolution of crucial issues that may arise. This technology can be put into use in a very short period and will be of assistance in the rescue of most individuals who are impacted by a disaster. In this research, we investigate the network performance of intelligent edge computing that is helped by UAVs to speed up SAR operations and functionality. When designing the new network that we have created, we have taken several different network metrics into consideration. These metrics include latency, throughput, traffic that is delivered and received, and route loss. It has also been shown that the recommended parameter optimization leads to a significant increase in network performance, which, in the end, results in SAR operations that are much more successful when dealing with hazardous situations or natural disasters.

Research institutions have focused a significant amount of attention on UAVs, throughout the course of the last several years. This is largely attributed to the enormous variety of applications that UAVs are capable of, which includes the observation and monitoring of natural disasters, rescue operations, military operations, civic activities, and many more. The most common reason for using UAVs is to fulfil the requirements of consumers while also providing them with the services they need. The provision of wireless communication capabilities to end users is one example of this, as is the role of acting as a relay node to improve the coverage of the core network, among a great number of other applications with a similar purpose. UAVs are meant to be more versatile and capable of covering larger territories than Tethered Balloons and Loon Balloons. In any catastrophe, regardless of whether it was caused by natural forces or by people, there is a large potential risk of damage to buildings, transportation networks, communication systems, and essential services. This risk exists regardless of whether the disaster was caused by humans or by natural forces. The typical terrestrial communication infrastructure is made unusable whenever significant natural disasters take place, such as landslides, forest fires, floods, earthquakes, and so on. Because of this, individuals are faced with a wide variety of obstacles. In this specific situation, UAVs show out to be the superior choice for offering sufferers with a wireless connection that is not just rapid and inexpensive but easy and basic to set up as well. However, there are several issues, such as interference between UAVs and other BSs, coordination between UAVs, requirements for quality of service, constraints on size, weight, and power, latency, coverage, position of UAVs, and so on [36].

The development of effective disaster management is one of the most critical challenges that must be surmounted on a global scale. Even if there is an immediate and vital need for help in the aftermath of a disaster, the inaccessibility of large sections of the terrain presents one of the most major challenges that develops throughout the management process. This is because one of the most significant impediments that arises is the inability to reach certain areas of the landscape. UAVs, which have grown more prevalent in recent years, are increasingly being used to aid in a range of disaster management tasks and to contribute to an increase in the efficiency of a disaster response. This is one of the ways that UAVs are helping to enhance the efficiency of a disaster response. The many varied ways in which UAVs might be used in disaster management are the primary emphasis of this chapter. To accomplish this, the various applications of UAVs that are currently being used as well as those that have the potential to be used in emergency planning and management have been divided into four categories. These categories are as follows: coverage and information collecting; resource dispatch and survivor rescue; and communication. In addition to that, the chapter dives into the several coverage algorithms, sensor systems, and communication structures that UAVs have the capability of using in the context of emergency management [37].

Traditional methods, which are thought to entail a large amount of time and effort, as well as carry a high risk for both the search team and the victim, are the major technique that SAR teams use while searching for victims of natural disasters. These methods also present a high risk for both the search team and the victim. Because of this problem, we made the decision to do research to contribute to the hunt for those who had been affected by the calamity. During this inquiry, we designed a system that makes use of passive infrared (PIR) sensors to determine and quantify the direction angle of the victim. This was accomplished as part of the system that we built. With the use of the triangulation method, we may be able to pinpoint the exact position of the victim by using the direction angle that is derived from the observations made at each of the several observation sites. In addition to this sensor set, we designed a quadcopter UAV so that it may be transported across the area that was impacted by the tragedy. A regional website that can display the data generated by the system has been built to make monitoring more straightforward. The results of the testing indicate that the system can determine the precise position of a victim to an accuracy of within a difference of less than 5 m in distance. The technology can search for victims in an empty space 35 m long and 15 m wide in 14 minutes and 20 seconds. The data monitoring system displayed not only the position of the victim but also the flight path of the UAV as well as the locations of the observation points in addition to showing the victim's location [38].

Estrada and Ndoma [39] in their paper examine the major role that UAVs, commonly known as drones, play in the context of giving aid to victims of humanitarian crises and reacting to natural disasters. UAVs are also known as drones. The primary objective of this research is to examine how UAVs (also known as drones) in the here and now or in the not-too-distant future may be able to give aid to those who have survived a natural disaster such as a tsunami, earthquake, floods, or any other kind of natural disaster. We are going to begin by making the premise that any natural catastrophe will always have a considerable danger of inflicting damage to the infrastructure, transportation systems, access to telecommunications networks, and key services quickly after it has taken place. This is the starting point for our discussion. This article outlines three possible uses for UAVs, commonly known as drones, in the context of giving assistance to humanitarian groups and reacting to natural disasters. UAVs are also known as drones. The following are examples of these: (1) the aerial evaluation after a natural disaster, (2) the natural disaster logistic and freight delivery, and (3) the aerial surveillance after a natural disaster.

The potential applications of UAVs, more often referred to as drones, in disaster management are investigated and explored in this chapter. There is strong evidence to support the strategic investment in UAV technology to generate new operational capabilities and enhance those that currently exist. These goals may be accomplished by both developing new capabilities and enhancing those that already exist. The current state of research, in addition to the breadth of it, provides the foundation for this body of data. In order to establish these skills, the authorities need to have a solid understanding of a wide range of factors, each of which will have an effect on the formulation of a strategy for dealing with a particular category of catastrophe. The formulation of a concept of operations for various kinds of catastrophes will be influenced by the aforementioned factors and variables.

Public safety authorities and organizations will need to collaborate closely with national airspace controllers to make sure that these actions are safe and do not endanger other aircraft because the technical abilities of UAVs go beyond the scope of their authorized use. This is because the technological capabilities of UAVs exceed the permitted use of this technology. This is due to the fact that it is anticipated that the technological capabilities of UAVs will continue to advance. It will also depend on the mentality of the community and how seriously they take the potential risks that exist in order to ascertain whether or not UAVs will be effective in disaster management. It is imperative that the relevant authorities address these concerns and institute preventative measures to guard against any invasions of personal privacy or violations of individual rights. It is also vital that authorities respond to the concerns that have been raised by the public.

In partially viewable and GPS-denied environments, the difficulty of multi-agent remote sensing to discover survivors or survey points of interest continues. In papers [30–39], a framework is proposed for multi-agent target-finding that combines DRL-based control with online POMDP-based planning. The framework is put into practice with the idea that planning and control are two distinct issues. Using a contemporary online POMDP solver, the planning problem is defined as a decentralized multi-agent graph search problem. Modern DRL approaches are used to address the control problem, which is categorized as a local continuous-environment exploration problem.

12.3 RESEARCH METHODOLOGY

This section describes the strategies used to provide a solution for a given problem in a logical and systematic way to achieve our aims and objectives.

12.3.1 Methodology

- **Path planning:** To determine an appropriate route from a predetermined start position to a goal position, the study used a path-planning algorithm. The algorithm took into account the map data that was given, which included details about the city block, and classified unidentified places as vacant. The goal of the path-planning algorithm was to allow the UAVs to navigate safely across the environment by avoiding obstacles.
- **Path interpolation:** The study used UAV Dubins connections to interpolate the path after determining the original path. The reference path for the UAVs was made smoother due to this interpolation technique. The interpolated path gave the UAVs a more consistent and efficient trajectory by linking waypoints along the original path using UAV Dubins connections.

- **Fixed-wing guidance model:** The study used a fixed-wing guidance model to simulate the UAV's trajectory and verify the routes. This model calculates the overall flying time while taking the UAV's airspeed into consideration and adding a buffer to accommodate for safety. Based on the discovered paths, the fixed-wing guidance model created a simulated UAV trajectory that could be assessed and contrasted with the reference paths.
- **Path smoothing:** In addition to the methods mentioned above, the study looked into path smoothing as a way to strengthen the initial path and enhance obstacle avoidance. The study created a smoother path employing UAV Dubins connections that allowed more effective navigation while taking environmental impediments into account. The path was rounded to reduce sharp curves and enhance the UAVs' overall path quality.

To implement above techniques, a **MATLAB** toolbox is used.

12.4 RESULTS AND DISCUSSION

12.4.1 MATLAB code

The MATLAB script performs path planning for a UAV in a city block environment. It begins by loading a pre-defined occupancy map from a file. The script sets the threshold for considering unknown spaces as unoccupied. Next, it selects a start pose and goal pose for the UAV, specifying their coordinates and orientation. The script creates a 3D plot of the map and marks the start and goal poses with red and green dots, respectively. The script then checks if a valid path exists between the start and goal poses. If a path is found, it creates another plot to visualize the map and adds the start and goal poses. An interpolated path object is generated by copying the original path object and adding 1,000 intermediate points to create a smooth trajectory. This interpolated path is shown as a green line in the plot, representing the desired path for the UAV. Using a fixed-wing guidance model, the script computes the total flight time required to traverse the interpolated path. Simulated waypoints are generated based on the interpolated path, airspeed, and time to reach the goal. The simulated UAV trajectory is plotted as a red line in the 3D plot. The script adds a legend to differentiate the reference (interpolated) path from the simulated trajectory. Finally, the plot's view is adjusted for optimal visualization. In summary, the script loads an occupancy map, selects start and goal poses, and generates an interpolated path for a UAV in a city block environment. It then simulates the UAV's trajectory and provides visual representations of the planned path and simulated flight.

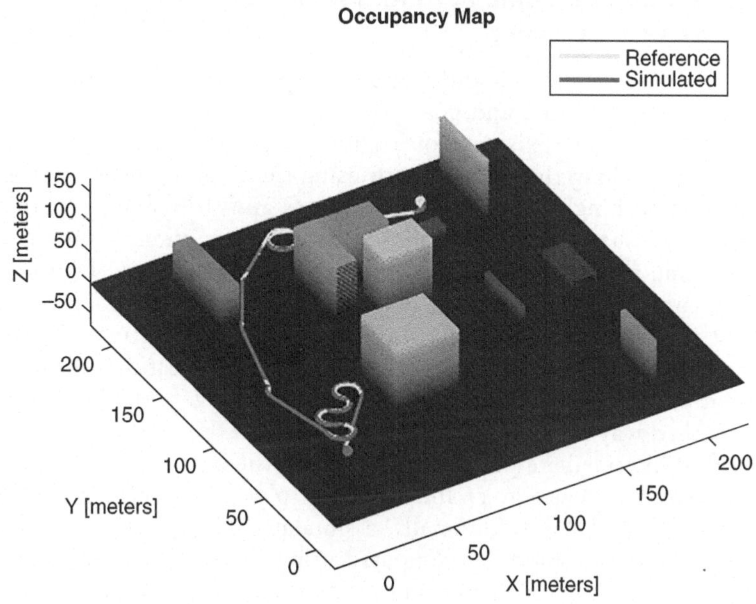

Figure 12.1 Occupancy Map-1.

If a valid path is not found between the start and goal poses, the script does not proceed to create the "OriginalPath" (Figure 12.1) or perform any further visualization or simulation. Assuming a path is found, the script displays the "OriginalPath" (Figure 12.1) with the 3D map, start and goal poses, interpolated path, and simulated UAV trajectory. This provides a comprehensive view of the planned path and how the UAV would navigate through the city block environment. The purpose of visualizing the interpolated path and the simulated UAV trajectory is to compare the planned path (reference) with the simulated trajectory and assess their consistency. The green line representing the interpolated path serves as a reference trajectory, while the red line depicts the simulated trajectory based on the fixed-wing guidance model. By plotting both the reference and simulated paths, it becomes possible to analyse any discrepancies between them and evaluate the effectiveness of the path-planning algorithm. The legend helps distinguish between the reference and simulated paths in the plot. The chosen view orientation of the 3D plot, set to [31 63], provides an optimal perspective for observing the entire path, start and goal poses, and any obstacles present in the environment.

In conclusion, the script allows for the visualization and evaluation of a planned path for a UAV in a city block environment. It provides insights into the path-planning process, the desired trajectory, and how the simulated UAV would navigate through the environment based on a fixed-wing guidance model.

12.4.2 Smooth Dubins path and simulate UAV trajectory

In the updated code, an additional step is performed if a valid path is found between the start and goal poses. Let's describe this new section. After confirming the existence of a path, the script proceeds to smooth the waypoints along the original path using the function (**exampleHelper UAVPathSmoothing**). This smoothing process aims to refine the path and eliminate any sharp turns or abrupt changes in direction, resulting in a more continuous and feasible trajectory for the UAV. A new Figure 12.2 window named "SmoothedPath" is created to display the smoothed path. The 3D map is plotted, and the start and goal poses are marked as red and green dots, respectively. Similar to the previous section, an interpolated path object (**interpolatedSmoothWaypoints**) is generated by copying the smoothed waypoints object (**smoothWaypointsObj**) and interpolating it with 1,000 intermediate points for smoother visualization. The smoothed path is then plotted as a green line in the 3D plot, representing the refined trajectory for the UAV. Additionally, a simulated flight path based on the fixed-wing guidance model is computed using the interpolated smoothed waypoints, airspeed, and estimated time to reach the goal. This simulated trajectory is plotted as a red line in the plot. A legend is added to differentiate between the smoothed reference path and the simulated trajectory. The legend labels are "Smoothed Reference" and "Simulated". The plot is

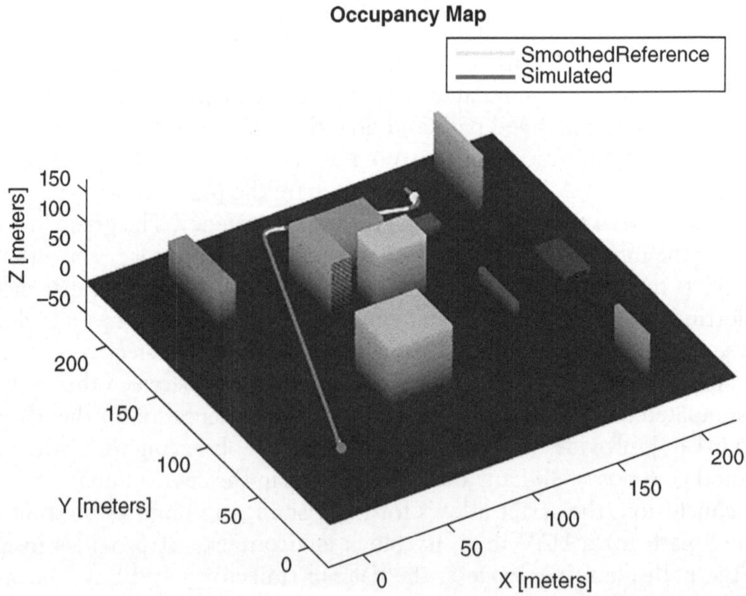

Figure 12.2 Occupancy Map-2.

finalized with a specific view orientation. In summary, this updated section of the code performs path smoothing on the original path and visualizes the smoothed path alongside the simulated flight trajectory. The smoothed path is displayed as a green line, while the simulated trajectory is shown as a red line. The purpose is to assess the effectiveness of the smoothing algorithm and compare the refined path with the simulated flight.

12.4.3 Discussion

The provided MATLAB script demonstrates the path-planning and visualization process for a UAV in a city block environment. The results obtained from running the script can be divided into two parts: the original path and the smoothed path.

A. **Original path:** The script initially checks for the existence of a path between the start and goal poses. If a path is found, it visualizes the original path in the "OriginalPath" (Figure 12.2). The 3D map is displayed, and the start and goal poses are marked with red and green dots, respectively. An interpolated path object is generated to create a smooth reference trajectory for the UAV, which is plotted as a green line. Additionally, a simulated UAV trajectory is computed using the interpolated path, airspeed, and time to reach the goal, and it is shown as a red line. The legend helps differentiate between the reference (interpolated) path and the simulated trajectory. The view of the 3D plot is set for optimal visualization.
B. **Smoothed path:** If a valid path is found, the script proceeds to smooth the original path. The "SmoothedPath" (Figure 12.2) is created to visualize the smoothed trajectory. The 3D map, start and goal poses, and interpolated smoothed waypoints are displayed. The smoothed path, represented by the interpolated smoothed waypoints, is plotted as a green line. The simulated flight path based on the smoothed waypoints, airspeed, and estimated time to reach the goal is shown as a red line. A legend is added to distinguish between the smoothed reference path and the simulated trajectory. The view of the plot is set to [–31 63].

The results obtained from this script provide valuable insights into the path-planning process for a UAV in a city block environment. The visualization allows for a clear understanding of the planned path, the effects of path smoothing, and the simulated trajectory of the UAV.

The comparison between the original path and the smoothed path can help evaluate the effectiveness of the path-smoothing algorithm. Smoothing the path aims to improve the UAV's navigation by reducing abrupt changes in direction and creating a more continuous trajectory. The visualization and simulation of the smoothed path provide a better understanding of the UAV's behaviour and feasibility. Additionally, the script enables the assessment of the UAV's trajectory based on a fixed-wing guidance model. The simulated flight path demonstrates how the UAV would navigate through the environment, taking into account factors such as airspeed and time to reach the goal. Overall, the results obtained from this script provide a comprehensive analysis of the path-planning process, path smoothing, and simulated UAV trajectory. These results can be used to evaluate and optimize the UAV's path-planning algorithm, enhancing its performance and safety in real-world scenarios.

Furthermore, the results obtained from the script can be used to analyse and address various aspects of the UAV's path planning and navigation in a city block environment. Here are some points for further discussion:

- **Path-planning efficiency:** The script's ability to find a valid path between the start and goal poses indicates the effectiveness of the path-planning algorithm. The presence of a path demonstrates that the algorithm successfully navigates the UAV through the given environment while avoiding obstacles. Assessing the computational efficiency of the path-planning algorithm can provide insights into its scalability and real-time applicability.
- **Path-smoothing impact:** The introduction of path smoothing in the script aims to enhance the UAV's trajectory by reducing sharp turns and improving the overall navigability of the path. The evaluation of the smoothed path, compared to the original path, helps assess the impact of path smoothing on the UAV's motion. Analysing factors such as path length, smoothness, and reduction of abrupt changes in direction can provide valuable insights into the benefits and trade-offs of path-smoothing techniques.
- **Simulation accuracy:** The simulation of the UAV's trajectory based on a fixed-wing guidance model allows for a realistic assessment of the UAV's behaviour during flight. The accuracy of the simulated trajectory can be examined by comparing it with the reference (interpolated) path. Evaluating factors such as trajectory tracking, adherence to the planned path, and overall stability of the UAV's motion can help determine the effectiveness of the guidance model and its suitability for real-world applications.
- **Obstacle avoidance:** The occupancy map used in the script serves as a representation of the environment, including obstacles and free spaces. The ability of the path-planning algorithm to successfully navigate the UAV through the city block environment while avoiding obstacles can be evaluated. By analysing the planned path and the UAV's simulated trajectory, it becomes possible to assess the algorithm's effectiveness in obstacle detection and avoidance.

- **Trade-offs and optimizations:** The script allows for various parameters and thresholds to be adjusted, such as the free and occupied space thresholds, interpolation resolution, and time buffer for UAV trajectory simulation. Exploring the effects of these parameters and optimizing them based on specific requirements can further improve the path-planning and navigation process. Finding an optimal balance between path smoothness, computational efficiency, obstacle avoidance, and trajectory accuracy is crucial for achieving high-performance UAV navigation.

The results obtained from the script offer a foundation for analysing and improving the path planning and navigation of UAVs in city block environments. By evaluating factors such as path-planning efficiency, impact of path smoothing, simulation accuracy, obstacle avoidance, and trade-offs in parameter tuning, researchers and engineers can enhance the performance, safety, and real-world applicability of UAV navigation systems.

The use of these methods in SAR operations for UAVs in an urban or city block environment results in enhanced efficiency, improved safety, and better resource utilization. The advantages gained include quicker response times, improved manoeuvrability, enhanced situational awareness, and optimized mission results, all of which eventually help SAR operations in urban or city block environment be more efficient and successful.

- **Collision avoidance:** UAVs can avoid collisions with buildings or structures, trees, power lines, and other obstacles in urban environments by using path-planning and path-smoothing algorithms. The UAV can manoeuvre safely and reduce the risk of accidents or destruction to the UAV and nearby infrastructure by carefully planning the path and optimizing it through smoothing.
- **Effective navigation:** UAVs can efficiently navigate through difficult urban areas with the help of path planning. The UAV can take the most efficient route while conserving time and energy by taking into account obstacles and limitations. The UAV can reach the desired place quickly due to this effective navigation, which reduces the response time in urgent circumstances.
- **Improved manoeuvrability:** Path interpolation and path-smoothing algorithms improve the UAV's manoeuvrability in confined locations and complicated urban structures. The UAV can navigate through narrow alleyways, around buildings, and in other difficult locations with better ease and accuracy by producing a smoother and continuous trajectory. This enhances the UAV's capability to enter difficult-to-reach spaces and completely search the target area.
- **Realistic trajectory simulation:** A fixed-wing guidance model that depends on the dynamics and guidance laws of the UAV enables

realistic trajectory simulation. This simulation aids to assess whether the proposed route is feasible and in making any necessary adjustments to the flight parameters. Operators may acquire information about potential difficulties and improve the UAV's performance for SAR missions by simulating the UAV's flight.

- **Enhanced situational awareness:** The optimized flight route made possible by these methods enhances the UAV's capability to gather essential information and give rescue workers real-time situational awareness. The UAV is capable of taking overhead photos, video footage, and sensor data by thoroughly covering the target region, which can help in discovering survivors, detecting dangers or hazards, and evaluating the overall situation. Rescue teams on the ground can use this information to make informed judgements and efficiently prioritize their efforts.
- **Reduced mission time and costs:** Reduced mission durations for UAVs during SAR missions are made possible by efficient navigation, shorter flying times, and optimized trajectories. UAVs can extend battery life, decrease fuel consumption (in the case of hybrid or gasoline-powered UAVs), and lower operational expenses or costs by reducing the amount of time needed to cover the target area. As a result, UAVs can cover bigger areas or complete greater numbers of missions in less time, resulting in extended operational endurance.

UAVs are now deployed to locate and provide coverage to survivors in city block or urban disaster areas after utilization of path-planning, interpolation, and path-smoothing strategies to create optimized paths; UAVs may navigate safely and effectively through the city block or urban environment using the optimized paths created by path planning, avoiding obstacles and arriving at their designated places in a timely manner. Further, Figure 12.3 shows UAVs' position.

Consider the size of urban disaster area is [1,000,1,000] m, the size of each grid cell is 100 m, altitude level is 0 to 50 m, and maximum flight time limit for each UAV is 30 minutes.

The plot's blue circles represent the UAVs' positions. Red squares on the layout represent the positions of the survivors trapped in the disaster area. The disaster area's boundary is shown by the red polygon. The spatial linkages between the UAVs, survivors, and the disaster area can be better understood by using a three-dimensional representation. By taking into consideration the X, Y, and Z coordinates, it enables us to see the locations in three dimensions. This can be especially helpful in situations when vertical location or altitude is crucial, such as when taking into account the height of UAVs or the presence of multi-story buildings in a disaster zone.

Position of number of users be: Users= [X, Y, Z] coordinates

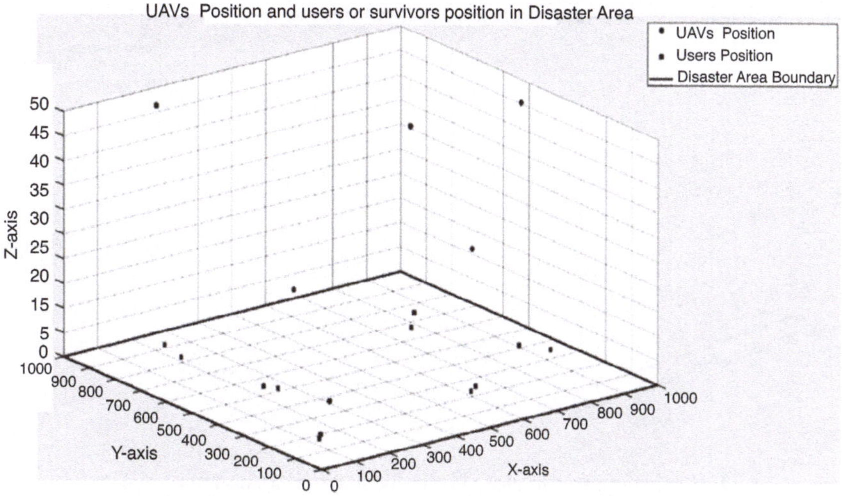

Figure 12.3 UAV position.

User1=[150, 200, 0], User2=[130, 180, 0], User3=[250, 550, 0], User4=[270, 520, 0], User5= [650, 250, 0], User6=[620, 230, 0], User7=[780, 670, 0], User8=[850, 750, 0], User9=[220, 830, 0], User10=[240, 920, 0], User11=[950, 350, 0], User12= [910, 420, 0]

Position of UAVs be: UAVs= [*X*, *Y*, *Z*] coordinates

UAV1=[100, 100, 10], UAV2=[300, 500, 20], UAV3=[600, 200, 30], UAV4=[800, 700, 40], UAV5=[200, 900, 50], UAV6=[900, 400, 50]

The maximum coverage range of each UAV is 100 m. The UAVs are positioned in such a way that they can cover the entire disaster area and can provide coverage to all users or survivors trapped in the city block or urban disaster area. When UAVs are within the range of 100 m from the survivors then UAVs automatically detect survivors and provide coverage to them. The UAVs use their onboard sensors and capabilities as part of the coverage system to locate and determine the presence of survivors in the urban disaster region. Cameras, thermal imaging devices, and even cutting-edge technology like LiDAR (light detection and ranging) may be used as these sensors. These sensors enable the UAVs to identify and locate survivors using their optical or thermal signatures. The UAVs start to prioritize their coverage and help once they find the survivors. Various strategies could be used by them in order to ensure thorough and efficient coverage. The UAVs have the ability to dynamically distribute their positions and trajectories to cover as many survivors as possible in the region.

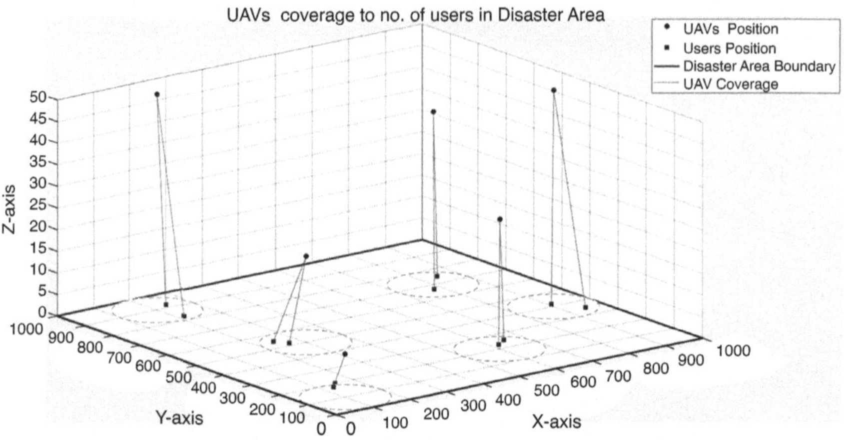

Figure 12.4 UAV coverage to survivors in city block or urban disaster area.

Figure 12.4 shows the UAVs coverage to survivors in city block or urban disaster area. UAVs can be extremely helpful in discovering and rescuing survivors during disaster SAR efforts in an area of a city block. Rapid deployment, a view from a high point, and the capacity to access difficult-to-reach regions are among the several benefits of using UAVs in such operations. The UAVs can carry out SAR operations by efficiently navigating across the city block environment after employing path-planning and path-smoothing algorithms.

- **Rapid deployment:** In disaster-stricken areas, UAVs can be immediately deployed to provide quick aid or help in SAR missions. They enable quick response times as they can be launched from a local base or alternatively from a mobile command centre.
- **Aerial surveillance:** Because UAVs have an elevated vantage point, they may observe a larger area of the city block or urban environment. They can take high-resolution photos and videos due to their cameras, infrared sensors, and other imaging equipment, giving the rescue crews a real-time awareness of the situation.
- **Search and localization:** UAVs can thoroughly search the city block area for survivors using the available map data and the intended path. Streets, alleyways, the tops of buildings and other sites that are challenging for ground troops to access can be flown over by UAVs. They can recognize signs of life, such as visual signals, heat signatures, or sounds, and communicate this knowledge to the ground-based rescue crews.
- **Mapping and assessment:** UAVs with innovative sensors can provide accurate maps of the disaster region. They are able to produce 3D models, spot structural flaws, and assess the overall condition of

structures, buildings, and infrastructure. This information aids in setting priorities for rescue efforts and ensuring the security of the teams conducting them.

- **Communication and coordination:** UAVs serve as communication intermediates between other UAVs, command centres, and ground personnel. They create wireless communication networks that act as a crucial link for communication and coordination. The efficiency and effectiveness of SAR efforts are increased overall due to this connectivity.
- **Delivery of supplies:** UAVs can deliver vital supplies, like medical kits, food, water, and communication supplies, to survivors or inhospitable locations. This ability ensures immediate assistance to individuals in need, particularly in circumstances where ground access is difficult or dangerous.
- **Remote sensing and hazard detection:** UAVs with specialized sensors can identify dangerous substances like gas leaks and hazardous materials in disaster zones. The rescue teams can use this information to determine safe paths and reduce potential risks.
- **Post-disaster assessment:** UAVs may continue to assist by conducting post-disaster assessments after the initial SAR efforts are finished. They can keep an eye out for secondary risks, assess how well the emergency response is working, and help with the organization of the recovery and reconstruction process.

12.5 CONCLUSION AND FUTURE DIRECTION

An occurrence or incident that substantially harms, devastates, or disturbs a particular region of a city or urban area is referred to as a disaster in the context of a city block or urban environment. In conclusion, this study examined the utilization of UAVs in disaster scenarios inside a city block or urban environment. The objective is to provide efficient and secure paths that enable UAVs move from a specified start position to a goal position while avoiding obstacles [40–52]. This study employed a path planning algorithm to identify an appropriate path from the starting position to the desired position. In the event that a path was discovered, the original path was shown on the map. To provide a smoother reference path, the path was also interpolated using UAV Dubins connections. Additionally, a simulated UAV trajectory with a fixed-wing guidance model was projected, taking into account the total flight time and including a safety margin. The study also investigated path-smoothing methods to improve the original path. UAV Dubins connections were used to create a smoother path, which more effectively avoided obstacles. Similar to the preceding phase, a simulated flight path based on the fixed-wing guidance model was projected following which the smoothed path had been visualized on the map. The study offered UAVs optimized paths to navigate safely and effectively in such difficult circumstances by utilizing path planning, interpolation, and path-smoothing approaches. These findings advance the subject of emergency preparedness and can help increase the efficiency of UAV deployment in urban or city block environments. We have also shown the positions of UAVs in city block or urban disaster area and UAVs positioned in such a way that they can cover the entire disaster area and can provide coverage to all users or survivors who are trapped in disaster area. To implement the above study, we have used MATLAB toolbox.

The UAVs can safely and efficiently move across the city block environment by using path planning and path-smoothing algorithms. The planned paths are designed to account for the hazards, obstacles, topography, and other restrictions inherent in the disaster region. Path smoothing improves the trajectories even more, providing that the UAVs travel continuously and smoothly. This leads to quicker reaction times, improved collision avoidance, improved situational awareness, and lower mission expenditures. In conclusion, UAVs are essential in SAR efforts following a tragedy when a city block is involved. They provide assistance in search and localization, aerial surveillance, map the region, set up communication systems, transport supplies, identify hazards, and take part in post-disaster evaluations. The use of path planning and path-smoothing techniques improves the efficacy and efficiency of UAV navigation, allowing them to successfully complete their missions.

REFERENCES

1. Barritt, Brian, et al. "Operating a UAV mesh & internet backhaul network using temporospatial SDN." In *2017 IEEE Aerospace Conference*. IEEE, 2017.
2. Khan, Amina, Sumeet Gupta, and Sachin Kumar Gupta. "Multi-hazard disaster studies: Monitoring, detection, recovery, and management, based on emerging technologies and optimal techniques." *International Journal of Disaster Risk Reduction* 47 (2020): 101642.
3. Xu, Jianwen, Kaoru Ota, and Mianxiong Dong. "Big data on the fly: UAV-mounted mobile edge computing for disaster management." *IEEE Transactions on Network Science and* Engineering 7.4 (2020): 2620–2630.
4. Khan, Amina, Sumeet Gupta, and Sachin Kumar Gupta. "Disaster management solution based on collaboration between SAR team and multi-UAV." In *International Conference on Futuristic Communication and Network Technologies*. Singapore: Springer Nature Singapore, 2021.
5. Hayat, Samira, Evşen Yanmaz, and Raheeb Muzaffar. "Survey on unmanned aerial vehicle networks for civil applications: A communications viewpoint." *IEEE Communications Surveys &* Tutorials 18.4 (2016): 2624–2661.

6. Khan, Amina, Sumeet Gupta, and Sachin Kumar Gupta. "UAV-enabled disaster management: Applications, open issues, and challenges." In *Advanced Computer Science Applications*, Karan Singh, Latha Banda, and Manisha Manjul, Eds. Apple Academic Press, New York, 2023, pp. 43–58.
7. Berie, Habitamu Taddese, and Ingunn Burud. "Application of unmanned aerial vehicles in earth resources monitoring: Focus on evaluating potentials for forest monitoring in Ethiopia." *European Journal of Remote Sensing* 51.1 (2018): 326–335.
8. Gupta, Akshita, et al. "Unmanned aerial vehicles integrated HetNet for smart dense urban area." *Transactions on Emerging Telecommunications Technologies* 33.10 (2022): e4123.
9. Shahmoradi, Javad, et al. "A comprehensive review of applications of drone technology in the mining industry." *Drones* 4.3 (2020): 34.
10. Yue, Xuejun, et al. "Software defined radio and wireless acoustic networking for amateur drone surveillance." *IEEE Communications* Magazine 56.4 (2018): 90–97.
11. Wang, Jian, et al. "Blockchain enabled verification for cellular-connected unmanned aircraft system networking." *Future Generation Computer Systems* 123 (2021): 233–244.
12. Recchiuto, Carmine Tommaso, and Antonio Sgorbissa. "Post-disaster assessment with unmanned aerial vehicles: A survey on practical implementations and research approaches." *Journal of Field Robotics* 35.4 (2018): 459–490.
13. Khan, Amina, Sumeet Gupta, and Sachin Kumar Gupta. "Unmanned aerial vehicle-enabled layered architecture based solution for disaster management." *Transactions on Emerging Telecommunications Technologies* 32.12 (2021): e4370.
14. Bähnemann, Rik, et al. "A decentralized multi-agent unmanned aerial system to search, pick up, and relocate objects." In *2017 IEEE International Symposium on Safety, Security and Rescue Robotics (SSRR)*. IEEE, 2017.
15. Khan, Amina, Sumeet Gupta, and Sachin *Kumar* Gupta. "Multi-UAV integrated HetNet for maximum coverage in disaster management." *Journal of Electrical Engineering* 73.2 (2022): 116–123.
16. Gupta, Lav, Raj Jain, and Gabor Vaszkun. "Survey of important issues in UAV communication networks." *IEEE Communications Surveys & Tutorials* 18.2 (2015): 1123–1152.
17. Arafat, Muhammad Yeasir, and Sangman Moh. "Routing protocols for unmanned aerial vehicle networks: A survey." *IEEE Access* 7 (2019): 99694–99720.
18. Khan, Amina, Sumeet Gupta, and Sachin Kumar Gupta. "Cooperative control between multi-UAVs for maximum coverage in disaster management: Review and proposed model." In *2022 2nd International Conference on Computing and* Information *Technology (ICCIT)*. IEEE, 2022.
19. Fu, Bing, et al. "An improved A* algorithm for the industrial robot path planning with high success rate and short length." *Robotics and Autonomous Systems* 106 (2018): 26–37.
20. Zhao, Leyang, et al. "Efficient and high path quality autonomous exploration and trajectory planning of UAV in an unknown environment." *ISPRS International Journal of Geo-Information* 10.10 (2021): 631.

21. Zhou, Xiaohui, et al. "Survey on path and view planning for UAVs." *Virtual Reality & Intelligent Hardware* 2.1 (2020): 56–69.
22. Zhong, Xunyu, et al. "Hybrid path planning based on safe A* algorithm and adaptive window approach for mobile robot in large-scale dynamic environment." *Journal of Intelligent & Robotic Systems* 99 (2020): 65–77.
23. Dhulkefl, Elaf Jirjees, and Akif Durdu. "Path planning algorithms for unmanned aerial vehicles." *International Journal Trend Scientific Research* Dev 3.4 (2019): 359–362.
24. Wan, Yuting, et al. "An accurate UAV 3-D path planning method for disaster emergency response based on an improved multiobjective swarm intelligence algorithm." *IEEE Transactions on Cybernetics*, 53: 2658–2671, 2022.
25. Nikhil, Naren, et al. "Unmanned aerial vehicles (UAV) in disaster management applications." In *2020 Third International Conference on Smart Systems and Inventive Technology (ICSSIT)*. IEEE, 2020.
26. Bayerlein, Harald, et al. "Multi-UAV path planning for wireless data harvesting with deep reinforcement *learning.*" *IEEE Open Journal of the Communications* Society 2 (2021): 1171–1187.
27. de Alcantara Andrade, Fabio Augusto, et al. "Autonomous unmanned aerial vehicles in search and rescue missions using real-time cooperative model predictive control." *Sensors* 19.19 (2019): 4067.
28. Panda, Kirtan Gopal, et al. "Design and deployment of UAV-aided post-disaster emergency network." *IEEE Access* 7 (2019): 102985–102999.
29. Monwar, Momena, Omid Semiari, and Walid Saad. "Optimized path planning for inspection by unmanned aerial vehicles swarm with energy constraints." In *2018 IEEE Global Communications Conference (GLOBECOM)*, 2018.
30. Hildmann, Hanno, and Ernö Kovacs. "Using unmanned aerial vehicles (UAVs) as mobile sensing platforms (MSPs) for disaster response, civil security and public safety." *Drones* 3.3 (2019): 59.
31. Zhao, Jinduo, et al. "Path planning research of a UAV base station searching for disaster victims' location information based on deep reinforcement learning." *Entropy* 24.12 (2022): 1767.
32. Khaghani, Mehran, and Jan Skaloud. "Autonomous vehicle dynamic model-based navigation for small UAVs." *Navigation: Journal of the Institute of* Navigation 63.3 (2016): 345–358.
33. Yu, Xiaobing, Chenliang Li, and JiaFang Zhou. "A constrained differential evolution algorithm to solve UAV path planning in disaster scenarios." *Knowledge-Based Systems* 204 (2020): 106209.
34. Gunetti, Paolo, Haydn Thompson, and Tony Dodd. "Simulation of a soar-based autonomous mission management system for *unmanned* aircraft." *Journal of Aerospace Information Systems* 10.2 (2013): 53–70.
35. Alsamhi, Saeed Hamood, et al. "UAV computing-assisted search and rescue mission *framework* for disaster and harsh environment mitigation." *Drones* 6.7 (2022): 154.
36. Khan, Amina, Sumeet Gupta, and Sachin Kumar Gupta. "Emerging UAV technology for disaster detection, mitigation, response, and preparedness." *Journal of Field Robotics* 39.6 (2022): 905–955.
37. Rajan, Jibin, et al. "Disaster management using unmanned aerial vehicles." In *Unmanned Aerial Systems*. Academic Press, 2021, pp. 129–155.

38. Hutabarat, Fajar Sam, et al. "Unmanned aerial vehicle design for disaster victim search and rescue operation using wireless sensor network localization." *IAES International Journal of Robotics and* Automation 11.4 (2022): 288.
39. Estrada, Mario Arturo Ruiz, and Abrahim Ndoma. "The uses of unmanned aerial vehicles–UAV's-(or drones) in social logistic: Natural disasters response and humanitarian relief aid." *Procedia Computer Science* 149 (2019): 375–383.
40. Sharma, Vijay Kumar. "CNTFET circuit-based wide fan-in domino logic for low power applications." *Journal Circuits Systems Computers* 31.02 (2022): 2250036.
41. Sharma, Vijay Kumar, and Manisha Pattanaik. "A reliable ground bounce noise reduction technique for nanoscale CMOS circuits." *International Journal Electronics* 102.11 (2015): 1852–1866.
42. Sharma, Vijay Kumar, Sumit Patel, and Manisha Pattanaik. "High performance process variations aware technique for sub-threshold 8T-SRAM cell." *Wireless Personal Communications* 78.1 (2014): 57–68.
43. Sharma, Vijay Kumar, and Manisha Pattanaik. "Techniques for low leakage nanoscale VLSI circuits: A comparative study." *Journal Circuits Systems Computers* 23.05 (2014): 1450061.
44. Sharma, Vijay Kumar. "A survey of leakage reduction techniques in CMOS digital circuits for nanoscale regime." *Australian Journal Electrical Electronics Engineering* 18.4 (2021): 217–236.
45. Kumar, Chanchal, Avinash Sharan Mishra, and Vijay Kumar Sharma. "Leakage power reduction in CMOS logic circuits using stack ONOFIC technique." In *2018 Second International Conference on Intelligent Computing and Control Systems (ICICCS)*, 2018.
46. Sharma, Vijay Kumar, and Manisha Pattanaik. "Design of low leakage variability aware ONOFIC CMOS standard cell library." *Journal Circuits Systems Computers* 25.11 (2016): 1650134.
47. Naz, Syed F., Sadat Riyaz, and Vijay Kumar Sharma. "A review of QCA nanotechnology as an alternate to CMOS." *Current Nanoscience* 18.1 (2022): 18–30.
48. Riyaz, Sadat, and Vijay Kumar Sharma. "Design of reversible Feynman and double Feynman gates in quantum-dot cellular automata nanotechnology." *Circuit World* 49.1 (2023): 28–37.
49. Raina, Bhoomika, Chetna Verma, Mehak Gupta, and Vijay Kumar Sharma. "Binary coded decimal (BCD) seven segment circuit designing using quantum-dot cellular automata (QCA)." In *2021 5th International Conference on Trends in Electronics and Informatics (ICOEI)*, 2021.
50. Mushtaq, Umayia, and Vijay Kumar Sharma. "Performance analysis for reliable nanoscaled FinFET logic circuits." *Analog Integrated Circuits Signal Processing* 107.3 (2021): 671–682.
51. Kajal and Vijay Kumar Sharma. "Design and simulation of FinFET circuits at different technologies." In *2021 6th International Conference on Inventive Computation Technologies (ICICT)*, 2021.
52. Sharma, Vijay Kumar. "Designing of face recognition system." In *2019 International Conference on Intelligent Computing and Control Systems (ICCS)*, Madurai, 2019.

Chapter 13

BHJ organic solar cells

Design and exploration of several modeling approaches

Syed Mohd Hussain, Md. Waseem Akram, Umayia Mushtaq, and Nazida Ansari

13.1 INTRODUCTION

The fast-paced growth today leads to a continuously rising energy need. This demand is exacerbated by the limited availability of traditional energy sources (i.e., coal, oil, petroleum, etc.), and the negative long-term consequences of their use force us to work on developing alternate renewable energy sources. In this case, photovoltaic (PV) technology, which harvests energy directly from the sun, is one of the best alternatives available for coming generations. As a result of the global energy crisis, the exhaustion of fossil fuel resources, and worries about greenhouse gas emissions, the popularity of generating renewable solar energy is steadily increasing. The market for PV systems is expanding quickly, and most commercially accessible solar cells are constructed of crystalline silicon (c-Si), which is readily available but has many limitations like costly, weight-sensitive, limited absorption spectrum, non-degradable, and complex to process material. Crystalline silicon (c-Si) solar cells are relatively brittle and can be harmed by wind, hail, and other environmental elements. The fragility of the solar panel may result in higher maintenance costs. Organic materials can develop a cheap alternative technology using eco-friendly materials that are abundantly available, making them feasible for generating power on a large scale. There is a good possibility of producing very thin solar cells with minimal material requirements using an organic semiconductor with extremely high optical absorption coefficients. Another attractive feature of organic photovoltaic (OPV) cells is their ability to produce thin, flexible devices using low-temperature methods and high-throughput printing methods using a Roll-to-Roll (R2R) method [1,2]. OPV cells, with their affordability, lightweight design, and mechanical flexibility, are ideally suited for widespread use in the renewable energy industry. These advantages enable their application in portable devices, wearable technology, and building-integrated PVs.

To understand how an organic solar cell (OSC) functions, we must be familiar with the phenomenon of exciton creation, exciton diffusion, and charge carriers transfer from one layer to another layer and their collection

 DOI: 10.1201/9781003643609-13

at electrodes. These processes are very important for the operation of a PV device and their understanding requires knowledge of carrier and electric field profile within the device and their variation with the applied voltage. In 1978, a single material (merocyanine) was sandwiched between Ag and Al electrodes to create the first OSC, which had an efficiency of less than half of a percent [3]. Two electrodes with district values of work functions form a bipolar device called a solar cell. The variation in work functions establishes an electric field that directs from the cathode, the lower work function electrode, to the anode, the higher work function electrode [4]. OSCs exhibit low dielectric constants, resulting in significant binding energy values within the range of (0.3–0.5) eV. Consequently, OSCs cannot achieve exciton dissociation solely through thermal energy. An OPV system needs at least two separate OSCs to get around this restriction. Dissociation of excitons occurs at the interface between these OSCs due to their energy-level offset, which surpasses the binding energy. The migration of electrons to the acceptor material at the contact leaves a hole in the donor. This process forms a charge-transfer state where charge carriers are still mutually attracted. Eventually, thermal energy prevails over the binding energy, leading to the state of charge separation. Electron–hole pairs in the charge-transfer state may undergo recombination at the contact between the two materials. The charge carriers then permeate via the appropriate interfacial layers and diffuse toward the respective electrodes. In the external circuit of the solar cell, the charge carriers are accumulated and used to perform work. Hence, resulted in the generation of current. Conjugated polymers and fullerene derivatives are the main components in the bulk heterojunction (BHJ) structure commonly utilized in most recent OPV cells. The bilayer or planar heterojunction (PHJ) is created by depositing the acceptor and donor materials on top of one other, forming a heterojunction interface. Tang developed the inaugural PHJ OSC, with a power conversion efficiency (PCE) of approximately 1.0% [5]. In 1995, the BHJ concept was introduced for polymer solar cells, exhibiting a PCE of below 1.5% [6,7]. Yet, the best OPV cell PCE is approximately 18.2%, which is extremely low [8]. Solar cells are designed in tandem configurations, meaning they are layered one on top of the other, to address the challenge of limited light absorption by a single OSC. This configuration enables the unabsorbed percentage of light from one cell to be captured by another cell. There are two distinct methods for modeling a PV cell. The initial model is an analytical one based on the device's structure, which describes its functioning through the PV effect. This model focuses on several physical aspects, including current density, potential distribution, charge distribution, and other relevant factors. The second kind of model is a circuit model that replicates the behavior of the solar cell structure. By investigating its current, voltage, and power, this model representation can be utilized to assess solar cell's electrical efficiency [9]. This chapter has been organized as an introduction in Section 13.1; circuit modeling and different types of circuit modeling including proposed tandem OSC circuit models in Section 13.2. In Section 13.3, analytical modeling is explored, including discussions on voltage-dependent dark current and photocurrent, along with the development of an analytical model for BHJ-OSC that incorporates a novel empirical expression for the generation of space-dependent photocarriers.

13.2 CIRCUIT MODELING

There are two distinct approaches to modeling a PV cell: type one focuses on the cell's structure and explains its operation based on the PV effect and it also explores the physical concepts like charge distribution, the cell's effective depth, and related factors. The second type of model analyses the electrical efficiency of a solar cell by examining its power, current, and voltage. It is analogous to an electrical circuit and is commonly referred to as "circuit modeling" because its outputs can be measured [9,10]. The process of modeling a circuit's behavior involves employing electronic circuits and fundamental electrical components such as sources of voltage and current, resistors, capacitors, and inductors. Before physically building complicated electronic systems, engineers and designers can use this approach to study and simulate the systems' behavior. Depending upon the level of precision and information needed for the study, circuit modeling can range from simple lumped-parameter models to more complex distributed-parameter models. These models can be made with a variety of simulation tools, including MATLAB/Simulink, SPICE (Simulation Program with Integrated Circuit Emphasis), and specialized software from semiconductor manufacturers. In the context of device modeling, such as for semiconductor devices like transistors, or diodes, creating mathematical formulas or equivalent circuits that precisely represent the electrical properties and operation of the device under various operating conditions is known as circuit modeling. After that, the behavior of circuits comprising these devices can be predicted by integrating these models into more extensive circuit simulators. All things considered, circuit modeling is an essential component of electronic design and analysis that helps engineers maximize performance, identify possible problems, and reduce costs and development time.

The behavior and performance of the circuit model for a solar cell are investigated using various circuit equations, including those governing the current–voltage relations, which predict and verify the behavior and performance of the device. There exist different types of circuit modeling of solar cells as given below:

1. 1-Diode Conventional Solar cell circuit mode
2. 2-Diode Conventional Solar cell circuit mode
3. 3-Diode Model of OSC

4. Mazhari's OSC model
5. Improved Mazhari's OSC model
6. OSC model proposed by G. del Pozo
7. OSC model proposed by Francisco
8. Degraded OSC model
9. Proposed tandem OSC model

The conventional circuit model is a simplified representation of the equivalent and parallel circuit for inorganic solar cells, whereas the 2-diode model is a modified version for inorganic solar cells. Unfortunately, the S-shaped current versus voltage characteristic of OSCs is not adequately explained by either of these theories. To address this limitation, the 3-Diode model for an OSC has been developed.

Mazhari's model successfully characterizes the electrical behavior of the S-shaped bend in the first quadrant of the current versus voltage curve for OSCs. However, it falls short in explaining the linear rise observed in the S-shaped bend within the third quadrant. To address this limitation, a refined model, based on Mazhari's work, is proposed. This enhanced model offers a more comprehensive explanation for the linear increase of the S-shaped bend in the third quadrant.

13.2.1 Conventional solar cell circuit model [11]

Figure 13.1 depicts a standard silicon solar cell, whose properties are commonly estimated by using the "Shockley diode equation" [12]. A silicon solar cell functions as a light detector, converting solar energy into electricity. Additionally, the light detector and an anti-parallel connected diode are used to control the current, as illustrated in Figure 13.2. Kirchhoff's law determines output current (denoted as I in the model below):

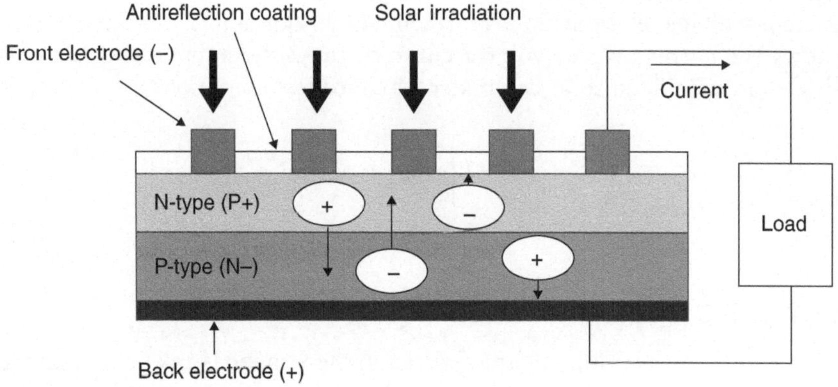

Figure 13.1 Single solar cell: p-n junction structure [13].

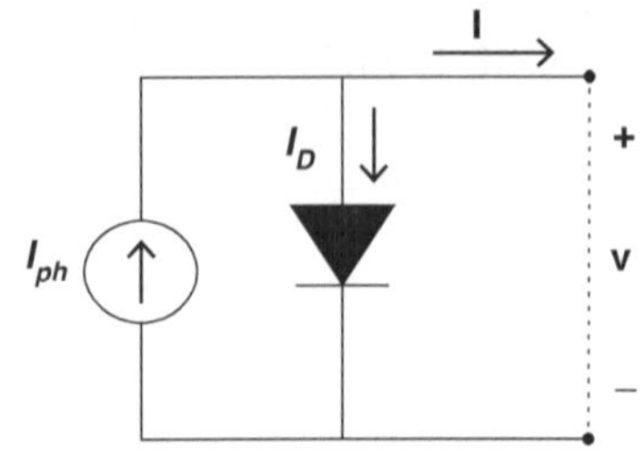

Figure 13.2 Single solar cell: electrical circuit representation [14].

$$I = I_{ph} - I_0\left[\exp\left(\frac{V}{AV_t}\right) - 1\right], \tag{13.1}$$

such that I_{ph} represents the photocurrent, I_0 shows the reverse saturation current of the diode, V is the diode voltage, and V_t is used for the thermal voltage of the diode.

The amount of irradiation has a significant impact on a solar cell's efficiency, and different types of solar cells perform similarly at different levels of irradiance (illustrated in Figure 13.3a and b).

13.2.2 2-Diode model of conventional solar cell [15]

Figure 13.2 depicts the electrical behavior of a typical solar cell using an ideal 1-diode equivalent circuit. This model doesn't apply to solar cells as believed by some researchers because it assumes perfect conditions and requires a lot of modification to fit experimental data better [16]. Researchers commonly rely on the 2-diode model to understand the behavior of crystalline silicon solar cells. As seen in Figure 13.4, this model describes the diffusion current (ID1) as well as the recombination current (ID2). In a 2-diode model, the current–voltage relationship is represented by Equation (13.2). Figure 13.5 shows the current versus voltage curve of the 2-diode circuit model of a single-crystal silicon solar cell under standard test conditions:

$$I = I_{ph} - I_{01}\left[\exp\left(\frac{V + R_sI}{A_1V_t}\right) - 1\right] - I_{01}\left[\exp\left(\frac{V + R_sI}{A_2V_t}\right) - 1\right] - \left(\frac{V + IR_s}{R_{sh}}\right) \tag{13.2}$$

13.2.3 3-Diode model of OSC [17]

The current versus voltage characteristics in the S-shaped kink of the 1-diode solar cell model or the basic 2-diode solar cell model cannot be explained by the circuits shown in Figures 13.2 and 13.4 [18], making them unsuitable

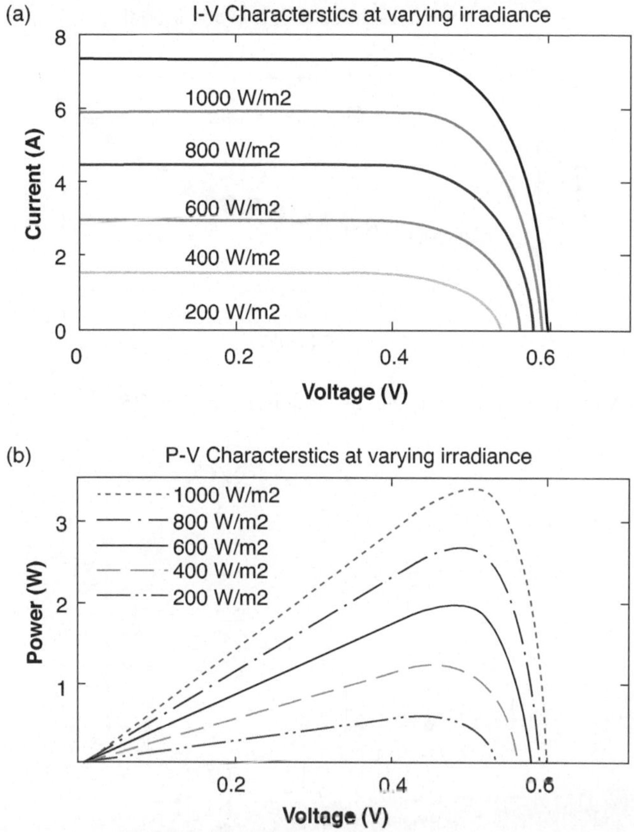

Figure 13.3 (a) Current versus voltage and (b) power versus voltage graphs for a standard solar cell at 25°C showing various irradiation levels [11].

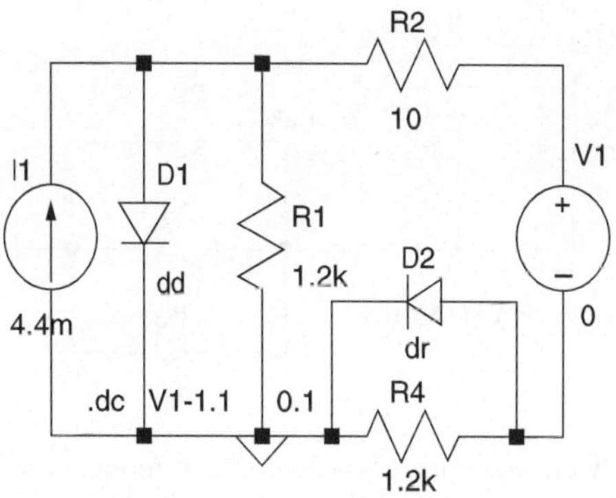

Figure 13.4 The standard equivalent solar cell circuit schematic, utilizing a 2-diode configuration [15].

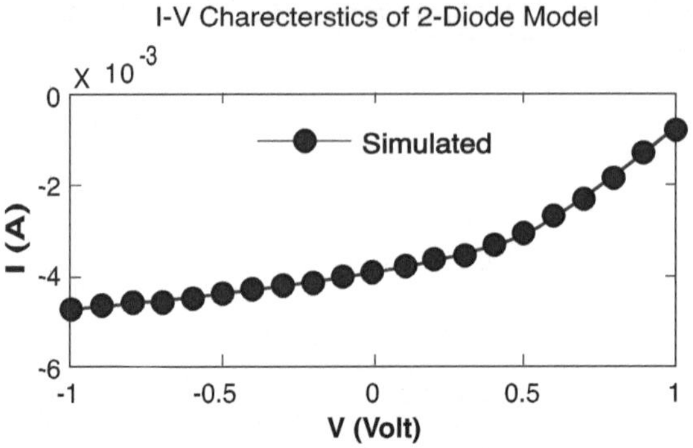

Figure 13.5 *I-V* behavior of 2-diode circuit model of solar cell [15].

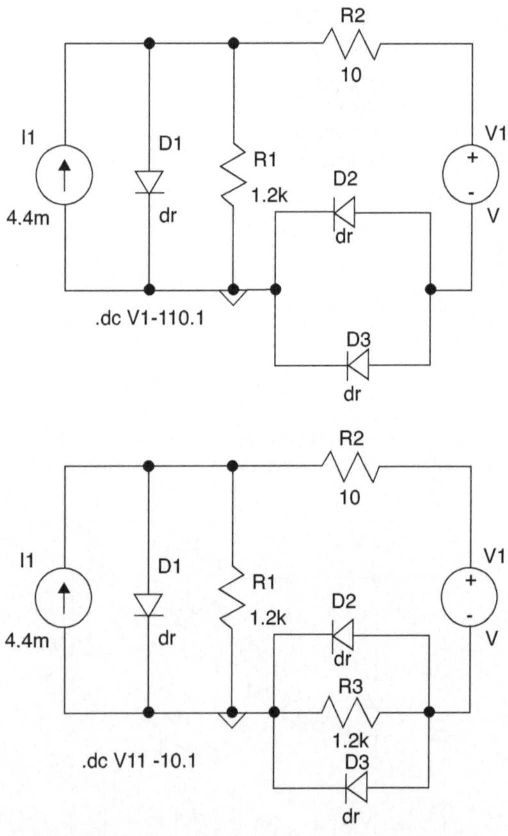

Figure 13.6 (a) 3-Diode conventional equivalent circuit model of solar cell [19]. (b) Solar cell model with a modified three-diode standard equivalent circuit [17]. (c) Comparison between 3-diode model and 3-diode modified model [17, 19]. (d) Comparison between 2-diode model and 3-diode modified model [15,17].

(Continued)

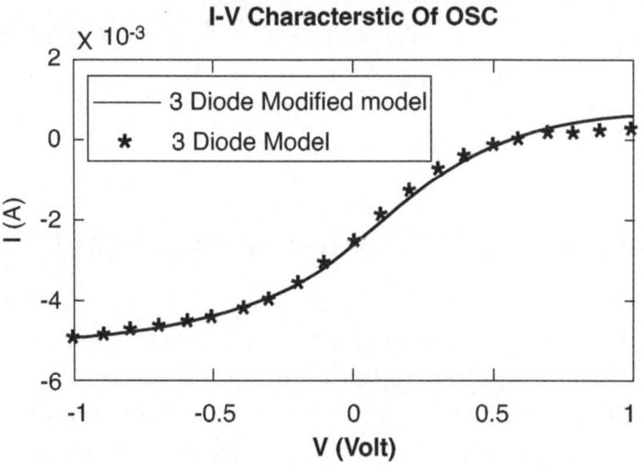

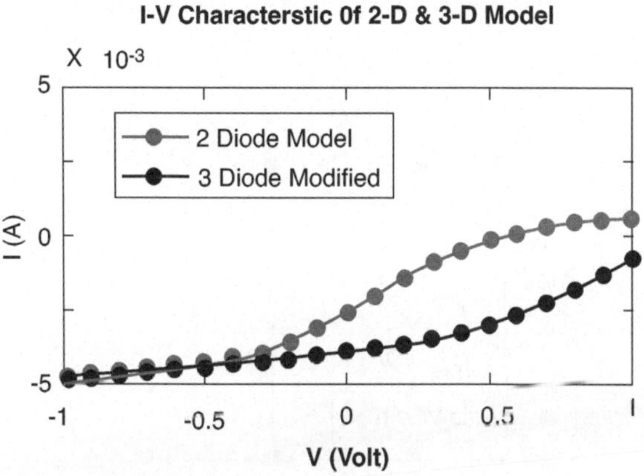

Figure 13.6 (Continued) (a) 3-Diode conventional equivalent circuit model of solar cell [19]. (b) Solar cell model with a modified three-diode standard equivalent circuit [17]. (c) Comparison between 3-diode model and 3-diode modified model [17, 19]. (d) Comparison between 2-diode model and 3-diode modified model [15,17].

for modeling OSCs. In Figure 13.6a, the 3-diode model represents the entire range of experimental S-shaped *I-V* curves [19]. To make this model better, a shunt resistance has been introduced between diode-D2 and diode-D3, as shown in Figure 13.6b. Figure 13.6c and d, respectively, demonstrates the comparison of *I-V* characteristics between the 2-diode model and 3-diode modified model and between the 3-diode model and 3-diode modified model [15,17,19]. Instead of a smooth curve, the *I-V* relationship exhibits a small "kink" or bend, resembling the letter "S" when plotted, hence termed an "S-shaped kink." The precise cause of this behavior varies with the specific solar cell technology, material properties, and device design. This kink can arise from factors such as trap-assisted recombination, interface effects, or changes in charge carrier mobility. Researchers investigate this phenomenon to optimize solar cell performance and comprehend the underlying physics. In summary, the S-shaped bend in the third quadrant of the current versus voltage curve is a distinctive feature that warrants further study to enhance solar cell efficiency and reliability.

13.2.4 Mazhari's organic solar cell model [20]

Figure 13.7a shows the PV response of OSC, which is described by the solar cell circuit model whose current versus voltage relationship is shown in Figure 13.7b. For a given light intensity, *I*1 represents the rate of net polaron generation, which is believed to be constant. Diode Dr models the polaron loss resulting from recombination, whereas diode De represents the free carrier extraction after polaron dissociation. When there is no light, the device's current–voltage characteristics are simulated using diode Dd.

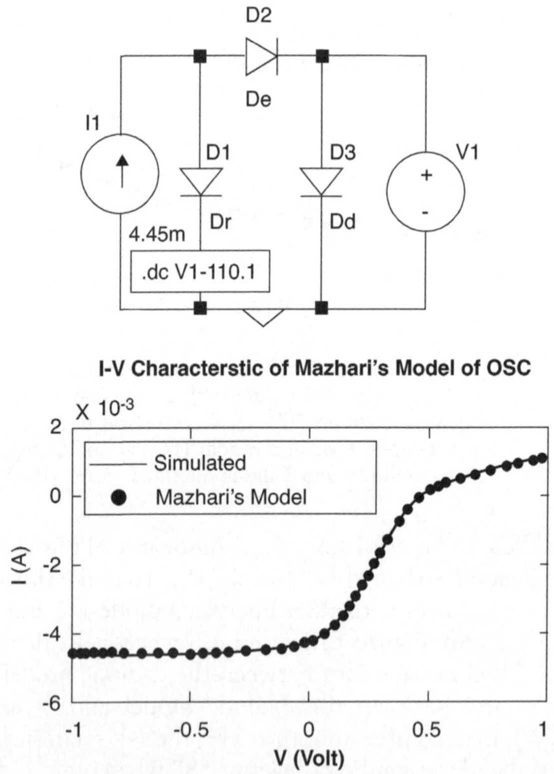

Figure 13.7 (a) Mazhari's equivalent circuit model with lumped parameters for OSC [20]. (b) *I-V* characteristics of bilayer OSC model proposed by Mazhari [20].

Additionally, D1 and D2 function as "ideal" diodes, behaving like a short circuit during forward bias and an open circuit during reverse bias [20].

13.2.5 Improved Mazhari's OSC model [21]

In the third quadrant, in an effort to increase the S-shaped curve's simulation accuracy, a slight adjustment has been made to Mazhari's existing model. This modification specifically targets the behavior observed in that quadrant, aiming for a more precise representation [21]. Mazhari's model, which is depicted in Figure 13.8a, consists of 3-Diodes as well as a photocurrent source I_{ph} with neither shunt resistances nor series resistance, mimics the dark current, recombination current, and extraction current in the diodes previously stated. A more refined model, depicted in Figure 13.8b, has been proposed based on Mazhari's model. This enhanced model includes a shunt resistance in the terminal current–voltage relationship, introducing a linear component necessary for explaining the S-shaped curve with a linear increase in the third quadrant. Figure 13.9 compares this refined model with Mazhari's model [21].

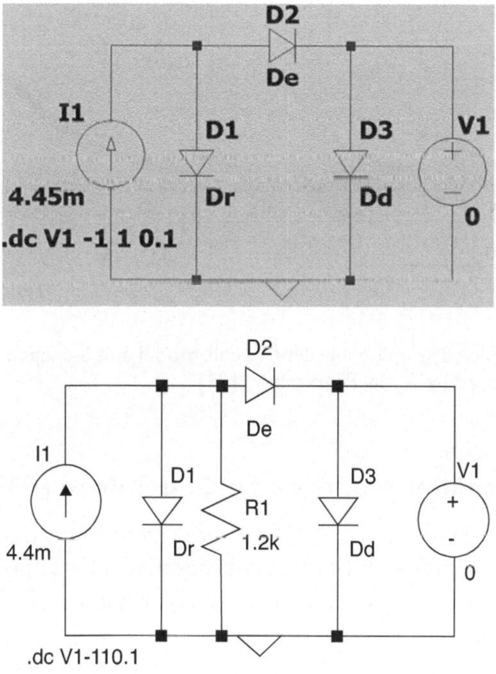

Figure 13.8 (a) Mazhari's proposed lumped-parameter equivalent circuit model for solar cells [20]. (b) An improved lumped-parameter equivalent circuit model for solar cells [22].

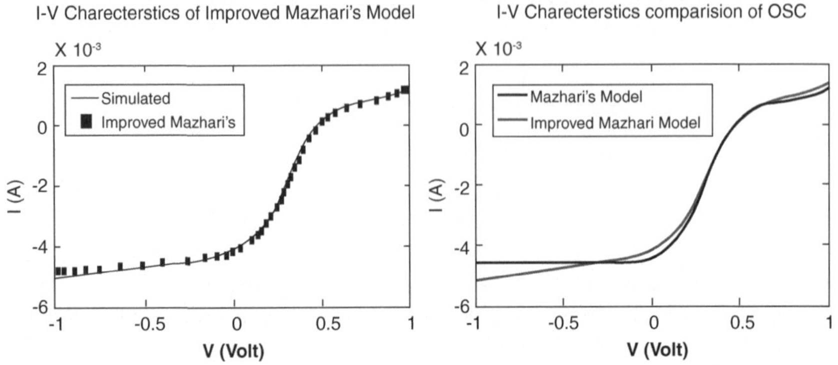

Figure 13.9 Improved Mazhari's model and its comparison analysis with Mazhari's model [21].

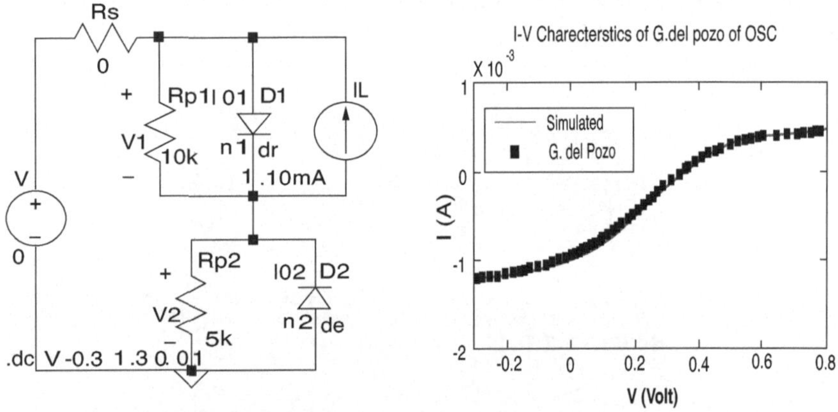

Figure 13.10 Organic solar cell equivalent circuit model and S-shaped *I-V* characteristic proposed by G. del Pozo model [23].

13.2.6 OSC model proposed by G. del Pozo [23]

The total terminal voltage of the OSC results from summing up the voltages across the series connection of three sub-circuits. This approach is outlined in the equivalent circuit model we previously discussed [23] and illustrated in Figure 13.10:

$$V = V_R + V_1 + V_2; \tag{13.3}$$

The calculation of terminal current flowing through series-connected sub-circuits involves using one of the following three equations:

$$I = \frac{V_R}{R_S}; \quad I = I_{RS}\left(e^{\frac{V_1}{n_1 V_{Th}}} - 1\right) + \frac{V_1}{R_{p1}} - I_{LP}; I = -I_{RP}\left(e^{-\frac{V_2}{n_2 V_{Th}}} - 1\right) + \frac{V_2}{R_{p2}} \tag{13.4}$$

In these equations, I_{RS} and I_{RP} represent the reverse currents, n_1 and n_2 are the ideality factors, thermal voltage is denoted by V_{Th}, and I_{LP} stands for the photogenerated current. In the past, researchers have done the calculation of terminal current or voltage using numerical methods or approximations [24].

13.2.7 OSC model proposed by Francisco [25]

OSCs in experimental stages often demonstrate an additional S-shaped current rise in the first quadrant at voltages significantly higher than V_{OC} [26–35]. This presents a challenge for existing models, such as the earlier equivalent circuit model by G. del Pozo depicted in Figure 13.10, which effectively captures the S-shaped segment in the current versus voltage relationship but falls short in explaining the extra current surge beyond the S-shaped bend in the first quadrant. To address this issue, Francisco suggests enhancing the model by substituting a shunt resistance R2 with an additional diode D3, incorporating the ideality factor n_3 and reverse current I_{03} as shown in Figure 13.11. This modification has been made to describe the observed current rise beyond the S-shaped bend in the first quadrant, providing a more accurate representation of how OSCs behave under experimental conditions, and making a significant contribution to the advancement of OSC technology.

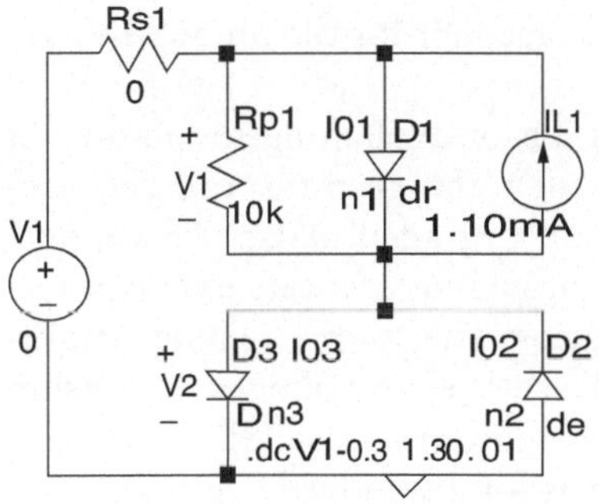

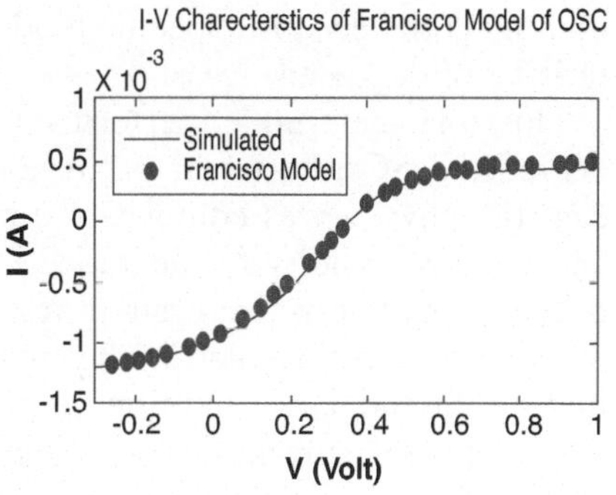

Figure 13.11 Organic solar cell lumped-parameter equivalent circuit model and S-shaped current versus voltage relationship proposed by Francisco [25].

13.2.8 Degraded OSC model [36]

An equivalent circuit model can aid us in our goal of comprehending the physics underlying OSC degradation. Deep insights into degradation can support the development of solar cells with extended lifespans. Numerous reports in the literature highlight the complex nature of degradation in OSCs, indicating that it remains a topic not fully understood. Degradation in OSCs is generally categorized into physical and chemical degradation. Chemical degradation occurs when electrode material, oxygen, and moisture interact with the organic molecules, forming new organic compounds with reduced electrical and optical properties. On the other hand, physical degradation includes issues like top electrode delamination, morphological changes in the active layer, electrode material diffusion into organic layers, and impairment of different interfaces. Additional chemical degradation processes include photooxidation, photoelectrochemical reactions, and low work function electrode oxidation at the top. Refer to reference [35] for an in-depth study of degradation in OSCs. In this study, we examine the model's suitability for degraded OSCs. A widely studied donor–acceptor pair known for efficient OSCs is poly-3-hexylthiophene (P3HT) and phenyl [6,6] C61 butanoic acid methyl ester (PCBM). As a result, investigations were conducted on OSCs utilizing the BHJ blend of PCBM and P3HT to validate the solar cell model [36].

The model effectively describes the behavior of degraded OSCs. However, due to distinct charge transport phenomena, the physics governing degraded OSCs differs significantly. The model's validity for degraded OSCs is discussed, accurately representing their behavior across various test conditions. Degradation effects are assessed by measuring current density-voltage (*J-V*) characteristics in both dark and illuminated states at different time intervals [37]. Under illumination, the current density resembles that of a conventional silicon p-n junction solar cell. This comparison aids in understanding how degradation impacts the performance of OSCs and highlights the need for specific models that account for the unique characteristics of degraded OSCs.

This solar cell model performs a good job of explaining the nature of degraded OSCs; however, because of distinct charge transport phenomena, the physics was completely different in degraded OSCs. This model addresses its validity for degraded OSCs, providing accurate explanations of their behavior in the different test scenarios. The study explained degradation effects by calculating the current density versus voltage relationship under light and dark conditions at different time durations [38]. Under light conditions, the current density is the same as for a standard Si p-n junction solar cell.

Degraded OSCs are effectively characterized by this model. However, because of distinct charge transport phenomena, the physics governing degraded OSCs differs significantly. The model's validity for degraded OSCs is discussed, accurately representing their behavior across various test conditions. Degradation effects are explained by calculating the current density versus voltage relationship under dark and in light condition at different time durations [38]. This comparison develops the understanding of how degradation impacts the performance of OSCs and highlights the need for specific models that account for the unique characteristics of degraded OSCs. Under illumination, the current density resembles that of a standard silicon p-n junction solar cell given by

$$J = J_s\left\{\exp\left(\frac{Q(V - JAR_s)}{nkT}\right) - 1\right\} + \frac{V - JAR_s}{AR_p} - J_{ph} \tag{13.5}$$

In the above equation, the critical variables and parameters for modeling the current–voltage relationship of a solar cell, particularly using the Shockley diode equation or the equivalent circuit model, are used. Key terms include current density (J), which quantifies electric current per unit area, and saturation current density (J_s), which measures reverse bias current. The elementary charge (Q) is a basic constant, while the applied potential (V) refers to the voltage across the cell. The active area (A) is the part exposed to light, generating current. Series resistance (R_s) and shunt resistance (R_P) influence the cell's efficiency and leakage, respectively. The ideality factor (n) shows deviation from ideal behavior, and the Boltzmann constant (k) connects particle energy to temperature, affecting current via thermal voltage. Absolute temperature (T) affects cell efficiency, and photogenerated current density ($J_{ph}(V)$) represents the light-induced current. These parameters collectively determine and optimize solar cell performance.

Figure 13.12a illustrates the equivalent circuit for traditional p-n junction solar cells [36,39]. J_{ph} is constant and unaffected by the applied potential for a given illumination intensity. In the condition of forward bias, the direction of photogenerated current flow is the opposite of the direction caused by the applied potential, and the above formula depends on the current density superposition. Although Equation (13.1) was first used to explain the features of OSCs, it is now widely recognized that Equation (13.1) is invalid for OSCs because of distinct physics [39].

In many respects, conventional solar cells and OSCs differ from one another. These factors comprise the structural configuration, and the presence of two different materials (in the active layer, bound electron–hole pairs—known as excitons—form from the blend of donor and acceptor materials). The limited mobility of charge carriers, photon absorption (with a binding energy of around 0.5 eV), and the potential degradation of materials all contribute to this. Thermal energy cannot dissociate excitons at normal ambient temperature into independent charge carriers. Since the energy difference between the vacant molecular orbitals with the lowest

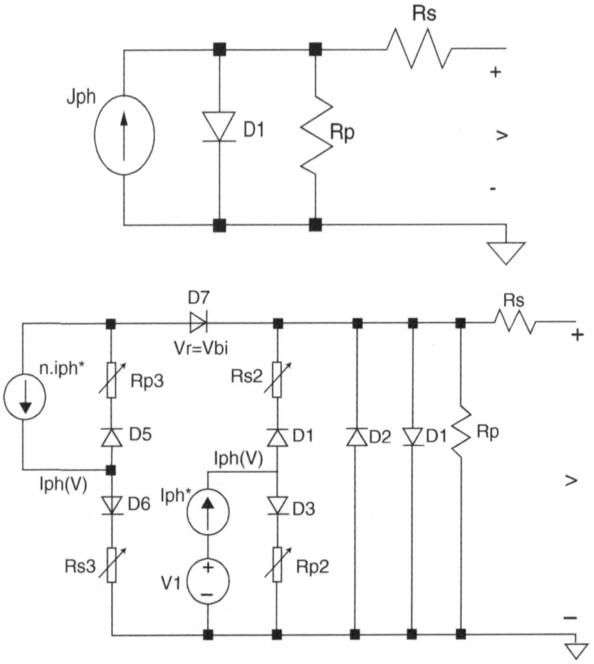

Figure 13.12 (a) Schematic for traditional Si p-n junction solar cell [38]. (b) Schematic for degraded organic solar cell [41].

energy levels of the acceptor and donor substances creates a strong driving force for exciton separation into free charge carriers, excitons formed far from the interface between the acceptor and donor materials must migrate toward this interface. Excitons produced directly at the acceptor and donor interface immediately dissociate [40]. Due to coulombic attraction during the dissociation process, electron–hole pairs develop across the contact. The internal electric field will provide the necessary driving force to separate the pair of electrons and holes into free charge carriers.

A reduction in photocurrent can occur from the recombination. The charge carriers are either separated at the interface or trapped in certain states. While being transported through a drift-diffusion process, the unutilized charge carriers are drawn from the cell to produce a photocurrent. Ultimately, the generated photocurrent relies on how charges move, excitons separate and diffuse, and excitons are generated in the acceptor and donor materials. Considering these aspects, we thoroughly analyzed the characteristics and developed a comparable circuit model for OSCs.

The behavior in dark conditions can be examined by assuming a pair of diodes named D1 and D2, each with the distinct value of ideality factors.

These diodes are connected in parallel with different polarities, along with parasitic shunt and series resistances. The total current density versus voltage relationship is expressed as [41]:

$$J = J_{s1}(V)\left\{\exp\left(\frac{q(V - JAR_s)}{n_1 kT}\right) - 1\right\} - J_{s2}(V)\left\{\exp\left(\frac{q(V - JAR_s)}{n2kT}\right) - 1\right\} + \frac{V - JAR_s}{AR_p}, \quad (13.6)$$

where J represents current density, J_{s2} represents the current density of D2 diode in the reversed bias, $J_{s1}(V)J_{n0}(V) + J_{p0}(V)$, for diode D1, the reverse bias current densities for holes and electrons are represented by J_{p0} and J_{n0}, respectively, while the ideality factors for diodes D1 and D2 are given by $n1$ and $n2$, respectively.

When illuminated, the J-V relationship would undergo modifications to

$$J = J_{s1}(V)\left\{\exp\left(\frac{q(V - JAR_s)}{n_1 kT}\right) - 1\right\} - J_{s2}(V)\left\{\exp\left(\frac{q(V - JAR_s)}{n2kT}\right) - 1\right\} + \frac{V - JAR_s}{AR_p} - J_{ph}(V), \quad (13.7)$$

where $J_{ph}(V)$ represents the current density generated by illumination in the solar cell. At a given illumination intensity, the conventional model operates under the assumption that J_{ph} remains constant, irrespective of the applied potential. However, OSCs do not hold this assumption. In OSCs, the photocurrent remains constant when the illumination intensity is constant but changes with variations in the applied voltage. We have derived a relationship that describes how $J_{ph}(V)$ varies with V. According to this relationship, if $(\mu\tau V_{bi}/d) > d$, $J_{ph}(V)$ can be expressed as given in [41]:

$$J_{ph}(V) = J_{ph}, \quad \text{for } V \text{ less than } 0,$$

$$J_{ph}(V) = J_{ph}\left\{1 - \exp\left(\frac{V - V_{bi} - J_{ph}(V)AR_{s2}}{n_3 kT}\right)\right\}, \quad \text{for } V \text{ less than } V_{bi} \text{ and greater than } 0,$$

$$J_{ph}(V) = J_{ph}\left\{1 - \exp\left(\frac{V - V_{bi} - J_{ph}(V)AR_{s2}}{n_4 kT}\right)\right\}, \quad \text{for } V \text{ greater than } V_{bi} \quad (13.8)$$

In the above equations, "n_3" and "n_4" are the terms that represent the extraction and recombination of electron–hole pairs. $J_{ph}(V)$ represents the current density generated by illumination in the solar cell which is independent of applied potential. The point where dark and illuminated curves intersect corresponds to zero photocurrents. Simultaneously, the resulting electric field $(V_{bi} - V/d)$ vanishes when applied potential is V_{bi}. Our investigations led us to design an equivalent circuit for OSC as illustrated in Figure 13.12b. The encircled region represents a dark condition.

For the comparison analysis of various solar cell models, it's important to evaluate how each model represents solar cell behavior under different conditions and their respective strengths and limitations. The conventional silicon solar cell model, based on the Shockley diode equation, provides a straightforward approach but oversimplifies non-ideal effects. The 2-diode model improves accuracy by incorporating recombination effects but adds complexity. For OSCs, the 3-diode model captures the S-shaped kink in the *I-V* curve, offering a more precise fit but with increased complexity. Mazhari's model and its improved version address different aspects of OSCs, including polaron dynamics and the S-shaped curve, though the latter's increased complexity requires careful calibration. G. del Pozo's model and Francisco's model focus on detailed sub-circuit behaviors and extra phenomena beyond the S-shaped kink, enhancing representation but complicating practical use. Lastly, the degraded OSC model explores the impact of degradation on performance, crucial for long-term durability but complex due to the varied nature of degradation processes. Each model serves specific purposes based on the solar cell type and desired accuracy, balancing between simplicity, detail, and practical application. Table 13.1 summarizes the comparison analysis of various solar cell models.

13.2.9 Proposed tandem OSC circuit model

One major obstacle in OPVs is the restricted light absorption by individual OSCs. OSCs typically use organic semiconductors to convert sunlight into electrical energy. However, organic materials often absorb a limited amount of incident light, particularly in the near-infrared and visible regions, owing to their low absorption coefficients. Some of the variables that contribute to this decreased absorption are as follows:

Electrode losses: OSCs commonly utilize transparent conducting electrodes, such as conductive polymers or indium tin oxide (ITO). To reduce the overall amount of light absorption, these electrodes can absorb some of the incident light before it reaches the active layer.

i. **Limited light harvesting:** Based on their molecular structures, organic materials often have limited light-harvesting capabilities. This may lead to ineffective utilization of incident photons.

Table 13.1 A comparative table for performance evaluation of different models of solar cell

Model	*Description*	*Strengths*	*Limitations*
Conventional silicon	Based on the Shockley diode equation, represents silicon solar cell behavior.	Straightforward approach, easy to implement.	Oversimplifies non-ideal effects, less accurate for detailed analysis.
2-Diode model	Incorporates recombination effects to improve accuracy over the conventional model.	More accurate representation of real solar cell behavior.	Increased complexity, harder to implement.
3-Diode model (OSCs)	Captures the S-shaped kink in the *I-V* curve, providing a more precise fit for organic solar cells.	Improved fit for organic solar cells, accounts for more detailed behavior.	High complexity, challenging to calibrate and implement.
Mazhari's model	Addresses polaron dynamics and the S-shaped curve for OSCs.	Detailed representation of OSC behavior, includes polaron dynamics.	Requires careful calibration, increased complexity.
Improved Mazhari's model	An enhanced version of Mazhari's model, with better handling of specific OSC behaviors.	More accurate than the original Mazhari's model, improved fit for OSC characteristics.	Even more complex, requires extensive calibration.
G. del Pozo's model	Focuses on detailed sub-circuit behaviors and extra phenomena beyond the S-shaped kink.	Captures a wide range of behaviors, very detailed.	Highly complex, difficult to use in practical applications.
Francisco's model	Similar to G. del Pozo's model, it provides detailed sub-circuit analysis and additional phenomena representation.	Enhanced accuracy, detailed representation.	Complicates practical use, very complex.
Degraded OSC model	Explores the impact of degradation on OSC performance over time.	Crucial for long-term durability analysis, specific to degradation processes.	Complex due to varied nature of degradation processes, requires extensive data for accurate modeling.

ii. **Absorption spectrum:** The solar spectrum may not be completely absorbed by the absorption spectra of organic materials. They may have peaks in certain regions, leaving other wavelengths less absorbed.

iii. **Thin active layer:** OSCs typically have limited light absorption due to their thin active layers, which restrict the amount of material available for absorbing light. While increasing the active layer thickness could improve absorption, doing so may also result in the recombination of more charge carrier and lead to the reduction in the device performance.

Researchers are exploring various methods to overcome these constraints and increase the efficiency of OSCs. Some of the methods are given below:

i. **Broadening absorption spectrum:** Researchers are trying to develop organic materials that can absorb the light from the sun at a larger spectrum of wavelengths. Development of innovative materials, optimizing structures for efficiency, and advancing manufacturing methods are focal points for both industry and scientific communities. These efforts aim to significantly enhance PCE, addressing the growing energy demand more effectively.
ii. **Light-trapping structures:** To improve light absorption and extend the optical path length, light-trapping structures or nanostructured materials are used.
iii. **Improving active layer morphology:** Controlling the active layer's shape to ensure efficient separation of charges and generation of exciton, this can improve the absorption of photons.
iv. **Tandem and multi-junction architectures:** Solar cells are constructed in tandem configuration, where many cells are stacked on top of one another, to overcome the problem of limited light absorption in single-junction OSCs. This design allows the unabsorbed light from one cell to be captured by the adjacent cell, hence in a tandem cell with a high value of PCE in comparison to individual cells.

Conjugated polymer and fullerene composite-based polymer solar cells present unique opportunities as a renewable energy source. They can be easily manufactured using cost-effective printing and coating methods on flexible and lightweight substrates [42–45]. Although PCE (η) of 5% has been reported in the literature [46–50], commercialization has been impeded by their limited efficiency. The multilayer "tandem cell" structure offers several advantages, similar to the series connection of two PV cells. Combining individual cells enhances the open circuit voltage (V_{OC}) by adding up their voltages. Employing two semiconductors with varying band gaps allows for effective photon absorption throughout the solar emission spectrum. Typically, the first cell features a wide-band-gap semiconductor, while the second cell utilizes a narrower-band-gap material [51]. Compared to single cells comprised of narrower-band-gap materials, tandem cells exhibit better PCE because they can produce more charge carriers from photons with energies above the energy gap. These charge carriers settle to their corresponding band edges quite rapidly. However, in polymer-fullerene composites, the poor mobility of charge carriers leads to lower fill factor and V_{OC}, thick active layer internal resistance, and higher internal resistance. [52]. Polymer-based solar cells see enhanced performance through the use of tandem cell architecture, which allows for greater optical density across a wider segment of the solar emission spectrum without an increase in internal resistance.

Figure 13.13 illustrates the schematic structure of the most efficient polymeric tandem OSC proposed by Kim et al. [53]. A thin transparent layer of wide-band-gap Ti*Ox* links the two cells, providing a robust foundation for another cell fabrication on the top. PEDOT:PSS serves as the hole-transport layer for both cells. The photogenerated holes in the front cell traverse through the PEDOT:PSS layer and gather at the anode of ITO, while the photogenerated electrons traverse through the Ti*Ox* layer and reach the Ti*Ox*–PEDOT: PSS interface. Similarly, holes from the back cell travel and

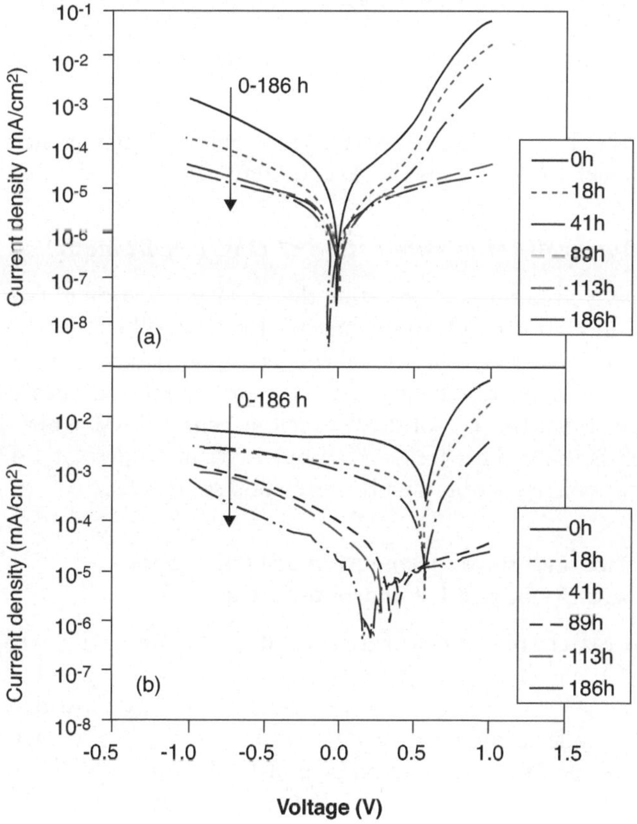

Figure 13.13 Illustration of the tandem solar cell structure [53].

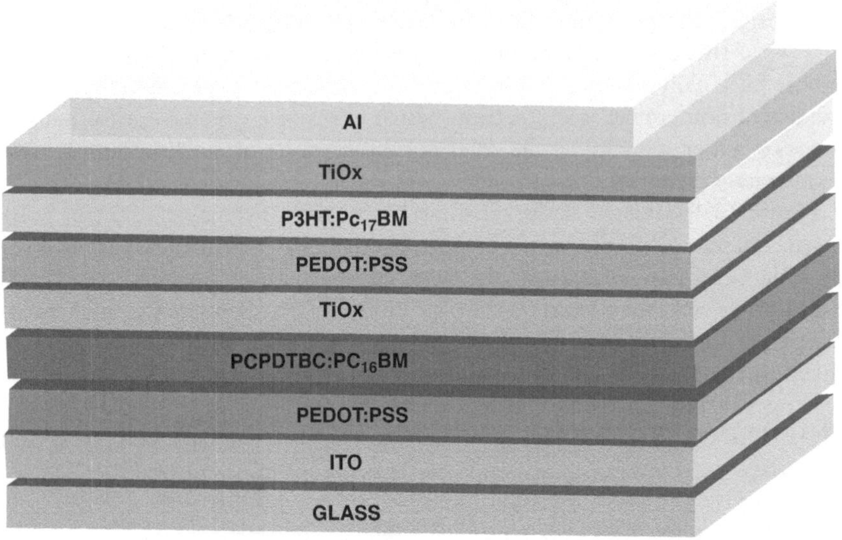

Figure 13.14 Proposed two-cell series solar cell circuit model and its *I-V* characteristics.

collect at the TiOx–PEDOT: PSS interface, while electrons move through the TiO*x* layer and collect at the Al cathode.

13.2.9.1 Two-cell series connection of OSC circuit model

We propose a circuit model depicting the series connection of two OSCs, as illustrated in Figure 13.14. In this model, the first cell exhibits a photocurrent (I_{ph1}) of 8 mA, while the second cell has a photocurrent (I_{ph2}) of 12 mA. The polaron loss due to recombination is represented by diodes D1 and D2. It is observed that the V_{OC} of the first cell measures 0.6 V, while that of the second cell is 0.8 V. This yields a net open circuit voltage of 1.4 V; the net open circuit voltage is obtained by individual solar cells.

13.2.9.2 Two-cell series connection of OSC model with S-shaped I-V characteristics

We propose a circuit model of a two-cell series connection of OSC with S-shaped *I-V* characteristics as depicted in Figure 13.15, in which both 1st cell and 2nd cell have photocurrent of 8 mA, the polaron loss due to recombination is modeled by diode dr (D1 and D2), while the extraction of free carriers after polaron dissociation is modeled by diode de (D3 and D4). In this connection, it is found that the VOC of the 1st cell is 0.6 V, and the 2nd cell is 0.6 V, and the net open circuit voltage is 1.2 V which is nothing but the addition of the open circuit voltage of the individual cell.

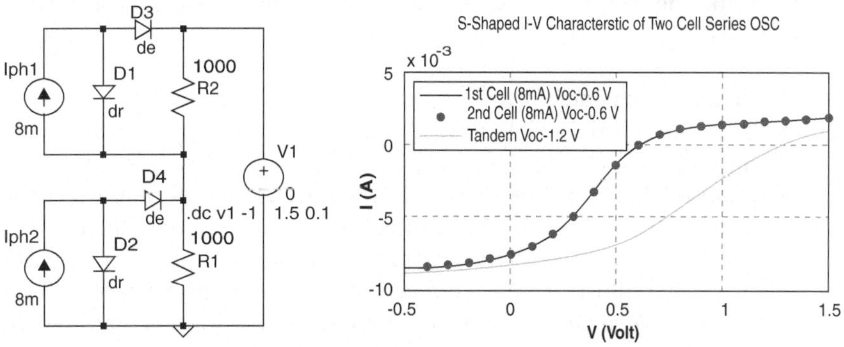

Figure 13.15 Proposed two-cell series OSC circuit model and its S-shaped *I-V* characteristics.

In certain cases, the corresponding electrode's grains and pinholes can facilitate water and molecular oxygen to enter the active layer. When exposed to oxygen and/or water, polymer solar cells are exposed to photooxidation, which causes charged species to arise in the active layer [54]. Oxygen and water near the electrodes may develop a doping profile around the contacts when illuminated. Mateker et al. observed the appearance of S-shaped current density versus voltage curves in the OSCs containing a blend of poly-[(di(2-ethylhexyloxy) benzo [1,2-b:4,5-b'] dithiophene)-co-(octylthieno [3,4-c] pyrrole-4,6-dione)] (PBDTTPD) and phenyl [6,6] C61 butanoic acid methyl ester (PCBM). The diffusion of the organic impurities which are accumulated at the cathode possibly has a reaction with the cathode as well resulting in S-shaped characteristics in series connections of two OSCs.

13.3 ANALYTICAL MODELING

Making mathematical representations of a system's, object's, or phenomenon's behavior is known as analytical modeling. Usually, a combination of fundamental concepts, physical rules, or empirical data is used to create these models. To shed light on a system's behavior, performance, or characteristics, analytical models seek to capture its most significant features. They frequently take the form of equations, mathematical expressions, or algorithms that are used in behavior analysis, design parameter optimization, and outcome prediction. In short, analytical modeling is the process of creating a mathematical model, which can help in the explanation of a system, the study of the effects of various components, and the prediction of behavior. Thus, an analytical model is a mathematical and linguistic description of a system.

The following steps are often involved in the analytical modeling process:

a. **Problem statement:** Clearly define the system or phenomenon that must be modeled as well as the aims and purposes of the modeling project.
b. **Finding relevant variables:** Identify the critical factors that affect the system's behavior and establish interconnections between them.
c. **Mathematical presentation:** Create expressions or equations in mathematics that illustrate the connections between the variables. This might involve using theoretical ideas, actual evidence, or physical laws.
d. **Solution:** To find solutions or to get insights into how the system behaves, solve the equations or mathematical models. This may involve applying analytical tools like linear algebra, calculus, differential equations, or optimization techniques.
e. **Validation:** Validate the analytical model's predictions or outcomes by comparing the results with data obtained from simulations, experiments, or recognized solutions.
f. **Verification:** Verify to determine whether the model faithfully captures the system's behavior in different scenarios.

Numerous disciplines, including engineering, physics, economics, biology, and social sciences, depend extensively on analytical modeling. It offers an effective tool for forecasting, generating new technologies, analyzing complex structures, and directing decision-making procedures. However, the accuracy and application of the conclusions may be affected by the constraints of analytical models, which could include simplifications or assumptions made during the modeling process. To guarantee analytical models' dependability and use, it is crucial to thoroughly validate and verify them.

13.3.1 BHJ-OSC: voltage-dependent photocurrent and dark current

By solving electron and hole continuity equations and investigating Shockley-Read-Hall recombination, a physics-based mathematical model is developed to explain the forward dark current that varies with external voltage, as seen in bulk heterojunction organic solar cells (BHJ-OSCs). On the other hand, recent work by Street et al. [55] suggests that mobile carriers recombining at SRH-type interface states close to the buried heterojunction is the primary recombination mechanism. A revised Braun model is used to assess the electric field-dependent dissociation efficiency of bound electron–hole pairs [56].

13.3.1.1 Dark current model

Two metals make contact with the active layer: one has a lower work function, ϕ_c (represents the work function of the cathode), while the other has a higher work function, ϕ_a (represents the work function of the anode). During thermal equilibrium, the energy band bends, resulting in a flat Fermi level. The built-in potential is created by the difference in work functions between two metals. When these metals are in contact, electrons move from the metal with the lower work function to the one with the higher work function until equilibrium is reached, generating an electric field and a potential difference at the interface. This potential difference is directly related to the difference in the work functions of the two metals:

$$eV_{bi} = \phi_a - \phi_c = E_g - \phi_1 - \phi_2, \tag{13.9}$$

where V_{bi} is the built-in potential and e is the elementary charge [57]. The dark current due to Shockley-Hall-Read recombination (SHR) recombination (shown in Figure 13.16) can be calculated as follows [56]:

$$J_{\text{dark}} = en_i \exp\left(-\frac{E_g}{2eV_t}\right)\left[\frac{1}{\tau_n}\exp\left(\frac{\phi_2}{eV_t}\right)+\frac{1}{\tau_p}\exp\left(\frac{\phi_1}{eV_t}\right)\right]$$
$$\times\left[\frac{LV_t}{V_{bi-V}}\exp\left(\frac{V}{V_t}\right)\left\{\exp\left(\frac{V_{bi}-V}{2V_t}\right)-1\right\}-\frac{LV_t}{V_{bi}}\left\{\exp\left(\frac{V_{bi}}{2V_t}\right)-1\right\}\right] \tag{13.10}$$

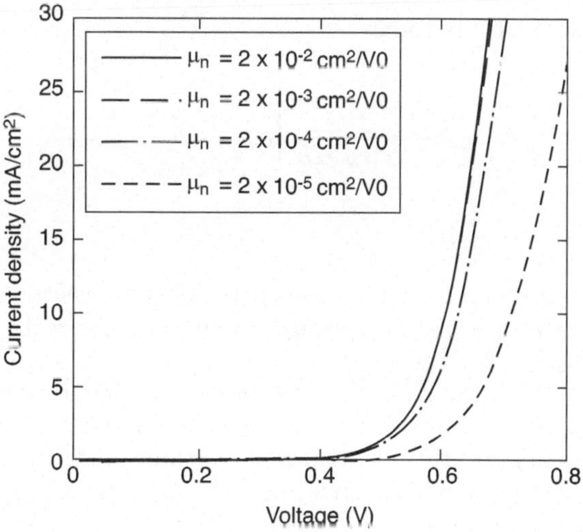

Figure 13.16 P3HT:PC61BM solar cells: Dark current–voltage characteristics for varying μ_n at constant μ_p [57].

Consequently, it has an inverse relationship with carrier lifetimes and increases exponentially with an increase in injection barrier heights (ϕ_1 & ϕ_2).

13.3.1.2 Braun's model for dissociation efficiency

Braun [55] presented a theoretical framework that considers geminate recombination with a non-zero reaction radius, aimed at determining the electric field-dependent escape probability. This reaction radius represents the less distance necessary for holes-bound electrons to recombine. Additionally, Braun [56] introduced an expression to compute dissociate rate dependent on the electric field:

$$M(F) = \frac{K_d(F)}{K_d(F) + K_f}, \tag{13.11}$$

where the recombination rate is K_f, the separation rate of bound electron-hole pair (EHP) is K_d, the electric field is F, and M represents escape probability.

13.3.1.3 Braun's model: modified version

According to Braun's model, this has been modified to

$$K_d(F) = \frac{Dr_c}{r_0^3 e^{r_c/r_0}} J_1\left(2\sqrt{-2b}\right)/\sqrt{-2b} \tag{13.12}$$

Now Equation (13.11) becomes

$$M(F) = \left[1 + \frac{K_f r_0^3 e^{r_c/r_0}}{Dr_c J_1\left(2\sqrt{-2b}\right)/\sqrt{-2b}}\right], \tag{13.13}$$

where

- D: Denotes the combined diffusion coefficients for electrons and holes in the respective media, illustrating how charge carriers disperse through the material.
- r_c: Refers to the Onsager radius, a fundamental length scale that represents the distance over which charge carriers interact.
- $\varepsilon_0\varepsilon_r$: Represents the effective dielectric constant of the blend.
- J_1: The Bessel function of the first kind and order, which appears in problems related to wave propagation and static potentials, describing its behavior under certain conditions.
- $b = e^3F/8\pi\varepsilon_0\varepsilon_r k^2T^2$ is reduced field.

e: is the elementary charge.
T: represents the absolute temperature.
K: represents the Boltzmann constant [56].

13.3.1.4 Photocurrent

In the blend layer, the EHP generation rate (G) can be stated as follows [58]:

$$G(\lambda,V)=\alpha(\lambda)\left[1-R(\lambda)\right]\lambda I_0(\lambda)M/hc, \tag{13.14}$$

such that G represents the generation rate, the wavelength of the incident photon is λ, $\alpha(\lambda)$ is the blend material absorption coefficient, R denotes the loss factor, I_0 is solar spectra's speed of the intensity, c denotes speed of light, and h represents Planck's constant. The integration of incident photon wavelengths across the entire solar spectrum provides the photocurrent density resulting from incident radiation:

$$J_{ph}(V)=\int_0^{\infty}\{J_n(\lambda,V)+J_p(\lambda,V)\}d\lambda. \tag{13.15}$$

13.3.1.5 Net external current

The solar cell's total external current density is [57]

$$J(V)=J_{\text{dark}}(V)+\frac{V-JR_s}{R_p}-J_{ph}(V), \tag{13.16}$$

where R_p and R_s are the shunt and series resistances, respectively. In the above formulas, V has been replaced with $(V-JR_s)$ to calculate the external current. In BHJ-OSC, current characteristics of both net current (the sum of dark and photocurrent) and dark current are investigated across varying external voltages. In P3HT:PCBM OSC under varying light intensities, the net external current versus voltage relationship is depicted in Figure 13.17 [57].

13.3.2 A new empirical expression capturing space-dependent photo-carrier generation has been integrated into an analytical model for BHJ-OSC

The study proposes an analytical model for BHJ-OSCs, taking into account space-dependent carrier production rates based on empirical expressions In our investigation, we use the BHJ-OSC setup depicted in Figure 13.18. The active layer in this configuration comprises P3HT and PCBM in the ratio of 1:1 mixture, with ITO and aluminum serving as anode and cathode

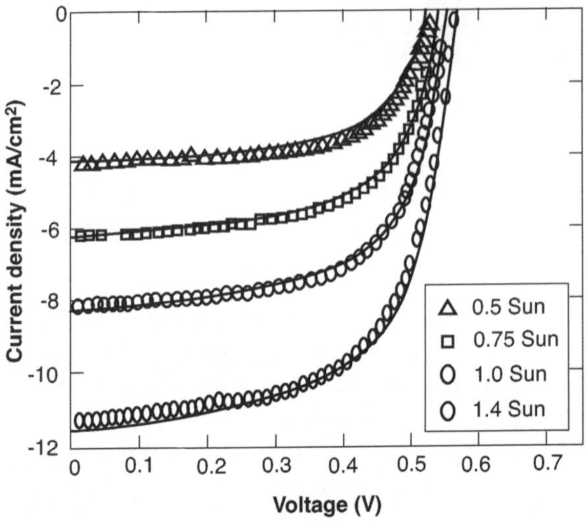

Figure 13.17 P3HT:PCBM organic solar cell: Net external *I-V* characteristics at various intensities of light [57].

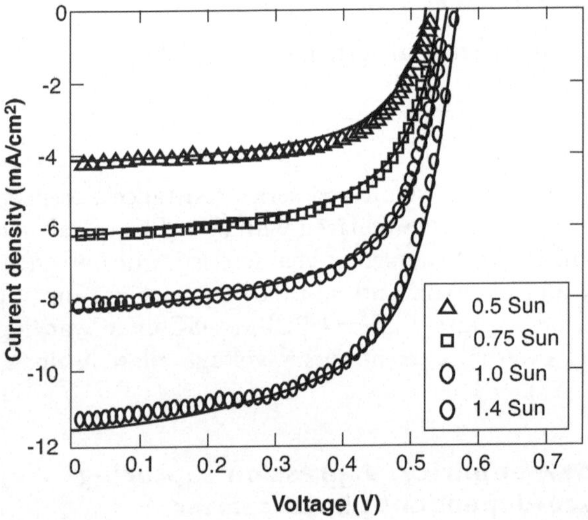

Figure 13.18 P3HT:PCBM-based BHJ organic solar cell configuration [62].

layers, respectively. The injecting layer consists of PEDOT:PSS, similar to previous research [59,60–62].

Following Equation (13.17), a combination of exponential functions is employed to model the exponential decline in carrier generation concerning position, as outlined in the established empirical formula for conventional solar cells, where thickness usually ranges in meters [62]:

$$G(x) = \sum_{i=1}^{n} a_i \exp(-b_i x)\,. \tag{13.17}$$

The constant terms a_i and b_i are calculated using the regression analysis method [63]. The carrier generation rate profile in OSCs due to the thin layers exhibits oscillations instead of following an exponential decay concerning position, as mentioned earlier [64]. As a result, Equation (13.17) is not suitable for fitting the profile of the generation rate. According to a thorough investigation of BHJ-OSC generation profiles, it is found that they can be approximated by the summation of the sine functions, as presented below:

$$G(x) = \sum_{m} a_m \sin(b_m x + c_m). \tag{13.18}$$

The constant terms a_m, b_m, and c_m can be calculated using a regression analysis method. The charge carrier production rate depicted in Figure 13.19 for the various blend thicknesses (200 nm, 140 nm, 100 nm, and 60 nm) is marked with red rectangles. The profile of the generation rate would exhibit an exponential decrease concerning position, as the active layer has a thickness measured in micro-meters. [65]. In such cases, Equation (13.18), based

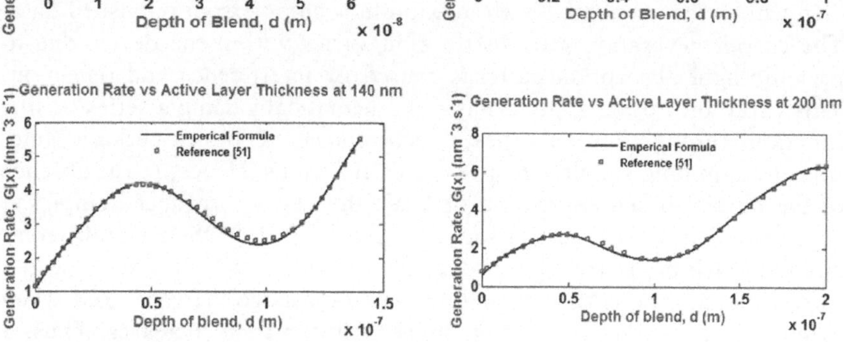

Figure 13.19 Carrier generation rate profile for various active layer thickness at 200, 140, 100, and 60 nm [66].

on Furlan and Amon's (1985) [63] expression, would provide a more accurate representation [62]. Therefore, the proposed Equation (13.18) is better suited for OSC's thin film-based structure.

Thus, till now, we have discussed circuit modeling and different types of circuit modeling; analytical modeling like voltage-dependent dark current and photocurrent and an analytical model have been integrated with a novel empirical expression that captures space-dependent photo-carrier generation in BHJ-OSC [67–79].

13.4 CONCLUSION

Detailed studies of different types of modeling including the circuit modeling of BHJ-OSC have been done, and a new circuit model of tandem structure OSC has also been proposed. The different circuit models of tandem OSCs conclude that when two cells are connected in series, their open circuit voltage combines. In OSCs utilizing BHJ structures, the net external current density results from combining the dark current and photocurrent densities. A detailed physics-based mathematical model explains this, illustrating how both currents change with external voltage. The dark current density occurs without light, while the photocurrent density arises from light absorption across the solar spectrum. The net current density, representing the total current per unit area, integrates these two components and fluctuates based on the applied voltage. An empirical method for expressing space-dependent carrier generation rate has demonstrated that the device's blend thickness affects the carrier generation rate in a mathematical framework for BHJ-OSCs. It increases up to a blend depth of 60 nm, then decreases up to 100 nm, and increases again up to 200 nm. The empirical expression has the advantage of taking into account the spatial distribution of the $G(x)$ profile rather than assuming a constant generation rate. The novel model's generation profile, $G(x)$, is identical to that extracted with maximum accuracy from published data. The carrier generation rate distribution varies within the device due to periodic light absorption patterns caused by interference and reflection. This variation can be approximated mathematically using a series of sine functions through Fourier series, which represent complex periodic functions by summing simple sine and cosine functions. Hence, in the absence of the physics-based expression of $G(x)$, the newly introduced empirical expression can be a practical tool for modeling BHJ-OSC. The different circuit models of tandem OSCs conclude that when two cells are connected in series connections then their open circuit voltage increases and when connected in parallel connection short circuit current increases. Thus, a tandem structure of OSCs has shown better performance as compared to single-junction OSCs.

REFERENCES

1. Shaheen, S.E., Radspinner, R., Peyghambarian, N., & Jabbour, G.E., Fabrication of bulk heterojunction plastic solar cells by screen printing. *Appl. Phys. Lett.*, 79(18), 2996–2998, 2001.
2. Gustafsson, G., Cao, Y., Treacy, G.M., Klavetter, F., Colaneri, N., & Heeger, A.J., Flexible light-emitting diodes made from soluble conducting polymers. *Nature*, 357(6378), 477–479, 1992.
3. Ghosh, A.K., & Feng, T., Merocyanine organic solar cells. *J. Appl. Phys.*, 49(12), 5982–5989, 1978.
4. Kumar, P., *Organic Solar Cells: Device Physics, Processing, Degradation, and Prevention*. CRC Press, Boca Raton, FL, 2016.
5. Tang, C.W., Two-layer organic photovoltaic cell. *Appl. Phys. Lett.*, 48(2), 183–185, 1986.
6. Yu, G., & Heeger, A.J., Charge separation and photovoltaic conversion in polymer composites with internal donor/acceptor heterojunctions. *J. Appl. Phys.*, 78(7), 4510–4515, 1995.
7. Halls, J.J.M., Walsh, C.A., Greenham, N.C., Marseglia, E.A., Friend, R.H., Moratti, S.C., & Holmes, A.B., Efficient photodiodes from interpenetrating polymer networks. *Nature*, 376(6540), 498–500, 1995.
8. National Renewable Energy Laboratory (NREL), Organic Photovoltaic Solar Cells. https://www.nrel.gov/pv/organic-photovoltaic-solar-cells.html. Accessed 2024.
9. Khatibi, A., Razi Astaraei, F., & Ahmadi, M.H., Generation and combination of the solar cells: A current model review. *Energy Sci. Eng.*, 7(2), 305–322. 2019.
10. Hiramoto, M., Suezaki, M., & Yokoyama, M., Effect of thin gold interstitial-layer on the photovoltaic properties of tandem organic solar cell. *Chem. Lett.*, 19(327), 327–330 1990.
11. Raugei, M., & Leccisi, E., A comprehensive assessment of the energy performance of the full range of electricity generation technologies deployed in the United Kingdom. *Energy Pol.*, 90, 46–59, 2016.
12. Rasool, F., Drieberg, M., Badruddin, N., & Mahinder Singh, B.S., PV panel modeling with improved parameter extraction technique. *Sol. Energy*, 153, 519–530, 2017.
13. Chin, V.J., Salam, Z., & Ishaque, K., Cell modelling and model parameters estimation techniques for photovoltaic simulator application: A review. *Appl. Energy*, 154, 500–519, 2015.
14. Bonkoungou, D., Koalaga, Z., & Njomo, D., Modelling and simulation of photovoltaic module considering single-diode equivalent circuit model in MATLAB – semantic scholar. *Int. J. Emerg. Technol. Adv. Eng.*, 3(3), 493–502, 2013.
15. Rajasekar, N., Krishna Kumar, N., & Venugopalan, R., Bacterial foraging algorithm based solar PV parameter estimation. *Sol. Energy*, 97, 255–265, 2013.
16. Pillai, D.S., Sahoo, B., Ram, J.P., Laudani, A., Rajasekar, N., & Sudhakar, N., Modelling of organic photovoltaic cells based on an improved reverse double diode model. *Energy Procedia*, 117, 1054–1061, 2017.

17. Mohammed, S.S., Modeling and simulation of photovoltaic module using MATLAB/Simulink. *Int. J. Chem. Environ. Eng.*, 2(5), 350–355, 2011.
18. Tsai, H.-L., Tu, C.-S., & Su, Y.-J., Development of generalized photovoltaic model using MATLAB/SIMULINK. In *Proceed World Congress Engineering Computer Science,* pp. 1–6, 2008.
19. Gilot, J., Wienk, M.M., & Janssen, R.A.J., Double and triple junction polymer solar cells processed from solution. *Appl. Phys. Lett.*, 90(14), 143512, 2007.
20. Mazhari, B., An improved solar cell circuit model for organic solar cells. *Sol. Energy Mater. Sol. Cells*, 90(7–8), 1021–1033, 2006
21. Xu, C., Yu, F., Lin, W., & Huang, G., An improved organic solar cell lumped-parameter equivalent circuit model. *Crystals*, 8(7), 277, 2018.
22. Huang, G., Yu, F., & Xu, C., An analytical solution to lumped parameter equivalent circuit model of organic solar cells. *Crystals*, 8, 224, 2018.
23. Del Pozo, G., Romero, B., & Arredondo, B., Evolution with annealing of solar cell parameters modeling the S-shape of the current–voltage characteristic. *Sol. Energy Mater. Sol. Cells*, 104, 81–86, 2012.
24. Araujo de Castro, F., Heier, J., Nüesch, F., & Hany, R., Origin of the kink in current-density versus voltage curves and efficiency enhancement of polymer-C60 heterojunction solar cells. *IEEE J. Sel. Top. Quantum Electron.*, 16(6), 1690–1699, 2010.
25. García-Sánchez, F.J., Lugo-Muñoz, D., Muci, J., & Ortiz-Conde, A., Lumped parameter modeling of organic solar cells' S-shaped I–V characteristics. *IEEE J. Photovolt.*, 3(1), 330–335, 2013.
26. Wagner, J., Gruber, M., Wilke, A., Tanaka, Y., Topczak, K., Steindamm, A., H¨ormann, U., Opitz, A., Nakayama, Y., Ishii, H., Pflaum, J., Koch, N., & Brütting, W., Identification of different origins for s-shaped current voltage characteristics in planar heterojunction organic solar cells. *J. Appl. Phys.*, 111(5), 054509, 2012.
27. Kumar, A., Sista, S., & Yang, Y., Dipole induced anomalous S-shape I-V curves in polymer solar cells. *J. Appl. Phys.*, 105, 094512, 2009.
28. Wagenpfahl, A., Rauh, D., Binder, M., Deibel, C., & Dyakonov, V., S-shaped current-voltage characteristics of organic solar devices. *Phys. Rev. B*, 82, 115306, 2010.
29. Wang, J.C., Ren, X.C., Shi, S.Q., Leung, C.W., & Chan, P.K L., Charge accumulation induced S-shape J–V curves in bilayer heterojunction organic solar cells. *Org. Electron.*, 12, 880–885, 2011.
30. He, L., Jiang, C., Wang, H., Lai, D., & Rusli, R., High efficiency planar Si/organic heterojunction hybrid solar cells. *Appl. Phys. Lett.*, 100(7), 073503, 2012.
31. Matsuo, Y., Ozu, A., Obata, N., Fukuda, N., Tanaka, H., & Nakamura, E., Deterioration of bulk heterojunction organic photovoltaic devices by a minute amount of oxidized fullerene. *Chem. Commun.*, 48, 3878–3880, 2012.
32. Ge, J., Liu, J., Guo, X., Qin, Y., Luo, H., Guo, Z.-X., & Li, Y., Photovoltaic properties of dimeric methanofullerenes containing hydroxyl groups. *Chem. Phys. Lett.*, 535, 100–105, 2012.
33. Wang, N., Yu, J., Zheng, Y., Guan, Z., & Jiang, Y., Organic photovoltaic cells based on a medium-bandgap phosphorescent material and C60. *J. Phys. Chem.* C, 116, 5887–5891, 2012.

34. Elumalai, N.K., Yin, L.M., Chellappan, V., Jie, Z., Peining, Z., & Ramakrishna, S., Effect of C60 as an electron buffer layer in polythiophene-methanofullerene based bulk heterojunction solar cells. *Phys. Status Solidi* A, 209, 1592–1597, 2012.
35. Jørgensen, M., Norrman, K., & Krebs, F.C., Stability/degradation of polymer solar cells. *Sol. Energy Mater. Sol. Cells*, 92(7), 686–714, 2008.
36. Kumar, P., & Gaur, A., Model for the J-V characteristics of degraded polymer solar cells. *J. Appl. Phys.*, 113(9), 094505, 2013.
37. Sze, S.M., *Physics of Semiconductor Devices*, pp. 1–832. Wiley, New York, 1981.
38. Kumar, P., Jain, S.C., Kumar, V., Chand, S., & Tandon, R.P., A model for the JV characteristics of P3HT: PCBM solar cells. *Appl. Phys.*, 105, 104507, 2009.
39. Schilinsky, P., Waldauf, C., Hauch, J., & Brabec, C.J., Simulation of light intensity dependent current characteristics of polymer solar cells. *J. Appl. Phys.*, 95, 2816, 2004.
40. Clarke, T.M., & Durrant, J.R., Charge photogeneration in organic solar cells. *Chem. Rev.*, 110, 6736, 2010.
41. Gaur, A., & Kumar, P., An improved circuit model for polymer solar cells. *Prog. Photovoltaics: Res. Appl.*, 22, 937–948, 2014.
42. Sariciftci, N.S., Smilowitz, L., Heeger, A.J., & Wudl, F., Photoinduced electron transfer from a conducting polymer to Buckminsterfullerene. *Science*, 258, 1474, 1992.
43. Yu, G., Gao, J., Hemmelen, J.C., Wudl, F., & Heeger, A.J., Polymer photovoltaic cells: Enhanced efficiencies via a network of internal donor-acceptor heterojunctions. *Science*, 270, 1789, 1995.
44. Brabec, C.J., Organic photovoltaics: Technology and market. *Sol. Energy Mater. Sol. Cells*, 83, 273, 2004.
45. Coakley, K.M., & McGehee, M.D., Conjugated polymer photovoltaic cells. *Chem. Mater.*, 16, 4533, 2004.
46. Ma, W., Yang, C., Gong, X., Lee, K., & Heeger, A.J., Thermally stable, efficient polymer solar cells with nanoscale control of the interpenetrating network morphology. *Adv. Funct. Mater.*, 15, 1617, 2005.
47. Reyes-Reyes, M., Kim, K., & Carroll, D.L., High-efficiency photovoltaic devices based on annealed poly(3-hexylthiophene) and 1-(3-methoxycarbonyl)-propyl-1-phenyl-(6,6)C61 blends. *Appl. Phys. Lett.*, 87, 083506, 2005.
48. Li, G., Shrotriya, V., Huang, J., Yao, Y., Moriarty, T., Emery, K., & Yang, Y., High-efficiency solution processable polymer photovoltaic cells by self-organization of polymer blends. *Nat. Mater.*, 4, 864–868, 2005.
49. Kim, Y., Cook, S., Tuladhar, S.M., Choulis, S.A., Nelson, J., Durrant, J.R., Bradley, D.D., Giles, M., McCulloch, I., Ha, C.S., & Ree, M., A strong regioregularity effect in self-organizing conjugated polymer films and high-efficiency polythiophene: Fullerene solar cells. *Nat. Mater.*, 5, 197–203, 2006.
50. Kim, J.Y., Kim, S.H., Lee, H.H., Lee, K., Ma, W., Gong, X., & Heeger, A.J., New architecture for high-efficiency polymer photovoltaic cells using solution-based titanium oxide as an optical spacer. *Adv. Mater.*, 18, 572–576, 2006.
51. Wanlass, M.W., Emery, K.A., Gessert, T.A., Horner, G.S., Osterwald, C.R., & Coutts, T.J., Practical considerations in tandem cell modeling. *Sol. Cells*, 27, 191–204, 1989.

52. Riedel, I., & Dyakonov, V., Influence of electronic transport properties of polymer-fullerene blends on the performance of bulk heterojunction photovoltaic devices. *Phys. Status Solidi A*, 201, 1332–1341, 2004.
53. Kim, J.Y., Lee, K., Coates, N.E., Moses, D., Nguyen, T.Q., Dante, M., & Heeger, A.J., Efficient tandem polymer solar cells fabricated by all-solution processing. *Science*, 317, 222–225, 2007.
54. Jørgensen, M., Norrman, K., Gevorgyan, S.A., Tromholt, T., Andreasen, B., & Krebs, F.C., Stability of polymer solar cells. *Adv. Mater.*, 24, 580, 2012.
55. Street, R.A., & Schoendorf, M., Interface state recombination in organic solar cells. *Phys. Rev. B*, 81, 205307, 2010.
56. Wojcik, M., & Tachiya, M., Accuracies of the empirical theories of the escape probability based on Eigen model and Braun model compared with the exact extension of Onsager theory. *J. Chem. Phys.*, 130, 104107, 2009.
57. Saleheen, M., Arnab, S., & Kabir, M., Analytical model for voltage-dependent photo and dark currents in bulk heterojunction organic solar cells. *Energies*, 9, 412, 2016.
58. Anjan, M., & Kabir, M.Z., Modeling of current-voltage characteristics of CdS/CdTe solar cells. *Phys. Status Solidi A*, 208, 1813–1816, 2011.
59. Mandoc, M., Koster, L., & Blom, P., Optimum charge carrier mobility in organic solar cells. *Appl. Phys. Lett.*, 90, 133504, 2007.
60. Cowan, S.R., Banerji, N., Leong, W.L., & Heeger, A.J., Charge formation, recombination, and sweep-out dynamics in organic solar cells. *Adv. Funct. Mater.*, 22, 1116–1128, 2006.
61. Lenes, M., Koster, L., Mihailetchi, V., & Blom, P., Thickness dependence of the efficiency of polymer: Fullerene bulk heterojunction solar cells. *Appl. Phys. Lett.*, 88, 243502, 2006.
62. Chowdhury, M.M., & Alam, M.K., An optoelectronic analytical model for bulk heterojunction organic solar cells incorporating position and wavelength dependent carrier generation. *Sol. Energy Mater. Sol. Cells*, 132, 107–117, 2015.
63. Sievers, D.W., Shrotriya, V., & Yang, Y., Modeling optical effects and thickness dependent current in polymer bulk-heterojunction solar cells. *J. Appl. Phys.*, 100, 114509, 2006.
64. Pettersson, L.A.A., Roman, L.S., & Inganas, O., Modeling photocurrent action spectra of photovoltaic devices based on organic thin films. *J. Appl. Phys.*, 86, 487–496, 1999.
65. Monestier, F., Simon, J.-J., Torchio, P., Escoubas, L., Flory, F., Bailly, S., de Bettignies, R., Guillerez, S., & Defranoux, C., Modeling the short-circuit current density of polymer solar cells based on P3HT: PCBM blend. *Sol. Energy Mater. Sol. Cells*, 91, 405–410, 2007.
66. Chowdhury, M.M., & Alam, M.K., An analytical model for bulk heterojunction organic solar cells using a new empirical expression of space dependent photocarrier generation. *Solar Energy*, 126, 64–72, 2016.
67. Sharma, V.K., CNTFET circuit-based wide fan-in domino logic for low power applications. *J. Circuits Syst. Comput.*, 31(02), 2250036, 2022.
68. Sharma, V.K., & Pattanaik, M., A reliable ground bounce noise reduction technique for nanoscale CMOS circuits. *Int. J. Electron.*, 102(11), 1852–1866, 2015.

69. Sharma, V.K., Patel, S., & Pattanaik, M., High performance process variations aware technique for sub-threshold 8T-SRAM cell. *Wirel. Pers. Commun.*, 78(1), 57–68, 2014.
70. Sharma, V.K., & Pattanaik, M., Techniques for low leakage nanoscale VLSI circuits: A comparative study. *Journal of Circuits, Systems, and Computers*, 23(05), 1450061, 2014.
71. Sharma, V.K., A survey of leakage reduction techniques in CMOS digital circuits for nanoscale regime. *Aust. J. Electr. Electron. Eng.*, 18(4), 217–236, 2021.
72. Kumar, C., Mishra, A.S., & Sharma, V.K., Leakage power reduction in CMOS logic circuits using stack ONOFIC technique. In *2018 Second International Conference on Intelligent Computing and Control Systems (ICICCS)*, 2018.
73. Sharma, V.K., & Pattanaik, M., Design of low leakage variability aware ONOFIC CMOS standard cell library. *J. Circuits Syst. Comput.*, 25(11), 1650134, 2016.
74. Naz, S.F., Riyaz, S., & Sharma, V.K., A review of QCA nanotechnology as an alternate to CMOS. *Curr. Nanosci.*, 18(1), 18–30, 2022.
75. Riyaz, S., & Sharma, V.K., Design of reversible Feynman and double Feynman gates in quantum-dot cellular automata nanotechnology. *Circuit World*, 49(1), 28–37, 2023.
76. Raina, B., Verma, C., Gupta, M., & Sharma, V.K., Binary coded decimal (BCD) seven segment circuit designing using quantum-dot cellular automata (QCA). In *2021 5th International Conference on Trends in Electronics and Informatics (ICOEI)*, 2021.
77. Mushtaq, U., & Sharma, V.K., Performance analysis for reliable nanoscaled FinFET logic circuits. *Analog Integr. Circuits Signal Process.*, 107(3), 671–682, 2021.
78. Kajal, & Sharma, V. K., Design and simulation of FinFET circuits at different technologies. In *2021 6th International Conference on Inventive Computation Technologies (ICICT)*, 2021.
79. Sharma, V. K., Designing of face recognition system. In *2019 International Conference on Intelligent Computing and Control Systems (ICCS)*, 2019.

Chapter 14

Design of high-performance SRAM cell

Balwinder Raj, Arun Kumar, Balwant Raj, and Shashi Bala

14.1 6T SRAM CELL

Static random-access memory (SRAM) cell design with complementary metal-oxide-semiconductor (CMOS) technology tends to have high-speed semiconductor memories with applied clock cycles which ranges from 2 to 30 ns [1–5]. SRAM cells hold the two stored values '0' and '1' in flops circuits while power is 'ON', that is, it is known as volatile memory cell. With advanced materials and device structures nowadays non-volatile SRAM cell designs are reported in the literature. It consists of two cross-coupled inverters for storage and two access transistors used for read and write operation, making a total of six transistors required for single-bit stage in SRAM cell. The word line is connected to the gate terminal of access transistors, and the bit line is attached to the source/drain terminals of transistors [6–11]. The word line needs to make high to select the cell while the bit line and bit line bar are responsible for read and write operations on the SRAM cell and further connected to the sense amplifier. Internally, the SRAM cells stored value on one side of the cell and its complement value on the other side of the cell, i.e. one side of cross-coupled cell holds '0' and other side holds '1' [12–18].

14.1.1 Working of 6T SRAM

The 6T SRAM cell design with nMOSFETs and pMOSFETs is presented in Figure 14.1. The cross-coupled inverters consisting of M1, M5 and M2, M6 behave as the storage elements to hold the digital values '0' and '1' [19–23]. Major design effort in SRAM cell is directed to reduce the cell area by scaling the dimensions of individual metal-oxide-semiconductor field-effect transistors (MOSFETs). In addition to this, power consumption and power dissipation need to be taken care of in order to place the millions of cells on a chip for high size memory [24–30].

 DOI: 10.1201/9781003643609-14

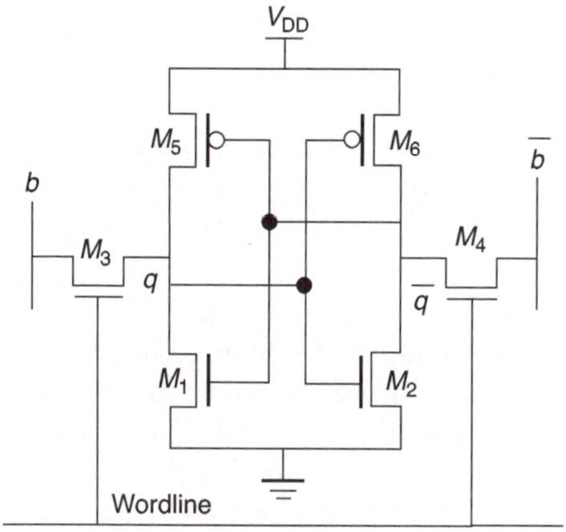

Figure 14.1 A basic SRAM cell.

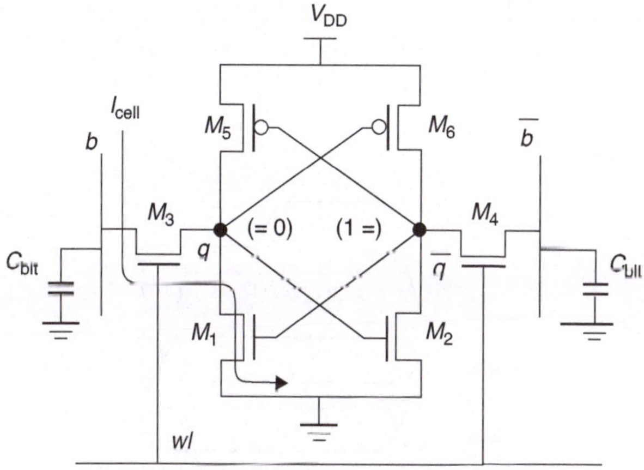

Figure 14.2 Read operation in 6T SRAM cell.

There are three basic operations in a working SRAM:

i. Read operation
ii. Write operation
iii. Hold operation

14.1.1.1 Read operation

Consider that a '0' is stored on the left side of crossed coupled inverter, and a '1' is stored on the right side of the SRAM cell to perform the read operation. Therefore, M_1 is 'ON' and M_2 is 'OFF' due to the complement input on their gate terminal [31–36]. Initially, q and $\overline{q}$ are kept at a high voltage nearly equal to V_{DD} by a pair of column pull-up transistors. The row selection line is raised to V_{DD} from low-voltage state leading to turn on access transistors M_3 and M_4 that are used for read and write operation. Current flows through M_3 and M_1 to ground, as shown in Figure 14.2, as current path is complete. Thereby, cell current slowly discharges the bit line capacitance C_{bit}. However, the voltage on $\overline{q}$ remains high since there is no path to ground through M_2 on the other side of the cell, for the flow of current [37–42]. The difference of q and $\overline{q}$ is fed back to a sense amplifier for the identification of stored bit '0' or '1', which is then stored in a data buffer as shown in Figure 14.3.

Once the read cycle is completed, the word line become low and stays at zero while the column lines are precharged to a high value for further operation [43–50]. During the scaling of MOSFET devices for read stability, it is required to ensure that voltage at the nodes of crossed coupled inverters are not changed during the read cycle. The SRAM cell retains its current state until one of the internal nodes crosses the switching threshold. The voltage at node q needs to be controlled by sizing of transistor M_1 and transistor M_3 to avoid alteration in the state of the cell while reading operation is performed [51–54]. This can be accomplished by making the conductance of transistor M_1 about 3–4 times that of transistor M_3 so that the drain voltage of M_1 does not go above Vth as shown in Figure 14.4.

$$W_1/W_3 \approx 1.5$$

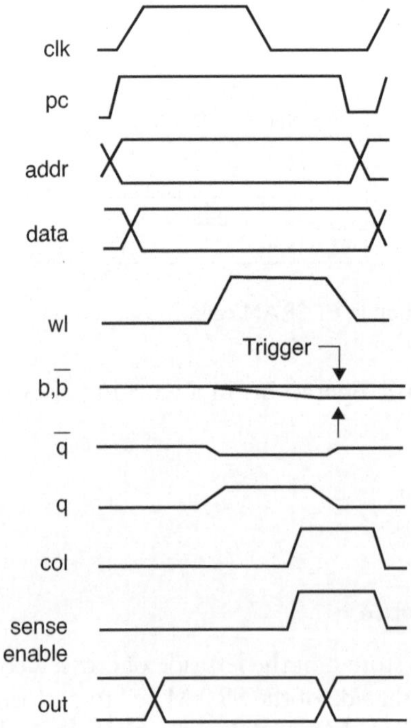

Figure 14.3 Timing diagram of read operation.

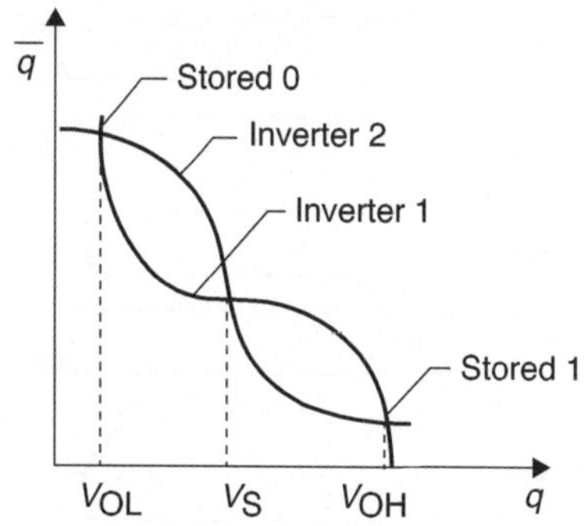

Figure 14.4 Stage flip w.r.t. voltage at output node.

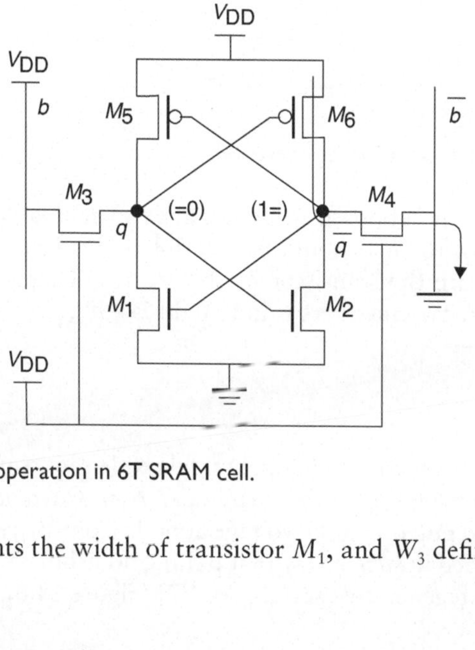

Figure 14.5 Write operation in 6T SRAM cell.

where W_1 presents the width of transistor M_1, and W_3 defined the width of transistor M_3.

14.1.1.2 Write operation

During the wire operation, one bit line, either bit line or bit line bar, is forced to keep low while the other bit line remains at higher potential (V_{DD}) to write 0 or 1 [55–58]. In Figure 14.5, to write 1, bit line bar is forced low, and to write 0, bit line is forced low. The conditions when writing 1 are shown in Figure 14.5.

Eventually, transistor M_1 turns 'OFF' and its drain voltage rise to V_{DD} subjected to the pull-up action of transistors M_5 and M_3. The transistor M_2 turns 'ON' simultaneously and assists transistor M_4 in pulling output to low value. When the SRAM cell finally flips to the new state, the row line is returned to its low standby level [59–63].

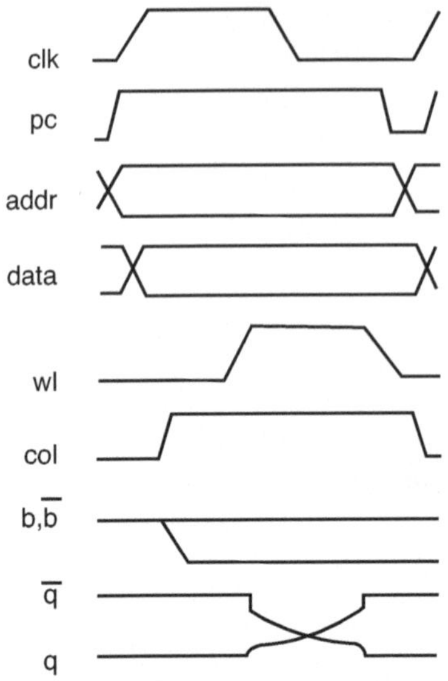

Figure 14.6 Timing diagram of write operation.

The SRAM cell is supposed to be designed in such a way that the conductance of transistor M_4 should be multiple times larger than transistor M_6 leading to drain the transistor M_2 below *Vss*. Thereby, a regenerative effect between the two inverters is initiated [64–68].

Here,

$$W_4/W_6 \approx 1.5$$

SRAM cell form a pseudo-n-channel metal oxide semiconductor (NMOS) inverter while turning on for the write operation. Current drains through the two devices to pull down the voltage at node from V_{DD} [69–73]. The size of devices is decided in such a way that pulling node below *Vs* forces the cell to switch via the regenerative action [74–77] (Figure 14.6).

14.1.1.3 Hold operation

For hold operation, word line is set to low making M_3 and M_4 to off state. Bit lines in this mode are charged to V_{DD}. The two cross-coupled inverters will keep on reinforcing each other as long as they are connected to power supply [78–82]. In this way, the memory element will keep the 1 or 0 intact as long as any read/write operation is being performed.

14.1.1.4 10T SRAM

Figure 14.7 depicts a schematic of a 10T SRAM with differential read bit lines (BL and BLB). NMOS (M_9 and M_7) for the BL and NMOS (M_8 and M_{10})

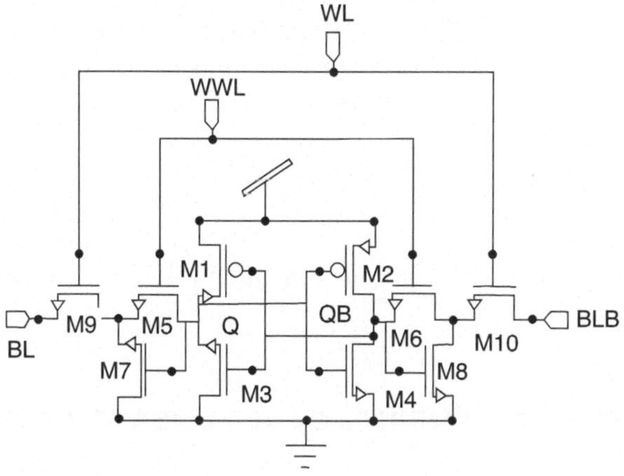

Figure 14.7 10T SRAM cell.

for BLB are added to the 6T SRAM [83–87]. The 10T cell allowed bit interleaving and exhibits superior sense margin with a differential read path.

14.1.1.5 Write operation

Firstly, write enable input is given to the write circuitry, which enables the inverters used to provide voltages at bit lines. Now, input is given to data input line to enable '1' and '0' at BL and BLB to write '1' and vice versa for writing '0' [88–94].

Now, wordline (WL) is asserted followed by write wordline (WWL) and the cell is connected to the BL and BLB line. Thus, current flows via M_6 and M_{10} and discharges through the BLB. This leads to a decrease in the potential at $\bar{q}$ point and thus it compels the p-channel metal oxide semiconductor (PMOS) transistors on q side to turn on. At one point of time, voltage rises above switching threshold value and thus value stored in cell flips. Thus, we have written data successfully in a cross-coupled back-to-back inverter network [95–98].

14.1.1.6 Read operation

Read operation of 10T cell is faster than that of 6T cell. Both the bit lines are asserted to high with the help of a precharge circuit at the columns. Now, whatever data is stored in the cell appears in form of gate voltage at the gate of M_7 and M_8 transistors. Thus, these transistors will be either on or off accordingly as gate potential is greater than the threshold potential of the device [99–103].

Now as the WL is asserted, BL will discharge through the M_9 and M_7 transistors. Time taken by bit line to discharge through transistors accounts for the read delay.

To reduce this delay, current from the cell is maintained at high value because in this case, it will not degrade the static noise margin (SNM) of cell during read operation and read operation is safe due to the isolated operation of the storage and read mechanisms.

14.1.1.7 Hold operation

After the data has been stored inside the cell, it is saved inside the cell due to back-to-back operation of storage cell, and in this case, we will not assert any word line or bit line.

14.2 MERITS AND DEMERITS OF 10T OVER 6T CELL

- Merits:
  - Read SNM improved.
  - Delay in reading is lowest (under 45 nm and 32 nm technology).
  - BL leakage is also reduced.
- Demerits:
  - Operates in sub-threshold.
  - Area inefficient.

14.3 UMC TECHNOLOGIES

We have used two UMC (United Microelectronics Corporations) 90 nm and 65 nm CMOS technologies for circuit designing.

14.3.1 UMC 90 nm technology

14.3.1.1 Technology parameters

$L = 90\,\text{nm}$
$\mu_o = 332.1\,\text{cm}^2/(\text{V-s})$
$V_{th} = 0.35\,\text{V}$ (Threshold Voltage) (NMOS)
$V_{th} = -0.35\,\text{V}$ (Threshold Voltage) (PMOS)
$\varepsilon_o = 8.85 \times 10^{-12}\,\text{F/m}$ (Permeability)
$C_{ox} = \varepsilon_o\ \varepsilon_r / T_{ox}$ (Oxide Capacitance)
$K_N = \mu_0{}^{*}\ C_{ox}$
$I_{ds} = K_N\ W\,(V_{gs} - V_T)^2/2\text{L}$ (Drain source Current)

We have used Cadence to make the circuit design. Cadence is EDA (Electronic Design Automation) tool which is the platform for custom IC design to enable the designers to harmonize the divergent world of analog and digital design.

14.3.1.2 Library device setup

MOSFETs: All MOSFETs in the library are four-terminal, with the body terminal explicitly connected. To confirm electromigration rules, customer may choose proper finger width, finger number, or source/drain metal according to device operating point.

14.3.1.3 Supported devices

Supported MOSFETs that are used are as follows:

- N_10_SPHVT (High Gate Threshold Voltage Transistor)
- N_10_SPLVT (Low Gate Threshold Voltage Transistor)
- P_10_SPHVT
- P_10_SPLVT

14.3.2 UMC 65 nm technology

14.3.2.1 Technology parameters

$L=90\,\text{nm}$
$\mu_o=332.1\,\text{cm}^2/(\text{V-s})$
$V_{th}=0.28\,\text{V}$ (Threshold Voltage) (NMOS)
$V_{th}=-0.28\,\text{V}$ (Threshold Voltage) (PMOS)
$\varepsilon_o=8.85\times10^{-12}\,\text{F/m}$ (Permeability)
$C_{ox}=\varepsilon_o\,\varepsilon_r/T_{ox}$ (Oxide Capacitance)
$K_N=\mu_0{}^*\,C_{ox}$
$I_{ds}=K_N\,W\,(V_{gs}-V_T)^2/2\,\text{L}$ (Drain source Current)

14.3.2.2 Library device setup

MOSFETS: Supported MOSFETs that are used are as follows:

- N_11_SPHVT (High Gate Threshold Voltage Transistor)
- N_11_SPLVT (Low Gate Threshold Voltage Transistor)
- P_11_SPHVT
- P_11_SPLVT

14.4 CIRCUIT SCHEMATIC

We have used Cadence to make the design. Cadence is an EDA (Electronic Design Automation) tool which is a platform for custom IC design to enable the designers to harmonize the divergent world of analog and digital design. For schematic and symbol creation, we use Composer (Figures 14.8–14.17).

While drawing the schematic of the SRAM cell, we use components from following libraries:

1. Analog Library
2. UMC90nm Library
3. UMC65sp Library

14.4.1 Components from analog library

- Vdc
- Gnd

14.4.2 Components from UMC90nm library

N_10_SPLVT
P_10_SPLVT

14.4.3 Components from UMC65sp library

N_11_SPLVT
P_11_SPLVT

14.4.4 Schematic implementation on composer

Figure 14.18 is the schematic of the circuit of 6T SRAM cell. The components are taken from ADD INSTANCE.

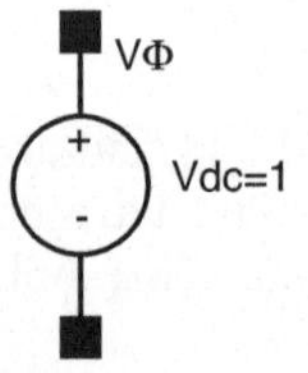

Figure 14.8 Schematic diagram of Vdc.

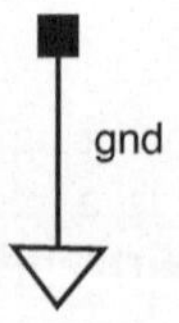

Figure 14.9 Schematic diagram of Gnd.

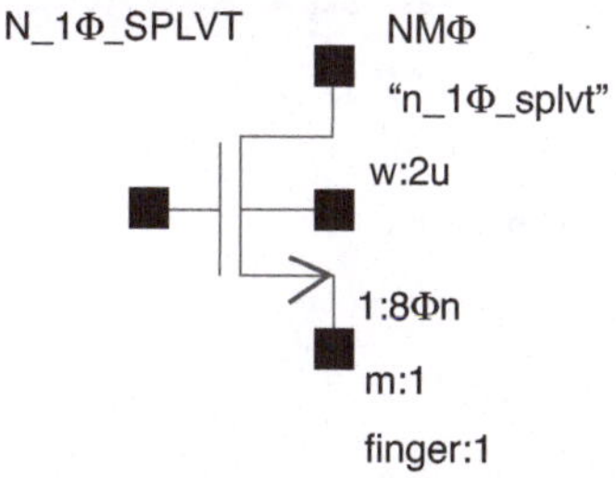

Figure 14.10 Schematic diagram of NMOS in 90 nm technology.

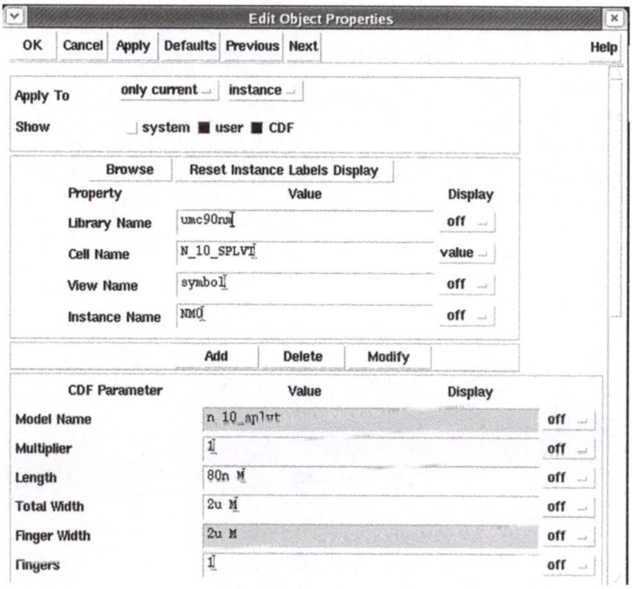

Figure 14.11 Parameters values for NMOS in 90 nm technology.

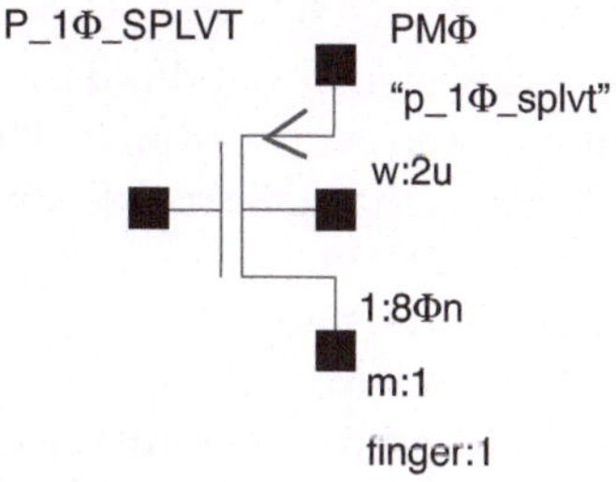

Figure 14.12 Schematic diagram of PMOS in 90 nm technology.

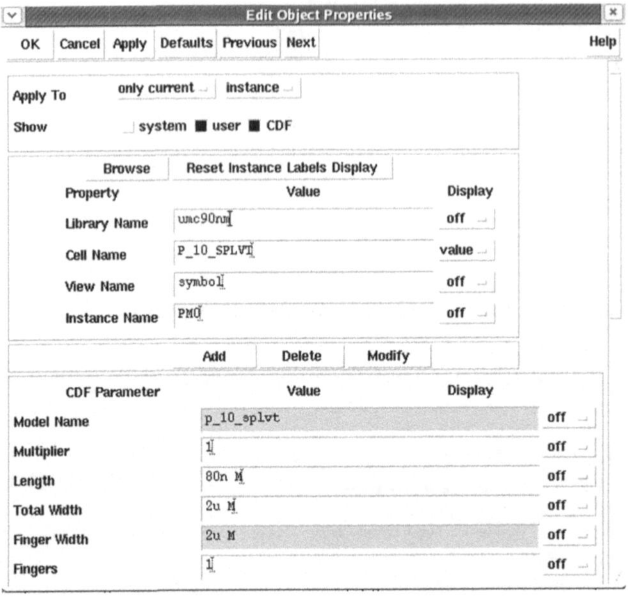

Figure 14.13 Parameters value of PMOS in 90 nm technology.

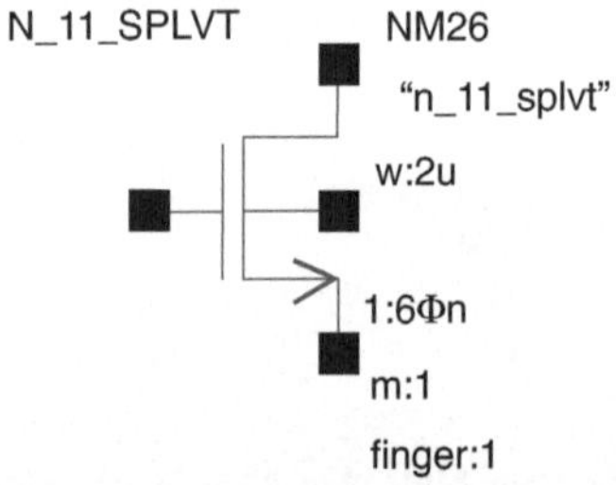

Figure 14.14 Schematic diagram of NMOS in 65 nm technology.

Input pins are writen (write enable), datain (data input), pre (precharge), sense (sense amplifier input), and wl (wordline). The input in the form of pulse wave is given to these pins through stimuli. Pins b and b_bar are activated according to writen and datain.

Output is observed at the nodes q and $\overline{q}$.

Bulks of all components from UMC65sp library and UMC90nm library are connected to ground for avoiding any bulk effect.

Figure 14.19 is the schematic of the circuit of 10T SRAM cell. The components are taken from ADD INSTANCE.

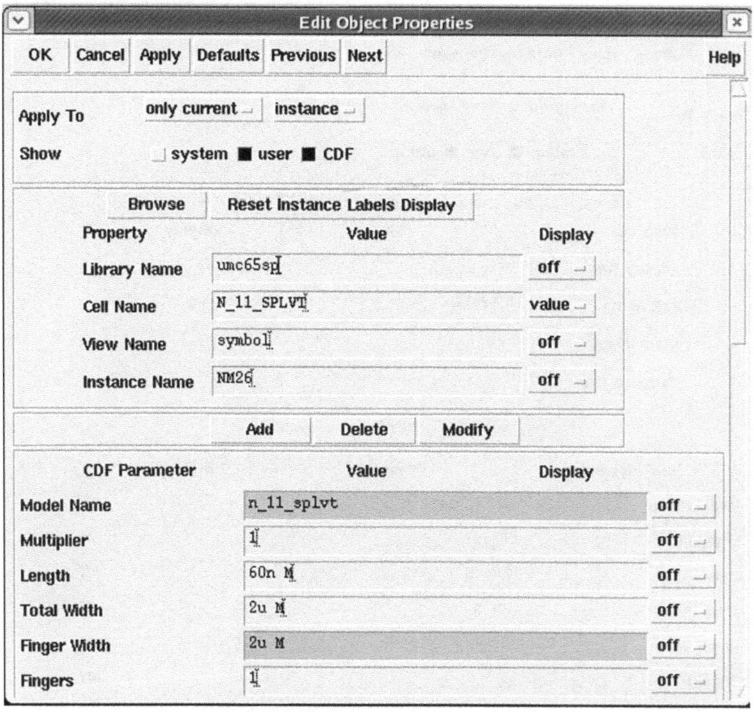

Figure 14.15 Parameters values of NMOS in 65 nm technology.

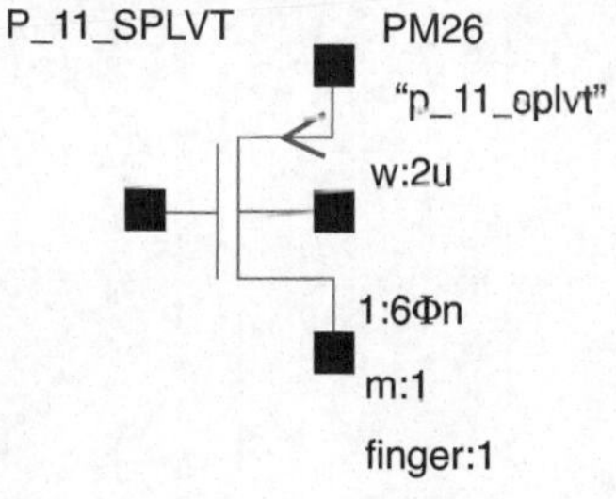

Figure 14.16 Schematic diagram of PMOS in 65 nm technology.

Input pins are writen (write enable), datain (data input), pre (precharge), sense (sense amplifier input), and wl (wordline) and another wwl. The input in the form of pulse wave is given to these pins through stimuli. Pins b and b_bar are activated according to writen and datain.

Output is observed at the nodes q and $\overline{q}$.

Bulks of all components from UMC65 sp library and UMC90nm library are connected to ground for avoiding any bulk effect

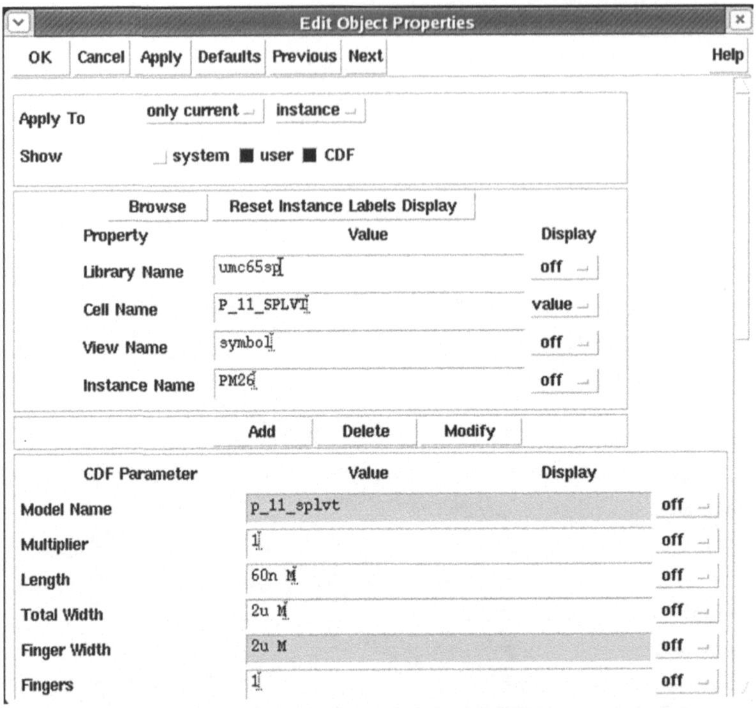

Figure 14.17 Parameter values of PMOS in 65 nm technology.

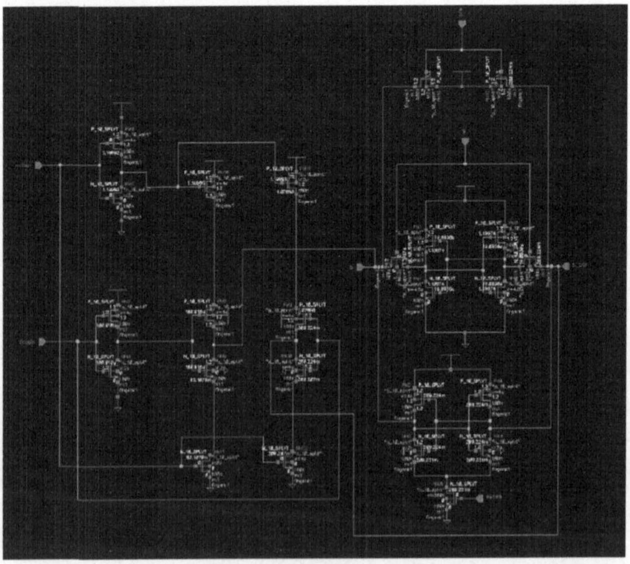

Figure 14.18 Schematic of 6T SRAM.

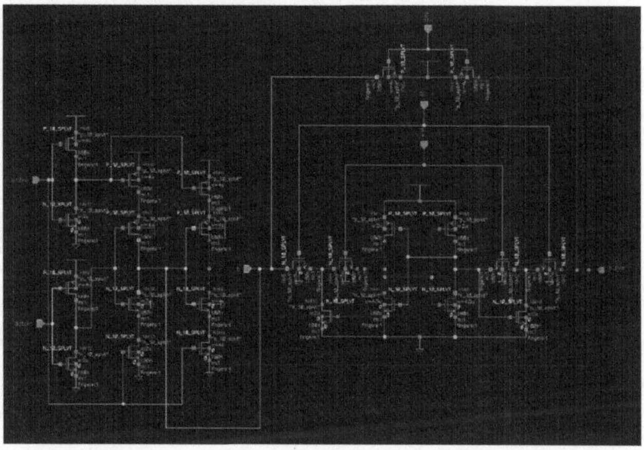

Figure 14.19 Schematic of 10T SRAM.

14.5 SIMULATION AND COMPARISON

The tool we are using is Cadence Virtuoso Schematic Editor. We used graphical interface for making our SRAM cell. An illustration is shown in Figure 14.20.

We can select instances from a library, and connect those using wires on the virtual breadboard. After the errors are removed, we open Analog Environment from the Tools menu. Three steps are done using analog environment:

i. **Stimuli:** The inputs to the pins are given in the form of pulses. These pulses have pulse width, time period, delays, rise time, and fall time. Such parameters are decided in the stimuli section.
ii. **Outputs to be plotted:** In analog environment, there's a column where we can define the outputs that are needed to be observed in the simulated result (waveform).
iii. **Analysis:** Whether transient or dc analysis, it is decided in this analog environment. Such an analysis is essential in determining the dc and transient characteristics of a circuit.

14.5.1 Analysis

We can choose for any of the analysis from the window shown in Figure 14.21.

After the analysis parameters are fixed, the outputs are determined that we want to observe. The output is not the output of the circuit, but all

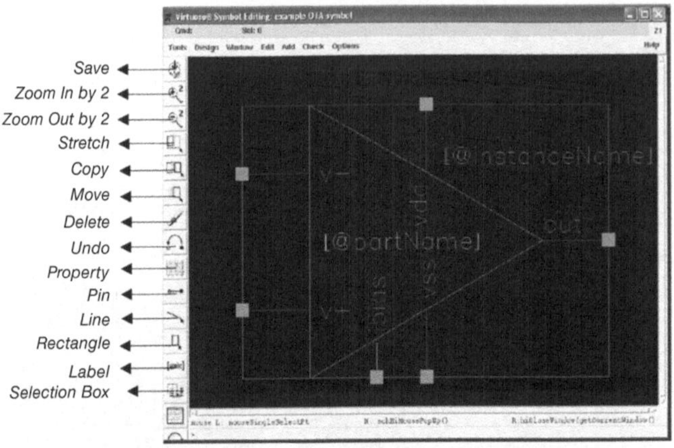

Figure 14.20 Graphical interface.

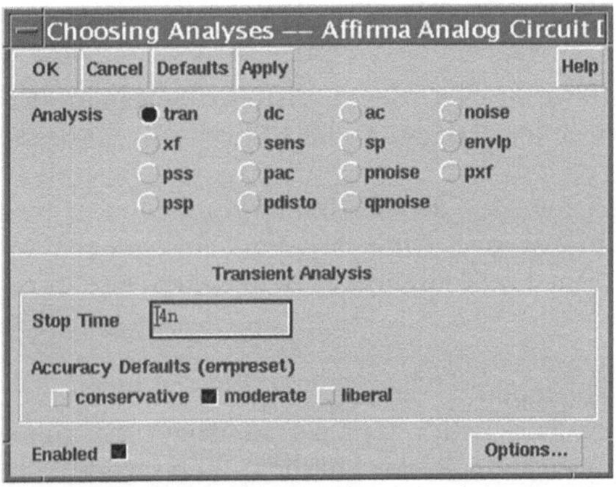

Figure 14.21 Analysis window.

the pins or nodes present in the circuit, whose variation along time axis is wished to be observed.

Finally, we click on the ***Netlist and Run*** button in the analog environment. And this runs the simulation, which displays on the screen the waveform depicting variations of the inputs and output of the circuit along time axis.

Using these waveforms, we can find out the read and write delays that occur during the process. Markers may be used to mark the 50% of transitions and the difference between two markers on different signals can help us in finding the delays [104–116].

14.5.2 Results for 6T SRAM cell

90 nm:

Write delay:

Voltage	*Write delay*
1.2	1.0544 ns
1	1.1834 ns
0.8	1.289 ns
0.6	1.6964 ns

Read delay:

Voltage	*Read delay*
1.2	527.75 ps
1	696.25 ps
0.8	743.77 ps
0.6	763.36 ps

SNM:

Voltage (V)	*Hold*	*Read*
1.2	446.6 mV	408 mV
0.8	344.4 mV	233.4 mV

65 nm:

Write delay:

Voltage	*Write delay*
1	1.0684 ns
0.9	1.2864 ns
0.8	1.2864 ns
0.7	1.2723 ns

Read delay:

Voltage	*Read delay*
1	180 ps
0.9	340.72 ps
0.8	435.73 ps
0.7	1.0169 ns

SNM:

Voltage	*Hold*	*Read*
1V	384 mV	58.2 mV
0.8V	328.3 mV	31.9 mV

14.5.3 Results for 10T cell

90 nm:

Write delay:

Voltage (V)	*Write delay*
0.6	535.24 ps
0.8	255.54 ps
1.0	174.492 ps
1.2	132.048 ps

Read delay:

Voltage (V)	*Read delay*
0.6	172 ps
0.8	68.056 ps
1.0	37.728 ps
1.2	32.449 ps

SNM:

Voltage (V)	*Hold*	*Read*
1.2	436.8 mV	73.8 mV
0.8	344.7 mV	55.9 mV

65 nm:

Write delay:

Voltage (V)	*Write delay*
1	262.4 ps
0.8	324.8 ps
0.7	371.33 ps
0.6	460 ps

Read delay:

Voltage (V)	*Read delay*
0.7	340.43 ps
0.8	170.36 ps
1.0	82.887 ps

SNM:

Voltage (V)	*Hold*	*Read*
1	410 mV	24.4 mV
0.8	348.1 mV	38.8 mV

14.6 CONCLUSION

This chapter presents the design of SRAM cell including the design of write-read circuitry along with design of sense amplifier for high-performance applications. The design of the storage cell is easy as there are just the back-to-back inverters, but apart from it, how data will get on to those storage points is critical and different ways are used to get it done. The way chosen will be fulfilling the requirements of the least leakage and least delay in the best optimum configuration to achieve high-performance design. Decoders used will be consisting of two stages, i.e. pre decode stage and final decode stage which will result in reducing the number of inputs for each gate. Cell connected like shown in the 10T configuration will provide us with the high speed or low delay. By using 10T cell read and write delay has been reduced and reduction in delay is 59.82%. This amount of delay has been decreased while we compare the 10T cell from the conventional 6T cell used, although it has problems of area efficiency.

REFERENCES

1. A. Bhardwaj, P. Kumar, B. Raj, and S. Anand, "Design and Performance Optimization of Doping Less Vertical Nanowire TFET Using Gate Stack Technique", *Journal of Electronic Materials (JEMS)*, Vol. 41, No. 7, pp. 4005–4013, 2022.
2. J. Singh, and B. Raj, "Tunnel Current Model of Asymmetric MIM Structure Levying Various Image Forces to Analyze the Characteristics of Filamentary Memristor", *Applied Physics A*, Vol. 125, No. 3, p. 203, 2019.
3. C. Goyal, J. S. Ubhi, and B. Raj, "Low Leakage Zero Ground Noise Nanoscale Full Adder Using Source Biasing Technique", *Journal of Nanoelectronics and Optoelectronics*, Vol. 14, pp. 360–370, 2019.

4. G. Singh, R. K.Sarin, and B.Raj, “A Novel Robust Exclusive-OR Function Implementation in QCA Nanotechnology with Energy Dissipation Analysis”, *Journal of Computational Electronics*, Vol. 15, No. 2, pp. 455–465, 2016.
5. T. Wadhera, D. Kakkar, G. Wadhwa, and B. Raj, “Recent Advances and Progress in Development of the Field Effect Transistor Biosensor: A Review”, *Journal of Electronic Materials*, Vol. 48, No. 12, pp. 7635–7646, 2019.
6. G. Wadhwa, P. Kamboj, and B. Raj, “Design Optimisation of Junctionless TFET Biosensor for High Sensitivity”, *Advances in Natural Sciences: Nanoscience and Nanotechnology*, Vol. 10, No. 7, p. 045001, 2019.
7. P. Bansal, and B. Raj, “Memristor Modeling and Analysis for Linear Dopant Drift Kinetics”, *Journal of Nanoengineering and Nanomanufacturing*, Vol. 6, pp. 1–7, 2016.
8. A. Singh, M. Khosla, and B. Raj, “Circuit Compatible Model for Electrostatic Doped Schottky Barrier CNTFET”, *Journal of Electronic Materials*, Vol. 45, No. 12, pp. 4825–4835, 2016.
9. D. Vaithiyanathan, Ashima, and B. Raj, “Performance Analysis of Charge Plasma induced Graded Channel Si Nanotube”, *Journal of Engineering Research*, Vol. 9, pp. 146–154, 2021.
10. A. S. Tomar, V. K. Magraiya, and B. Raj, “Scaling of Access and Data Transistor for High Performance DRAM Cell Design”, *Quantum Matter*, Vol. 2, pp. 412–416, 2013.
11. N. Jain, and B. Raj, “Parasitic Capacitance and Resistance Model Development and Optimization of Raised Source/Drain SOI FinFET Structure for Analog Circuit Applications”, *Journal of Nanoelectronics and Optoelectronics*, Vol. 13, pp. 531–539, 2018.
12. S. Singh, S. K. Vishvakarma, and B. Raj, “Analytical Modeling of Split-Gate Junction-Less Transistor for a Biosensor Application”, *Sensing and Bio-Sensing*, Vol. 18, pp. 31–36, 2018.
13. M. Gopal, and B. Raj, “Low Power 8T SRAM Cell Design for High Stability Video Applications”, *ITSI Transaction on Electrical and Electronics Engineering*, Vol. 1, No. 5, pp. 91–97, 2013.
14. B. Raj, J. Mitra, D. K. Bihani, V. Rangharajan, A. K. Saxena, and S. Dasgupta, “Analysis of Noise Margin, Power and Process Variation for 32 nm FinFET Based 6T SRAM Cell”, *Journal of Computer*, Vol. 5, No. 6, pp. 1–8, 2010.
15. D. Sharma, R. Mehra, and B. Raj, “Comparative Analysis of Photovoltaic Technologies for High Efficiency Solar Cell Design”, *Superlattices and Microstructures*, Vol. 153, p. 106861, 2021.
16. P. Kaur, A. S. Buttar, and B. Raj, “A Comprehensive Analysis of Nanoscale Transistor Based Biosensor: A Review”, *Indian Journal of Pure and Applied Physics*, Vol. 59, pp. 304–318, 2021.
17. D. Yadav, B. Raj, and B. Raj, “Design and Simulation of Low Power Microcontroller for IoT Applications”, *Journal of Sensor Letters*, Vol. 18, pp. 401–409, 2020.
18. S. Singh, and B. Raj, “A 2-D Analytical Surface Potential and Drain Current Modeling of Double-Gate Vertical t-Shaped Tunnel FET”, *Journal of Computational Electronics*, Vol. 19, pp. 1154–1163, 2020.
19. J. Singh, and B. Raj, “An Accurate and Generic Window Function for Non-linear Memristor Model”, *Journal of Computational Electronics*, Vol. 18, No. 2, pp. 640–647, 2019.

20. M. Kaur, N. Gupta, S. Kumar, B. Raj, and A. K. Singh, "Comparative RF and Crosstalk Analysis of Carbon Based Nano Interconnects", *IET Circuits, Devices & Systems*, Vol. 15, No. 6, pp. 493–503, 2021.
21. N. Kandasamy, F. Ahmad, D. Ajitha, B. Raj, and N. Telagam, "Quantum Dot Cellular Automata Based Scan Flip Flop and Boundary Scan Register", *IETE Journal of Research,* Vol. 69, pp. 535–548, 2023.
22. S. K. Sharma, B. Raj, and M. Khosla, "Enhanced Photosensitivity of Highly Spectrum Selective Cylindrical Gate In1-xGaxAs Nanowire MOSFET Photodetector", *Modern Physics Letter-B*, Vol. 33, No. 12, p. 1950144, 2019.
23. J. Singh, and B. Raj, "Design and Investigation of 7T2M NVSARM with Enhanced Stability and Temperature Impact on Store/Restore Energy", *IEEE Transactions on Very Large Scale Integration Systems*, Vol. 27, No. 6, pp. 1322–1328, 2019.
24. A. K. Bhardwaj, S. Gupta, B. Raj, and A. Singh, "Impact of Double Gate Geometry on the Performance of Carbon Nanotube Field Effect Transistor Structures for Low Power Digital Design", *Computational and Theoretical Nanoscience*, Vol. 16, pp. 1813–1820, 2019.
25. N. Jain, and B. Raj, "Thermal Stability Analysis and Performance Exploration of Asymmetrical Dual-k Underlap Spacer (ADKUS) SOI FinFET for Security and Privacy Applications", *Indian Journal of Pure & Applied Physics (IJPAP)*, Vol. 57, pp. 352–360, 2019.
26. A. Singh, M. Khosla, and B. Raj, "Design and Analysis of Dynamically Configurable Electrostatic Doped Carbon Nanotube Tunnel FET", *Microelectronics Journal*, Vol. 85, pp. 17–24, 2019.
27. N. Jain, and B. Raj, "Dual-k Spacer Region Variation at the Drain Side of Asymmetric SOI FinFET Structure: Performance Analysis towards the Analog/RF Design Applications", *Journal of Nanoelectronics and Optoelectronics*, Vol. 14, pp. 349–359, 2019.
28. J. Singh, S. Sharma, B. Raj, and M. Khosla, "Analysis of Barrier Layer Thickness on Performance of In1-xGaxAs Based Gate Stack Cylindrical Gate Nanowire MOSFET", *Journal of Nanoelectronics and Optoelectronics*, Vol. 13, pp. 1473–1477, 2018.
29. N. Jain, and B. Raj, "Analysis and Performance Exploration of High-k SOI FinFETs over the Conventional Low-k SOI FinFET toward Analog/RF Design", *Journal of Semiconductors*, Vol. 39, No. 12, p. 124002, 2018.
30. C. Goyal, J. S. Ubhi, and B. Raj, "A Reliable Leakage Reduction Technique for Approximate Full Adder with Reduced Ground Bounce Noise", *Journal of Mathematical Problems in Engineering*, Vol. 2018, p. 3501041, 2018.
31. J. S. Anuradha, B. Raj, and M. Khosla, "Design and Performance Analysis of Nano-Scale Memristor-Based Nonvolatile SRAM", *Journal of Sensor Letter,* Vol. 16, pp. 798–805, 2018.
32. G.Wadhwa,andB.Raj,"ParametricVariationAnalysisofCharge-Plasma-Based Dielectric Modulated JLTFET for Biosensor Application", *IEEE Sensor Journal*, Vol. 18, No. 15, pp. 6070–6077, 2018.
33. J. Singh, and B. Raj, "Comparative Analysis of Memristor Models for Memories Design", *Journal of Semiconductors*, Vol. 39, No. 7, p. 074006, 2018.
34. D. Yadav, S. S. Chouhan, S. K. Vishvakarma, and B. Raj, "Application Specific Microcontroller Design for IoT Based WSN", *Sensor Letter*, Vol. 16, pp. 374–385, 2018.

35. G. Singh, R. K. Sarin, and B. Raj, "Fault-Tolerant Design and Analysis of Quantum-Dot Cellular Automata Based Circuits", *IEEE/IET Circuits, Devices & Systems*, Vol. 12, pp. 638–664, 2018.
36. J. Singh, and B. Raj, "Modeling of Mean Barrier Height Levying Various Image Forces of Metal Insulator Metal Structure to Enhance the Performance of Conductive Filament Based Memristor Model", *IEEE Nanotechnology*, Vol. 17, No. 2, pp. 268–275, 2018.
37. A. Jain, S. Sharma, and B. Raj, "Analysis of Triple Metal Surrounding Gate (TM-SG) III-V Nanowire MOSFET for Photosensing Application", *Opto-Electronics Journal*, Vol. 26, No. 2, pp. 141–148, 2018.
38. A. Jain, S. Sharma, and B. Raj, "Design and Analysis of High Sensitivity Photosensor Using Cylindrical Surrounding Gate MOSFET for Low Power Sensor Applications", *Engineering Science and Technology, an International Journal*, Vol. 19, No. 4, pp. 1864–1870, 2016.
39. A. Singh, M. Khosla, and B. Raj, "Analysis of Electrostatic Doped Schottky Barrier Carbon Nanotube FET for Low Power Applications", *Journal of Materials Science: Materials in Electronics*, Vol. 28, pp. 1762–1768, 2017.
40. G. Saiphani Kumar, A. Singh, and B. Raj, "Design and Analysis of Gate All around CNTFET Based SRAM Cell Design", *Journal of Computational Electronics*, Vol. 17, No.1, pp. 138–145, 2018.
41. G. P. Singh, B. S. Sohi, and B. Raj, "Material Properties Analysis of Graphene Base Transistor (GBT) for VLSI Analog Circuits", *Indian Journal of Pure & Applied Physics (IJPAP)*, Vol. 55, pp. 896–902, 2017.
42. S. Kumar and B. Raj, "Estimation of Stability and Performance Metric for Inward Access Transistor Based 6T SRAM Cell Design Using n-type/p-type DMDG-GDOV TFET", *IEEE VLSI Circuits and Systems Letter*, Vol. 3, No. 2, pp. 25–39, 2017.
43. S. Sharma, A. Kumar, M. Pattanaik, and B. Raj, "Forward Body Biased Multimode Multi-Threshold CMOS Technique for Ground Bounce Noise Reduction in Static CMOS Adders", *International Journal of Information and Electronics Engineering*, Vol. 3, No. 3, pp. 567–572, 2013.
44. H. Singh, P. Kumar, and B. Raj, "Performance Analysis of Majority Gate SET Based 1-bit Full Adder", *International Journal of Computer and Communication Engineering(IJCCE)*, Vol. 2, No. 4, pp. 1–8, 2013.
45. A. K. Bhardwaj, S. Gupta, and B. Raj, "Investigation of Parameters for Schottky Barrier (SB) Height for Schottky Barrier Based Carbon Nanotube Field Effect Transistor Device", *Journal of Nanoelectronics and Optoelectronics*, Vol. 15, pp. 783–791, 2020.
46. P. Bansal, and B. Raj, "Memristor: A Versatile Nonlinear Model for Dopant Drift and Boundary Issues", *Journal of Computational and Theoretical Nanoscience*, Vol. 14, No. 5, pp. 2319–2325, 2017.
47. N. Jain, and B. Raj, "An Analog and Digital Design Perspective Comprehensive Approach on Fin-FET (Fin-Field Effect Transistor) Technology-A Review", *Reviews in Advanced Sciences and Engineering (RASE)*, Vol. 5, pp. 1–14, 2016.
48. S. Sharma, B. Raj, and M. Khosla, "Subthreshold Performance of In1-xGaxAs Based Dual Metal with Gate Stack Cylindrical/Surrounding Gate Nanowire MOSFET for Low Power Analog Applications", *Journal of Nanoelectronics and Optoelectronics*, Vol. 12, pp. 171–176, 2017.

49. B. Raj, A. K. Saxena, and S. Dasgupta, "Analytical Modeling for the Estimation of Leakage Current and Subthreshold Swing Factor of Nanoscale Double Gate FinFET Device" *Microelectronics International*, Vol. 26, pp. 53–63, 2009.
50. S. S. Soniya, G. Wadhwa, and B. Raj, "An Analytical Modeling for Dual Source Vertical Tunnel Field Effect Transistor", *International Journal of Recent Technology and Engineering (IJRTE)*, Vol. 8, No. 2, pp. 603–608, 2019.
51. S. Singh, and B. Raj, "Design and Analysis of Hetrojunction Vertical T-Shaped Tunnel Field Effect Transistor", *Journal of Electronics Material*, Vol. 48, No. 10, pp. 6253–6260, 2019.
52. C. Goyal, J. S. Ubhi, and B. Raj, "A Low Leakage CNTFET Based Inexact Full Adder for Low Power Image Processing Applications", *International Journal of Circuit Theory and Applications*, Vol. 47, No. 9, pp. 1446–1458, 2019.
53. B. Raj, A. K. Saxena, and S. Dasgupta, "A Compact Drain Current and Threshold Voltage Quantum Mechanical Analytical Modeling for FinFETs", *Journal of Nanoelectronics and Optoelectronics (JNO)*, Vol. 3, No. 2, pp. 163–170, 2008.
54. G. Wadhwa, and B. Raj, "An Analytical Modeling of Charge Plasma Based Tunnel Field Effect Transistor with Impacts of Gate Underlap Region", *Superlattices and Microstructures*, Vol. 142, p. 106512, 2020.
55. S. Singh, and B. Raj, "Modeling and Simulation Analysis of SiGe Heterojunction Double GateVertical t-Shaped Tunnel FET", *Superlattices and Microstructures*, Vol. 142, p. 106496, 2020.
56. A. Singh, D. K. Saini, D. Agarwal, S. Aggarwal, M. Khosla, and B. Raj, "Modeling and Simulation of Carbon Nanotube Field Effect Transistor and Its Circuit Application", *Journal of Semiconductors (JoS)*, Vol. 37, p. 074001, 2016.
57. N. Jain, and B. Raj, "Device and Circuit Co-Design Perspective Comprehensive Approach on FinFET Technology - A Review", *Journal of Electron Devices*, Vol. 23, No. 1, pp. 1890–1901, 2016.
58. S. Kumar, and B. Raj, "Analysis of ION and Ambipolar Current for Dual-Material Gate-Drain Overlapped DG-TFET", *Journal of Nanoelectronics and Optoelectronics*, Vol. 11, pp. 323–333, 2016.
59. N. Anjum, T. Bali, and B. Raj, "Design and Simulation of Handwritten Multiscript Character Recognition", *International Journal of Advanced Research in Computer and Communication Engineering*, Vol. 2, No. 7, pp. 2544–2549, 2013.
60. S. Sharma, B. Raj, and M. Khosla, "A Gaussian Approach for Analytical Subthreshold Current Model of Cylindrical Nanowire FET with Quantum Mechanical Effects", *Microelectronics Journal*, Vol. 53, pp. 65–72, 2016.
61. K. Singh, and B. Raj, "Performance and Analysis of Temperature Dependent Multi-Walled Carbon Nanotubes as Global Interconnects at Different Technology Nodes", *Journal of Computational Electronics*, Vol. 14, No. 2, pp. 469–476, 2015.
62. S. Kumar, and B. Raj, "Compact Channel Potential Analytical Modeling of DG-TFET Based on Evanescent Mode Approach", *Journal of Computational Electronics*, Vol. 14, No. 2, pp. 820–827, 2015.

63. K. Singh, and B. Raj, "Temperature Dependent Modeling and Performance Evaluation of Multi-Walled CNT and Single-Walled CNT as Global Interconnects", *Journal of Electronic Materials*, Vol. 44, No. 12, pp. 4825–4835, 2015.
64. V. K. Sharma, M. Pattanaik, and B. Raj, "INDEP Approach for Leakage Reduction in Nanoscale CMOS Circuits", *International Journal of Electronics*, Vol. 102, No. 2, pp. 200–215, 2014.
65. K. Singh, and B. Raj, "Influence of Temperature on MWCNT Bundle, SWCNT Bundle and Copper Interconnects for Nanoscaled Technology Nodes", *Journal of Materials Science: Materials in Electronics*, Vol. 26, No. 8, pp. 6134–6142, 2015.
66. N. Anjum, T. Bali, and B. Raj, "Design and Simulation of Handwritten Gurumukhi and Devanagri Numerical Recognition", *International Journal of Computer Applications*, Vol. 73, No. 12, pp. 16–21, 2013.
67. S. Khandelwal, V. Gupta, B. Raj, and R. D. Gupta, "Process Variability Aware Low Leakage Reliable Nano Scale DG-FinFET SRAM Cell Design Technique", *Journal of Nanoelectronics and Optoelectronics*, Vol. 10, No. 6, pp. 810–817, 2015.
68. V. K. Sharma, M. Pattanaik, and B. Raj, "ONOFIC Approach: Low Power High Speed Nanoscale VLSI Circuits Design", *International Journal of Electronics*, Vol. 101, No. 1, pp. 61–73, 2014.
69. S. Khandelwal, B. Raj, and R. D. Gupta, "FinFET Based 6T SRAM Cell Design: Analysis of Performance Metric, Process Variation and Temperature Effect", *Journal of Computational and Theoretical Nanoscience*, Vol. 12, pp. 2500–2506, 2015.
70. S. Singh, Y. Shekhar, R. Jagdeep, S. Anurag, and B. Raj, "Impact of HfO_2 in Graded Channel Dual Insulator Double Gate MOSFET", *Journal of Computational and Theoretical Nanoscience*, Vol. 12, No. 6, pp. 950–953, 2015.
71. V. K. Sharma, M. Pattanaik, and B. Raj, "PVT Variations Aware Low Leakage INDEP Approach for Nanoscale CMOS Circuits", *Microelectronics Reliability*, Vol. 54, pp. 90–99, 2014.
72. B. Raj, A. K. Saxena and S. Dasgupta, "Quantum Mechanical Analytical Modeling of Nanoscale DG FinFET: Evaluation of Potential, Threshold Voltage and Source/Drain Resistance", *Journal of Material Science in Semiconductor Processing*, Vol. 16, No. 4, pp. 1131–1137, 2013.
73. M. Gopal, S. S. D. Prasad, and B. Raj, "8T SRAM Cell Design for Dynamic and Leakage Power Reduction", *International Journal of Computer Applications*, Vol. 71, No. 9, pp. 43–48, 2013.
74. M. Pattanaik, B. Raj, S. Sharma, and A. Kumar, "Diode Based Trimode Multi-Threshold CMOS Technique for Ground Bounce Noise Reduction in Static CMOS Adders", *Advanced Materials Research*, Vol. 548, pp. 885–889, 2012.
75. B. Raj, A. K. Saxena, and S. Dasgupta, "Nanoscale FinFET Based SRAM Cell Design: Analysis of Performance Metric, Process Variation, Underlapped FinFET and Temperature Effect", *IEEE Circuits and System Magazine*, Vol. 11, No. 2, pp. 38–50, 2011.
76. V. K. Sharma, M. Pattanaik, and B. Raj, "Leakage Current ONOFIC Approach for Deep Submicron VLSI Circuit Design", *International Journal of Electrical, Computer, Electronics and Communication Engineering*, Vol. 7, No. 4, pp. 239–244, 2013.

77. T. Chawla, M. Khosla, and B. Raj, "Design and Simulation of Triple Metal Double-Gate Germanium on Insulator Vertical Tunnel Field Effect Transistor", *Microelectronics Journal*, Vol. 114, p. 105125, 2021.
78. P. Kaur, S. S. Gill, and B. Raj, "Comparative Analysis of OFETs Materials and Devices for Sensor Applications", *Journal of Silicon*, Vol. 14, pp. 4463–4471, 2022.
79. S. K. Sharma, P. Kumar, B. Raj, and Balwinder Raj, "In1-xGaxAs Double Metal Gate-Stacking Cylindrical Nanowire MOSFET for Highly Sensitive Photo Detector", *Journal of Silicon*, Vol. 14, pp. 3535–3541, 2022.
80. B. Raj, A. K. Saxena, and S. Dasgupta, "Analytical Modeling of Quasi Planar Nanoscale Double Gate FinFET with Source/Drain Resistance and Field Dependent Carrier Mobility: A Quantum Mechanical Study", *Journal of Computer (JCP)*, Vol. 4, No. 9, pp. 1–8, 2009.
81. S. Bhushan, S. Khandelwal, and B. Raj, "Analyzing Different Mode FinFET Based Memory Cell at Different Power Supply for Leakage Reduction", in *Seventh International Conference on Bio-Inspired Computing: Theories and Application, (BIC-TA 2012), Organized IIITM Gwalor*, pp. 89–100, 14–16 Dec 2012.
82. J. Singh, and B. Raj, "Temperature Dependent Analytical Modeling and Simulations of Nanoscale Memristor", *Engineering Science and Technology, an International Journal*, Vol. 21, pp. 862–868, 2018.
83. S. Singh, S. Bala, B. Raj, and B. Raj, "Improved Sensitivity of Dielectric Modulated Junctionless Transistor for Nanoscale Biosensor Design", *Sensor Letter*, Vol. 18, pp. 328–333, 2020.
84. V. Kumar, S. K. Vishvakarma, and B. Raj, "Design and Performance Analysis of ASIC for IoT Applications", *Sensor Letter*, Vol. 18, pp. 31–38, 2020.
85. A. Jaiswal, R. K. Sarin, B. Raj, and S. Sukhija, "A Novel Circular Slotted Microstrip-fed Patch Antenna with Three Triangle Shape Defected Ground Structure for Multiband Applications", *Advanced Electromagnetic (AEM)*, Vol. 7, No. 3, pp. 56–63, 2018.
86. G. Wadhwa, and B. Raj, "Label Free Detection of Biomolecules Using Charge-Plasma-Based Gate Underlap Dielectric Modulated Junctionless TFET", *Journal of Electronic Materials (JEMS)*, Vol. 47, No. 8, pp. 4683–4693, 2018.
87. G.Singh, R. K. Sarin, and B. Raj, "Design and Performance Analysis of a New Efficient Coplanar Quantum-Dot Cellular Automata Adder", *Indian Journal of Pure & Applied Physics (IJPAP)*, Vol. 55, pp. 97–103, 2017.
88. A. Singh, M. Khosla, and B. Raj, "Design and Analysis of Electrostatic Doped Schottky Barrier CNTFET Based Low Power SRAM", *International Journal of Electronics and Communications*, Vol. 80, pp. 67–72, 2017.
89. P. Kaur, V. Pandey, and B. Raj, "Comparative Study of Efficient Design, Control and Monitoring of Solar Power Using IoT", *Sensor Letter*, Vol. 18, pp. 419–426, 2020.
90. G. Wadhwa, P. Kamboj, J. Singh, and B. Raj, "Design and Investigation of Junctionless DGTFET for Biological Molecule Recognition", *Transactions on Electrical and Electronic Materials*, Vol. 22, pp. 282–289, 2020.
91. T. Chawla, M. Khosla, and B. Raj, "Optimization of Double-Gate Dual Material GeOI-Vertical TFET for VLSI Circuit Design", *IEEE VLSI Circuits and Systems Letter*, Vol. 6, No. 2, pp. 13–25, 2020.

92. S. K. Verma, S. Singh, G. Wadhwa and B. Raj, "Detection of Biomolecules Using Charge-Plasma Based Gate Underlap Dielectric Modulated Dopingless TFET", *Transactions on Electrical and Electronic Materials (TEEM)*, Vol. 21, pp. 528–535, 2020.
93. N. Jain, and B. Raj, "Impact of Underlap Spacer Region Variation on Electrostatic and Analog/RF Performance of Symmetrical High-k SOI FinFET at 20 nm Channel Length", *Journal of Semiconductors (JoS)*, Vol. 38, No. 12, p. 122002, 2017.
94. S. Singh, and B. Raj, "Analytical Modeling and Simulation Analysis of T-Shaped III-V Heterojunction Vertical T-FET", *Superlattices and Microstructures*, Vol. 147, p. 106717, 2020.
95. G. Singh, R. K. Sarin, and B. Raj, "Design and Analysis of Area Efficient QCA Based Reversible Logic Gates", *Journal of Microprocessors and Microsystems*, Vol. 52, pp. 59–68, 2017.
96. A. Singh, M. Khosla, and B. Raj, "Compact Model for Ballistic Single Wall CNTFET under Quantum Capacitance Limit", *Journal of Semiconductors (JoS)*, Vol. 37, p. 104001, 2016.
97. S. Singh, M. Khosla, G. Wadhwa, and B. Raj, "Design and Analysis of Double-Gate Junctionless Vertical TFET for Gas Sensing Applications", *Applied Physics A*, Vol. 127, No. 16, pp. 883–890, 2021.
98. I. Singh, B. Raj, M. Khosla, and B. K. Kaushik, "Potential MRAM Technologies for Low Power SoCs", *SPIN*, Vol. 10, No. 04, p. 2050027, 2020.
99. S. Singh, and B. Raj, "Parametric Variation Analysis on Hetero-Junction Vertical t-Shape TFET for Suppressing Ambipolar Conduction", *Indian Journal of Pure and Applied Physics*, Vol. 58, pp. 478–485, 2020.
100. S. S. Soniya, G. Wadhwa, and B. Raj, "Design and Analysis of Dual Source Vertical Tunnel Field Effect Transistor for High Performance", *Transactions on Electrical and Electronics Materials*, Vol. 21, pp. 74–82, 2019.
101. M. Kaur, N. Gupta, S. Kumar, B. Raj, and A. K. Singh, "RF Performance Analysis of Intercalated Graphene Nanoribbon Based Global Level Interconnects", *Journal of Computational Electronics*, Vol. 19, pp. 1002–1013, 2020.
102. G. Wadhwa, B. Raj, "Design and Performance Analysis of Junctionless TFET Biosensor for High Sensitivity", *IEEE Nanotechnology*, Vol. 18, pp. 567–574, 2019.
103. J. Singh, and B. Raj, "Enhanced Nonlinear Memristor Model Encapsulating Stochastic Dopant Drift", *Journal of Nanoelectronics and Optoelectronics*, Vol. 14, pp. 958–963, 2019.
104. V. K. Sharma, "CNTFET Circuit-based Wide Fan-In Domino Logic for Low Power Applications", *Journal of Circuits Systems and Computers*, Vol. 31, No. 02, p. 2250036, 2022.
105. V. K. Sharma and M. Pattanaik, "A Reliable Ground Bounce Noise Reduction Technique for Nanoscale CMOS Circuits", *International Journal of Electronics*, Vol. 102, No. 11, pp. 1852–1866, 2015.
106. V. K. Sharma, S. Patel, and M. Pattanaik, "High Performance Process Variations Aware Technique for Sub-Threshold 8T-SRAM cell", *Wireless Personal Communications*, Vol. 78, No. 1, pp. 57–68, 2014.
107. V. K. Sharma, and M. Pattanaik, "Techniques for Low Leakage Nanoscale VLSI Circuits: A Comparative Study", *Journal of Circuits Systems and Computers*, Vol. 23, No. 05, p. 1450061, 2014.

108. V. K. Sharma, "A Survey of Leakage Reduction Techniques in CMOS Digital Circuits for Nanoscale Regime", *Australian Journal of Electrical and Electronics Engineering*, Vol. 18, No. 4, pp. 217–236, 2021.
109. C. Kumar, A. S. Mishra, and V. K. Sharma, "Leakage Power Reduction in CMOS Logic Circuits Using Stack ONOFIC Technique", in *2018 Second International Conference on Intelligent Computing and Control Systems (ICICCS)*, 2018.
110. V. K. Sharma and M. Pattanaik, "Design of Low Leakage Variability Aware ONOFIC CMOS Standard Cell Library", *Journal of Circuits Systems Computers*, Vol. 25, No. 11, p. 1650134, 2016.
111. S. F. Naz, S. Riyaz, and V. K. Sharma, "A Review of QCA Nanotechnology as An alternate to CMOS", *Current Nanoscience*, Vol. 18, No. 1, pp. 18–30, 2022.
112. S. Riyaz and V. K. Sharma, "Design of Reversible Feynman and Double Feynman Gates in Quantum-dot Cellular Automata Nanotechnology", *Circuit World*, Vol. 49, No. 1, pp. 28–37, 2023.
113. B. Raina, C. Verma, M. Gupta, and V. K. Sharma, "Binary Coded Decimal (BCD) Seven Segment Circuit Designing Using Quantum-Dot Cellular Automata (QCA)", in *2021 5th International Conference on Trends in Electronics and Informatics (ICOEI)*, 2021.
114. U. Mushtaq and V. K. Sharma, "Performance Analysis for Reliable Nanoscaled FinFET Logic Circuits", *Analog Integrated Circuits Signal Processing*, Vol. 107, No. 3, pp. 671–682, 2021.
115. Kajal and V. K. Sharma, "Design and Simulation of FinFET Circuits at Different Technologies", in *2021 6th International Conference on Inventive Computation Technologies (ICICT)*, 2021.
116. V. K. Sharma, "Designing of Face Recognition System", in *2019 International Conference on Intelligent Computing and Control Systems (ICCS)*, 2019.

Index

Reichsgerichts-Entscheidungen
in kurzen Auszügen

Zivilsachen — Band 165

1942

J. Schweitzer Verlag, Berlin und München

Reichsgerichts-Entscheidungen

in kurzen Auszügen

Zivilsachen Band 165

1942

J. Schweitzer Verlag, Berlin und München

Printed in Germany

Druck von Dr. F. P. Datterer & Cie., Freising-München.

Von dieser Sammlung erschienen folgende Bändchen:

I. Zivilsachen:	Bd.	76—100	. . .	je	RM.	0.80
	„	101—140	. . .	je	RM.	1.—
	„	141—155	. . .	je	RM.	2.—
Serien:	„	76—155	mit 3 Reg.	zus.	RM.	76.—
	„	81—155	83—119 120—130	zus.	RM.	71.—
	„	91—155	131—140	zus.	RM.	61.—
	„	101—155		zus.	RM.	53.—
	„	111—155		zus.	RM.	43.—
	„	121—155	mit 4 Reg.	zus.	RM.	33.—
	„	131—155	mit 4 Reg.	zus.	RM.	28.—
Gesamtregister zu	Bd.	83—119	. . . .		RM.	6.—
Gesamtregister zu	Bd.	120—130	. . . .		RM.	1.80
Gesamtregister zu	Bd.	131—140	. . . .		RM.	1.50
Gesamtregister zu	Bd.	141—150	. . . .		RM.	1.50
II. Strafsachen:	Bd.	45—55	. . .	je	RM.	0.80
	„	56—64	. . .	je	RM.	1.—
	„	65—72	. . .	je	RM.	2.—

Serie: Bd. 45—72 mit Ges.-Reg. zu Bd. 45—60 zus. RM. 29.—

Gesamtregister zu Band 45 60 . . . . RM. 3.70

Jedes Bändchen entspricht einem Bande der amtlichen Sammlung.

1. Langfristiger Mietvertrag. Knebelung. Gute Sitten. (BGB. §§ 138, 567.) In einer Fabrik wurden Maschinen zur Anfertigung von Schuhen hergestellt. Bestimmte Arten von Maschinen, die unter Patentschutz standen, wurden nur mietweise abgegeben. Die Mietverträge waren in Formblättern festgelegt. Der Mietzins setzte sich regelmäßig zusammen aus einer Jahresgrundgebühr und einem monatlich zu zahlenden Betrag, der sich nach dem Umfang der Benutzung der Maschinen (Umdrehungszahl der Hauptwelle, Zahl der mit der Maschine hergestellten Schuhe) richtete; für gewisse Maschinen war lediglich eine feste Jahresmiete, für andere lediglich eine Umdrehungsmiete zu zahlen. Die Dauer der Mietzeit wurde auf 20 Jahre festgesetzt; sie verlängerte sich um je ein Jahr, wenn der Mieter nicht 60 Tage vor Ablauf kündigte. Im Jahr 1928 räumte die Gesellschaft, der die Fabrik gehörte, den Mietern das Recht ein, gegen Zahlung einer Ablösungssumme vorzeitig zu kündigen. Ein Fabrikant, der seit dem Jahr 1904 mit der Gesellschaft in Geschäftsverbindung stand, klagte auf Feststellung der Nichtigkeit der Verträge. Er beanstandete folgende Bestimmungen: Die Verträge legten dem Mieter die Verpflichtung auf, die gemieteten Maschinen bis zur völligen Ausnutzung ihrer Leistungsfähigkeit für alle in seinem Betriebe vorkommenden Arbeiten, die damit ausgeführt werden konnten, zu benutzen (Benutzungspflicht). Der Mieter war verpflichtet, in dem Raum, in dem die Maschine stand, kein Schuhwerk mit anderen Maschinen bestimmter Art zu bearbeiten als solchen, die er von der Gesellschaft gemietet hatte (Raumklausel). Der Mieter durfte endlich die gemieteten Maschinen nur zur Verarbeitung solchen Schuhwerks benutzen, bei dessen Herstellung keine anderen Maschinen bestimmter Art Verwendung gefunden hatten als solche der Gesellschaft (Verkettungsklausel). Begründet wurde die Klage damit, daß schon die lange Laufzeit der Verträge eine übermäßige Bindung bedeute, da bei Schuhmaschinen nur mit einer Lebensdauer von etwa 10 Jahren gerechnet werde. Durch die anderen Klauseln werde der Mieter gezwungen, nacheinander immer neue Maschinen auf 20 Jahre zu mieten. Das sei um so schädlicher, als andere Fabriken gleichwertige und sogar bessere Maschinen herstellten und ein Kaufmaschinenbetrieb vorteilhafter arbeite als ein Mietmaschinenbetrieb. Die im Jahr 1928 eingeräumte Lösungsmöglichkeit komme nicht in Betracht, weil die Ablösungssumme übermäßig hoch sei. Es bestehe auch ein auffälliges Mißverhältnis zwischen Leistung und Gegenleistung, da der vom Kläger zu zahlende Mietpreis jährlich 70000 bis 75000 Mark, der Kaufpreis für gleichwertige Maschinen anderer Fabriken aber nur 309000 RM betrage. Die beklagte Gesellschaft erreiche durch ihr Vertragssystem eine Verlängerung des Patentschutzes auf unbestimmte Zeit, was dem Patentgesetz widerspreche. Das Landgericht sprach dem Kläger das Recht zu, jeden Mietvertrag 30 Jahre nach seinem Inkrafttreten zu kündigen; dieses Recht wurde für den Fall des Ineinandergreifens mehrerer Mietverträge eingehend geregelt. Das Berufungsgericht wies die Klage ab. Das Reichsgericht bestätigte das Urteil.

I. Das Berufungsgericht hatte zunächst dargelegt, daß es sich nur um Mietverträge, nicht um Verträge mit gesellschaftsrechtlichem oder arbeitsrechtlichem Einschlag oder um Lizenzverträge handle, ferner daß die einzeln abgeschlossenen Verträge keine rechtliche Einheit bildeten und kein Gesamtrechtsverhältnis begründeten. Zur Begründung eines auffälligen Mißverhältnisses zwischen den beiderseitigen Leistungen hatte das Berufungsgericht eine Gegenüberstellung der bedungenen Mieten mit den Kaufpreisen ähnlicher Maschinen für ungeeignet gehalten, da es sich zumeist um Maschinen unter Patentschutz handle und die Herstellungskosten durch die Mietbeträge nur in geringem Maße belastet würden. Von einer unlösbaren Bindung auf ewige Zeiten könne keine Rede sein. Der Kläger sei aus freiem Entschluß mit der Beklagten in Verbindung getreten; die Maschinen hätte er auch anderwärts beziehen können; er müsse nun auch die beim Abschluß der Verträge erkennbar gewesenen Bindungen und Nachteile hinnehmen. Einen Verstoß gegen das Patentgesetz hatte das Berufungsgericht verneint.

II. Mit Recht hatte das Berufungsgericht die Verträge als Mietverträge angesehen; ebenso wurde seine Auffassung gebilligt, daß es sich bei jedem Abschluß um einen neuen Mietvertrag handelte und daß die einzelnen Verträge nicht zu einem einheitlichen Gesamtrechtsverhältnis zusammenzufassen waren.

III. Das Berufungsgericht hatte nicht außer acht gelassen, daß auch ein rein tatsächlicher Zwang zum Mieten immer neuer Maschinen und zur Aufrechterhaltung des Gesamtverhältnisses mit der beklagten Gesellschaft eine sittenwidrige Knebelung des Klägers in sich schließen konnte. Da es sich um formblattmäßig abgeschlossene Verträge handelte, kam es nicht so sehr darauf an, wie sie sich bei einem der zahlreichen Mieter auswirkten; vielmehr war in erster Reihe eine Nachprüfung der Verträge auf ihre Wirkung im allgemeinen vorzunehmen. Daß bei dem Kläger irgendwelche besondere Verhältnisse vorlagen, welche die bindende Wirkung der Mietverträge über das sich aus ihrem Inhalt ergebende Maß hinaus verstärkt hätten, war nirgends hervorgetreten.

IV. Dafür, daß zwischen den beiderseitigen Leistungen ein Mißverhältnis bestand, war der Kläger beweispflichtig. Die in früheren Jahren vereinbarten Mieten mit dem Preise zu vergleichen, zu dem die Maschinen jetzt zu kaufen sind, gestattete keinen Rückschluß auf die Angemessenheit der Mieten, weil die Maschinen früher unter Patentschutz standen. Daraus war früher eine Bemessung der Mieten über den Betrag hinaus gerechtfertigt, der sich allein auf Grund der Herstellungskosten als angemessen ergeben mochte. Auch daraus, daß die Miete nur einen sehr geringen Teil des Preises der damit hergestellten Schuhe darstellte, konnte ein wesentliches Anzeichen dafür gewonnen werden, daß sie nicht unangemessen hoch war. Für eine Vernichtung der Verträge aus dem Gesichtspunkt eines Mißverhältnisses zwischen Leistung und Gegenleistung wäre zudem noch notwendig gewesen, daß der beklagten Gesellschaft nach irgendeiner Richtung eine verwerfliche Gesinnung vorzuwerfen gewesen wäre.

V. Aus der Verkettungsklausel ergab sich nicht von vornherein ein Zwang zum Mieten weiterer Maschinen; wohl aber konnte sich ein solcher Zwang aus dem Laufen der früheren Verträge ergeben, wenn wirtschaftliche Verhältnisse die Einstellung neuer Maschinen veranlaßten. In Verbindung mit der langen Laufzeit der Verträge konnte diese Klausel dazu führen, daß der Kläger immer neue Verträge mit der beklagten Gesellschaft abschließen mußte und so in eine Bindung an sie gelangte, die aus sich selbst heraus kein Ende fand. Eine Beschränkung in der wirtschaftlichen Freiheit ist aber nicht ohne weiteres sittenwidrig; es kommt auf die gesamten Umstände an und sind die Belange beider Vertragsteile in Betracht zu ziehen. Die vom Kläger beanstandeten Verpflichtungen gingen hienach nicht zu weit. Die Benutzungspflicht hing damit zusammen, daß die Miete sich vielfach nach dem Umfang der Benutzung richtete. Die Raum- und Verkettungsklauseln ließen sich daraus rechtfertigen, daß die einzelnen Maschinen der beklagten Gesellschaft aufeinander abgestimmt waren und daß durch die Verwendung geschlossener Maschinensätze Herstellungsfehler vermieden werden konnten, was die beklagte Gesellschaft schon deshalb anstreben mußte, weil sie den Mieter die Verwendung ihres Warenzeichens gestattet hatte. Auch wenn die Beklagte mit diesen Vertragsbindungen den Zweck verfolgte, sich auf lange Zeit einen festen Abnehmerkreis zu sichern, war das nicht an sich verwerflich. War die Bindung des Klägers an die beklagte Gesellschaft durch die große Zahl der abgeschlossenen Verträge auch schwer lösbar geworden, so war eine Lösung doch nicht unmöglich. Gewisse dabei vom Kläger zu bringende Opfer waren nicht unbillig, nachdem er es für vorteilhaft erachtet hatte, sich auf lange Zeit zu binden und immer wieder neue Verträge zu schließen. Die im Jahr 1928 den Mietern eingeräumte Möglichkeit der Kündigung galt auch für die früher geschlossenen Verträge. Selbst wenn diese wegen zu starker Bindung nichtig gewesen wären, wäre doch ihre Fortsetzung mit der neu eingeführten Änderung als Neuabschluß mit neuem Inhalt anzusehen gewesen. Die im Falle der Kündigung zu zahlende Summe, die nach der Berechnung des Berufungsgerichts etwa $1/5$ der noch zu erwartenden Mietbeträge ausmachte, war nicht unbillig. Wenn auch der Kläger für die Ablösung und für den Ankauf neuer Maschinen große Summen aufwenden müßte, deren Beschaffung ihm schwierig wäre, so konnte er sich doch hierauf nicht berufen, weil er sich selbst in diese Lage gebracht und früher das immer wiederholte Mieten für zweckmäßig gehalten, auch unter Ausnutzung dieser Verträge sein Unternehmen zu hoher wirtschaftlicher Blüte gebracht hatte.

VI. Von einer sittenwidrigen Ausnutzung einer Vormachtstellung konnte schon deshalb keine Rede sein, weil die beklagte Gesellschaft nie eine solche Vormachtstellung besessen hatte. Es lag auch nichts Sittenwidriges darin, daß die beklagte Gesellschaft den für ihre Maschinen geltenden Patentschutz ausnutzte. Allerdings kann auch bei Verträgen, die gültig abgeschlossen worden sind, eine Veränderung der Verhältnisse, namentlich ein Wegfall der Geschäftsgrundlage, dazu führen, daß die Ausübung vertraglich begründeter Rechte unzulässig wird und nach § 242 BGB. unterbleiben muß. Darauf hatte sich aber der Kläger selbst nicht berufen.

VII. Die Zulassung einer Kündigung der Mietverträge nach 30jähriger Dauer, die das Landgericht dem Kläger zugebilligt hatte, beruhte auf der Annahme, daß die sämtlichen Verträge ein einheitliches Ganzes bildeten. Das traf nicht zu. Jeder neue Vertrag beruhte auf einem neuen Entschluß des Klägers. Für diesen mochten wirtschaftliche Notwendigkeiten bestimmend sein, die sich im Zusammenhang mit den früheren Verträgen ergaben; nirgends aber bestand eine Verpflichtung, eine neue Maschine zu mieten. Deshalb konnte keine Zusammenrechnung der Zeiträume aus den einzelnen Verträgen stattfinden. Es kam weiter auch in Betracht, daß seit dem Jahr 1928 jederzeit eine Lösung durch Kündigung unter Zahlung einer Abstandssumme möglich war. (V, 29. Juli 1940.) Amtl. Sammlg. S. 1—22.
Vgl. Bd. 115 S. 218; Bd. 130 S. 143; Bd. 132 S. 305; Bd. 150 S. 1; Bd. 152 S. 251.

2. **Rekurs.** (VO. vom 28. Februar 1939, §§ 4, 13.) Mit dem Widerspruch gegen einen Beschluß in einer Entmündigungssache wurde ein Antrag auf Ablehnung aller Richter des Landgerichts, das über den Widerspruch zu entscheiden hatte, verbunden. Das Oberlandesgericht wies den Antrag ab. Der Rekurs gegen diesen Beschluß wurde für unzulässig erklärt. Die Verordnung zur weiteren Überleitung der Rechtspflege im Lande Österreich bestimmt im § 4, daß Rekurse gegen Beschlüsse der Oberlandesgerichte in bürgerlichen Rechtsstreitigkeiten ausgeschlossen sind; das gilt auch für Beschlüsse, die das Oberlandesgericht als Erstgericht erlassen hat. Eine ähnliche ausdrückliche Bestimmung fehlt im dritten Abschnitt der Verordnung, der das Verfahren außer Streitsachen regelt. Im § 13 wird nur die Frage behandelt, ob gegen einen Beschluß, den das Oberlandesgericht als Rechtsmittelgericht

erlassen hat, ein weiterer Rechtszug zulässig ist; der Fall, daß es als Erstgericht entschieden hat, wird nicht erwähnt. Der Sinn der Verordnung geht aber dahin, daß Beschlüsse der Oberlandesgerichte nicht anfechtbar sein sollen, soweit nicht die Verordnung selbst eine Ausnahme schafft. Es kann nicht angenommen werden, daß in bürgerlichen Rechtsstreitigkeiten jedes Rechtsmittel gegen Beschlüsse der Oberlandesgerichte versagt, im Verfahren außer Streitsachen ein Rekurs zwar gegen die Rechtsmittelentscheidung der Oberlandesgerichte unzulässig, gegen ihre im ersten Rechtsgang erlassenen Beschlüsse aber zulässig sein sollte, zumal gerade die Fälle, in denen solche Beschlüsse ergehen, sehr selten sind und minder wichtige Angelegenheiten betreffen. (VIII, 30. September 1940.)

Amtl. Sammlg. S. 22—24.

3. **Ehescheidung. Neuerungsverbot.** (EheG. § 61; DurchfVO. § 76; ÖstZPO. §§ 482, 483; VO. vom 22. Dezember 1938, § 42.) Aus § 55 EheG. wurde auf Scheidung geklagt. Die beklagte Ehefrau widersprach der Scheidung und beantragte Abweisung der Klage. Das Landgericht schied die Ehe ohne Schuldausspruch. Die Beklagte legte Berufung ein und beantragte, den Kläger für schuldig zu erklären. Das Berufungsgericht bestätigte das Urteil; es nahm an, daß der Antrag der Beklagten im Berufungsverfahren nicht mehr zulässig sei. Das Reichsgericht verwies die Sache zurück. In dem Antrag auf Abweisung des Scheidungsbegehrens in Verbindung mit dem auf das Verschulden des Klägers gestützten Widerspruch war ein Antrag auf Schuldigerklärung des Klägers zu erblicken; anders wäre die Sachlage nur zu beurteilen gewesen, wenn sich aus dem Verhalten der Beklagten ergeben hätte, daß sie einen solchen Antrag nicht stellen wolle. Das Neuerungsverbot der Österreichischen Zivilprozeßordnung hätte die Stellung des Antrags im zweiten Rechtszug nicht zugelassen. Die Verordnung vom 22. Dezember 1938 zur Einführung des großdeutschen Eherechts in den sudetendeutschen Gebieten und die Durchführungsverordnung vom 27. Juli 1938 haben das Neuerungsverbot wesentlich eingeschränkt; diese Einschränkung gilt nicht nur für die Klageänderung, sie gewinnt vielmehr aus dem Gedanken der Einheit des Ehescheidungsverfahrens weitere Bedeutung. Diese Einheit kommt auch im sachlichen Recht

dadurch zum Ausdruck, daß die beklagte Partei neben der Widerklage oder statt ihrer den Antrag auf Mitschulderklärung stellen kann, so daß dieser jederzeit zu berücksichtigen ist. (IV, 3. Oktober 1940.)

Amtl. Sammlg. S. 24—26.

Vgl. Bd. 162 S. 402.

4. Ehescheidung. Unterhaltsanspruch. Verwirkung. Gute Sitten. (BGB. §§ 826, 1578; EheG. §§ 66, 68, 74; ZPO. § 323.) Eine Ehe wurde wegen Ehebruchs des Ehemanns geschieden. Das Urteil wurde durch beiderseitigen Rechtsmittelverzicht sofort rechtskräftig. Noch am gleichen Tage gab der Ehemann vor einem Notar eine Erklärung ab, worin er sich als im Scheidungsurteil schuldig erklärter Teil zur Unterhaltsgewährung verpflichtet anerkannte und der Frau ein Unterhaltsgeld von monatlich 200 *RM* zusicherte; die Verpflichtung sollte bis zum Lebensende der Frau oder bis zu ihrer Wiederverheiratung gelten. Ausdrücklich verzichtete der Ehemann darauf, wegen etwaiger Änderungen seiner persönlichen Verhältnisse eine Herabsetzung der Unterhaltssumme zu beanspruchen. Nachdem er eine Reihe von Jahren hindurch der Verpflichtung nachgekommen war, klagte er auf Feststellung, daß die Verpflichtung nicht bestehe. Die Klage wurde auf Verstoß gegen die guten Sitten gestützt. Ein solcher sollte schon darin bestanden haben, daß die Beklagte, obwohl auch sie selbst sich des Ehebruchs schuldig gemacht hatte, den Kläger veranlaßte, sich als allein schuldig zu bekennen, daß sie in einer überstürzten Form den Unterhaltsanspruch geltend gemacht und ihn trotz Bewußtseins des eigenen Unrechts fortgesetzt festgehalten habe; des weiteren wurde ihr vorgeworfen, daß sie einen unsittlichen Lebenswandel führe, indem sie mit einem bei ihr in Miete wohnenden Manne in dauerndem Geschlechtsverkehr stehe. Das Berufungsgericht wies die Klage ab. Die Revision des Klägers hatte keinen Erfolg. Von einem Verstoß gegen die guten Sitten hätte gesprochen werden können, wenn die Beklagte das Scheidungsurteil durch unlauteres Verhalten herbeigeführt oder das rechtskräftige Urteil in Kenntnis seiner Unrichtigkeit ausgenutzt hätte. Ein solcher Verstoß lag aber nicht schon deshalb vor, weil die Feststellung der Alleinschuld des Klägers unrichtig war. Daß die Gewährung von Unterhalt an einem mitschuldigen Ehegatten nicht grundsätzlich gegen die guten Sitten verstößt, ergibt sich schon daraus, daß das Ehegesetz (abweichend vom früheren Recht) dem mitschuldigen Ehegatten sogar einen gesetzlichen Unterhaltsanspruch einräumt (§ 68). Damit die Benutzung eines solchen Urteils unsittlich erscheine, müßten besondere Umstände hinzukommen. Solche waren nicht nachgewiesen. Daß das Unterhaltsabkommen im unmittelbaren Anschluß an die Verkündung des Scheidungsurteils geschlossen worden war, konnte um so weniger beanstandet werden, als die Verhandlungen hierüber schon vor der Scheidung stattgefunden hatten. Eine Verwirkung des Unterhaltsanspruchs wegen unsittlichen Lebenswandels hatte das Berufungsgericht abgelehnt, obwohl die Beklagte nicht bestritten hatte, mit ihrem Mieter wie Mann und Frau zusammen zu leben; sie hatte sich damit verteidigt, daß das Verhältnis nirgends Anstoß errege, da es in keiner Weise nach außen in die Erscheinung trete. Das Reichsgericht trat dieser Auffassung bei. Nach Auflösung der Ehe kann von dem unterhaltsberechtigten Teil nicht ein Verhalten gefordert werden, das auf eine Fortdauer der Treuepflicht bis zum Abschluß einer neuen Ehe hinausliefe. Ein geschlechtsvertrauliches Zusammenleben in unauffälliger Gemeinschaft reicht zur Annahme eines unsittlichen Lebenswandels im Sinne des § 74 EheG. nicht aus. Nach Abschluß des Vertrags war eine Änderung der Gesetzgebung eingetreten. Nach früherem Recht (BGB. § 1378) kam es für die Anrechnung des Ertrags der Arbeit einer schuldlos geschiedenen Frau darauf an, ob nach den Verhältnissen, in denen die Ehegatten gelebt hatten, Erwerb der Frau üblich war; nach neuem Recht (EheG. § 66) ist maßgebend, ob eine Erwerbstätigkeit von ihr nach den Umständen erwartet werden darf. In dem Vertrag war der Fall einer Änderung der Gesetzgebung nicht vorgesehen; die Verpflichtung sollte vielmehr bis zum Tode oder bis zur Wiederverheiratung der Frau fortbestehen. Ersichtlich waren die Parteien damals davon ausgegangen, daß die Beklagte nicht genötigt sein sollte, ihren Lebensunterhalt durch Arbeit zu erwerben. § 323 ZPO. sieht nicht schlechthin die Möglichkeit einer Änderung vollstreckbarer Verpflichtungserklärungen vor, wenn die Verhältnisse sich wesentlich geändert haben, sondern besagt nur, daß die Unterwerfung unter die Zwangsvollstreckung einer Abänderung nicht im Wege steht, wenn die Vereinbarung die Klausel der gleichbleibenden Verhältnisse enthält. Wenn auch Unterhaltsverträgen der hier in Betracht kommenden Art eine solche Klausel regelmäßig innewohnt, entscheidet doch der durch Auslegung zu ermittelnde Vertragswille der Parteien darüber, in welchem Umfang eine Änderung der Verhältnisse die übernommene Verpflichtung beeinflussen soll. Da anzunehmen war, daß beim Vertragsschluß an eine Änderung der Gesetzgebung nicht gedacht wurde, mußte eine ergänzende Auslegung stattfinden und gefragt werden, wie die Parteien den eingetretenen Fall bei Erkennung seiner Möglichkeit geregelt haben würden. Das Berufungsgericht war zu dem Ergebnis gekommen, daß es dem Willen der Parteien entsprochen hätte, der Änderung keinen Einfluß auf die übernommene Verpflichtung einzuräumen. Das beruhte auf tatsächlichen Erwägungen, die Rechts- oder Erfahrungsgrundsätzen nicht zuwiderliefen. (IV, 7. Oktober 1940.)

Amtl. Sammlg. S. 26—32.

Vgl. Bd. 155 S. 55; Bd. 156 S. 265; Bd. 163 S. 287.

5. **Ehescheidung. Wiederholte Klage. Widerspruch.** (EheG. § 55; ZPO. § 616.) Eine auf § 55 EheG. gestützte Scheidungsklage wurde abgewiesen; das Urteil erlangte die Rechtskraft. Nach Ablauf von zwei Jahren erhob der Ehemann neuerdings Klage auf Scheidung. Das Berufungsgericht gab der Klage statt und erklärte den Kläger für schuldig. Das Reichsgericht wies die Klage ab. Das Berufungsgericht hatte angenommen, § 616 ZPO. stehe der neuen Klage nicht entgegen, weil aus Briefen der Beklagten, die der Kläger vorlegte, zu entnehmen sei, daß ihr die zur Aufrechterhaltung der Ehe erforderliche eheliche Gesinnung fehle und daß lediglich wirtschaftliche Erlägungen sie veranlaßten, an der Ehe festzuhalten; es hatte sich auf den Standpunkt gestellt, daß unheilbar zerrüttete Ehen als für die Volksgemeinschaft wertlos zu scheiden seien. Das Reichsgericht betonte, daß nach Abweisung der früheren Klage für eine neue Klage nur dann Raum wäre, wenn in der Folgezeit neue Tatsachen hinzugekommen wären, die ein anderes Gesamtbild ergäben. In dem früheren Urteil war die Abweisung der Klage darauf gestützt worden, daß der Kläger

allein an der Zerrüttung der Ehe schuld sei und daß der Beklagten kein irgendwie erheblicher Vorwurf gemacht werden könne. Daß die Beklagte nachträglich ein Verhalten gezeigt hätte, daß ihr erheblich zur Last zu legen wäre, war nicht nachgewiesen. Selbst wenn die vorgelegten Briefe den vom Berufungsgericht gezogenen Schluß zuließen, blieb die Möglichkeit offen, daß schon im früheren Rechtsstreit das Gericht mit einer solchen Einstellung der Beklagten gerechnet hatte; eine wesentliche Änderung der Verhältnisse war also nicht dargetan. Ob eine dreijährige Fortdauer der Trennung eine Grundlage dafür geboten hätte, ohne Rücksicht auf die frühere Entscheidung den Sachverhalt neu zu würdigen, konnte dahingestellt bleiben, da noch nicht drei Jahre seit der letzten mündlichen Verhandlung im damaligen Rechtsstreit verstrichen waren. (IV, 7. Oktober 1940.)

Amtl. Sammlg. S. 32—35.

Vgl. Bd. 164 S. 249.

6. **Geschäftskauf. Pachtvertrag.** (BGB. §§ 242, 556, 581; RG. zum Schutz des Einzelhandels; DurchfVO. vom 23. Juli 1934.) Durch einen Vertrag vom Jahr 1925 wurden Räume, in denen ein Lebensmittelgeschäft betrieben wurde, auf die Dauer von 10 Jahren verpachtet; außer dem Pachtzins war eine einmalige Abfindung von 4000 *RM* zu zahlen. Nach Ablauf der Pachtzeit wurde die Vertragsdauer um 5 Jahre verlängert. Der Verpächter zog in eine andere Stadt und eröffnete dort ein gleiches Geschäft. Im Jahr 1931 erwarb der Pächter das Nachbarhaus, in dem ein Möbelgeschäft betrieben wurde; im Jahr 1938 verlegte er das Lebensmittelgeschäft dorthin. Der Verpächter klagte auf Zurückverlegung des Geschäfts in die Pachträume oder Schadenersatz. Die Klage hatte keinen Erfolg. Wenn auch im Fall eines bloßen Geschäftsverkaufs der Käufer beim Fehlen einer ausdrücklichen oder stillschweigenden anderen Abrede mit dem gekauften Geschäft nach seinem Belieben schalten und walten kann, entsteht doch in einem Fall, in dem der Käufer zugleich Pächter der Räume ist, in denen bisher das Geschäft betrieben wurde, die Frage, ob ihm der Pachtvertrag verbietet, das gekaufte Geschäft aus den gepachteten Geschäftsräumen heraus zu verlegen, wenn dadurch die Verwertung der Geschäftsräume nach

Ablauf der Pachtzeit zum Nachteil des Verpächters wesentlich beeinträchtigt würde. Nach dem Stande der Gesetzgebung zur Zeit des Vertragsschlusses konnte der Kläger damit rechnen, daß er nach Beendigung der Pachtzeit selbst wieder ein Lebensmittelgeschäft in seinem Hause werde betreiben können und daß er, wenn der Beklagte das Geschäft bis dahin tüchtig geführt hatte, davon Nutzen haben würde, anderseits war es aber auch dem Beklagten unbenommen, das gekaufte Geschäft nach Ablauf der Pachtzeit in einem nahegelegenen Hause weiterzuführen. Diese Sachlage hatte sich aber durch gesetzliche Vorschriften, die während der Dauer des Pachtverhältnisses ergingen, wesentlich geändert. Die Errichtung und Übernahme, unter Umständen auch die Verlegung einer Verkaufsstelle ist nunmehr an eine Genehmigung geknüpft; diese kann versagt werden, wenn durch sie in der betreffenden Gegend eine außergewöhnliche Übersetzung eintreten würde. Nach Lage der Sache war ausgeschlossen, daß der Kläger die Genehmigung erhalten hätte, nach Ablauf des Pachtvertrags in seinem Haus ein Lebensmittelgeschäft zu betreiben, solange der Beklagte ein solches im Nebenhause betrieb. Die von den Parteien beim Vertragsschluß nicht vorausgesehene Entwicklung der Verhältnisse hatte es demnach mit sich gebracht, daß der Betrieb beider Geschäfte nebeneinander nicht möglich war und daß eine Partei der anderen weichen mußte. Die Frage war also, ob der Beklagte nach Treu und Glauben verpflichtet war, seine eigenen Belange denen des Klägers hintanzusetzen. Als Grundsatz hatte zu gelten, daß es dem Pächter nicht gestattet ist, aus der Stellung heraus, die ihm der Pachtvertrag gegeben hat, dem Verpächter die künftige Verwertung der Pachträume in der bisherigen Weise unmöglich zu machen. Für den vorliegenden Fall war aber weiter zu berücksichtigen, daß der Kläger das Geschäft an den Beklagten verkauft hatte. Nach den Feststellungen des Berufungsgerichts war damals der Gang des Geschäftes verlustbringend. Der Beklagte hatte durch seine Tüchtigkeit den Umsatz um ein Mehrfaches gesteigert. Diesen Erfolg seiner Arbeit würde der Beklagte zum Vorteil des Klägers verlieren, wenn er den Geschäftsbetrieb in seinem Hause aufgeben müßte. Der Kläger konnte sein von ihm geführtes gleichartiges

Geschäft in der anderen Stadt weiterführen und die freiwerdenden, an den Beklagten verpachteten Räume durch anderweitige Vermietung oder Verpachtung, wenn auch nicht gerade zum Betrieb eines Lebensmittelgeschäfts, angemessen nutzen. Bei Berücksichtigung dieser Umstände war dem Kläger zuzumuten, sich damit abzufinden, daß er sein Eigentum, das er durch einen langfristigen Vertrag gegen Entgelt aus der Hand gegeben hatte, heute nicht mehr so nutzen konnte, wie er es möchte (II, 4. September 1940.) Amtl. Sammlg. S. 35—41.

Vgl. Bd. 158 S. 180.

7. **Werklieferungsvertrag. Gewährleistung. Vertragsauslegung. Verjährung.** (BGB. §§ 157, 633, 635, 638.) Eine Maschinenfabrik übernahm gegenüber einer G. m. b. H. die Herstellung einer neu erfundenen Maschine (Mutternpresse). Das Schreiben, in dem sie die Übernahme des Auftrags bestätigte, führte genau auf, welche Arbeit die Maschine leisten und welche Vorteile gegenüber anderen Maschinen dadurch erzielt werden sollten. In gedruckten allgemeinen Lieferungsbedingungen, die sie der Bestätigung des Auftrags beilegte, übernahm sie die Haftung für Mängel der Lieferung in der Weise, daß sie alle Teile, die innerhalb von 6 Monaten nach der Lieferung unbrauchbar oder schadhaft würden, auszubessern oder neu zu liefern zusicherte. Weitere Ansprüche waren ausgeschlossen; den zum Betrieb der Maschine gehörenden Wärmeofen sowie die notwendigen Werkzeuge hatte die Bestellerin zu beschaffen. Die Maschine wurde geliefert. Zu einer Abnahme kam es nicht, weil sich bei den Versuchen ergab, daß die Werkzeuge zu Bruch gingen; sie hätten gegenüber den Ansprüchen, welche die Maschine an sie stellte, aus so hartem Stahl bestehen müssen, wie er derzeit nicht geliefert werden kann. Die Fabrik nahm die Maschine zurück, klagte aber gegen die G. m. b. H. auf Zahlung von Wechseln, die sie von ihr erhalten hatte. Die Beklagte erhob Widerklage auf Schadenersatz. Das Berufungsgericht sprach der Beklagten einen Betrag von 45 000 *RM* nebst Zinsen zu auf Grund der Feststellung, daß sie bei ordnungsmäßiger Vertragserfüllung in der Lage gewesen wäre, eine oder mehrere Maschinen mit einem Gewinn von dieser Höhe zu verkaufen. Das Reichsgericht verwies die Sache zurück. Es handelte sich um

einen Werklieferungsvertrag über die Herstellung einer nicht vertretbaren Sache (§ 651 BGB.). Die Haftung der Klägerin für etwaige Mängel wurde daher durch die §§ 633 ff. BGB. bestimmt. Eine Abnahme des Werkes durch die Bestellerin war nicht Voraussetzung des Gewährleistungsanspruchs; es genügte, daß die Klägerin das von ihr hergestellte Werk der Beklagten erfüllungshalber geliefert hatte. Eine bei Abschluß eines Werklieferungsvertrags übernommene Gewähr kann eine dreifache Bedeutung haben:

a) Es kann sich um die gewöhnliche Zusicherung einer Eigenschaft des Werkes handeln ohne die ausdrückliche Zusage, für seine Vertragsmäßigkeit unbedingt einstehen zu wollen. In diesem Falle steht dem Besteller ein Schadenersatzanspruch nur zu, wenn das Fehlen der zugesicherten Eigenschaft auf einem von dem Unternehmer zu vertretenden Umstande beruht.

b) Der Unternehmer kann die Haftung dafür, daß das Werk die zugesicherten Eigenschaften hat, in der Weise übernehmen, daß er für die Vertragsmäßigkeit seines Werkes unbedingt einzustehen sich verpflichtet.

c) Endlich kann er auch die selbständige Gewähr für einen über die Vertragsmäßigkeit des Werkes hinausgehenden Erfolg übernehmen.

Das Berufungsgericht hatte den letzten Fall als gegeben erachtet, weil in dem Bestätigungsschreiben die Leistungen, die mit der Maschine erzielt werden sollten, genau aufgezählt waren und in Verbindung mit dem Umstande, daß die Bestellerin den Ofen und die Werkzeuge zu beschaffen hatte, über die Leistungsfähigkeit der Maschine selbst hinaus eine bestimmte Leistungsfähigkeit der ganzen Anlage und damit ein bestimmter wirtschaftlicher Erfolg im Sinn einer unbedingten Gewähr für diesen Erfolg habe zugesagt werden sollen. Dieser Auffassung trat das Reichsgericht nicht bei. Gewiß kann darin, daß eine bestimmte Leistungsfähigkeit nicht nur des zu liefernden Werkes als solchen, sondern der gesamten Anlage, für die das Werk bestimmt ist, zugesichert wird, die besondere Gewähr für einen über die Vertragsmäßigkeit des Werkes hinausgehenden Erfolg liegen; das ist aber nur dann der Fall, wenn der zugesicherte Erfolg erkennbar ein anderer und weiterer ist als die bloße Vertragsmäßigkeit des Werkes selbst, mag auch die zugesicherte Eigenschaft in dessen Leistungsfähigkeit im Rahmen einer Anlage bestehen. So wurde eine selbständige Gewährleistung in einem Falle anerkannt, in dem bei Lieferung eines Ringofens zugesichert worden war, daß bei ordnungsmäßigem Betrieb eine bestimmte Menge von Steinen mit einem bestimmten jährlichen Verdienst hergestellt werden könne; die Zusicherung ging wesentlich über die Leistungsfähigkeit des Ofens hinaus, weil der zugesicherte Erfolg nicht nur von diesem selbst, sondern von besonderen, nicht ohne weiteres feststehenden wirtschaftlichen Verhältnissen abhing. In einem anderen Falle war als besonderer Erfolg des Einbaus des zu liefernden Werkes (8 Generatoren) in eine schon vorhandene Anlage ein gewisser, gegen früher ermäßigter Kohlenverbrauch versprochen worden. Im vorliegenden Falle waren aber nur gewisse Leistungen der zu liefernden Maschine zugesichert worden und es war nicht zu ersehen, daß der zu erreichende Leistungserfolg noch von anderen Umständen abhing. Die Annahme des Berufungsgerichts, die Klägerin habe einen über die Vertragsmäßigkeit ihres Werkes hinausgehenden besonderen Erfolg gewährleistet, entbehrte also einer ausreichenden Grundlage. Aber selbst wenn es sich um eine selbständige Gewährübernahme gehandelt hätte, wäre daraus nicht zu folgern gewesen, daß die Klägerin im Falle der Nichterreichung des zugesicherten Erfolges der Beklagten zum Schadenersatz und sogar zum uneingeschränkten, auch auf den entgangenen Gewinn ausgedehnten Schadenersatz verpflichtet wäre. Wenn das auch in der Regel als die Folge einer Gewährsübernahme anzusehen ist, kann sich doch aus den Umständen des Falles etwas anderes ergeben. Für die Beurteilung dieser Auslegungsfrage kann vor allem auch den zum Vertragsinhalt erhobenen allgemeinen Lieferungsbedingungen, welche die Haftung für Mängel des Werkes einschließlich zugesicherter Eigenschaften einschränken, eine besondere Bedeutung zukommen. Das Berufungsgericht hatte die Berücksichtigung der Lieferungsbedingungen mit der Begründung abgelehnt, daß eine so außergewöhnliche Abrede wie die Übernahme der unbedingten Gewährleistung durch besonderen Vertrag nicht durch gedruckte Lieferungsbedingungen außer Kraft gesetzt werden könne. Eine Gewährleistung, besonders eine selbständige, löst aber, falls die Zusage sich nicht verwirklicht, nicht notwendig einen Schadenersatzanspruch aus; ihre Wirkung kann beschränkt werden auf die Zubilligung eines Rücktrittsrechts oder eines Anspruchs auf Ersatz von Aufwendungen. Von einem Widerspruch der Lieferungsbedingungen mit den aus dem Bestätigungsschreiben ersichtlichen Erklärungen über die von der Maschine zu leistende Arbeit konnte somit keine Rede sein. Es mußte also geprüft werden, ob bei einer den Anforderungen des § 157 BGB. entsprechenden Vertragsauslegung anzunehmen war, daß der Beklagten auch ein Anspruch auf entgangenen Gewinn zugebilligt werden sollte. Es handelte sich, wie auch der Beklagten bekannt war, um eine noch nicht erprobte Maschine; demnach wäre, selbst bei einem Interesse der Klägerin an der Erlangung des Auftrags, es mit ihren Belangen schwer zu vereinbaren gewesen, wenn sie die weitgehende, in ihren Folgen kaum übersehbare Gefahr auf sich genommen hätte, der Beklagten einen Gewinn zu gewährleisten. Dagegen sprach gerade der Umstand, daß die Klägerin es trotz der Eigenart des Sonderauftrags für nötig hielt, die ihre Haftung wesentlich einschränkenden allgemeinen Lieferungsbedingungen ausdrücklich zum Vertragsinhalt zu machen. Falls das Berufungsgericht auf Grund erneuter Prüfung zu dem Ergebnis gelangte, die Klägerin habe eine Gewähr nur für die eigene Leistung und deren Erfolg übernommen, war zu berücksichtigen, daß die Lieferungsbedingungen Schadenersatzansprüche wegen Mängel des Werkes ausdrücklich auch insoweit ausschlossen, als diese Mängel in dem Fehlen zugesicherter Eigenschaften bestanden und auf der Bauart beruhten. Derartige Ansprüche unterliegen der sechsmonatigen Verjährung des § 638 BGB. Für diese Frage kam es darauf an, ob die Maschine abgenommen worden war oder ob die Abnahme deshalb abgelehnt worden war, weil weitere Erprobungsversuche und Nachbesserungen aussichtslos erschienen. (II, 28. September 1940.)

Amtl. Sammlg. S. 41—54.

Vgl. Bd. 58 S. 179; Bd. 71 S. 174; Bd. 128 S. 214; Bd. 146 S. 120; JW. 1919 S. 241; 1921 S. 828; 1926 S. 2526; 1939 S. 38.

8. **Kraftfahrzeugversicherung. Führerscheinabrede.** (Allg. KraftVersBed. § 3). Ein Spediteur nahm für seinen

Lastkraftwagen Versicherungsschutz gegen Haftpflicht und Fahrzeugbeschädigung. Die Gewährung des Versicherungsschutzes war an die Voraussetzung geknüpft, daß der berechtigte Fahrer des Fahrzeugs bei Eintritt des Schadens den zur Führung des Fahrzeugs vorgeschriebenen Führerschein hatte oder der Versicherungsnehmer das entschuldbarerweise annehmen konnte. Ein Sudetendeutscher, der einen tschechoslowakischen Führerschein besaß, trat in den Dienst des Versicherungsnehmers und wurde von diesem mit der Führung des Wagens betraut. Er stieß mit einem Personenkraftwagen zusammen; dieser wurde zertrümmert, der Führer schwer verletzt. Auch der Wagen des Spediteurs wurde beschädigt. Er klagte gegen die Versicherungsgesellschaft auf Feststellung, daß diese verpflichtet sei, ihm sowohl die Geldbeträge, die er auf Grund seiner Verantwortung für den Unfall zu leisten hatte, als auch seinen eigenen Schaden zu ersetzen. In zwei Rechtszügen drang die Klage durch. Das Reichsgericht wies sie ab. Der Kläger hatte gewußt, daß sein Fahrer keinen vorgeschriebenen Führerschein besaß; der tschechoslowakische Führerschein gab ihm keine Erlaubnis, innerhalb des Deutschen Reiches einen Lastkraftwagen zu führen. Die Untergerichte hatten aber angenommen, daß er in entschuldbarer Weise annehmen konnte, daß er eine solche Erlaubnis besitze. Er hatte sich bei der Polizeibehörde hienach erkundigt; dort war ihm nach Anfrage beim Bezirksamt mitgeteilt worden, der Führer könne weiter fahren, müsse aber eine Ergänzungsprüfung machen. Zur Ablegung einer solchen Prüfung war es nicht gekommen. Das Reichsgericht entschied, daß diese Sachlage nicht ausreiche, um den Versicherungsschutz zu begründen. Dieser besteht, wenn der Führer den Führerschein nicht besitzt, nur dann, wenn der Versicherungsnehmer das in entschuldbarer Weise annehmen konnte. Diese Voraussetzung, die einen Ausnahmetatbestand schafft, darf nicht ausdehnend für den Fall angenommen werden, daß der Versicherungsnehmer zwar weiß, der berechtigte Führer besitze den vorgeschriebenen Führerschein nicht, aber gutgläubig und auch entschuldbarerweise der Meinung ist, die zuständige Behörde habe ihm ausnahmsweise eine Fahrerlaubnis erteilt. Eine derartige Ausdehnung würde den Umfang des bestimmt abgegrenzten Versicherungsschutzes wesentlich

erweitern und die Versicherungsgefahr zu Lasten des Versicherers beträchtlich verschieben, (VII, 15. Oktober 1940.)
Amtl. Sammlg. S. 54—62.
Vgl. Bd. 158 S. 284; Bd. 160 S. 220; JW. 1933 S. 765.

9. **Ehescheidung. Mitschuldantrag. Rechtskraft.** (EheG. § 60; ÖstZPO. § 182.) Im Gebiet des österreichischen Rechts wurde Klage auf Scheidung einer Ehe wegen Verschulden der Ehefrau erhoben. Diese beantragte, die Klage abzuweisen und die Ehe wegen Verschulden des Klägers zu scheiden; als Widerklage war der Antrag nicht bezeichnet. Das Landgericht wies beide Scheidungsanträge ab. Auf die Berufung des Klägers wurde das Urteil aufgehoben und die Sache unter Rechtskraftvorbehalt an das Landgericht zurückverwiesen. Gegen den Aufhebungsbeschluß legte der Kläger Rekurs ein, soweit noch über sein Verschulden zu verhandeln sei. Er hatte keinen Erfolg. Der Frau war nachgewiesen, daß sie nach der Trennung vom Ehegatten die Ehe gebrochen hatte. Das Landgericht hatte angenommen, daß der Kläger diese Verfehlung nicht als ehezerstörend empfunden habe. Das Berufungsgericht war auf Grund der von ihm erhobenen Beweise zur entgegengesetzten Auffassung gelangt; es hatte hienach eine Prüfung des Verschuldens des Klägers für notwendig erachtet. Der Rekurs hielt eine solche Prüfung für überflüssig, weil in dem nicht angefochtenen Teile des Urteils endgültig dahin entschieden sei, daß ein Verschulden des Klägers nicht bestehe. Das erklärte das Reichsgericht für rechtsirrig. Gegen die Abweisung ihres Scheidungsbegehrens hatte die Beklagte allerdings kein Rechtsmittel eingebracht. Damit war aber dieser Teil des Urteils nicht rechtskräftig geworden. Die Frage des Bestandes einer Ehe und ihrer Lösung bilden, auch wenn verschiedene Begehren als Klage und Widerklage einander gegenüberstehen, eine Einheit. Zur Frage der Lösung der Ehe gehört als untrennbarer Teil auch die Frage der Schuld. Über Klage und Widerklage auf Scheidung oder Aufhebung einer Ehe und auch über die Frage des Verschuldens muß daher auf Grund eines einheitlichen Verfahrens und in einem einheitlichen Urteil erkannt werden. Das sachliche Ergebnis des Rechtsstreits in Ehesachen läßt darüber nur ein einheitliches Urteil zu, das nur im ganzen der Rechtskraft fähig ist. Wenn ein Rechtsmittel auch nur einen

Teil des den Rechtsstreit in einem Rechtsgang erledigenden Urteils ergreift, wird damit die Rechtskraft des ganzen Urteils gehemmt. Auch die vom Rechtsmittel nicht erfaßten Teile des Urteils bleiben nicht bestehen; sie behalten aber insofern Bedeutung, als sie das höhere Gericht derart binden, daß das Urteil die Partei, die kein Rechtsmittel eingelegt hat, nicht besser stellen darf, als es das von ihr nicht angefochtene Urteil getan hatte. Der Beklagten war es also versagt, auf ihre abgewiesene Widerklage nochmals zurückzukommen. Sie war aber nicht gehindert, sich der Klage gegenüber zu verteidigen und dabei auf ein Verschulden des Klägers zu berufen. Das Landgericht, an das die Sache zurückverwiesen wurde, konnte nicht mehr dazu gelangen, die Ehe auf Grund der Widerklage der Beklagten zu scheiden; wohl aber konnte es das Verschulden des Klägers unter dem Gesichtspunkt seiner Mitschuld prüfen. Das muß auch im Gebiet des österreichischen Rechts gelten, weil der Grund der Einheitlichkeit der Entscheidung in Ehesachen nicht auf der verfahrensrechtlichen Gestaltung, sondern auf der nunmehr gleichen sachlichen Regelung des Ehescheidungsrechts beruht. Die Beklagte konnte sich auch auf Verfehlungen des Klägers berufen, wegen welcher sie das Klagerecht schon verloren hatte, sofern deren Berücksichtigung der Billigkeit entsprach. (IV, 17. Oktober 1940.)

Amtl. Sammlg. S. 62—64.

Vgl. Bd. 162 S. 402.

10. Kostenentscheidung. Anfechtbarkeit. (ZPO. §§ 93, 99.) Durch ein Versäumnisurteil wurde die Beklagte zur Zahlung der eingeklagten Forderung verurteilt. Sie legte Einspruch ein und erklärte, in der Hauptsache keinen Antrag stellen zu können, da die Forderung während des Verfahrens fällig geworden sei; dagegen beantragte sie, die Kosten der Klägerin aufzuerlegen. Das Landgericht hielt das Versäumnisurteil aufrecht mit der Maßgabe, daß die Klägerin die Kosten des Rechtsstreits zu tragen habe mit Ausnahme der durch die Säumnis entstandenen Kosten, die der Beklagten zur Last gelegt wurden. Die Klägerin legte Berufung ein mit der Bitte, das Rechtsmittel im Falle der Unzulässigkeit der Berufung als sofortige Beschwerde zu behandeln. Das Oberlandesgericht verwarf sowohl die Berufung als auch die sofortige Beschwerde als unzulässig und legte die Kosten der Klägerin auf. Sie legte gegen den Beschluß sofortige Beschwerde ein, hatte aber keinen Erfolg. Der Beschluß des Oberlandesgerichts enthielt zwei Entscheidungen verschiedenen Inhalts: eine über die Zulässigkeit der Berufung (§ 519 ZPO.) und eine über die sofortige Beschwerde (§ 99 ZPO.). Die gegen den Beschluß eingelegte sofortige Beschwerde war nur zur ersten Entscheidung zulässig (§§ 519 b, 547 ZPO.), nicht aber zur zweiten Entscheidung (§ 567 ZPO.). Soweit die Beschwerde zulässig war, entbehrte sie des inneren Grundes. Die Kostenentscheidung eines Urteils kann regelmäßig nicht selbständig angefochten werden. Das Urteil des Landgerichts konnte nicht als eine Entscheidung in der Hauptsache angesehen werden, weil sich der Anspruch der Beklagten nur auf den Kostenpunkt beschränkte. In Frage kam nur, ob die Berufung deshalb zulässig war, weil die Hauptsache durch eine auf Grund eines Anerkenntnisses ausgesprochene Verurteilung erledigt wurde (§ 99 Abs. 2 ZPO.). Ein solches Anerkenntnisurteil war aber hier nicht ergangen, sondern nur ein Versäumnisurteil, das in der Hauptsache deshalb rechtskräftig wurde, weil die Beklagte, da sie die während des Verfahrens eingetretene Berechtigung des Klageanspruchs anerkennen mußte, keinen Einspruch gegen ihre Verurteilung eingelegt hatte. Dieser Tatbestand kam zwar sachlich dem der Verurteilung auf Grund eines Anerkenntnisses nahe; das rechtfertigte aber doch nicht, diese verfahrensrechtliche Ausnahmevorschrift über die im Gesetz klar und ausdrücklich getroffene Regelung hinaus anzuwenden, zumal die einer Urteilsanfechtung im Kostenpunkt allein grundsätzlich abgeneigte Einstellung des Gesetzes unverkennbar ist. (V, 10. Oktober 1940.)

Amtl. Sammlg. S. 65—65.

Vgl. Bd. 152 S. 248; JW. 1936 S. 2544.

11. Gesellschaft mit beschränkter Haftung. Aufsichtsrat. Entsendungsrecht. Stimmrechtsbeschränkung. Auslegung der Satzung. (GmbHG. § 52.) Nach der Satzung einer G. m. b. H., die sich mit der Herstellung und dem Vertrieb von Düngemitteln und chemischen Erzeugnissen aller Art befaßte, war ein Gesellschafter, der auf seinen Namen mindestens 25% des Gesellschaftskapitals besaß, berechtigt, eine Stelle im Aufsichtsrat zu beanspruchen. Auf Grund dieser Bestimmung war ein Kaufmann, der Geschäftsanteile in dieser Höhe in seiner Hand vereinigte, wiederholt in den Aufsichtsrat gewählt worden. Als er seine Geschäftsanteile an eine Aktiengesellschaft verkaufte, beantragte diese in einer Gesellschafterversammlung, ein Mitglied ihres Vorstandes in den Aufsichtsrat zu wählen. Der Antrag wurde von allen anderen Mitgliedern der G. m. b. H. abgelehnt. Die Aktiengesellschaft klagte gegen die G. m. b. H. und die übrigen Mitglieder auf Feststellung, daß das von ihr benannte Mitglied ihres Vorstandes Mitglied des Aufsichtsrats der G. m. b. H. sei. In zwei Rechtszügen wurde die Klage abgewiesen. Das Reichsgericht gab ihr statt. Das Berufungsgericht hatte die maßgebende Vorschrift der Satzung dahin ausgelegt, daß durch sie den übrigen Mitgliedern der Gesellschaft untersagt würde, gegen den Wahlvorschlag des begünstigten Gesellschafters zu stimmen, sofern nicht Gründe vorlägen, aus denen die Stimmrechtsbeschränkung entfiel; solche Gründe hatte es darin gefunden, daß die Aktiengesellschaft im Wettbewerb mit der G. m. b. H. stand und diese befürchten mußte, von ihr abhängig gemacht, unter Umständen sogar stillgelegt zu werden. Das Reichsgericht erkannte an, daß die zur Prüfung stehende Vorschrift der Satzung ihrem Wortlaut nach nicht klar und eindeutig sei und daß zur Behebung dieser Mehrdeutigkeit alle Auslegungsbehelfe herangezogen werden müßten, besonders solche, die sich aus den der Öffentlichkeit zugänglichen Akten des Handelsregisters ergaben und darum den über den Sinn der Vorschrift streitenden Gesellschaftern bekannt sein mußten; Nebenabreden, die in der Satzung keinen Ausdruck gefunden hatten, mußten unberücksichtigt bleiben, da die Satzung sich an einen unbestimmten Personenkreis wandte, der über die aus ihr nicht erkennbaren Gedanken und Absichten der Vertragschließenden nicht unterrichtet war. Ausdrücklich hob das Reichsgericht hervor, daß es in der Lage sei, die Auslegung des Berufungsgerichts frei nachzuprüfen. Es erblickte den Zweck der streitigen Vorschrift darin, daß einem Gesellschafter, der mindestens 25% des Gesellschaftskapitals besaß, die Befugnis gewährt werden sollte, dem Aufsichtsrat der Gesellschaft als Mitglied anzugehören oder durch eine Person seines Vertrauens darin vertreten zu sein. Nachdem die Höhe der Kapitalbeteiligung alleinige Voraussetzung für das Entstehen und die Ausübung des Rechtes war, vertrug es sich damit nicht, daß seine Verwirklichung von anderen Umständen, insbesondere der Zustimmung der Gesellschafter, abhängig sein sollte. Ziel der Auslegung konnte hienach nur sein, aus der Vorschrift jene Rechtsform als gewollt herzuleiten, die ihrem Zweck am nächsten kam und die Absichten der Vertragschließenden am vollkommensten in die Erscheinung treten ließ. Das war der Fall, wenn man dem begünstigten Gesellschafter ein Entsendungsrecht zubilligte. Daß ein solches Recht möglich ist, kann für den Bereich der G. m. b. H. nicht bezweifelt werden. Nach § 52 GmbHG. unterliegt die Frage, ob ein Aufsichtsrat bestellt werden soll, und die Art seiner Bestellung dem Willen der Gesellschafter. Es steht nichts entgegen, daß die Satzung das Ernennungsrecht für einen Teil oder alle Mitglieder des Aufsichtsrats einem Gesellschafter einräumt, die Befugnis hiezu auch von der Erfüllung gewisser Voraussetzungen abhängig macht. Nur das Entsendungsrecht gibt dem Berechtigten die Gewähr, seinen Anspruch durchsetzen zu können, ohne gewärtigen zu müssen, damit am Widerspruch der Mitgesellschafter zu scheitern. Selbst wenn man eine gesellschaftsrechtliche Bestimmung dahin, daß die übrigen Gesellschafter verpflichtet seien, für die Wahl eines vom Berechtigten vorgeschlagenen Aufsichtsrates zu stimmen, für rechtlich zulässig und wirksam ansehen wollte, könnte das nicht zur Folge haben, daß eine dieser Verpflichtung zuwiderlaufende Stimmabgabe gegenstandslos und der Vorgeschlagene trotz Fehlens einer Mehrheit Mitglied des Aufsichtsrats geworden wäre; der Berechtigte sähe sich vor die Notwendigkeit gestellt, gegen den seinen Vorschlag ablehnenden Beschluß vorzugehen, um sein Recht zur Geltung zu bringen. Das wäre ebenso der Fall, wenn sein Vorschlag unter Verletzung einer den Gesellschaftern auferlegten Stimmrechtsbeschränkung der Ablehnung verfiele. Das Recht konnte allerdings nicht uneingeschränkt und ohne Rücksicht auf die Belange der Gesellschaft ausgeübt werden. Es liefe der dem Gesellschafter obliegenden Treupflicht gegenüber der Gesellschaft zuwider, wenn er von seiner Befugnis in einer Weise Gebrauch machte, durch welche diese der Gefahr ausgesetzt wäre, einen ihr nicht zumutbaren Nachteil zu erleiden. Die Tatsache, daß die klagende Aktiengesellschaft mit der G. m. b. H. in Wettbewerb stand, begründete keine ernste Gefährdung der letzteren, die als wichtiger Grund gegen die Zulassung des benannten Aufsichtsratsmitglieds angesehen werden konnte. Die Aktiengesellschaft konnte eigensüchtige und den Belangen der G. m. b. H. abträgliche Bestrebungen nur zur Geltung bringen, wenn sie hiezu einen entsprechenden kapitalmäßigen Einfluß in dieser hatte. Hiefür hätte sie noch weitere Geschäftsanteile erwerben müssen. Die bloße Möglichkeit einer solchen Entwicklung hatte aber mit der Entsendung des benannten Aufsichtsratsmitglieds nichts zu tun und ließ eine Bekämpfung dieser Entsendung mit dem Hinweis auf eine danach drohende Gefahr nicht zu. Daß die Klägerin auf ihr Entsendungsrecht verzichtet habe, indem sie das von ihr benannte Mitglied zur Wahl stellte, konnte nicht angenommen werden; ebensowenig ließ sich gegenüber dem eindeutigen Zweck der Vorschrift die Auffassung rechtfertigen, die Entsendungsbefugnis sei zufolge eines durch die ständige anderweitige Handhabung geschaffenen körperschaftlichen Gewohnheitsrecht außer Kraft gesetzt worden. Wenn der Kaufmann, von dem die Aktiengesellschaft ihre Geschäftsanteile erworben hatte, sich in den Aufsichtsrat hatte wählen lassen, statt diesem kraft seines Entsendungsrechts ohne weiteres beizutreten, konnte das seinen Grund darin haben, daß er mit Widerständen seitens der Mitgesellschafter nicht zu rechnen brauchte. (II, 12. Oktober 1940.)

Amtl. Sammlg. S. 68—85.

Vgl. Bd. 112 S. 273; Bd. 119 S. 368; Bd. 133 S. 90; Bd. 137 S. 305; Bd. 140 S. 303; Bd. 141 S. 204; Bd. 159 S. 272, 321; Bd. 160 S. 257; Bd. 164 S. 129; JW. 1939 S. 354.

12. **Teilverzicht. Berufungssache.** (BGB. § 397; ZPO. §§ 306, 511 a, 523.) In einer Klage auf Schadenersatz war auch Zahlung eines Schmerzensgeldes von mindestens 500 *RM* verlangt. Das Landgericht erklärte den Anspruch dem Grunde nach für gerechtfertigt. Der Beklagte legte wegen des Anspruchs auf Schmerzensgeld Berufung ein. Im Berufungsverfahren erklärte der Kläger, daß er auf seinen Schmerzensgeldanspruch verzichte, soweit dieser über 500 *RM* hinausgehe. Der Beklagte beantragte, die Klage wegen dieses Anspruchs abzuweisen und hinsichtlich

des Mehrbetrags ein Verzichtsurteil gegen den Kläger zu erlassen. Das Berufungsgericht verwarf die Berufung als unzulässig. Das Reichsgericht erklärte sie für zulässig. Auch nach der Beschränkung der Berufung auf das Schmerzensgeld war der Streitwert der Beschwerung des Berufungsführers mit mehr als 500 RM anzunehmen, da der Kläger mindestens 500 RM als Schmerzensgeld verlangte, wenn er auch im übrigen die Bemessung dem Gericht überlassen hatte. Da der Kläger bei der mündlichen Verhandlung auf den geltend gemachten Schmerzensgeldanspruch, soweit er über 500 RM hinausging, verzichtete, mußte das Berufungsgericht auf den vom Beklagten gestellten Antrag den Kläger auf Grund seines Verzichts mit dem Anspruch abweisen. Diese Entscheidung dem Landgericht zu überlassen, war nicht zulässig. Soweit der Verzicht reichte, war eine weitere Verhandlung der Sache nicht erforderlich. Das Revisionsgericht war nicht in der Lage, das Verzichtsurteil zu erlassen, weil beim Fehlen der Revisionssumme die Revision nur die Frage der Zulässigkeit der Berufung betraf. Die Annahme der Verzichtserklärung lag schon in dem Antrag des Beklagten, den Kläger auf Grund seines Verzichts mit dem Anspruch abzuweisen. Obwohl hienach infolge des Verzichts des Klägers der Anspruch, soweit er über 500 RM hinausgegangen war, auch sachlichrechtlich erloschen war, wurde dadurch die Berufung doch nicht unzulässig. Der Kläger konnte nicht dadurch, daß er im Laufe des Berufungsverfahrens auf einen Teil seines Anspruchs verzichtete, die zulässigerweise eingelegte Berufung des Beklagten unzulässig machen. Es ist etwas ganz anderes, ob der Rechtsmittelkläger durch Zurücknahme eines Teils seines Rechtsmittels sich selbst die Rechtsmittelsumme entzieht und dadurch die Weiterverfolgung seines Rechtsmittels unmöglich macht oder ob der Gegner des Rechtsmittelklägers diesem sein Rechtsmittel durch einen Teilverzicht oder ein Teilanerkenntnis aus der Hand zu schlagen sucht. Dem Gegner des Rechtsmittelklägers diese Möglichkeit zu geben, würde zu unerträglichen Folgen führen. (VI, 25. September 1940.)

Amtl. Sammlg. S. 85—91.

Vgl. Bd. 5 S. 387; Bd. 76 S. 292; Bd. 107 S. 53; Bd. 113 S. 246; Bd. 139 S. 221; Bd. 161 S. 350;

JW. 1897 S. 601; 1899 S. 830; 1902 S. 422; 1911 S. 988; 1920 S. 558; 1936 S. 2712; 1938 S. 2909.

13. **Krankenkasse. Vertrauensarzt. Beamter. Amtspflichtverletzung. Schadenersatz.** (WeimVerf. Art. 131; BGB. §§ 276, 328, 839; RVO. §§ 368, 368 d, 369 b; Gesetz über den Aufbau der Sozialversicherung vom 5. Juli 1934 Abschn. II Art. 1, Art. 2 § 1, Art. 3 §§ 1, 2, Art. 7 §§ 1, 2; 3. VO. vom 18. Dezember 1934 Nr. 3; 9. VO. vom 13. Juli 1935 Art. 2 §§ 2, 3; 16. VO. vom 9. Juni 1938 §§ 1, 3). Ein Arbeiter, der Mitglied einer Innungskrankenkasse war, wurde von dieser auf ein Gutachten des Kassenarztes in ein Krankenhaus zur Beobachtung eingewiesen. Der Arzt des Krankenhauses, der ihn behandelte, erklärte ihn für etwa 10 Tage arbeitsunfähig. Zur Feststellung, ob nach Ablauf dieser Zeit Arbeitsfähigkeit eintreten würde, veranlaßte die Kasse eine Untersuchung durch den Vertrauensarzt der vertrauensärztlichen Dienststelle. Dieser sprach sich dahin aus, daß ein Berufswechsel nötig sei. Inzwischen hatte der Krankenhausarzt der Kasse ein Nachtragsgutachten übersandt, worin eine chirurgische Behandlung für notwendig bezeichnet und volle Arbeitsunfähigkeit bestätigt wurde. Die stellvertretende Leiterin der Kasse sandte dieses Nachtragsgutachten nicht an den Vertrauensarzt, sondern an den Kassenarzt, der den Arbeiter zuerst behandelt hatte; dieser führte die Behandlung fort, fand aber eine Behandlung im Krankenhaus nicht für notwendig. Das Befinden des Arbeiters verschlechterte sich immer mehr; schließlich trat völlige Lähmung ein. Mit der Behauptung, daß die Verschlechterung durch eine Behandlung im Krankenhaus hätte abgewendet werden können, klagte der Arbeiter gegen die Kasse auf Schadenersatz. Während des Rechtsstreits starb er; seine Erben traten in den Rechtsstreit ein. Die Kasse bestritt allgemein einen Rechtsanspruch ihrer Mitglieder auf Krankenhausbehandlung, nahm hiefür vielmehr ihr freies Ermessen in Anspruch; sie stellte auch ein Verschulden ihrer stellvertretenden Leiterin und des Vertrauensarztes in Abrede. Das Berufungsgericht wies die Klage ab. Das Reichsgericht verwies die Sache zurück. Die Revision war ohne Rücksicht auf den Wert des Beschwerdegegenstandes zulässig, da die Klage auf Amtspflichtverletzung gestützt war. Es handelte sich nicht etwa um einen Anspruch

gegen die Kasse als Träger der Krankenversicherung, über den die Versicherungsbehörden zu entscheiden hätten, sondern um einen Schadenersatzanspruch, der sich nicht in diesen Leistungen erschöpfte und für den die ordentlichen Gerichte zuständig waren. Die Frage, ob in schuldhaften Handlungen oder Unterlassungen des Leiters einer Krankenkasse eine Amtspflichtverletzung gefunden werden kann, war zu bejahen. Unter Ausübung öffentlicher Gewalt im Sinne des Art. 131 WeimVerf. ist nicht nur die Ausübung obrigkeitlicher Zwangsgewalt, sondern auch die Betätigung öffentlichen Schutzes und öffentlicher Fürsorge zu begreifen. Die Krankenkassen haben eine gewisse Zwangsgewalt gegenüber ihren Mitgliedern insofern, als diese den für die Krankenversicherung bestehenden Vorschriften und den darauf gestützten Anordnungen der Kassen unterworfen sind; aber auch die fürsorgerische Tätigkeit der Kassen ist, weil auf öffentlich-rechtlichem Gebiete sich bewegend, zumal nach der neueren Gestaltung des Versicherungsrechts, hoheitlicher Art. Die Angestellten der Krankenkassen sind zwar nicht Beamte im staatsrechtlichen Sinne, wohl aber im Sinne des Art. 131 WeimVerf. und des § 839 BGB.; als Beamte in diesem Sinne gelten alle Personen, die der Staat oder eine dazu befugte Körperschaft mit öffentlicher Gewalt umkleidet. Die stellvertretende Leiterin der Kasse verletzte ihre Pflicht gegenüber dem Arbeiter, indem sie das Nachtragsgutachten des Krankenhausarztes nicht an den Vertrauensarzt, der um Untersuchung des Arbeiters angegangen worden war, weitergab. Allerdings war die Kasse nicht verpflichtet, ein Gutachten des Vertrauensarztes zu erholen; wenn sie sich aber an ihn wandte und damit zu erkennen gab, daß sie seine sachverständige Ansicht für die weitere Behandlung des Krankheitsfalles benötige, mußte es als ihre Pflicht erachtet werden, dabei nicht nur die eigenen Belange, sondern auch jene des Versicherten, die ihr anvertraut waren, zu beachten und zur Vermeidung eines Fehlurteils ihm alle ihr zur Verfügung stehenden Unterlagen über den Fall zu unterbreiten. Durch das Nachtragsgutachten des Krankenhausarztes, das die Notwendigkeit chirurgischer Behandlung bescheinigte, änderten sich, auch für die stellvertretende Leiterin der Krankenkasse erkennbar, die dem Vertrauensarzt zur Verfügung gestellten Unterlagen für die ärztliche Beurteilung der Arbeitsfähigkeit in entscheidender Weise. Der Vertrauensarzt hatte auch im Laufe des Verfahrens die Erklärung abgegeben, daß er bei Kenntnis des Nachtragsgutachtens sich für eine Krankenhausbeobachtung ausgesprochen hätte. Es genügte nicht, daß das Nachtragsgutachten an den behandelnden Kassenarzt gegeben wurde; dadurch konnte bei diesem (wie es nach der Feststellung im angefochtenen Urteil auch der Fall war) die irrige Auffassung hervorgerufen werden, der Vertrauensarzt habe im Gegensatz zu dem Nachtragsgutachten eine Krankenhausbehandlung nicht für notwendig gehalten. Einen solchen Irrtum zu verhüten und den Kassenarzt davon, daß der Vertrauensarzt das Nachtragsgutachten nicht gekannt hatte, zu unterrichten, wäre gleichfalls eine Pflicht der stellvertretenden Leiterin der Kasse gewesen. Diese Unterlassungen stellten eine Amtspflichtverletzung dar, welche die Haftung der Kasse für den daraus entstandenen Schaden begründete. Dagegen konnte der Klageanspruch nicht auf ein schuldhaftes Verhalten des Vertrauensarztes gestützt werden. Dieser stand nach der Neuregelung des sozialen Versicherungsrechts im Dienste der Landesversicherungsanstalt, nicht mehr, wie früher, im Dienste der Krankenkasse. Bei Ausübung seiner Tätigkeit, zu der auch die Erstattung von Gutachten auf Ersuchen von Krankenkassen gehört, erfüllt er hoheitliche Aufgaben. Das gilt jedenfalls dann, wenn er, wie im vorliegenden Falle, von einer Krankenkasse als Organ des versicherungsrechtlichen Gemeinschaftsdienstes in Anspruch genommen wird. Für ein etwaiges Verschulden hiebei hätte also nicht die beklagte Kasse, sondern die Landesversicherungsanstalt als die Körperschaft, in deren Dienst er stand, eintreten müssen. Ob ihm wirklich ein Verschulden zur Last zu legen war, konnte bei dieser Rechtslage ungeprüft bleiben. Selbst wenn ein solches Verschulden anzunehmen gewesen wäre, hätte das die Frage der Haftung der Beklagten für das schuldhafte Verhalten ihrer stellvertretenden Leiterin nicht berührt. In diesem Falle hätte es sich um zwei ursächlich zusammenwirkende schuldhafte Unterlassungen gehandelt, von denen die eine die Haftung der Krankenkasse, die andere die Haftung der Landesversicherungsanstalt begründet hätte. Es wäre dann eine Gesamthaftung anzunehmen gewesen; in solchen Fällen kann der eine Haftpflichtige sich nicht auf die Möglichkeit berufen, daß der Geschädigte von dem anderen Haftpflichtigen Ersatz des Schadens erlangen könne. Im übrigen gehört allerdings das Nichtbestehen einer anderweitigen Ersatzmöglichkeit zur Grundlage einer Klage aus Art. 131 WeimVerf. in Verbindung mit § 839 BGB. und ist vom Kläger darzutun. In Frage kam, ob der Kassenarzt dem Erblasser der Kläger zu Schadenersatz verpflichtet gewesen wäre. Der zwischen einer Krankenkasse und dem Kassenarzt bestehende Vertrag ist im Hinblick auf die Versicherten als ein Vertrag zugunsten Dritter zu beurteilen. Danach erwirbt der Versicherte einen unmittelbaren, im ordentlichen Rechtsweg verfolgbaren Anspruch gegen den Kassenarzt auf sachgemäße ärztliche Behandlung wie ein anderer Kranker gegen den ihn behandelnden Arzt. Im Verlauf des Verfahrens war ausgeführt, daß der Kassenarzt, als ihm das Nachtragsgutachten zugleich mit dem Gutachten des Vertrauensarztes zuging, angenommen habe, der Vertrauensarzt habe das Nachtragsgutachten gekannt, und daß er bei Kenntnis des wahren Sachverhalts wahrscheinlich den Antrag auf Überweisung des Arbeiters in ein Krankenhaus gestellt hätte. In der neuen Verhandlung hatte das Berufungsgericht dazu Stellung zu nehmen, ob das ausreichte, um eine Haftung des Kassenarztes zu begründen. (III, 27. September 1940.) Amtl. Sammlg. S. 91—107. Vgl. Bd. 51 S. 258; Bd. 130 S. 401; Bd. 131 S. 67; Bd. 137 S. 20, 38; Bd. 139 S. 149, 343; Bd. 140 S. 126, 415; Bd. 142 S. 140; Bd. 144 S. 262; Bd. 145 S. 182; Bd. 151 S. 385; Bd. 152 S. 385; Bd. 154 S. 117; JW. 1915 S. 916; 1932 S. 1325; 1937 S. 1321.

14. Rekurs. (ÖstZPO. §§ 471, 507, 513, 514, 519, 528; ÜberlVO. §§ 4, 7, 9; VO. vom 1. September 1939 §§ 9, 29.) Gegen die Zurückweisung einer unzulässigen Berufung durch das Berufungsgericht ist nach österreichischem Recht der Rekurs zulässig. Für die Zurückweisung einer unzulässigen Revision durch das Berufungsgericht fehlt eine entsprechende Vorschrift. § 519 ÖstZPO. trifft nicht zu, da er nur die Zulässigkeit von Rekursen gegen die im Berufungsverfahren ergangenen Beschlüsse des Berufungsgerichts regelt, ein die Revision zurückweisender Beschluß aber schon im Revisionsverfahren ergeht. Sinngemäß muß aber auch gegen einen solchen Beschluß der Rekurs zugelassen werden. Es bedarf dazu nicht der Erwägung, daß es nicht angeht, durch einen unrichtigen Beschluß eines Untergerichts eine zulässige Revision zu verhindern. Daran ist auch durch die Verordnung vom 28. Februar 1939 zur weiteren Überleitung der Rechtspflege im Lande Österreich und in den sudetendeutschen Gebieten (Überleitungsverordnung) nichts geändert worden. § 7 läßt die Revision dann zu, wenn die Unzulässigkeit der Berufung ausgesprochen wird. Wenn aber eine zulässige Berufung durch das Berufungsgericht nicht endgültig verwehrt werden kann, ist nicht anzunehmen, daß das Berufungsgericht eine Revision rechtskräftig zurückweisen darf. Dadurch wird der gleiche Rechtszustand wie im Altreiche hergestellt, wo das Berufungsgericht gar nicht in die Lage kommt, über die Zulässigkeit einer Revision zu entscheiden, vielmehr die Entscheidung dem Reichsgericht vorbehalten ist. (VIII, 14. Oktober 1940.)

Amtl. Sammlg. S. 107—109.

15. **Ehescheidung. Ehezerrüttung. Verzeihung.** (EheG. § 49.) Ein Mann, der in der Zeit von 1918 bis 1934 nicht weniger als 16 Strafen erhalten hatte, schloß im Jahr 1934 eine Ehe; im gleichen Jahr wurde er wieder zu 3 Jahren Zuchthaus verurteilt. Nach Verbüßung dieser Strafe beging er neue Straftaten und wurde hiewegen im Sommer 1938 verhaftet. Darauf klagte seine Frau auf Scheidung. Während des Verfahrens wurde der Mann zu zweieinhalb Jahren Zuchthaus und Sicherungsverwahrung verurteilt. In zwei Rechtszügen drang die Klage durch. Das Reichsgericht wies sie ab. Als gesetzliche Grundlage für die Klage konnte nur § 49 EheG. in Betracht kommen. Danach konnte die Klage nur durchdringen, wenn die Straftaten des Beklagten die Ehe der Parteien so tief zerrüttet hatten, daß die Wiederherstellung einer dem Wesen der Ehe entsprechenden Lebensgemeinschaft nicht erwartet werden konnte. Der Tatbestand erfordert ein Doppeltes: der beklagte Ehegatte muß schuldhaft etwas begangen haben, was sich als schwere Eheverfehlung oder als ehrloses oder unsittliches Verhalten darstellt; zum anderen muß als Folge dieser Verfehlung oder dieses Verhaltens die im Gesetz vorgeschriebene Zerrüttung der Ehe eingetreten sein. Die erste

Frage ist nach sachlichen Gesichtspunkten zu entscheiden und hängt nicht von den Empfindungen des anderen Ehegatten ab. Im Gegensatz dazu ist das zweite Tatbestandsmerkmal, die Ehezerrüttung und ihr Maß, davon abhängig, wie die Verfehlung oder das Verhalten des einen Ehegatten auf den anderen gewirkt hat. Nur wenn dieser sich durch den Vorgang verletzt fühlt, kann als Folge der Ereignisse der Ehezerrüttung eingetreten sein, deren Tiefe zugleich von dem Maß des Empfindens der Verletzung abhängen wird. Für dieses Tatbestandsmerkmal ist also die sittliche Wertung des Verhaltens des einen Ehegatten durch den anderen maßgebend; es kommt darauf an, ob der andere Ehegatte gerade infolge des ehrlosen oder unsittlichen Verhaltens oder der sonstigen Eheverfehlung eine erhebliche Einbuße an seinem ehelichen Gefühl gegenüber dem schuldigen Ehegatten erlitten und deshalb den Willen zur Fortsetzung der Ehe verloren hat. Erst wenn in solcher Weise ein Scheidungsrecht des verletzten Ehegatten entstanden ist, kann er es durch Verzeihung wieder verlieren. Das Verhalten der Klägerin stellte aber außer Zweifel, daß die Ehe durch die Straftaten des Beklagten nicht zerrüttet worden war. Selbst wenn sie bei der Eheschließung die große Zahl der Vorstrafen ihres Mannes nicht kannte, erfuhr sie davon doch gelegentlich der nachfolgenden Verhandlung, der sie beiwohnte; sie entnahm daraus aber keine Veranlassung, sich vom Beklagten zu lösen. Während des Scheidungsverfahrens schrieb sie an das Gericht, sie habe sich mit dem Beklagten versöhnt und bitte nur, ihn möglichst schnell auf freien Fuß zu setzen; in einem Brief an ihren Mann erklärte sie, sie sterbe aus Sehnsucht nach ihm, sie liebe ihn bis in den Tod, nur der Tod könne sie von ihm trennen. Auch in späteren Zuschriften äußerte sie sich ähnlich. Der Sachverhalt ergab eindeutig, daß die Klägerin ihr Scheidungsverlangen nur deshalb aufrecht erhielt, weil sie damit rechnen mußte, daß der Beklagte, wenn überhaupt, so erst nach langer Zeit zu ihr zurückkehren werde. Ein Ehegatte, der trotz Kenntnis des wirklichen Sachverhalts die Eheverfehlungen oder das ehrlose oder unsittliche Verhalten des anderen Ehegatten nicht als ehezerstörend empfunden, also auch kein Scheidungsrecht daraus gewonnen hat, kann aber Scheidung zu einer späteren Zeit nicht deshalb begehren, weil er nunmehr die Folgen erkannt hat, die ihm unangenehm oder untragbar erscheinen, mit der sittlichen Wertung der Tat selbst aber nichts zu tun haben. Davon ist wesentlich verschieden die Frage, ob ein entstandenes Scheidungsrecht durch Verzeihung wieder wirksam aufgegeben worden ist. Die Verzeihung beseitigt das Recht auf Ehescheidung nur dann, wenn der Verzeihende den Willen zur Aufgabe dieses Rechts in Würdigung aller Umstände gefaßt und geäußert hat. Deshalb hat die Rechtsprechung der Verzeihung die Wirkung versagt, sobald sie in Unkenntnis wesentlicher Auswirkungen der Eheverfehlung gewährt worden ist; es ist durchaus verständig, wenn der Scheidungsberechtigte bei der Prüfung, ob er verzeihen will oder nicht, auch die Folgen berücksichtigt, die aus der Verfehlung für die weitere Ehedauer, für die Familie, für seine eigene Lebensstellung, insbesondere auch in wirtschaftlicher Hinsicht, zu erwarten sind. Solche umfassende Beachtung aller Umstände steht mit den Geboten der Sittlichkeit und dem Wesen der Ehe durchaus im Einklang. Diese Übereinstimmung würde dagegen fehlen, wenn es zugelassen würde, daß ein Ehegatte, der sich zunächst durch Verfehlungen des anderen gar nicht verletzt gefühlt und seine eheliche Gesinnung ihm gegenüber unverändert behalten hat, später die Scheidung doch verlangen könnte, weil unliebsame Folgen der längst abgeschlossenen und ihm völlig bekannten Verfehlungen in anderer Hinsicht zutage getreten sind. (IV, 21. Oktober 1940.)

Amtl. Sammlg. S. 109--116.

16. Grundstückverkehr in Österreich. Rekurs. (ÖstZPO. § 526; ÖstABGB. §§ 380, 424, 425, 426, 431, 432, 433; GrundstVerkBek. vom 26. Januar 1937; VO. über Einführung dieser Bekanntmachung in Österreich vom 20. Juli 1938.) Über ein Grundstück in Österreich wurde mündlich ein Kaufvertrag geschlossen. Die Verkäuferin behauptete, der Vertrag sei nachträglich wieder aufgehoben worden. Die Käufer erhoben Klage und verlangten mit dieser, daß die Verkäuferin eine auf Kosten der Käufer zu errichtende Kaufurkunde beglaubigt unterfertige und einwillige, daß das Eigentum nach erteilter Genehmigung durch den Landrat für die Kläger in das Grundbuch einverleibt werde; hilfsweise beantragten die den Ausspruch, daß die Beklagte verpflichtet sei, gemeinsam mit ihnen beim zuständigen Landrat um die Genehmigung des mündlich abgeschlossenen Kaufvertrags nachzusuchen. Das Landgericht wies den Hauptantrag ab und erklärte den Hilfsantrag für unzulässig. Das Berufungsgericht ließ den Hilfsantrag zu, hob das Urteil des Landgerichts unter Vorbehalt der Rechtskraft auf und entschied, daß klargestellt werden müsse, ob der Kaufvertrag wieder aufgelöst worden sei. Das Reichsgericht verwies die Sache an das Berufungsgericht zurück. Der Antrag der Beklagten ging auf Wiederherstellung des Urteils des Landgerichts. Die Anfechtung eines Aufhebungsbeschlusses kann jedoch niemals zu einer Sachentscheidung in Urteilsform führen. Aber auch hievon abgesehen war eine sofortige Wiederherstellung des landgerichtlichen Urteils nicht möglich. Über ein Hilfsbegehren ist nur zu entscheiden, wenn das Hauptbegehren abgewiesen wird. Das Landgericht hatte das Hauptbegehren abgewiesen, über das Hilfsbegehren aber nicht entschieden, weil es dieses für unzulässig hielt. Das Berufungsgericht erklärte zwar das Hilfsbegehren für zulässig, entschied aber darüber ebenfalls nicht, weil es das Hauptbegehren unter bestimmten, noch festzustellenden Voraussetzungen für begründet hielt. Der Rekurs der Beklagten erstrebte die Abweisung des Hauptbegehrens. Im Falle seines Erfolges mußte also die Sache an das Berufungsgericht zur Entscheidung über das Hilfsbegehren zurückverwiesen werden. Der Auffassung des Berufungsgerichts, daß der noch nicht genehmigte Vertrag die Verkäuferin verpflichte, die Vertragsurkunde beglaubigt zu unterfertigen, trat das Reichsgericht nicht bei. Nach österreichischem Recht bedarf es zur Übertragung des Eigentums an einem Grundstück einerseits eines dahingehenden Verpflichtungsgeschäfts, andererseits der Übergabe des Grundstücks in der Form der Eintragung in den öffentlichen Büchern. An diesem Rechtszustande ist durch die Verordnung vom 20. Juli 1938, durch welche die Bekanntmachung über den Verkehr mit Grundstücken vom 26. Januar 1937 in Österreich eingeführt wurde, nichts geändert worden. Das Verpflichtungsgeschäft kann nach österreichischem Recht auch mündlich abgeschlossen werden. Der Vertrag bleibt bis zur Erteilung der Genehmigung des Landrats in Schwebe; beide Teile sind aber verpflichtet, zur Herbeiführung dieser Genehmigung zusammenzuwirken. Der Antrag an den Landrat bedarf keiner bestimmten Form. Die Kläger konnten den Antrag auch für sich allein und ohne Vorlage einer Urkunde über den Kaufvertrag stellen; aus der Tatsache, daß ein mündlicher Vertrag geschlossen worden war, ergab sich also keine Verpflichtung der Beklagten, darüber eine Urkunde auszustellen oder gemeinsam mit den Klägern um die Genehmigung nachzusuchen. Die Genehmigung setzt die Feststellung voraus, daß ein wirksamer Vertrag vorliegt. Wenn der Verkäufer das bestreitet, muß der Käufer durch das Gericht feststellen lassen, daß der Vertrag zu Recht besteht. Eine Klage auf Mitwirken bei dem Ansuchen um die Genehmigung ersetzt die Feststellungsklage nicht, da durch den Urteilsspruch über das Bestehen des Vertrags nicht entschieden würde; eine solche Klage ist auch überflüssig, weil ein gemeinsamer Antrag nicht notwendig ist. Hienach waren sowohl des Hauptbegehren als das Hilfsbegehren verfehlt. Unrichtig war auch die Ansicht des Berufungsgerichts, daß die Beklagte zur Ausstellung einer Urkunde über den Vertrag verpflichtet sei, sofern dieser noch zu Recht bestehe. Die Sache bedurfte ganz neuer Prüfung. (VIII, 21. Oktober 1940.)

Amtl. Sammlg. S. 117—120.

17. **Eheaufhebung. Bestätigung.** (EheG. §§ 37, 38, 52.) Eine Frau klagte auf Aufhebung ihrer Ehe wegen arglistiger Täuschung, weil der Mann verschwiegen hatte, daß er an Lungentuberkulose litt, daß auch seine Mutter an dieser Krankheit gelitten hatte und daß zwei ihrer Brüder in jungen Jahren daran gestorben waren; hilfsweise stützte sie ihr Verlangen auf Irrtum und begehrte äußerstenfalls Scheidung der Ehe aus § 52 EheG. In zwei Rechtszügen wurde die Klage abgewiesen. Das Reichsgericht hob die Ehe auf. Das Berufungsgericht hatte entscheidendes Gewicht darauf gelegt, daß die Klägerin mit ihrem Manne nach Entdeckung seines Gesundheitszustandes noch geschlechtlich verkehrt und damit ihren Willen, die Ehe fortzusetzen, kundgegeben habe. Es war aber festgestellt worden, daß ein solcher Verkehr nur einmal stattgefunden hatte und daß die Frau dazu nach anfänglicher Ablehnung nur auf Drängen des Mannes bereit gewesen war, als sie von einem langem Badeaufenthalt zurückgekehrt war. Das Gesetz schließt aber die Auf-

hebung der Ehe, wenn dazu Gründe vorliegen, nur dann aus, wenn der berechtigte Ehegatte ein Verhalten an den Tag gelegt hat, das unzweideutig auf seinen Willen schließen läßt, die Ehe fortzusetzen. Es ist nicht ausgeschlossen, daß ein solcher Fortsetzungswille durch den Geschlechtsverkehr hinreichend kundgegeben wird; es kommt aber auf die Umstände an. Die Ehe trotz der Gefahr fortzusetzen, die aus der Krankheit des Beklagten für die Klägerin selbst und ihre Nachkommen drohte, bedeutete einen Entschluß von großer Tragweite für die Klägerin. Es mußten schon Bedenken dagegen bestehen, einen einmaligen Geschlechtsverkehr unmittelbar nach der Heimkehr von langer Abwesenheit als Beweis für diesen Willen zu werten; völlig ausgeschlossen war es aber, nachdem feststand, daß die Klägerin das Verlangen des Beklagten zunächst abgelehnt und erst nach längerem Drängen nachgegeben hatte. Daß der Beklagte selbst bei Eingehung der Ehe den Aufhebungsgrund kannte, war nicht nachgewiesen, so daß eine Aufhebung wegen arglistiger Täuschung nicht möglich war. Auf das Scheidungsbegehren der Klägerin brauchte nicht eingegangen zu werden, da es nur hilfsweise vorgebracht worden war. (IV, 7. November 1940.) Amtl. Sammlg. S. 121—125.
Vgl. Bd. 164 S. 372.

18. **Ehescheidung. Wiederholung der Klage.** (EheG. § 55; ZPO. § 616.) Eine im Jahr 1908 geschlossene Ehe ging im Jahr 1935 in die Brüche. Der Ehemann verließ seine Familie und zog zu einer Frau, mit der er schon seit 1932 ehebrecherische Beziehungen unterhalten hatte. Seine im Jahr 1938 erhobene, auf § 55 EheG. gestützte Klage wurde im Oktober 1938 abgewiesen; das Urteil erlangte die Rechtskraft. Auf die im Mai 1939 wiederholte Klage schied das Berufungsgericht die Ehe und erklärte den Kläger für schuldig. Das Reichsgericht wies die Klage ab. Das Berufungsgericht hatte die Auffassung vertreten, daß § 616 ZPO. auf die Wiederholung einer Klage aus § 55 EheG. nicht anzuwenden sei. Zu dem Scheidungsanspruch auf Grund des § 55 EheG. gehöre eine mehr als dreijährige Trennung, eine unheilbare Zerrüttung der Ehe und die mangelnde Zulässigkeit oder mangelnde Rechtfertigung eines etwa erhobenen Widerspruchs; alle diese Tatsachen könnten durch Zeitablauf eine andere Bedeu-

tung gewinnen, so daß eine neue Prüfung zu einem anderen Ergebnis führen könne. Diese Auffassung erklärte das Reichsgericht für irrig. Die rechtskräftige Abweisung einer Scheidungsklage hat die endgültige Feststellung zur Folge, daß sich aus der Tatsachenlage, wie sie bei Abschluß der Tatsachenverhandlung des früheren Rechtsstreits bestand, kein Scheidungsrecht ergebe und daß für eine abweichende Beurteilung durch den Richter eines späteren Rechtsstreits kein Raum sei. Die Lage kann sich allerdings dadurch ändern, daß in der Folge neue Tatsachen hinzukommen, die unter den nach § 55 EheG. maßgeblichen Gesichtspunkten ein anderes Gesamtbild ergeben. Das frühere Urteil hatte die Klage wegen Widerspruchs der Frau abgewiesen, also die ausreichende Dauer der Trennung, die unheilbare Zerrüttung der Ehe und die Schuld des Klägers an dieser Zerrüttung angenommen. Eine rechtserhebliche Veränderung der Tatsachenlage war seitdem nicht eingetreten. Das Berufungsgericht hatte sich nur über den Widerspruch der Beklagten eine andere Meinung gebildet. Es hatte ihn für zulässig anerkannt, ihm aber die Beachtung mit der Begründung verweigert, daß zerstörte Ehen grundsätzlich geschieden werden sollten, daß die Belange der längst erwachsenen Kinder einer Scheidung nicht im Wege ständen und daß der Kläger auch den künftigen Unterhalt der Beklagten gesichert habe. Mochte diese Beurteilung auch im Vergleich zu jener des früheren Urteils den Vorzug verdienen, so konnte es hierauf doch nicht ankommen, solange nicht durch eine Veränderung der Tatsachenlage eine neue Beurteilung des Scheidungsanspruchs zulässig geworden war. Das traf nicht zu. Ob eine Änderung des Bildes nur durch den Ablauf weiterer Zeit eingetreten ist, muß für die Frage, ob die Klage wiederholt werden kann, ohne daß sich sonst etwas Neues ereignet hat, ohne Bedeutung bleiben. Die entgegengesetzte Auffassung würde zu einer uneingeschränkten Wiederholung der Rechtsstreitigkeiten um denselben Gegenstand führen, ein Ergebnis, welches das Gesetz gerade ausschließen will. Die in § 55 EheG. zur Grundlage des Scheidungsanspruchs gemachte Ehezerrüttung ist zumeist ein Dauerzustand und kann sich ihre Bedeutung durch Zeitablauf wandeln. Hienach kann ein Bedürfnis dafür bestehen, die Wiederholung der Klage in geeigneten Fällen auch dann zuzulassen, wenn nur die Zeit ändernd gewirkt haben kann. Es genügt aber, die Klage nach drei Jahren seit der Rechtskraft des früheren Urteils wieder zuzulassen. Diese Zeitspanne war hier noch nicht verstrichen. (IV, 7. November 1940.) Amtl. Sammlg. S. 125—128.

Vgl. Bd. 164 S. 249.

19. Gesellschaft mit beschränkter Haftung. Gründervorteil. Sonderrecht. Umdeutung. Rechtsmißbrauch. (BGB. § 140; GmbHG. §§ 5, 53.) Ein Frauenarzt führte mehrere Jahre hindurch eine Privatklinik. Im Jahr 1910 gründete er mit einem anderen Arzt eine G. m. b. H., der er die Klinik zur Fortführung überließ. Er wurde zum Geschäftsführer und leitenden Arzt bestimmt. Im Jahr 1923 legte er sein Amt als Geschäftsführer nieder. Da er den Standpunkt vertrat, bei Gründung der Gesellschaft sei vereinbart worden, daß er allein das Recht habe, frauenärztliche Kassenkranke dritter Klasse in der Klinik zu behandeln und zu operieren, und daß andere Ärzte frauenärztliche und geburtshilfliche Privatfälle nur mit seiner Genehmigung dort behandeln dürften, wurde in einer Mitgliederversammlung im März 1924 ein Beschluß gefaßt, der dieses Recht ausdrücklich anerkannte. Im Jahr 1933 beschloß die G. m. b. H., der inzwischen verschiedene andere Personen durch Erwerb von Geschäftsanteilen als Mitglieder beigetreten waren, ihre Umwandlung in ein gemeinnütziges Unternehmen. Von da an ließ der Geschäftsführer auch Ärzte, die nicht Mitglieder waren, zur Ausübung ihrer Tätigkeit in der Anstalt zu. Im Jahr 1937 wurde ein Antrag des Aufsichtsrats angenommen, durch den der Beschluß vom März 1924 als mit der neuen Fassung der Satzung nicht mehr vereinbar aufgehoben wurde. Der hiedurch benachteiligte Arzt klagte auf Feststellung, daß dieser Beschluß unwirksam sei. Das Berufungsgericht wies die Klage ab. Die Revision des Klägers blieb erfolglos. Bei dem vom Kläger geltend gemachten Recht handelte es sich um einen Gründer-Sondervorteil, der ihm für seine Person bei Gründung der Gesellschaft zugesagt worden war. Eine solche Vereinbarung hätte, um wirksam zu sein, der Aufnahme in den Gesellschaftsvertrag bedurft. Der Beschluß vom März 1924 reichte zu ihrer Begründung nicht aus, da darin eine Satzungsänderung zu finden war, wofür eine Mehrheit von drei Vierteln der abgegebenen Stimmen und gerichtliche oder notarische Beurkundung erforderlich gewesen wären; beide Erfordernisse fehlten. Der Kläger hatte hiegegen vorgebracht, daß die Berufung auf die Nichtigkeit der Vereinbarung wegen Formmangel gegen Treu und Glauben verstoße. Das Reichsgericht erkannte an, daß es, nachdem das vom Kläger beanspruchte Recht viele Jahre hindurch von der beklagten Gesellschaft anerkannt worden war, an sich gegen Treu und Glauben verstoßen konnte, die Ansprüche des Klägers an dem Formmangel scheitern zu lassen; diese für das reine Vertragsrecht zur Herrschaft gelangten Grundsätze konnten aber auf den vorliegenden Fall keine Anwendung finden, weil es sich nicht um ein Recht handelte, das den Gegenstand eines zwischen dem Kläger und der beklagten Gesellschaft formgültig abgeschlossenen Vertrages bildete, sondern um ein bei Gründung der Gesellschaft zwischen den Gründern bedungenes Gründersonderrecht. Solche Rechte müssen mit Rücksicht auf andere, insbesondere auf später eintretende Gesellschafter, und wegen der Belange Dritter, die ebenfalls durch die Satzung geschützt werden sollen, aus der Satzung selbst ersichtlich sein. Der Einwand des Rechtsmißbrauchs war also nicht gerechtfertigt. Es ging auch nicht an, die Vereinbarung als rein schuldrechtlichen Vertrag umzudeuten. Wenn auch wirtschaftlich der Erfolg, der sich aus dem Vorrecht für den Kläger ergab, auch durch einen schuldrechtlichen Vertrag gleichen Inhalts zwischen ihm und der Gesellschaft zu erreichen gewesen wäre, so war doch die Umdeutung deshalb ausgeschlossen, weil es begrifflich unmöglich war, an Stelle eines bei Gründung der Gesellschaft zwischen den Gründern formungültig vereinbarten Rechts den Anspruch aus einem Vertrag zwischen dem Kläger und der Gesellschaft zu setzen. Anders wäre die Sache zu beurteilen gewesen, wenn die Gesellschaft selbst nach ihrer Entstehung mit dem Kläger nicht in seiner Eigenschaft als Mitglied, sondern davon unabhängig einen solchen Vertrag geschlossen hätte, so daß das damit begründete Recht nicht mit dem Geschäftsanteil des Klägers verbunden, sondern davon unabhängig gewesen wäre. Im vorliegenden Falle stand aber die Zubilligung des Sonderrechts in so engem und unmittelbarem Zusammenhang mit der Gründertätigkeit des Klägers, daß die Aufnahme in den Gesellschaftsvertrag nötig gewesen wäre. Das muß auch gelten, wenn ein solcher Vorteil, der eine Abgeltung für eine Tätigkeit bei der Gründung der Gesellschaft darstellt, erst später eingeräumt wird; hier muß die für die Änderung der Satzung vorgesehene Form eingehalten werden. Die Frage, ob der Kläger, wenn ihm das Recht zugestanden haben sollte, sich auch noch darauf berufen könnte, nachdem mit seiner Zustimmung die Umwandlung der Gesellschaft in ein gemeinnütziges Unternehmen beschlossen worden war, brauchte nicht erörtert zu werden. (II, 26. Oktober 1940.)

Amtl. Sammlg. S. 129—136.

Vgl. Bd. 113 S. 241.

20. **Aufwertungsvertrag. Goldklausel.** (ÖstABGB. §§ 988, 989; TschechoslG. vom 9. Oktober 1936 über die Neuregelung der tschechoslowakischen Währung.) Durch einen außergerichtlichen Vergleich wurde eine Forderung auf 580 000 čK. festgesetzt; sie sollte in bestimmten Teilzahlungen abgetragen werden. Nachträglich wurde vereinbart, daß die Forderung wertbeständig sein sollte; alle Zahlungen sollten in solcher Höhe erfolgen, daß sie dem Wert der Tschechenkrone im damaligen Kurs (gleich 37,15 mg Feingold) entsprachen. Durch ein Gesetz vom 9. Oktober 1936 und eine Regierungsverordnung vom gleichen Tage wurde der Goldgehalt der Tschechenkrone auf 31,21 mg festgesetzt. Die Klage auf Berechnung der Teilzahlungen nach der Vereinbarung drang in allen Rechtszügen durch. Das Oberste Gericht in Brünn hatte ursprünglich die Auffassung vertreten, daß Aufwertungsvereinbarungen der öffentlichen Regelung der Währungsverhältnisse, die zwingender Natur sei, zu weichen hätten. An dieser Auffassung wurde aber nicht festgehalten, nachdem die Gründeberichte zu dem Abwertungsgesetz vom Jahr 1936 ausdrücklich anerkannt hatten, daß durch das Gesetz Verabredungen unberührt blieben, die auf Tschechenkronen, ausgedrückt durch einen bestimmten Goldgehalt, lauteten. Durch die Vereinbarung war allerdings nur festgelegt worden, was nach § 988 ABGB. für Darlehen allgemein Geltung hatte. Daraus konnten aber die Beklagten nichts für sich ableiten. Wenn die Parteien eine Bestimmung nachgiebigen Gesetzesrechts ausdrücklich zum Inhalt ihres Vertragswillens machten, gaben sie da-

durch zu erkennen, daß sie diesen Inhalt ihres Willens unabhängig von dem Schicksal der gesetzlichen Bestimmung ihrem Rechtsverhältnisse zugrunde gelegt wissen wollten, so daß das, was sie vereinbart hatten, für die Auslegung und Ordnung ihres Rechtsverhältnisses maßgebend sein und bleiben solle, ohne Rücksicht darauf, ob das Gesetz selbst, dem sie jene Bestimmung nachgebildet oder entnommen hatten, eine Änderung erfuhr oder nicht. Eine Ausnahme kann nur gelten, wenn zwingendes Recht der Vereinbarung entgegensteht. Das traf hier nicht zu. Durch das Gesetz vom 9. Oktober 1936 war die Vorschrift des § 988 ABGB. nicht aufgehoben, vielmehr waren nur die gesetzlichen Ansprüche von Gläubigern aus Rechtsverhältnissen, die mit der Tschechenkrone rechneten, beseitigt worden. Auf vertragliche Aufwertungsansprüche bezog sich diese Bestimmung nicht. (VIII, 28. Oktober 1940.)

Amtl. Sammlg. S. 136—140.

21. **Schiedsspruch. Vereinssatzung. Aufhebung. Gute Sitten. Knebelungsvertrag.** (ZPO. §§ 99, 280, 1027, 1041, 1048; BGB. § 138.) Innerhalb einer Aktiengesellschaft schlossen sich Aktionäre zu einer Schutzgemeinschaft zusammen. Nach deren Satzung war die Veräußerung von Aktien an Nichtmitglieder verboten; für Zuwiderhandlungen waren hohe Vertragsstrafen vorgesehen; über Streitigkeiten aus dem Gemeinschaftsverhältnis hatte ein Schiedsgericht zu entscheiden. Zwei Mitglieder veräußerten ihre Aktien an Nichtmitglieder; die Übereignung unterblieb, da der Aufsichtsrat der Aktiengesellschaft die nach deren Satzung erforderliche Genehmigung versagte. Andere Mitglieder der Schutzgemeinschaft riefen das Schiedsgericht zwecks Festsetzung der Vertragsstrafe an. Das Schiedsgericht wies den Antrag ab und legte die Kosten des Verfahrens beiden Parteien je zur Hälfte auf. Die Mitglieder, gegen die sich das Verfahren gerichtet hatte, klagten auf Aufhebung des Schiedsspruchs, soweit ihnen dadurch Kosten auferlegt waren, mit der Begründung, daß der Schiedsvertrag der Form des § 1027 ZPO. ermangele, auch die Beklagten (die damaligen Antragsteller) im Schiedsverfahren nicht nach Vorschrift der Gesetze vertreten gewesen seien. Das Landgericht wies die Klage ab. Im Berufungsverfahren verlangten die Klägerinnen die Aufhebung des Schiedsspruchs schlechthin und

begehrten ferner die Feststellung, daß die Schutzgemeinschaft, hilfsweise, daß der Schiedsvertrag nichtig sei. Das Berufungsgericht hob den Schiedsspruch auf und stellte die Nichtigkeit des Schiedsvertrags fest, wies aber den Antrag auf Feststellung der Nichtigkeit der Schutzgemeinschaft ab. Das Reichsgericht wies die Klage ab. Die Aufhebungsklage war zulässig, wenn auch die Klägerinnen nur durch die Auferlegung von Kosten beschwert waren. § 99 ZPO. war nicht anwendbar, weil die Aufhebungsklage kein Rechtsmittel im Sinne der Zivilprozeßordnung ist; gegen eine Übertragung des in § 99 enthaltenen Rechtsgedankens aus Gesichtspunkten der Rechtsähnlichkeit auf den Streitfall sprach, daß die Klage sich nicht nur gegen die Kostenverteilung wandte, sondern das ganze Schiedsverfahren als widerrechtlich angriff. Sachlich war die Klage nicht gerechtfertigt. Die Schutzgemeinschaft war als Verein anzusehen; das ergab sich aus ihrer körperschaftlichen Verfassung, ihrer Benennung und ihrer Unabhängigkeit vom Wechsel der Mitglieder. Schiedsabreden in Vereinssatzungen sind an die Formvorschrift des § 1027 ZPO. nicht gebunden. Daß der Verein nicht rechtsfähig war, machte nichts aus. Auch beim rechtsfähigen Verein hat die Satzung vor der die Rechtspersönlichkeit begründenden Eintragung nicht die Eigenschaft eines Willensaktes des Vereins, sondern vertragliche Wesensart. In beiden Fällen aber gilt die Satzung, sobald der Verein ins Leben getreten ist, als die von der Persönlichkeit seiner Mitglieder losgelöste Verfassung seines Eigenlebens und damit als eine nicht auf Vereinbarung beruhende Verfügung im Sinne des § 1048 ZPO. Die urkundliche Einsetzung eines die Streitigkeiten der Mitglieder mit dem Verein betreffenden Schiedsgerichts in der Vereinssatzung macht eine besondere, von anderen Abmachungen losgelöste schriftliche Niederlegung der Schiedsabrede entbehrlich. Das Vorbringen der Klägerinnen, daß die Beklagten im Schiedsverfahren nicht nach der Vorschrift der Gesetze vertreten gewesen seien, konnte der Klage nicht zum Erfolge verhelfen, weil den Mangel der Vertretung nur die davon selbst betroffene Partei geltend machen kann. Den im zweiten Rechtszug gestellten Antrag auf Nichtigerklärung der Schutzgemeinschaft hatten die Klägerinnen damit begründet, daß die Satzung eine Knebelung der Mitglieder in einem gegen die guten Sitten verstoßenden Ausmaße bedeute. Die Beklagten hatten den Einwand unzulässiger Klageänderung erhoben; die Zulässigkeit des Antrags ergab sich aber aus § 280 ZPO. Das Vorliegen des angeführten Nichtigkeitsgrundes hatte das Berufungsgericht mit Recht verneint. Derartige Schutzgemeinschaften von Aktionären zur Wahrnehmung ihrer gemeinschaftlichen Belange sind von grundsätzlichen Gesichtspunkten aus nicht zu beanstanden; der Zusammenschluß war freiwillig und ohne Druck geschehen; das Verbot, Aktien an Nichtmitglieder zu veräußern, diente dem Zweck der Verhütung einer Überfremdung der Gesellschaft; die Sicherung der Einhaltung der Satzung durch hohe Vertragsstrafen war ebensowenig zu beanstanden wie die unbeschränkte Zeitdauer der Schutzgemeinschaft und ihre Erstreckung auf die Erben verstorbener Mitglieder. In der Revision hatten die Klägerinnen ausgeführt, die ursprünglich nicht zu beanstandende Satzung sei zufolge veränderter Personenverhältnisse in der Leitung der Schutzgemeinschaft in ihrer Auswirkung und Handhabung sittenwidrig geworden. Dabei war übersehen, daß die Schutzgemeinschaft, wenn sie nicht von vornherein durch den Inhalt ihrer Satzung nichtig war, nicht nachträglich durch die Handhabung der Satzung von seiten bestimmter Personen nichtig werden konnte. Die Gefahr von Kursverlusten bei der Veräußerung von Aktien war für die Rechtswirksamkeit der Satzung ohne jeden Belang. Fühlte sich ein Mitglied der Schutzgemeinschaft durch die Handhabung der Satzung benachteiligt, so stand ihm der Austritt aus der Gemeinschaft frei. (VII, 29. Oktober 1940.)

Amtl. Sammlg. S. 140—146.

Vgl. Bd. 60 S. 96; Bd. 119 S. 248; Bd. 153 S. 267.

22. **Sippenurkunden.** (BGB. § 810.) Urkunden, die sich auf eine Sippe bezogen, befanden sich im Alleinbesitz eines Mitglieds der Sippe. Ein anderes Mitglied klagte auf Gewährung der Einsicht in die Urkunden. In zwei Rechtszügen wurde die Klage abgewiesen. Das Reichsgericht erkannte sie als begründet an. Der Bedeutung, welche die nationalsozialistische Volksgemeinschaft der Erforschung und Pflege der blutmäßigen Beziehungen der Volksgenossen beimißt, entspricht es, daß die Mittel zur **Feststellung der Sippenzusammenhänge und Sippenge-**

schichte keinem Volksgenossen durch die Eigennützigkeit eines anderen vorenthalten werden dürfen. In rechtsschöpferischer Erstreckung des in § 810 BGB. enthaltenen Grundsatzes auf diesen Lebensbereich muß heute jedem Volksgenossen gegen den Besitzer einer Urkunde, die sippenkundlichen Einzigkeitswert für ihn hat, ein Anspruch auf die Vorlegung zur Einsicht zugebilligt werden. Dieser Anspruch umfaßt allerdings nicht standesamtliche Bescheinigungen und ähnliche Unterlagen, die der Volksgenosse sich selbst beschaffen kann, oder Auszüge, Zusammenstellungen und dergleichen, die er selbst anzufertigen oder durch einen Sachverständigen anfertigen zu lassen in der Lage ist; wohl aber bezieht er sich auf alle Urkundenurschriften und solche Abschriften, deren Unterlagen verloren gegangen, nicht auffindbar oder sonstwie nicht erreichbar sind, sofern deren Einsicht zur Ermittelung der Vorfahren und ihrer Sippe von Wert sein kann. Dieser Anspruch war dem Kläger auch nicht dadurch verloren gegangen, daß er, wie der Beklagte einwandte, alle Bande zwischen sich und den anderen Sippenangehörigen zerschnitten hatte; denn er setzt keine Familienbeziehungen zu dem Urkundenbesitzer voraus. Da Ort, Zeit und Umfang der Vorlegung erst noch mit den Parteien zu erörtern und deren Voraussetzungen zu prüfen waren, wurde die Sache zurückverwiesen. (IV, 2. November 1940.)

Amtl. Sammlg. S. 146—148.

23. **Ehescheidung. Ehezerrüttung. Staatsangehörigkeit.** (EheG. § 55; EGzBGB. Art. 17; VO. vom 20. Juli 1939 über die Anwendung deutschen Rechts auf deutsche Staatsangehörige im Protektorat Böhmen und Mähren). Die Parteien, die damals tschechische Staatsangehörige waren, schlossen im Jahr 1926 in Prag miteinander die Ehe. Im Jahr 1936 siedelte der Ehemann, der mehr als zwei Jahre arbeitslos gewesen war, nach Deutschland über, wo er eine Stellung gefunden hatte. Die Frau, die vor der Ehe als Bankbeamtin tätig gewesen war und diese Stellung auch während der Ehe beibehalten hatte, blieb in Prag zurück. Beide Ehegatten besuchten einander wiederholt und standen miteinander in regem Briefwechsel. Im Jahr 1937 trat der Ehemann in Beziehungen zu einer anderen Frau und erklärte seiner Frau im August 1937, daß er die Ehe mit ihr nicht fortsetzen wollte. Im Sep-

tember 1937 brachte die Frau ein Kind zur Welt. Der Klage auf Scheidung der Ehe setzte sie Widerspruch entgegen. In zwei Rechtszügen wurde die Klage abgewiesen. Das Reichsgericht verwies die Sache zurück. Durch den Erlaß des Führers über das Protektorat Böhmen und Mähren vom 16. März 1939 war der Kläger deutscher Staatsangehöriger geworden; die Beklagte teilte diese Staatsangehörigkeit als seine Ehefrau. Seitdem war auf die Parteien das deutsche Eherecht anzuwenden. Eine Tatsache, die sich ereignet hat, während der Ehemann einem anderen Staat angehört, kann als Scheidungsgrund nur geltend gemacht werden, wenn sie nach den Gesetzen dieses Staates ein Scheidungsgrund ist. Diese Vorschrift konnte aber nicht in Anwendung kommen. § 55 EheG. fordert als Scheidungsgrund, daß die Ehe unheilbar zerrüttet und die häusliche Gemeinschaft der Ehegatten seit mindestens drei Jahren aufgehoben ist. Für die Entscheidung, ob beides vorliegt, ist nur der Zeitpunkt der letzten mündlichen Verhandlung vor dem Tatrichter maßgebend. Die Tatsachen, die als Grundlage des Scheidungsanspruchs vom Gesetz gefordert werden, sind also diese Zerrüttung und neben ihr nicht der Ablauf der Trennungsdauer, sondern nur die Vollendung dieses Ablaufs. Diese Vollendung fiel aber in die Zeit, da der Kläger schon Deutscher war. Der eigentliche Scheidungsgrund ist die Zerrüttung der Ehe; wenn das Gesetz daneben noch die Aufhebung der häuslichen Gemeinschaft seit drei Jahren fordert, soll damit die Anwendung des neuen Scheidungsgrundes eingeschränkt und Vorsorge getroffen werden, daß Störungen der Ehe nur dann zu ihrer Lösung führen, wenn eine Besonderheit dieser Ehe durch die Trennungsdauer in die Erscheinung getreten ist, sei es, daß die Aufhebung der häuslichen Gemeinschaft im Einzelfall ein Anzeichen für die Zerrüttung, nämlich ihre Folge ist, sei es, daß sie in anderen Fällen die Zerrüttung gefördert oder sogar verursacht hat. Hienach bedurfte es keiner Feststellung, ob der Scheidungsgrund des § 55 EheG. auch dem tschechischen Recht bekannt war. Ebenso entfiel die Prüfung, ob die Vorschrift überhaupt noch Bedeutung haben konnte, nachdem die Tschechoslowakei zu bestehen aufgehört hat. Das Berufungsgericht hatte die unheilbare Zerrüttung der Ehe verneint, weil die Beklagte ernstlich bereit war, die Ehe fortzusetzen; nur der Kläger beharrte auf der Trennung. Dieses einseitige Lossagen von der Gemeinschaft wurde nicht als Beweis für eine Zerrüttung der Ehe angesehen, zumal der Kläger viele Jahre lang gut mit seiner Frau gelebt hatte. Es genüge nicht, daß ein Ehegatte aus Laune, Willkür oder anderen Gründen die eheliche Gemeinschaft verweigere, sondern hiefür seien Gründe zu fordern, die bei der Berücksichtigung der Persönlichkeit des anderen Ehegatten und der Entwicklung, die das Zusammenleben genommen habe, Beachtung verdienten. Das erklärte das Reichsgericht für rechtsirrig. Es kommt nicht darauf an, aus welchem Grunde die Ehe zerrüttet ist, ob irgendwelche Vorgänge den Verlust der ehelichen Gesinnung bei dem einen oder anderen Ehegatten rechtfertigen. Die Scheidung konnte nicht deshalb versagt werden, weil der Verlust der ehelichen Gesinnung, aus dem sich die Zerrüttung ergab, nur auf der Laune oder Willkür des Klägers beruhte und er bei rein sachlicher Würdigung aller Umstände zur Beklagten zurückkehren sollte. Für eine andere Auffassung fehlt im Gesetz jeder Anhalt. Ausreichende Anhaltspunkte für die Annahme, daß die Lebensgemeinschaft der Ehegatten hätte wiederhergestellt werden können, hatte das Berufungsgericht nicht dargelegt. In einer Eingabe des Anwalts des Klägers, mit der eine Aussetzung des Verfahrens beantragt wurde, weil der Kläger seine Einberufung zum Heer erwartete, war allerdings erwähnt, daß die Möglichkeit einer Aussöhnung unter Berücksichtigung des Ernstes der allgemeinen Lage nicht ganz ausgeschlossen sei; selbst wenn man aber annahm, daß dieser Satz nicht nur vom Anwalt herrührte, hatte der Kläger die Äußerung doch offensichtlich nur unter dem Eindruck seiner bevorstehenden Einberufung getan. Seitdem war nichts mehr in der Richtung auf eine Versöhnung geschehen. Da der Kläger die Ehezerrüttung allein verschuldet hatte, stand der Beklagten das Recht zum Widerspruch zu. Das Berufungsgericht hatte den Widerspruch beachtet, weil die Beklagte durch eigene Arbeitstätigkeit für sich und ihren Mann den Unterhalt verdient habe in Zeiten, in denen der Kläger arbeitslos war, weil die Ehe über 11 Jahre lang ohne wesentliche Störung verlaufen war und weil es nicht angehe, daß der Kläger sich von der Erziehungspflicht

gegenüber dem erst $2^1/_2$ Jahre alten Kinde befreie. Das Reichsgericht fand, daß diese Begründung den gesetzlichen Anforderungen an die Entscheidung über den Widerspruch nicht genüge. Grundsätzlich sind unheilbar zerrüttete Ehen zu scheiden und nur ausnahmsweise ist der Widerspruch zu beachten, wenn besondere Gründe das gebieten. Schon das Alter der Parteien (der Kläger war 44, die Beklagte 41 Jahre alt) sprach dagegen, sie für den Rest ihres Lebens an ein wertloses Eheband zu ketten, obwohl eine neue Ehe, mindestens eine solche des Klägers, zu erwarten war und für die Volksgemeinschaft noch wertvoll sein konnte. Auch die Rücksicht auf das Kind wurde nur gering gewertet, da dieses den persönlichen Zusammenhang mit dem Vater und seine Einwirkung bei der Erziehung auch im Falle der Ablehnung der Scheidung entbehren mußte. Besonderes Gewicht wurde darauf gelegt, daß die Beklagte ein Ruhegeld von 135 *RM* monatlich bezog und leichter als manche andere Frau ihres Alters eine Erwerbstätigkeit aufnehmen konnte, nachdem sie bis vor wenigen Jahren als Bankbeamtin tätig gewesen war. (IV, 2. November 1940.) Amtl. Sammlg. S. 149—155.
Vgl. Bd. 160 S. 95.

24. **Miethaus. Verkehrssicherung. Mitverschulden.** (BGB. §§ 254, 539, 545, 823.) Mit Genehmigung der Stadtgemeinde wurde in einer erst geplanten Straße, mit deren Herstellung überhaupt noch nicht begonnen worden war, zwei Häuser erbaut und nach der Gebrauchsabnahme alsbald vermietet. Die Frau eines Mieters verließ das Haus an dem Abend des Tages, an dem sie eingezogen waren, um ihre Kinder, die während des Umzugs anderweitig untergebracht waren, abzuholen. In der Dunkelheit stürzte sie auf dem Wege bis zur nächsten Straße so unglücklich, daß sie eine Gehirnerschütterung erlitt, das rechte Augenlicht verlor und in ihrer Arbeitsfähigkeit erheblich beeinträchtigt wurde. Ihr Ehemann klagte gegen die Stadtgemeinde und gegen die Vermieterin auf Schadenersatz. Er verlangte, da er eine Hausgehilfin einstellen müsse, eine Rente von monatlich 40 *RM* und für seine Ehefrau ein angemessenes Schmerzensgeld, außerdem die Feststellung, daß die Beklagten als Gesamtschuldner ihm und seiner Ehefrau allen weitergehenden Schaden zu ersetzen hätten. Das Landgericht gab der Klage statt und

bemaß das Schmerzensgeld auf 2000 *RM*. Das Berufungsgericht wies die Klage gegen die Stadt ab, nahm Mitverschulden der Ehefrau des Klägers an und verurteilte die Vermieterin nur zur Hälfte. Auf die Revision des Klägers wurde die Sache zurückverwiesen. Für die Beurteilung war entscheidend, daß aus Wohnungsmangel die Häuser vor Herstellung einer zu ihnen führenden Straße für den Gebrauch freigegeben, vermietet und bezogen worden waren, und daß es sich nur darum handelte, für die Übergangszeit einen vorläufigen, einigermaßen brauchbaren Zugang zu schaffen. Bei der Eigentümlichkeit des Falles konnte es nicht darauf ankommen, welche Zumutungen unter gewöhnlichen Umständen an eine Stadt in Hinsicht auf eine Straßenanlage oder an einen Vermieter in Hinsicht auf den Zugang zum Mietgrundstück zu stellen sind. Außergewöhnliche Gefahrenlagen verlangen auch außergewöhnliche Maßnahmen. Hienach war die Stadt für das Fehlen jeder Sicherung und Beleuchtung auf dem Wege von den Häusern bis zur Straße verantwortlich, weil sie die Erlaubnis zum Bezuge der Häuser erteilt hatte. Ihre verfassungsmäßigen Vertreter mußten wissen, daß die Häuser alsbald bezogen werden würden. Dann war es ihre Pflicht, behufs Herstellung eines wenigstens vorläufigen Zugangs und ausreichender Beleuchtung das Erforderliche zu veranlassen. Die Grundlage der Klage gegen die Vermieterin bildete nicht der Mietvertrag, der ja keinen Anspruch auf Schmerzensgeld hätte begründen können, sondern der Umstand, daß sie durch Vermietung der Häuser einen, wenn auch beschränkten, Verkehr zu und von ihnen eröffnet hatte. Die §§ 539, 515 BGB. waren daher für die Entscheidung bedeutungslos. Wenn auch die Verkehrssicherungspflicht des Vermieters sich in der Regel nicht über das Mietgrundstück hinaus erstreckt, war das doch unter den außergewöhnlichen Umständen, die hier vorlagen, anders. Die Vermieterin mußte sich Gedanken darüber machen, wie sich der Verkehr zu und von den Mietgrundstücken bewerkstelligen lasse. Sie durfte sich nicht damit beruhigen, daß sie Kosten für die Straßenherstellung an die Stadt entrichtet hatte. Daß beim Einzug der Mieter noch nichts für die Schaffung eines ungefährdeten Zugangs zur nächsten Straße geschehen war, lag offen zutage. Unter diesen

Umständen mußte die Vermieterin, nötigenfalls im Einvernehmen mit der Stadt, zum mindesten für eine einigermaßen ausreichende Beleuchtung bis zur Straße sorgen. Durch eine Polizeiverordnung der Stadt war es für die Herstellung von Bauten den Bauherrn zur Pflicht gemacht, einen nach der Ansicht der Polizeibehörde genügenden Zufuhrweg vom Baugrundstück bis zum nächsten Fahrweg zu beschaffen und zu unterhalten. Danach war sogar durch ein Schutzgesetz die Haftung der Vermieterin aus § 823 Abs. 2 BGB. begründet. Diese Haftung ergab sich aber schon aus § 823 Abs. 1 BGB. Die Haftung der Stadt war durch die Polizeiverordnung nicht beseitigt, weil der Zufuhrweg unter mindestens prüfender Mitwirkung ihrer Polizei zu beschaffen gewesen wäre. Ungenügend begründet war auch die Annahme eines Mitverschuldens der verletzten Ehefrau. Diese hatte allerdings schon auf dem Heimweg zur Wohnung einen Eindruck davon bekommen, wie es mit dem Zugang beschaffen war. Das Berufungsgericht hatte angenommen, daß sie sich mit einer Taschenlampe hätte versehen müssen; es war aber nicht einmal behauptet, geschweige denn festgestellt, daß sie eine solche im Besitz hatte. (VI, 6. November 1940.)

Amtl. Sammlg. S. 155—160.

Vgl. JW. 1909 S. 161; 1910 S. 618.

25. **Grundstückveräußerung. Genehmigung. Vergleich.** (BGB. §§ 313, 317; GrundstVerkVO. § 2.) Zum Zweck der Aufhebung einer Erbengemeinschaft wurde die Zwangsversteigerung eines zum Nachlaß gehörigen Bauernhofes beantragt. Das Amtsgericht gab dem Antrag statt, stellte aber auf Grund einer Notverordnung das Verfahren auf die Dauer von 6 Monaten ein. Hiegegen erhoben die Antragsteller Beschwerde. Im Verfahren vor dem Landgericht schlossen die Erben einen Vergleich, durch den eine Schätzungskommission eingesetzt wurde; diese sollte den Wert des Hofes ermitteln und die Lage, Art und Größe des einer der Erbinnen zuzuteilenden Landes, das ein Fünftel des Hofes ausmachen sollte, feststellen; der Rest des Hofes sollte an ihren Bruder aufgelassen werden; zwei weitere Erbinnen sollten ihren Anteil in Geld erhalten. Der Landrat genehmigte die Auflassung der Grundstücke. Die Erbin, der bestimmte Grundstücke zugeteilt worden waren, klagte gegen ihren Bruder auf Auflassung

dieser Grundstücke Zug um Zug gegen Auflassung des Restbesitzes. Die Klage drang in allen Rechtszügen durch. Die Entscheidung hing davon ab, ob der vor dem Landgericht im Beschwerdeverfahren abgeschlossene Vergleich einem gerichtlich oder notarisch beurkundeten Vertrag, wie ihn § 313 BGB. für die Verpflichtung zur Übertragung des Eigentums an einem Grundstück verlangt, gleichzustellen war. Das Reichsgericht bejahte die Frage. Daß ein im Erkenntnisverfahren abgeschlossener, in der vorgeschriebenen Form beurkundeter Vergleich einem solchen Vertrage gleichsteht, ist anerkannten Rechtens; es liegt kein Grund vor, einem im Zwangsvollstreckungsverfahren aufgenommenen Vergleich dieselbe Wirkung nicht zuzuerkennen, am wenigsten dann, wenn der Vollstreckungsrichter nicht eigentlich als solcher tätig wird, sondern eine entscheidende Tätigkeit ausübt. Im Beschwerdeverfahren kann eine mündliche Verhandlung stattfinden; für ihre Niederschrift gelten dieselben Vorschriften wie für eine Niederschrift im Erkenntnisverfahren. Es sind also für die Festlegung des Willens der Parteien dieselben Sicherheiten gegeben wie beim Abschluß eines Vergleichs im Erkenntnisverfahren. Der Vergleich enthielt allerdings weder die Bezeichnung der Grundstücke, die an die Klägerin, noch jener, die an den Beklagten aufgelassen werden sollten; es genügte aber, daß gesagt war, das der Klägerin zuzuteilende Fünftel solle durch die Schätzungskommission festgelegt werden. Wie es zulässig ist, in einem formgerechten Vertrag zu bestimmen, einer der Vertragschließenden solle nach den darin angegebenen Richtlinien das Grundstück bezeichnen, zu dessen Übertragung sich der andere Teil verpflichte, so ist es auch zulässig zu vereinbaren, daß ein Dritter oder mehrere Dritte das zu übertragende Grundstück festlegen. Dem Gesetz ist Genüge getan, wenn die Verpflichtung zur Übertragung des Eigentums von Grund und Boden festgelegt wird; die nähere Bezeichnung des Grundstücks ist nicht notwendig. Die Genehmigung des Landrats war bis zur Auflassung widerruflich. Der Beklagte hatte vor dem Berufungsgericht vorgetragen, daß der Landrat auf seine Eingabe hin neue Ermittlungen angestellt habe, die möglicherweise zu einem Widerruf der Genehmigung führen könnten. Die Möglichkeit eines Widerrufs berührt aber die Wirksamkeit des genehmigten Kaufvertrags nicht; der vor der Auflassung erklärte Widerruf macht nur die Erfüllung des an sich verbindlichen Vertrags nachträglich unmöglich. Die Genehmigung war nur der Klägerin, nicht auch dem Beklagten zugestellt worden. Das genügte. Nur für die Versagung der Genehmigung ist vorgeschrieben, daß sie dem Antragsteller und seinem Vertragsgegner zuzustellen ist. (VII, 8. November 1940.)

Amtl. Sammlg. S. 161—165.

Vgl. Bd. 48 S. 183; Bd. 64 S. 82; Bd. 76 S. 191; Bd. 106 S. 148; Bd. 142 S. 303; Bd. 143 S. 100; Bd. 146 S. 300.

26. Große Haverei. Rettung aus Seenot. Gemeinsame Gefahr. (HGB. §§ 527, 535, 630, 641, 700, 740.) Im Kaiser-Wilhelm-Kanal stieß ein Dampfer mit einem anderen zusammen und sank. Die sofort durch den Kapitän benachrichtigte Reederei beauftragte eine Bergungsgesellschaft mit der Rettung von Schiff und Ladung. Auf dem Achterdeck des Schiffes befanden sich 412 Eisenbahnschienen, die an eine Gesellschaft in Kiel als Empfängerin abgeliefert werden sollten. Mit Taucherhilfe wurden 411 Schienen gehoben und auf Leichtern nach Kiel gebracht, wo sie der Empfängerin ausgehändigt wurden. Die Hebung des Dampfers gestaltete sich sehr schwierig; es mußte ein hölzerner Umbau errichtet werden, um das Auspumpen zu ermöglichen. Erst nach mehreren Wochen konnte die Arbeit beendet und das Schiff abgeschleppt werden. In der über den Havereifall aufgemachten Dispache wurde die Empfängerin der Schienen mit einem Beitrag von 15000 *RM* belastet. Sie verweigerte die Zahlung mit der Begründung, für die auf Deck verladenen Schienen habe keine Gefahr bestanden, weil sie im Wasser nicht hätten verderben und ohne erhebliche Kosten jederzeit hätten gehoben werden können; ihre Hebung stehe mit jener des Dampfers in keinem Zusammenhang. Die Klage der Reederei gegen die Empfängerin der Schienen drang in zwei Rechtszügen durch. Das Reichsgericht verwies die Sache zurück. Maßgebend waren die Vorschriften des Handelsgesetzbuchs, da es sich um ein Seeschiff handelte, das auf einer Seefahrt in dem zwei Meere verbindenden, als Seeschiffahrtsweg dienenden Kaiser-Wilhelm-Kanal gesunken war. Voraussetzung für die Anwendung der Vorschriften über große Haverei ist, daß vom Schiffer oder auf dessen Geheiß zum Zwecke der Rettung von Schiff und Ladung aus gemeinsamer Gefahr Kosten aufgewendet worden sind. Die beklagte Gesellschaft hatte sich darauf berufen, daß der Auftrag nicht vom Schiffer, sondern von der Eigentümerin des Schiffes erteilt worden sei; mit dem Sinken des Schiffes sei die Aufgabe des Schiffers beendet gewesen. Der Schiffer handelt aber in solchen Fällen in Vertretung der Reederei wie der Ladungsbeteiligten; diese Vertretungsbefugnis besteht beim Untergang des Schiffes fort, wenn Aussicht zur Hebung von Schiff und Ladung besteht und der Schiffer kraft seiner Fürsorgepflicht zu diesem Zwecke tätig wird. Ein die Reise beendigender und die Frachtverträge wegen dauernder Unmöglichkeit der Ausführung auflösender Verlust des Schiffes stand nicht in Frage. Mit Rücksicht auf die Möglichkeit der Hebung von Schiff und Ladung war die Interessengemeinschaft zwischen beiden auch nach dem Untergange bestehen geblieben; der Kapitän des Schiffes hatte auch das Rettungswerk eingeleitet und seine Ausführung überwacht. Unerheblich war, ob er noch die tatsächliche Gewalt über das Schiff hatte und demzufolge das Rettungswerk nicht als Bergung, sondern als Hilfeleistung zu betrachten war. Die Durchführung der Hebungsarbeiten im einzelnen war nicht vom Kapitän, sondern von sachkundigen Angestellten der Bergungsgesellschaft geleitet worden; das schloß aber nicht aus, daß die Arbeiten trotzdem als auf Geheiß des Schiffers geleistet anzusehen waren, da hiefür nicht erforderlich ist, daß jede einzelne Rettungsmaßnahme von ihm angeordnet oder geleitet wird. Unter den gegebenen Umständen konnte er sich darauf beschränken, den von der Bergungsgesellschaft aufgestellten Plan zu billigen und die Ausführung zu beaufsichtigen; den Abschluß des Vertrags mit der Bergungsgesellschaft konnte er der Reederei überlassen. Die Einrichtung der großen Haverei beruht auf dem Gedanken, daß mit Rücksicht auf die zwischen Schiff und Ladung bestehende Interessengemeinschaft die zur Rettung beider aus gemeinsamer Gefahr gebrachten außerordentlichen Opfer von allen Beteiligten gleichmäßig getragen werden sollen. Die Gefahr muß für das Schiff und die ganze Ladung, nicht nur für einen Teil bestanden haben; es muß sich um eine Gefahr gehandelt haben, die schon eingetreten war, nicht erst drohte. Unter Gefahr ist eine Sachlage zu verstehen, die notwendigerweise oder doch regelmäßig Schaden verursacht. Der Unterschied zwischen einer gegenwärtigen und einer erst in Zukunft zu befürchtenden Gefahr ist darin zu finden, daß Maßregeln, die dem von einem noch ungewissen Ereignis (z. B. schlechtem Wetter) abhängigen Eintritt einer Gefahr vorbeugen sollen, nicht zur großen Haverei gehören. Das Vorhandensein einer gegenwärtigen Gefahr ist nicht lediglich nach sachlichen Maßstäben zu beurteilen; es genügt vielmehr, wenn eine die sofortige Anwendung von Rettungsmaßregeln rechtfertigende Gefahr verständigerweise vom Schiffer als vorhanden angenommen werden durfte. Im vorliegenden Falle war der Dampfer nach dem Untergang in fast stromlosem Wasser einer Ausweichstelle des Kanals gelegen. Es war nicht ohne weiteres zu ersehen, in welcher Gefahr sich die Deckladung Eisenbahnschienen befand. Eine solche Gefahr lag nicht vor, wenn die Schienen jederzeit von dem Deck des gesunkenen Dampfers geholt werden konnten. Ließ sich also nicht sagen, daß sich das Schiff und die ganze Ladung gemeinsam in Seenot befunden hatten, so war es ungerechtfertigt, einen für die schwierige Rettung des Dampfers besonders hoch bemessenen Berge- oder Hilfslohn in großer Haverei unter Einbeziehung der Beklagten zu verteilen. In der Behauptung der Beklagten, daß die Eisenbahnschienen infolge ihrer günstigen Lage auf dem Achterdeck des gesunkenen Dampfers nicht gefährdet gewesen seien, lag auch das Bestreiten, daß sie sich sich mit den übrigen gesunkenen Werten in Seenot befunden hätten. Seenot ist eine der Seeschiffahrt eigentümliche Gefahr, die von der Schiffsmannschaft nicht mit eigenen Mitteln überwunden werden kann. Es genügt, wenn bei vernünftiger Beurteilung der gegebenen Verhältnisse die Beteiligten eine Entwicklung der Sachlage annehmen konnten, die ohne Dazwischentreten der Rettungsmaßregeln zur Beschädigung oder zum Untergang des Schiffes oder der Ladung führen würde. Gemeinsame Gefahr ist auch schon dann anzunehmen, wenn die Gefährdung auf die gleiche Ursache zurückgeht, mag auch für die Ladung nur eine Beschädigung, für das Schiff dagegen der gänzliche Verlust zu befürchten sein. Eine aus der Schwierigkeit des Rettungswerkes hervorgehende Steigerung der Gefahr für Schiff oder Ladung schließt deren Gemeinsamkeit nicht aus. Bestimmte Regeln lassen sich in dieser Beziehung nicht aufstellen. Maßgebend ist die Absicht des Schiffers, der unter mehreren möglichen Mitteln das für die Gesamtheit der Beteiligten mit den geringsten Nachteilen verbundene zu wählen hat. Auch wenn Ladung und Schiff durch verschiedene Handlungen gerettet worden sind, handelt es sich bei der Frage, ob diese als große Haverei zu beurteilen sind, nur um Ermittelung ihres Zweckes; es kommt also darauf an, ob sie auf die Rettung des Schiffes und der Ladung gerichtet und durch diese Absicht zu einer Einheit verknüpft sind. Das Erfordernis sachlicher Angemessenheit der vom Schiffer oder auf dessen Geheiß ergriffenen Maßregeln hat das Gesetz mit Absicht nicht aufgestellt; auch ein Verschulden des Schiffers oder der Retter bei der Ausführung der Rettung schließt große Haverei nicht aus. Wenn die Löschung der Eisenbahnschienen zur Erleichterung des Schiffes notwendig war, fiel sie in den Gesamtrettungsplan, auch wenn sie lange vor Bendigung der Hebung des Dampfers entfernt und an ihr Ziel gebracht worden waren. Anders wäre der Fall nur gelegen gewesen, wenn eine die Einheit des Rettungsunternehmens ausschließende Sonderrettung beabsichtigt gewesen wäre. Die Beklagte hatte sich auch darauf berufen, daß die Rettungsarbeiten nach Wegbringen der Schienen durch verschiedene Zwischenfälle erschwert worden seien. Mit solchen Möglichkeiten ist aber bei schwierigen Schiffshebungen immer zu rechnen. (I, 8. November 1940.) Amtl. Sammlg. S. 166—185. Vgl. Bd. 5 S. 00; Bd. 08 S. 1; Bd. 44 S. 136; Bd. 57 S. 23; Bd. 70 S. 274; Bd. 96 S. 316; Bd. 89 S. 171.

27. Vaterschaftsanerkenntnis. Erbbiologische Begutachtung. (ÖstABGB. §§ 163, 164.) Zu einem unehelich geborenen Kinde erkannte ein Mann, der die Mutter zu heiraten beabsichtigte, die Vaterschaft an. Nach mehreren Jahren klagte das Kind gegen einen anderen Mann auf Feststellung, daß dieser sein Vater sei, und auf Leistung von Unterhalt. Das Amtsgericht gab der Feststellungsklage statt, wies aber die Klage auf Unterhalt ab, weil der Beklagte zur Unterhaltszahlung außerstande sei. Die Berufung und die Revision des Beklagten hatten keinen

Erfolg. Nach ständiger Rechtsprechung des Reichsgerichts war die Revision zulässig, obwohl im ersten Rechtsgang ein Amtsgericht entschieden hatte. Wäre der Mann, der gegenüber der Klägerin die Vaterschaft anerkannt hatte, durch ein rechtskräftiges Urteil als Vater der Klägerin festgestellt worden, so hätte die Klage abgewiesen werden müssen; die urteilsmäßige Feststellung, daß ein Mann der Vater ist, schließt dieselbe Feststellung gegen einen anderen Mann aus. Anders liegt es, wenn der erste Mann die Vaterschaft nur im außerstreitigen Verfahren anerkannt hat. Ein solches Anerkenntnis entbehrt der Rechtskraft. Es kann nach den Regeln der Anfechtung einer bürgerlichrechtlichen Erklärung angefochten werden; auch ohne Anfechtung steht es einer Klage gegen einen anderen Mann nicht im Wege. Der Beklagte kann allerdings darauf verweisen, daß ein anderer Mann die Vaterschaft anerkannt habe; dieser Hinweis hat aber keine andere Bedeutung als die Berufung auf sonstige, gegen die Vaterschaft sprechende Beweismittel. Wird trotz dieses Hinweises rechtskräftig festgestellt, daß der Beklagte der Vater ist, so wird dadurch die auf das frühere Anerkenntnis gestützte Vermutung beseitigt; das Anerkenntnis wird wirkungslos. Im Rechtsstreit war außer Zweifel gestellt worden, daß der Mann, der die Vaterschaft anerkannt hatte, niemals mit der Kindsmutter in Geschlechtsverkehr gekommen war und mit der Anerkennung ihr nur eine Gefälligkeit hatte erweisen wollen. Damit stand es fest, daß er nicht der Vater der Klägerin sein konnte. Da kein Beweis dafür vorlag, daß die Kindsmutter während der Empfängniszeit mit anderen Männern geschlechtlich verkehrt hatte, lag auch kein Anlaß vor, ein erbbiologisches Gutachten zu erholen. Das Amtsgericht hatte festgestellt, daß zwischen der Klägerin und dem Beklagten auffallende Ähnlichkeit bestehe; der Einwand der Revision, daß das Gericht zu einer solchen Feststellung nicht berufen und daß die Bewertung dieser Feststellung bei der Entscheidung rechtsirrig sei, entbehrte der Berechtigung. Auch der Körper der Parteien kann Gegenstand einer gerichtlichen Augenscheinseinnahme sein; ihr Ergebnis kann das Gericht wie das Ergebnis einer anderen Beweisaufnahme nach freier Überzeugung würdigen. Zur Anordnung einer erbbiologischen Begutachtung wäre nur dann Anlaß gegeben gewesen, wenn Anhaltspunkte für eine so erhebliche Unähnlichkeit zwischen der Klägerin und dem Beklagten gegeben gewesen wäre, daß durch sie die Möglichkeit, daß die Klägerin vom Beklagten erzeugt sei, ausgeschlossen worden wäre. (VIII, 12. Dezember 1940.)

Amtl. Sammlg. S. 186—192.

Vgl. Bd. 164 S. 45.

28. **Offene Handelsgesellschaft. Kommanditgesellschaft. Rücktritt. Anfechtung. Fortfall der Geschäftsgrundlage. Rechtsschein.** (BGB. §§ 119, 123, 142, 154, 155, 723; HGB. §§ 133, 161.) Die offene Handelsgesellschaft B. & C. befaßte sich mit der Herstellung und dem Vertrieb von Fischkonserven. Sie sollte in eine Kommanditgesellschaft umgewandelt werden, an der K. und L. als persönlich haftende Gesellschafter beteiligt sein sollten; L. sollte für das nötige Betriebskapital sorgen, insbesondere einen Betrag von 500000 *RM* in englischer Sperrmark zur Verfügung stellen. In einer gemeinsamen Erklärung legten die Beteiligten fest, die Abmachung sollte unter dem Vorbehalt getroffen sein, daß K. und L. die ihnen angebotenen englischen Sperrmark in Höhe von 500000 *RM* erhalten und dem Unternehmen zur Verfügung stellen würden. B. und C. verkauften an die Kommanditgesellschaft das gesamte Vermögen der offenen Handelsgesellschaft B. & C. für 100000 *RM* und beließen ihr den Kaufpreis als Einlage. Für den Fall, daß der Sperrmarkbetrag bis zum 1. April 1938 nicht freigegeben wäre, wurde vereinbart, daß der Geschäftsbetrieb trotzdem aufgenommen und das nötige Geld durch einen Zwischenkredit beschafft werden sollte. Die Kommanditgesellschaft wurde am 23. Mai 1938 in das Handelsregister eingetragen und eröffnete ihre Tätigkeit im August 1938. Der Sperrmarkbetrag wurde nicht beigebracht. Die offene Handelsgesellschaft B. & C. sowie B. und C. für sich setzten K. und L. eine Frist bis zum 31. Januar 1939 unter Androhung des Rücktritts für den Fall ihrer Nichteinhaltung. Als die Frist erfolglos verstrich, klagten sie gegen die Kommanditgesellschaft sowie gegen K. und L. mit dem Antrag, die Kommanditgesellschaft zur Herausgabe des ihr überlassenen Vermögens der offenen Handelsgesellschaft B. & C., K. und L. aber zur Einwilligung in die Löschung der Kommanditgesellschaft und zur Abgabe der ent-

sprechenden Erklärungen zu verurteilen. Das Landgericht wies die Klage ab; das Berufungsgericht gab ihr statt. Das Reichsgericht stellte das Urteil des Landgerichts wieder her. Die Kläger hatten vor allem Unwirksamkeit des Gesellschaftsvertrags geltend gemacht, weil die wesentliche Voraussetzung der Bereitstellung des Sperrmarkbetrags nicht eingetreten sei. Hilfsweise hatten sie sich auf ihren Rücktritt vom Vertrag berufen. Endlich fochten sie den Vertrag wegen arglistiger Täuschung an. Das Berufungsgericht hatte einen versteckten Einigungsmangel angenommen. Gegenüber dem klaren und eindeutigen Wortlaut des Gesellschaftsvertrags ließ sich ein solcher aber, auch bei Berücksichtigung der Vorgeschichte, nicht feststellen. Der Auffassung des Berufungsgerichts, daß die Bindung der Kläger an den Gesellschaftsvertrag und damit dieser selbst durch die Beibringung des Sperrmarkguthabens bedingt sein sollte, stand die Vereinbarung entgegen, daß für den Fall der Nichtbeschaffung der Sperrmark trotzdem der Betrieb begonnen werden sollte. Weder für eine aufschiebende noch für eine auflösende Bedingung lagen Anhaltspunkte vor. Es ist feststehender Grundsatz, daß jeder Erklärende sich beim Worte nehmen und es sich gefallen lassen muß, daß seine Erklärung so verstanden wird, wie die Allgemeinheit sie auffaßt. Anderseits muß jeder Teil die Erklärung des Gegners gegen sich so gelten lassen, wie sie nach Treu und Glauben mit Rücksicht auf die Verkehrssitte zu verstehen ist. Gegenüber eindeutigen Vertragserklärungen kann kein Teil sich darauf berufen, daß er seinen Erklärungen einen anderen Sinn beigelegt habe; für die Anwendung des Grundsatzes vom versteckten Einigungsmangel war kein Raum. Auch ein Rücktrittsrecht stand den Klägern nicht zu. Beim Abschluß des Vertrags wußten sie, daß das Sperrmarkguthaben nicht bereit stand. Wenn sie den späteren Eingang des Guthabens zur Geschäftsgrundlage gemacht hätten, kam ein nachträglicher Fortfall der Geschäftsgrundlage, nicht aber das Fehlen der Geschäftsgrundlage beim Vertragsschluß in Frage. Der nachträgliche Fortfall der Geschäftsgrundlage begründet regelmäßig ein Rücktrittsrecht, wenn die vertraglich bedungene Leistung nach Treu und Glauben dem davon betroffenen Vertragsteil nicht mehr zuzumuten ist. Für das Gesellschaftsrecht gilt das aber nicht. Durch ständige

Rechtsprechung ist festgelegt, daß bei einer in Vollzug gesetzten Gesellschaft das Rücktrittsrecht wegen Unmöglichwerden der Leistung oder wegen Leistungsverzugs nicht geltend gemacht werden kann, vielmehr durch das Recht zur fristlosen Kündigung aus wichtigem Grunde oder durch das Recht, die Auflösung der Gesellschaft durch gerichtliche Entscheidung zu verlangen, ausgeschlossen wird. Diese Rechtssätze müssen auch auf den Fortfall der Geschäftsgrundlage angewendet werden. § 723 BGB. und § 133 HGB. enthalten Sonderbestimmungen, die den dem Gesellschaftsrecht eigentümlichen Verhältnissen angepaßt sind und hinter denen das sonst gewährte Recht auf Rücktritt vom Vertrag zurückstehen muß. Die Kläger konnten also wegen etwaigen Fortfalls der Geschäftsgrundlage nur auf Auflösung der Kommanditgesellschaft klagen, aber nicht vor der Auflösung die Herausgabe ihrer Einlagen fordern oder von den Beklagten verlangen, daß sie in die Löschung der Kommanditgesellschaft einwilligen sollten. Wenn die Kläger, wie sie behaupteten, von den Beklagten K. und L. arglistig getäuscht worden waren, konnten sie möglicherweise Schadenersatzansprüche geltend machen, nicht aber gegen die nicht aufgelöste Kommanditgesellschaft auf Herausgabe der von ihnen eingebrachten Gegenstände klagen oder die Einwilligung der Beklagten K. und L. in die Löschung der Kommanditgesellschaft verlangen. Die Anfechtung des Gesellschaftsvertrags wegen Irrtums oder arglistiger Täuschung erklärte das Reichsgericht für unzulässig. Für die Kapitalgesellschaften (Aktiengesellschaft, Gesellschaft mit beschränkter Haftung) sowie die eingetragenen Genossenschaften hat die Rechtsprechung schon bisher daran festgehalten, daß nach Eintragung dieser Körperschaften im Handels- oder Genossenschaftsregister eine Anfechtung des Gesellschaftsvertrags wegen Willensmängeln nicht mehr möglich ist, diese Rechtsbehelfe vielmehr in Abweichung von dem sonst geltenden Recht entfallen. Die offene Handelsgesellschaft und die Kommanditgesellschaft besitzen allerdings keine eigene Rechtspersönlichkeit; die Verselbständigung des Gesellschaftsvermögens ist aber auch bei ihnen in erheblichem Umfang anerkannt. Eine rückwirkende Vernichtung dieser Gesellschaften wäre mit den Bedürfnissen eines geordneten Handels- und Geschäftsverkehrs schlechthin unvereinbar, da dieser sich auf die nach außen verlautbarten Erklärungen und den so hervorgerufenen Rechtsschein muß verlassen können. An die Stelle der Anfechtung wegen Willensmängeln tritt hier die Auflösung der Gesellschaft aus wichtigem Grunde. Im Gegensatz zu der Rechtslage bei den Kapitalgesellschaften kommen auf diese Weise Willensmängel immerhin in einer nach außen und innen erträglichen Weise zur Auswirkung und werden nach beiden Richtungen klare und durchsichtige Rechtsverhältnisse geschaffen, wie sie im Handels- und Geschäftsverkehr zum richtig verstandenen Besten aller Beteiligten besonders notwendig sind. Die Belange des von dem Willensmangel betroffenen Gesellschafters werden über die Auflösung und Abwicklung der Gesellschaft hinaus dadurch gewahrt, daß er gegen die Mitgesellschafter Schadenersatzansprüche erheben kann (§§ 122, 823, 826 BGB.). Es erscheint auch billiger und zweckdienlicher, die Frage nach der Auflösung einer einmal in Vollzug gesetzten Gesellschaft nach Lage der Verhältnisse in der Gegenwart zu beurteilen, wie das bei der Auflösungsklage geschieht, als aus Gründen, die in der Vergangenheit liegen, die Gesellschaft ohne Beachtung der Entwicklung, die sie genommen hat, rückwirkend zu vernichten. (II, 13. November 1940.) Amtl. Sammlg. S. 193—207. Vgl. Bd. 78 S. 303; Bd. 81 S. 303; Bd. 89 S. 333, 398; Bd. 100 S. 134; Bd. 105 S. 209; Bd. 112 S. 280; Bd. 127 S. 186; Bd. 162 S. 177; JW. 1933 S. 1996; 1935 S. 2617; 1937 S. 1242.

29. **Pflichtteil.** (ÖstABGB. §§ 773, 783, 820, 821.) Eine Frau entzog ihrer einzigen Tochter den Pflichtteil mit der Begründung, daß sie verschuldet und verschwenderisch sei. Als Erben setzte sie die beiden Kinder ihrer Tochter zu je $^3/_8$ und eine Nichte zu $^2/_8$ ein; den beiden Enkelinnen vermachte sie noch verschiedene Schmuckstücke. Die Tochter klagte gegen die drei Erben als Gesamtschuldner auf Auszahlung ihres Pflichtteils in Höhe der Hälfte des Nachlasses. Die Untergerichte wiesen die Klage gegen die Nichte ab, weil nach dem mutmaßlichen Willen der Erblasserin der Pflichtteil der Klägerin von ihren Kindern, die an ihrer Stelle Erben geworden seien, allein getragen werden müsse. Das Reichsgericht verwies die Sache zurück. Da die Erben die Erbschaft bedingt angenommen hatten und diese ihnen schon eingeantwortet war, hafteten

sie für die Nachlaßschulden nur nach Verhältnis ihrer Erbteile, die Nichte also zu $^{2}/_{8}$. Insoweit konnte aber auch von ihr der Pflichtteil begehrt werden. Die Auffassung der Untergerichte hätte zur Voraussetzung gehabt, daß die Erblasserin den Pflichtteil der Klägerin deren Töchtern zugewandt hätte; sie hatte aber darüber hinaus eine andere Verteilung der Erbschaft vorgenommen. (VIII, 14. November 1940.) Amtl. Sammlg. S. 207—208.

30. Patent. Anmeldung. Kombination. Einschränkung. Bindung des Gerichts. (PatG. §§ 6, 47.) Für eine Vorrichtung zur Abnahme des Filterkuchens von Zellendrehfiltern wurde ein Patent angemeldet. Die Anmeldung wurde mit folgendem Anspruch bekannt gemacht: Vorrichtung zur Abnahme des Filterkuchens von Zellendrehfiltern, bestehend aus einer Walze mit rauher Oberfläche, die dauernd mit einer Filterkuchenschicht bedeckt ist, dadurch gekennzeichnet, daß die Erhebungen und Vertiefungen auf der mit einer Gegenwalze zusammenarbeitenden Abnahmewalze derart unterschnitten sind, daß der Filterkuchen auf der Walzenoberfläche festgehalten wird. Ein Fabrikant, der ähnliche Vorrichtungen herstellte, erhob Einspruch gegen die Erteilung des Patents; als es trotzdem erteilt wurde, legte er dagegen Beschwerde ein. Der Beschwerdesenat des Patentamts schränkte dann das Paten in der Weise ein, daß es nur noch galt für eine Kombination folgender Merkmale: einer Abnahmewalze, deren rauhe Oberfläche durch unterschnittene Erhebungen und Vertiefungen, z. B. durch um ihre Oberfläche gelegte Schnüre oder Drahtgeflecht, gebildet ist, und einer Gegenwalze, die an Stelle eines Schabers die Abnahme des Filterkuchens von der Abnahmewalze übernimmt. Schon vor Erlaß dieses Beschlusses hatte der Patentinhaber gegen den Fabrikanten, der die Beschwerde eingelegt hatte, Klage auf Unterlassung, Rechnungslegung und Schadenersatz erhoben. Das Landgericht hatte der Klage stattgegeben, das Berufungsgericht hatte sie nach Erlaß des Beschlusses des Beschwerdesenats des Patentams abgewiesen. Die Revision des Klägers hatte keinen Erfolg. Von der Kombination, wie sie im Beschluß des Beschwerdesenats des Patentamts beschrieben war, machte der Beklagte keinen Gebrauch. Allerdings verwendete er zur Ablösung des Filterkuchens von der Abnahmewalze eine Gegenwalze; dagegen entsprach die Oberfläche der Abnahmewalze nicht dem im Beschluß aufgeführten Merkmal, vielmehr war ihre Oberfläche lediglich aufgerauht, um das Haftvermögen des Filterkuchens zu vergrößern. In Frage kam hienach, ob durch das Patent der allgemeine Gedanke geschützt war, zur Erhaltung einer gleichmäßigen Filterkuchenschicht auf der Abnahmewalze die von dieser abzuführende Filterkuchenschicht statt durch einen Schaber durch eine Gegenwalze abzulösen. Der endgültig erteilte Patentanspruch der die Eigenschaft der Erfindung als Kombination betonte, deckte sich nicht mit dem bekanntgemachten Anspruch. Es konnte nicht zweifelhaft sein, daß der Beschwerdesenat des Patentamts im sachlichen Ergebnis den Patentschutz für den Fall der Anwendung einer Gegenwalze mit einer nur aufgerauhten Abnahmewalze versagte. Eine solche Beschränkung des Patentschutzes mußte schon deshalb als zulässig anerkannt werden, weil der Anmelder sich im voraus mit den Änderungen der Unterlagen einverstanden erklärt hatte, die der Beschwerdesenat für notwendig erachten sollte. Auch ohne das ergibt sich die Befugnis des Patentamts, den Umfang des Patentschutzes im Erteilungsverfahren festzulegen, soweit es dies im Hinblick auf die Unterlagen der Anmeldung oder den Stand der Technik für geboten erachtet, schon aus seiner Stellung als Erteilungsbehörde. Die Gerichte haben als Grundlage der Auslegung das Patent so hinzunehmen, wie es erteilt worden ist. Auch einschränkende Verfügungen des Patentamts sind für die Gerichte verbindlich. Insbesondere kann danach ein einzelnes Merkmal außerhalb einer patentierten Kombination keinen selbständigen Schutz genießen, wenn das Patentamt diese Frage im Erteilungsverfahren geprüft und verneint hat. Unerheblich war, ob durch die in der Formel des Beschlusses festgelegten Änderungen des Anspruchs und der Beschreibung der Beschränkungswille schon mit hinreichender Deutlichkeit zum Ausdruck kam; es genügte vielmehr, daß die Gründe des Beschlusses über die Grenzen, die das Patentamt dem Patentschutz gegeben hatte, sichere Auskunft gaben. Allerdings ist die Anmeldung zurückzuweisen, wenn der Anmelder sich mit einer vom Patentamt verlangten Beschränkung des Schutzbegehrens nicht einverstanden erklärt. Das ändert aber nichts daran, daß die den Patentschutz unzwei-

deutig beschränkende Entscheidung des Patentamts nicht nachträglich durch die Behauptung, es sei zu Unrecht das Einverständnis des Anmelders angenommen worden, ihrer Bedeutung entkleidet werden kann. (I, 15. November 1940.) Amtl. Sammlg. S. 209—219.
Vgl. Bd. 80 S. 54; Bd. 153 S. 47; Bd. 159, S. 1; JW. 1928 S. 1212.

31. Unterhalt. Zahlung ins Ausland. (BGB. §§ 242, 1612, 1613; RichtlDevBewirtsch. vom 22. Dezember 1938 Abschn. II Nr. 36, 37; Abschn. IV Nr. 49, 61.) Eine Ehe wurde im Jahr 1930 aus Verschulden des Ehemannes geschieden. Die Frau wanderte mit ihren zwei Kindern im Jahr 1933 nach Frankreich aus. Im Jahr 1936 wurde einer Klage der Kinder auf Zahlung von je 200 *RM* Unterhalt für den Monat September in der Weise stattgegeben, daß der Beklagte zur Zahlung dieses Betrags auf ein bei der Dresdner Bank zu errichtendes Kreditsperrmarkkonto verurteilt wurde; das Urteil erlangte die Rechtskraft. Im November 1939 erhoben die Kinder Unterhaltsansprüche in der Höhe von je 200 *RM* monatlich für die Zeit vom 1. Oktober 1936 bis zum 31. Dezember 1939 und verlangten die Verurteilung des Beklagten zur Einzahlung dieser Beträge auf ein Vorzugssperrguthabenkonto für die Kläger bei der Dresdner Bank. Das Amtsgericht gab der Klage statt. Das Berufungsgericht erklärte die Ansprüche dem Grunde nach für gerechtfertigt. Das Reichsgericht wies die Klage ab. Die Unterhaltspflicht ist grundsätzlich durch Zahlung einer Geldrente, also durch Übereignung von Bargeld, zu erfüllen. Hiefür war, da die Kläger im Ausland wohnten, die Genehmigung der Devisenstelle erforderlich. Diese hatte die Genehmigung zur Einzahlung auf ein bei der Dresdner Bank zu errichtendes Kreditsperrmarkkonto erteilt. Die Kläger hatten Einzahlung auf dieses Konto verlangt, sich also bereit erklärt, diese Einzahlung an Stelle der geschuldeten Bargeldübereignung an Erfüllungs statt anzunehmen. In der Rechtsprechung ist anerkannt, daß der Schuldner nach Treu und Glauben die durch die Devisengesetzgebung erforderlich werdenden Änderungen in der Art der geschuldeten Leistung auf sich nehmen muß, sofern er dadurch nicht beschwert wird. In der Regel werden die Belange des Schuldners keine erhebliche Beeinträchtigung dadurch erfahren, daß er die Leistung statt an den Gläubiger selbst an eine Bank auf ein für den Gläubiger bestimmtes Sperrguthaben bewirkt. Im vorliegenden Falle verhielt es sich aber anders. Der Beklagte hatte mit Recht eingewendet, daß seine Leistung zweckgebunden sei; sie war dazu bestimmt, den Lebensunterhalt der Kläger zu decken. Dieser Zweck konnte durch die Zahlung an eine inländische Bank auf ein Vorzugssperrguthaben der Kläger nicht erreicht werden. Zur Verfügung über dieses Guthaben waren die Kläger nur nach Maßgabe der Richtlinien für die Devisenbewirtschaftung in der Lage. Hienach kam für sie im wesentlichen nur die Verwendung zur Kapitalanlage im Inland in Betracht. Zur Schaffung von Vermögensansammlungen und Geldrücklagen sind aber Unterhaltsleistungen ihrem Wesen nach nicht bestimmt. Nach Treu und Glauben ließe es sich nicht rechtfertigen, dem Beklagten eine Leistung zuzumuten, die für ihn als Unterhaltsverpflichteten eine erhebliche Belastung bedeutete, aber den Zweck nicht zu erfüllen vermochte, den sie für die Kläger als Unterhaltsberechtigte haben sollte. Die Zahlung auf ein Sperrguthaben konnte dem Beklagten hienach auch dann nicht angesonnen werden, wenn für die Kläger die Möglichkeit bestand, ihre Sperrguthabenforderung im Auslande börsenmäßig zu verwerten. Die Kläger hatten selbst vorgetragen, daß dies nur mit erheblichen Verlusten möglich wäre; der Zweck der Leistung, die Deckung des standesmäßigen Unterhalts der Kläger, ließ sich dadurch nicht erreichen. Nach den Richtlinien für die Devisenbewirtschaftung sind die Devisenstellen berechtigt, die Genehmigung zur Überweisung von Unterhaltsrenten auf gesetzlicher Grundlage in das Ausland zu erteilen, wenn der Empfänger auf die Bezüge für seinen bescheidenen Lebensunterhalt dringend angewiesen ist und ihm eine Rückkehr in das Inland nicht zugemutet werden kann. Nur auf diesem Wege und in dem hienach zu bemessenden Umfange war der Beklagte zur Zeit in der Lage, seiner Unterhaltspflicht gegenüber den Klägern nachzukommen; während der Dauer ihres Aufenthalts im feindlichen Auslande war hiefür auch noch die Genehmigung des Reichswirtschaftsministers erforderlich. Die Kläger hatten nicht vorgebracht, daß sie um eine solche nachgesucht hatten. Für den Beklagten bestand keine

Veranlassung, selbst darum nachzusuchen. (IV, 18. November 1940.) Amtl. Sammlg. S. 219—223.

Vgl. Bd. 151 S. 39, 121; Bd. 152 S. 359; Bd. 164 S. 71; RAG. Bd. 17 S. 210.

32. **Gläubigeranfechtung. Vermögensverschleuderung.** (TschechoslAnfG. §§ 2, 3, 4.) Grundstücke wurden um einen Preis verkauft, der unter ihrem halben Werte lag; der Käufer war ein Schwager des Verkäufers. Eine Gläubigerin focht die Veräußerung wegen Gläubigerbenachteiligung an und verlangte Verurteilung des Käufers zur Duldung der Vollstreckung in die Grundstücke. Das Kreisgericht wies die Klage ab. Das Obergericht nahm ein gemischtes Geschäft (negotium mixtum cum donatione) an und gab der Klage teilweise statt. Das Reichsgericht verurteilte den Beklagten nach dem Klageantrag. Anfechtbar sind nach der tschechoslowakischen Anfechtungsordnung unentgeltliche Verfügungen des Schuldners. Bei der Beurteilung der Unentgeltlichkeit kommt nicht schlechthin der wahre Wert von Leistung und Gegenleistung, sondern zugleich auch deren Wertung durch die Beteiligten in Betracht. Ging deren Wille auf einen billigen Verkauf, so war keine Unentgeltlichkeit gegeben. War dagegen der Hauptzweck des Geschäfts Freigebigkeit, so unterlag es der Anfechtung. In der Regel ist dann das ganze Geschäft anfechtbar; nur wenn es sich in einen entgeltlichen und einen unentgeltlichen Teil zerlegen läßt, kommt eine Teilanfechtung in Frage. Die Untergerichte hatten darauf Wert gelegt, daß zur Zeit des Geschäftsabschlusses weder der Verkäufer noch der Käufer von dem Vorhandensein der Gläubigerin wußten. Darauf kam es aber nicht an. Die im letzten Jahr vor der Anfechtung vom Schuldner eingegangenen Kaufverträge sind anfechtbar, wenn der andere Teil in dem Geschäft eine die Gläubiger des Schuldners benachteiligende Vermögensverschleuderung erkannte oder erkennen mußte. Ob er weiß, daß der Veräußerer Schulden hat, ist ohne Belang. Das Mißverhältnis zwischen dem Kaufpreis und dem Wert der Grundstücke hatte der Erwerber, wenn er es nicht tatsächlich erkannte, bei Anwendung der im Verkehr erforderlichen Sorgfalt erkennen müssen. VIII, 25. November 1940.)

Amtl. Sammlg. S. 223—225.

33. **Preisprüfung. Aufrechnung. Verschulden beim Vertragsschluß. Schadenersatz.** (VierJahrPlG. § 2.) Der Eigentümer eines Gasthofes schloß im Jahr 1934 mit den Gaswerken der Stadt einen Vertrag auf 10 Jahre ab; nach einem zwölfmonatigen Probebetriebe sollte beiden Teilen gestattet sein, neue Verhandlungen über die Preisbildung anzubahnen. Der Vertrag begann am 23. Mai 1935 zu laufen. Am 20. Mai 1936 teilte die Leitung der Gaswerke dem Vertragsgegner mit, daß sie zu dem vereinbarten Preis nicht mehr weiter liefern könne, vielmehr für das Kubikmeter Gas mindestens 4.5 *Rpf* berechnen müsse. Sie klagte auf Feststellung, daß sie nicht verpflichtet sei, dem Beklagten das Gas um den vereinbarten Betrag zu liefern. Das Landgericht gab der Klage statt. Während des Berufungsverfahrens setzte die Preisprüfungsstelle den Gaspreis auf 4.5 *Rpf* für das Kubikmeter fest. Die Berufung und die Revision des Beklagten blieben erfolglos. Der tatsächliche Streit der Parteien über den Inhalt ihres Vertrages (ob dieser für 10 Jahre bindend geschlossen war oder nicht) war für die Entscheidung der Sache bedeutungslos geworden durch die Preisfestsetzung des Preisbildungskommissars; diese Festsetzung war für die Gerichte bindend und hatte auch keine Entschädigung für einen durch die Anordnung entstandenen oder noch entstehenden Schaden zur Folge. In Frage kam nur ob der Preis mit Rückwirkung für die Zeit vor dem Inkrafttreten des Gesetzes über den Vierjahresplan (29. Oktober 1936) rechtswirksam festgesetzt werden konnte. Auch die Anwendung dieser Befugnis auf die zur Zeit des Inkrafttretens des Gesetzes noch unabgewickelten und den Gegenstand von Meinungsstreitigkeiten der Beteiligten bildenden langdauernden Vertragsbeziehungen ist der Nachprüfung durch die Gerichte entzogen. Eine offenbare Überschreitung der gesetzlichen Ermächtigung lag nicht vor. Die Hauptangriffe der Revision richteten sich dagegen, daß das Berufungsgericht einem vom Beklagten behaupteten Verschulden beim Vertragsschluß nicht Rechnung getragen habe. Dieses sollte darin bestehen, daß die Klägerin durch unrichtige Berechnung über die Kosten der Gasfeuerung den Beklagten dazu gebracht habe, die Kohlenheizung aufzugeben. Das Berufungsgericht hatte in dieser Richtung ausgeführt, jeder Nachteil, den der Be-

klagte gegenüber der Vereinbarung, auch wenn diese auf 10 Jahre Geltung haben sollte, erlitte, sei ausschließlich auf Grund des Eingreifens des von ihm selbst angerufenen Reichskommissars und seiner Preisfestsetzung entstanden, so daß also ein ursächlicher Zusammenhang zwischen irgendwelchem Verschulden der Klägerin und einem ihm erwachsenen oder erwachsenden Schaden jeglicher Art fehle. Demgegenüber erklärte der Beklagte, daß der Schaden schon durch die Umstellung auf die Gasfeuerung entstanden sei und daß er sich dazu nicht entschlossen hätte, wenn er von der Klägerin pflichtgemäß aufgeklärt worden wäre; daher könne auch keine Rede davon sein, daß auf einem Umwege nichts anderes als die Umgehung der Anordnung des Reichskommissars erstrebt würde. Der Beklagte hatte aber eine Aufrechnung mit einer bestimmten Schadenersatzforderung gegenüber der Geldforderung der Klägerin nie erklärt. Diese Erklärung, die den Feststellungsantrag nicht berührt hätte, wäre in den unteren Rechtszügen möglich gewesen; im Revisionsverfahren konnte sie nicht nachgeholt werden. Mit der Anordnung des Reichskommissars und mit der Entscheidung des Gerichts über die Bezahlung des bezogenen Gases war allerdings nicht ohne weiteres darüber entschieden, ob nicht der Beklagte einen Schadenersatzanspruch aus Verschulden beim Vertragsschluß erheben konnte. Es bestanden aber erhebliche Bedenken dagegen. Wenn auch gesagt werden konnte, daß über diese Schadenersatzfrage der Reichskommissar nicht befunden habe, da es sich um einen selbständigen, außerhalb seiner Zuständigkeit liegenden bürgerlichrechtlichen Anspruch handelte, war damit noch nicht dargetan, daß gegenüber seiner Ordnung der Preiszahlung eine Aufrechnung überhaupt oder gerade aus solcher Gegenforderung stattfinden darf. Mittels eines Schadenersatzanspruchs kann nicht die endgültige Anordnung der Verwaltungsbehörde über ein im allgemeinen Interesse notwendiges Geschehen zwischen Vertragsparteien außer Wirkung gesetzt werden. Die Anordnung des Reichskommissars konnte sehr wohl dahin ausgelegt werden, daß der festgesetzte Preis auch wirklich zu bezahlen war und nicht wegen einer Gegenforderung, zumal einer streitigen, vorenthalten werden durfte; diese Auslegung lag um so näher, als der festgesetzte Preis nur die eigenen Kosten der Klägerin deckte.

(II, 6. November 1940.) Amtl. Sammlg. S. 226—236. Vgl. Bd. 152 S. 12; RAG. Bd. 14 S. 95; JW. 1936 S. 3175.

34. Unfallversicherung. Schadenersatz. Feststellungsklage. Unzulässige Rechtsausübung. (RVO. § 1542; ZPO. § 256.) Bei einem Zusammenstoß zwischen einem Straßenbahnwagen und einem Lastkraftwagen wurde ein Angestellter der Straßenbahn verletzt. Die zuständige Berufsgenossenschaft bewilligte die ihm nach der Reichsversicherungsordnung zustehenden Leistungen; auf Grund der Feststellung, daß er zu 50% erwerbsbeschränkt sei, erhielt er eine Unfallrente in entsprechender Höhe. Er wurde von der Aktiengesellschaft, welche die Straßenbahn betrieb, wieder eingestellt, und zwar mit den gleichen Bezügen, die er vor dem Unfall gehabt hatte. Die Berufsgenossenschaft verlangte von dem Halter und dem Führer des Kraftwagens, der den Unfall verursacht hatte, Ersatz dieser Rente. Diese erwiderten, daß der Verletzte nach seiner Wiedereinstellung mit den früheren Bezügen keinen Schaden mehr habe und daß darum auch kein Schadenersatzanspruch auf die Unfallberufsgenossenschaft habe übergehen können. Die Klage der Berufsgenossenschaft hatte keinen Erlog. Ein Schadenersatzanspruch des Verletzten gegen die Beklagten wegen Minderung seiner Erwerbsfähigkeit hätte zur Voraussetzung, daß ihm ein Vermögensnachteil entstanden wäre; das war aber nicht der Fall. Der Verletzte war wieder bei seiner früheren Arbeitgeberin im Dienst und bezog auf Grund seines Arbeitsvertrages denselben Lohn wie vor dem Unfall. Demgemäß konnte auch auf die Klägerin ein Anspruch des Verletzten nicht übergegangen sein. Es handelte sich bei der Bewilligung des früheren Arbeitslohnes auch nicht um eine freiwillige Leistung der Aktiengesellschaft, die dem Verletzten aus rein fürsorgerischen Erwägungen gewährt wurde; die Entlohnung richtete sich vielmehr nach den Vorschriften der Tarifordnung, die für die Arbeitgeberin verbindlich war. Die Berufsgenossenschaft hatte ausgeführt, es könne nicht angehen, daß ein öffentlichrechtlicher Versicherungsträger auf Grund der Reichsversicherungsordnung Leistungen an den Arbeitnehmer zu bewirken habe, während der Schädiger frei ausgehe. **Dieses Ergebnis findet aber seine Erklärung in der Verschiedenheit der Voraussetzun-**

gen, die für die Leistungen der Träger der Unfallversicherung und die Schadenersatzpflicht des Schädigers bestehen. Die Leistungen der Berufsgenossenschaft sind davon abhängig, ob und in welchem Umfange die Erwerbsfähigkeit des Verletzten an sich beeinträchtigt ist; die Schadenersatzpflicht des Schädigers setzt dagegen voraus, daß dem Verletzten wirklich ein Schaden erwachsen ist. Die bloße Minderung der Erwerbsfähigkeit an sich ist insoweit ohne Bedeutung. Die Berufsgenossenschaft hatte auch Feststellung verlangt, daß die Beklagten verpflichtet seien, ihr alle Aufwendungen zu ersetzen, die sie künftig an den Verletzten machen müsse. Das Berufungsgericht hatte angenommen, daß es an einem Interesse der Berufsgenossenschaft an dieser Feststellung fehle. Hiegegen hatte die Berufsgenossenschaft eingewendet, daß sich dieses Interesse schon aus der kurzen Verjährung ergebe, deren Verlängerung durch Vereinbarung unzulässig sei. Auch dieser Einwand war unbegründet. Die Beklagten hatten die Klägerin wegen ihrer sonstigen Leistungen aus dem Unfall befriedigt und im Rechtsstreit ausdrücklich ihre Verpflichtung anerkannt, der Klägerin aus dem Unfall zu haften und ihr, falls Schadenersatzansprüche des Verletzten auf sie übergegangen seien, die entsprechenden Beträge zu zahlen. Unter solchen Umständen war ein Interesse der Klägerin an alsbaldiger Feststellung dieser Verpflichtung nicht anzuerkennen. Die Verjährung war schon durch das Anerkenntnis der Beklagten unterbrochen; es war anzunehmen, daß es auch in Zukunft möglich sein würde, solche Anerkenntnisse herbeizuführen. Überdies würde die Klägerin den Beklagten, wenn sie sich später auf Verjährung berufen würden, im Hinblick auf ihr bisheriges Verhalten, besonders auf ihre Erklärung im Rechtsstreit, den Einwand unzulässiger Rechtsausübung entgegenhalten können. (VI, 16. November 1940.)

Amtl. Sammlg. S. 236—242.

35. **Leistungen für kirchliche Zwecke. Widmung. Zuständigkeit. Aufwertung.** (PrLandGemO. für die östlichen Provinzen § 88.) Im Jahr 1895 wurde in der Gemeinde A. durch Auspfarrung aus einer Stadtkirchengemeinde eine selbständige Kirchengemeinde errichtet. Die Gemeinde verpflichtete sich, für die Aufbringung des Gehalts des neu anzustellenden Predigers und des Orga-

nisten jährlich 1366 *RM* zu zahlen und die Entlohnung des Kirchendieners ebenfalls vorläufig zu übernehmen. Die Gemeindebeschlüsse wurden vom Kreisausschuß genehmigt. Als im Jahr 1920 die Zahlung eingestellt wurde, erhob die Kirchengemeinde Klage auf Zahlung des fälligen Jahresbetrages und der in Zukunft fällig werdenden Jahresbeträge von 1666 *RM*. Die Stadtgemeinde B., in welche die Gemeinde A. übernommen worden war, erkannte den Anspruch an und wurde entsprechend verurteilt. Sie kam dem Urteil mehrere Jahre hindurch nach. Auf Grund der Notverordnung vom 8. Dezember 1931 kürzte sie die Zahlungen auf 1350 *RM*. Am 1. April 1937 stellte sie die Zahlungen ein. Die Klage auf Fortsetzung der Zahlungen zu dem gekürzten Betrage drang durch. Da die Klage ihren Ausgang von dem Anerkenntnisurteil nahm, konnte die Zulässigkeit des Rechtswegs nicht in Frage kommen. Für Ansprüche auf vermögensrechtliche Leistungen zu kirchlichen Zwecken ist übrigens, sofern nicht besondere Bestimmungen im Wege stehen, der Rechtsweg selbst dann zulässig, wenn diese Ansprüche nach heutiger Auffassung ihren Ursprung im öffentlichen Recht haben. Die Bereiterklärung der Gemeinde A., die Mittel aufzubringen, die nach der Feststellung des Konsistoriums zur Deckung des laufenden Geldbedarfs der neu zu errichtenden Kirchengemeinde erforderlich waren, sollte die Schöpfung dieser Gemeinde ermöglichen und ihren Bestand sichern. Sie bildete die Grundlage für die Errichtung der Kirchengemeinde durch die kirchliche Behörde und für deren Genehmigung durch die Staatsregierung. Ein mit solcher Zweckbestimmung gegebenes und zum Bestandteil des Gründungsplanes eines öffentlichrechtlichen Verbandes gewordenes Leistungsversprechen entzieht sich der bürgerlichrechtlichen Beurteilung; es handelt sich vielmehr um die Begründung einer öffentlichrechtlichen Verpflichtung. Bei diesem als Widmung bezeichneten Rechtsvorgange schöpft das einseitige Leistungsversprechen seine verpflichtende Kraft aus dem Zweck, dem die Leistung nach Bestimmung der für die Errichtung des neuen Verbandes öffentlichrechtlich berufenen Organe gewidmet ist. Es kann deshalb nicht darauf ankommen, ob das Versprechen nach den Grundsätzen des bürgerlichen Schuldrechts eine Verpflichtung zu erzeugen vermag. Ebensowenig ist es geboten, daß, wenn es sich um das Versprechen einer politischen Gemeinde handelt, dieses Versprechen den Vorschriften genügt, von deren Beachtung ihre rechtsgeschäftliche Bindung gemeinderechtlich sonst abhängig ist. Die verpflichtende Kraft der Widmung setzt nur voraus, daß die Gemeinde ihren Willen, zum Besten des zu bildenden Verbandes eine Leistung zu übernehmen, durch ihre Vertretung ordnungsgemäß und mit der erforderlichen Genehmigung der Aufsichtsbehörde gefaßt und diesen Willensbeschluß der für die Körperschaftsgründung zuständigen Behörde zur Kenntnis gebracht hat. Diese Erfordernisse waren gegeben. Demgemäß konnte die ins Leben gerufene Kirchengemeinde die Ansprüche, die dieser Verpflichtung entsprachen, als eigene geltend machen. Die Nichtbeachtung der Vorschrift des § 88 Abs. 4 Ziff. 7 LGO. stand der rechtswirksamen Entstehung der Verpflichtung nicht im Wege. Es handelte sich nur noch darum, ob der Einwand der Beklagten, die Zuschußverpflichtung sei vermöge der ihr innewohnenden Beschränkung durch die wirtschaftliche Erstarkung der klagenden Kirchengemeinde erloschen, als berechtigt anzuerkennen war. Aus den Verhandlungen, die den Beschlüssen vorhergegangen waren, und aus den Beschlüssen selbst ergab sich aber eine solche Beschränkung nicht. Die Verpflichtung wurde also dadurch nicht beseitigt, daß durch die starke Zunahme der Mitglieder der Kirchengemeinde deren Steuerkraft eine wesentliche Steigerung erfahren hatte. Das Berufungsgericht hatte unter Anwendung der Grundsätze des Aufwertungsrechts eine niedrigere Summe zugesprochen. Das Reichsgericht erklärte, daß für die Aufwertung der Zuschußpflicht deren Zweckgebundenheit bestimmend sei. Danach erschien es naheliegend, daß die Zahlung eines dem Markbetrage des Zuschusses entsprechenden Reichsmarkbetrages unter Berücksichtigung der von der Klägerin hingenommenen Kürzung die angemessene Aufwertung darstellte. Ein solches Aufwertungsabkommen fiel in die laufende Verwaltung der Gemeinde und bedurfte keiner besonderen Form. (VII, 26. November 1940.)

Amtl. Sammlg. S. 242—248.

Vgl. Bd. 10 S. 242; Bd. 111 S. 211; JW. 1927 S. 785.

36. Uneheliche Vaterschaft. Vaterschaftsfeststellung. Abstammungsstreit. Revision. (BGB. §§ 130, 1717, 1719,

1720, 1723, 1735 a; EheG. §§ 6, 7; ZPO. §§ 546, 640, 644.) Zu einem am 1. Juni 1936 unehelich geborenen Kind erkannte der in Anspruch genommene Mann die Vaterschaft an und verpflichtete sich zur Unterhaltsleistung. Nunmehr klagte er auf Feststellung, daß er nicht der Vater des Kindes sei. Zur Begründung behauptete er, die Kindsmutter habe während der Empfängniszeit auch anderen Männern die Beiwohnung gestattet und ihn durch arglistiges Verschweigen dieser Tatsache zur Anerkennung der Vaterschaft veranlaßt; es sei auch den Umständen nach unmöglich, daß das Kind von ihm stamme. In zwei Rechtszügen wurde die Klage abgewiesen. Das Reichsgericht verwies die Sache zurück. Die Zulässigkeit der Revision hing davon ab, welcher Art der vom Kläger geltend gemachte Anspruch war, ob es sich insbesondere bei der von ihm begehrten verneinenden Feststellung um das Nichtbestehen der Zahlvaterschaft oder um die blutmäßige Abstammung der Beklagten von ihm handelte. Im ersten Falle wäre die Revision unzulässig gewesen. Das Rechtsverhältnis der unehelichen Vaterschaft ist zwar bisher als ein solches nichtvermögensrechtlicher Art angesehen worden, weil es sich nicht in dem vermögensrechtlichen Unterhaltsanspruch des Kindes erschöpfte, sondern auch Folgen nach sich zog, die nicht dem Vermögensrecht angehörten. Diese Auffassung kann aber, nachdem die neuere Rechtsentwicklung eine scharfe Grenze zwischen der blutmäßigen Abstammung und der bloßen Zahlvaterschaft gebracht hat, nicht mehr als zutreffend anerkannt werden. Soweit der unehelichen Vaterschaft im Gesetz andere als rein vermögensrechtliche Wirkungen beigelegt sind, ist für sie nicht mehr die Zahlvaterschaft, sondern die blutmäßige Abstammung entscheidend, so bei dem Ehehindernis der Verwandtschaft (EheG. §§ 6, 7), bei der Ehelichkeitserklärung (BGB. §§ 1723, 1735 a), bei der Legitimation eines unehelichen Kindes durch nachfolgende Ehe der Eltern (§ 1719 BGB.); die gesetzlichen Vermutungen des § 1720 können jederzeit durch den Nachweis der Unmöglichkeit der Abstammung entkräftet werden, so daß es auch hier letzten Endes auf die blutmäßige Abstammung ankommt. Rechtsstreitigkeiten, welche die blutmäßige Abstammung betreffen, sind immer nichtvermögensrechtlicher Art; die Revision ist deshalb bei ihnen ohne Rücksicht auf ihren Wert zulässig. Bei der großen Bedeutung, die der blutmäßigen Abstammung nach neuem Recht zukommt, kann im allgemeinen davon ausgegangen werden, daß bei Klagen, welche die uneheliche Abstammung betreffen, nicht die uneheliche Vaterschaft aus § 1717 BGB., sondern die blutmäßige Abstammung gemeint ist, sofern nicht die Umstände das Gegenteil ergeben. Da hienach im vorliegenden Fall ein nichtvermögensrechtlicher Anspruch Gegenstand des Rechtsstreits war, bestanden gegen die Zulässigkeit der Revision keine Bedenken. Klagen auf Feststellung des Bestehens oder Nichtbestehens blutmäßiger Abstammung sind im sog. Statusverfahren der §§ 640ff. ZPO. durchzuführen. Der Kläger hatte einen Mann als Zeugen benannt, mit dem die Kindsmutter geschlechtlichen Verkehr gehabt haben sollte; dieser hatte das auf Eid in Abrede gestellt. Das Berufungsgericht hatte hienach angenommen, daß der Kläger den ihm obliegenden Beweis, daß die Beklagte nicht von ihm abstamme, nicht erbracht habe. Hilfsweise hatte der Kläger Feststellung verlangt, daß er möglicherweise nicht der Vater sei. Eine solche Feststellung ist nicht zulässig. Die Ablehnung der verneinenden Abstammungsklage enthält ohnehin nicht die Feststellung der blutmäßigen Abstammung, läßt also andere Möglichkeiten offen. Die Zurückverweisung erfolgte, weil des Berufungsgericht weitere Beweisangebote abgelehnt, insbesondere von einer eidlichen Vernehmung der Kindsmutter abgesehen hatte. Die allgemeine Lebenserfahrung beweist, daß die Mutter eines unehelichen Kindes, wenn es sich um die Abstammung des Kindes mit ihren schwerwiegenden Folgen für alle Beteiligten handelt, bei einer uneidlichen Vernehmung häufig mit der Wahrheit zurückhält und erst unter dem Eindruck des Eides zu weiteren Angaben bereit ist. Ihre Beeidigung muß deshalb im Abstammungsstreit, sofern nicht der Sachverhalt bereits anderweit völlig geklärt ist, die Regel bilden; ohne diese Beeidigung kann von einer erschöpfenden Ausnutzung aller Bezugsquellen, die dem Gericht als Pflicht obliegt, nicht gesprochen werden. Insoweit ist der sonst geltende Grundsatz, daß die Beeidigung eines Zeugen dem im Revisionsverfahren nicht nachprüfbaren Ermessen des Gerichts überlassen ist, im Abstammungsstreit eingeschränkt. (IV, 28. November 1940.) Amtl. Sammlg. S. 248—254.

Vgl. Bd. 161 S. 277; Bd. 163 S. 321; Bd. 164 S. 62, 281.

37. **Revisionssumme. Anspruchhäufung.** (ZPO. §§ 5, 546; ÜberlVO. § 6; ÖstZPO. § 502; ÖstJurN. § 55.) Nach österreichischem Recht ist im Falle der Häufung mehrerer Ansprüche in einer Klage zu unterscheiden, ob diese Ansprüche selbständig sind oder ob sie in einem rechtlichen und tatsächlichen Zusammenhange stehen. Im zweiten Falle werden die sämtlichen Ansprüche zusammengerechnet und gelten als ein einziger Anspruch; im ersten Falle werden sie einzeln für sich bewertet. Diese Bestimmungen sind aber durch die Überleitungsverordnung außer Kraft gesetzt worden. Der Gesetzgeber wollte die Zulässigkeit der Revision im ganzen Gebiete Großdeutschlands einheitlich, und zwar so gestalten, wie sie in der deutschen Zivilprozeßordnung geregelt ist. Dieser ist eine solche Unterscheidung fremd. Als Wert des Beschwerdegegenstandes im Revisionsverfahren ist der Wert anzusehen, um den das Urteil des Berufungsgerichts hinter dem Begehren der Revision zurückbleibt. Bezieht sich die Revision auf mehrere Ansprüche, so ergibt sich die Beschwerdesumme in allen Fällen aus der Summe der Änderungen, die nach dem Antrage der Revision am Urteil des Berufungsgerichts vorgenommen werden sollen. Dabei ist es unerheblich, ob die vom Revisionsverfahren betroffenen Ansprüche in rechtlichem oder tatsächlichem Zusammenhange stehen oder ob sie voneinander unabhängig sind. (VIII, 5. Dezember 1940.) Amtl. Sammlg. S. 255—256.

38. **Hauszinssteuer. Amtspflicht. Grundsteuerkataster.** (BGB. § 839; WeimVerf. Art. 131.) Eine Stadt zog in staatlichem Auftrag die Hauszinssteuer aus den in ihr belegenen Häusern ein. Bei der Veranlagung wurden zwei nebeneinander gelegene Grundstücke verwechselt; demzufolge wurde von dem einen nur eine monatliche Steuer von 5.40 *RM* statt 14.40 *RM* erhoben. Das Haus ging an die Kreissparkasse über und wurde von dieser weiterverkauft. In der Folgezeit wurde der Irrtum festgestellt und berichtigt. Die Käufer verlangten von der Sparkasse die Rückgängigmachung des Kaufvertrags mit der Begründung, daß die Annahme einer Hauszinssteuer in der bisher eingeforderten Höhe für die Bemessung des Kaufpreises maßgebend gewesen sei. Die Sparkasse wurde ver-

urteilt und leistete dem Urteil Folge. Für den ihr dadurch entstandenen Schaden machte sie die Stadt haftbar. Ihre Klage wurde in allen Rechtszügen abgewiesen. Die Führung der Steuerrolle ist eine Ausübung öffentlicher Gewalt; die damit betrauten Beamten sind dienstlich verpflichtet, in der Steuerrolle wie in den Steuerzetteln den Sollbetrag richtig anzugeben. Diese Pflicht besteht aber nur gegenüber dem Staat als dem Steuergläubiger. Der Steuerpflichtige kann allerdings verlangen, daß er nicht durch eine unbegründete Steuererhebung betroffen wird. Ob in einer unrichtigen Veranlagung und Anforderung die Verletzung einer dem Steuerschuldner gegenüber obliegenden Amtspflicht erblickt werden könnte, brauchte aber nicht entschieden zu werden, da weder der Steuerrolle noch dem Steuerzettel die Bedeutung einer unter allen Umständen verbindlichen amtlichen Erklärung zukommt. Daher besteht keine Amtspflicht, durch die Führung der Steuerrolle und die Übersendung der Steuerzettel verbindliche Unterlagen für den Grundstückverkehr zu schaffen. Die bloße Übung, bei Ermittlung der Grundstückslasten die Hauszinssteuer nach den Angaben der Steuerzettel zu berücksichtigen, ist nicht geeignet, die Amtspflichten der mit der Einziehung der Steuern betrauten Beamten zu erweitern und ihnen eine besondere Fürsorgepflicht gegenüber den Steuerschuldnern aufzuerlegen. Anders liegt die Sache bei den Grundsteuerkatastern; diese bilden die gesetzliche Grundlage für die dem Rechtsverkehr dienenden Eintragungen in das Grundbuch, und die Verpflichtung ordnungsmäßiger Führung besteht darum auch gegenüber den am Grundstückverkehr beteiligten Personen. (III, 1. November 1940.)

Amtl. Sammlg. S. 257—260.

Vgl. Bd. 108 S. 102; Bd. 118 S. 241; Bd. 148 S. 375; JW. 1938 S. 4.

39. **Offene Handelsgesellschaft. Kommanditgesellschaft. Stille Gesellschaft. Innengesellschaft. Gestaltungsurteil. Schadenersatz. Wiederherstellung.** (HGB. §§ 105, 133, 140, 142, 161, 331; BGB. §§ 249, 262, 286.) Der Kaufmann J. geriet in Zahlungsschwierigkeiten. Er schloß deshalb im Februar 1935 mit R. einen Vertrag, wonach ihm dieser ein Betriebsdarlehen von 20000 *RM* zur Verfügung stellte; dafür bot er ihm an, daß er bis zum 1. Januar

1937 als Teilhaber in seine Firma, die dann in eine offene Handelsgesellschaft oder Kommanditgesellschaft umgewandelt werden sollte, eintreten könne. In einem zweiten Vertrag wurde bestimmt, daß R. sofort als Prokurist bei J. angestellt werde und als solcher Gehalt sowie Anteil am Reingewinn erhalten solle. R. erklärte rechtzeitig die Annahme des im ersten Vertrag gemachten Angebots; da aber eine Einigung über die Einzelheiten nicht zustande kam, blieb er weiter als Prokurist tätig. Im Jahr 1939 wurde J. wegen Vergehen nach § 175 StGB. zu einer Gefängnisstrafe von fünf Monaten verurteilt. Als R. davon erfuhr, verlangte er die Auflösung der Gesellschaft und die Zustimmung des J. zur Übernahme des Geschäfts. Seine Klage auf Anerkennung seines Rechtes, das Geschäft zu übernehmen, und Verurteilung des Beklagten, ihm das Geschäft zu überlassen, wurde in zwei Rechtszügen abgewiesen; das Reichsgericht bestätigte das Urteil hinsichtlich des ersten Antrags, verwies aber hinsichtlich des zweiten Antrags die Sache zurück. Die Klage war darauf gestützt, daß eine Handelsgesellschaft, sei es eine offene Handelsgesellschaft oder eine Kommanditgesellschaft, zwischen den Parteien als Gesellschaftern schon bestehe und daß zufolge der Verurteilung des Beklagten der Kläger das Recht habe, dessen Ausscheiden zu verlangen. Mit Recht hatte das Berufungsgericht diese Auffassung abgelehnt. Nach den abgeschlossenen Verträgen bestand Übereinstimmung darüber, daß der Beklagte vorerst Alleininhaber des Geschäfts bleibe; wenn der Kläger auch Anteil am Reingewinn erhielt, wurde damit nur eine stille Gesellschaft begründet. Die Revision führte demgegenüber aus, daß zwar im Außenverhältnis der Beklagte noch weiter als Geschäftsinhaber erschienen sei, daß aber nach der Annahme des Angebots durch den Kläger eine Innengesellschaft zwischen den Parteien bestanden habe. Hierin lag eine Verkennung des Wesens der offenen Handelsgesellschaft. Eine solche kann nur in der Weise errichtet werden, daß das Gesellschaftsverhältnis nach außen hervortritt und daß das Betriebsvermögen allen Gesellschaftern zur gesamten Hand gehört. Allerdings ist eine Innengesellschaft in der Weise möglich, daß vereinbart wird, im Innenverhältnis solle es so gehalten werden, als ob eine offene Handelsgesellschaft bestünde. Eine solche Vereinbarung würde aber nur eine schuldrechtliche Verpflichtung der Gesellschafter gegeneiander begründen und keinen Anspruch auf Auflösung der Gesellschaft und Überlassung des Gesellschaftsvermögens an den einen Teilhaber ergeben. Eine Abrede dieser Art kam im vorliegenden Falle auch gar nicht in Betracht, da der Kläger sich die Entscheidung darüber vorbehalten hatte, in welcher Form er sich an dem Geschäft des Beklagten beteiligen wollte. Auch die Frage, ob durch die Annahme des Angebots eine Kommanditgesellschaft begründet worden sei, war zu verneinen; hiefür hätte es noch eines besonderen Vertrages bedurft, da es im abgeschlossenen Vertrage noch offen gelassen war, ob der Kläger in einer Kommanditgesellschaft persönlich haftender Gesellschafter oder Kommanditist sein sollte. Der Vertrag war nur als ein Vorvertrag anzusehen; zum Abschluß des eigentlichen Gesellschaftsvertrages war es nicht gekommen. Zu Unrecht hatte aber das Berufungsgericht angenommen, daß der Beklagte zwar verpflichtet sei, dem Kläger den durch seinen Verzug entstandenen Schaden zu ersetzen, daß aber die Herstellung des Gesellschaftsverhältnisses für einen in der Vergangenheit liegenden Zeitpunkt unmöglich sei und daß der Kläger demgemäß nur Geldersatz beanspruchen könne. Unter Wiederherstellung im Sinne des § 249 BGB. ist nicht die Herstellung genau des gleichen Zustandes zu verstehen, wie er vor dem Eintritt des schädigenden Ereignisses bestanden hat oder ohne dessen Eintritt bestehen würde. Der Schadenersatzberechtigte kann vielmehr unter Umständen auch etwas fordern, was vorher nicht dagewesen ist, falls er es nur ohne das zum Schadenersatz verpflichtende Verhalten des Schädigers nach Erfahrungsgrundsätzen mit aller Wahrscheinlichkeit erlangt haben würde. Der Verzug des Beklagten konnte allerdings dem Kläger nicht ohne weiteres einen Anspruch aus § 142 HGB. verschaffen; das schloß aber nicht aus, daß der Beklagte verpflichtet war, dem Kläger die Rechtsstellung einzuräumen, die dieser ohne den Verzug gehabt hätte, also ihm das Geschäft zu überlassen, falls er ohne Verzug des Beklagten einen entsprechenden Gestaltungsanspruch gehabt hätte. Auf diese Weise ist als Folge des Verzugs auch eine Umwandlung des ursprünglichen Anspruchs (auf Aufnahme als Gesellschafter) in einen anderen Anspruch (auf

Überlassung des Geschäfts) möglich. Daneben bleibt noch der ursprüngliche Erfüllungsanspruch bestehen, da es in das Belieben des Gläubigers gestellt ist, ob er Erfüllungsfolgen geltend machen will. Eine andere Frage war, ob das Ereignis, das den Kläger berechtigt haben würde, das Übernahmeverlangen zu stellen, diese schwerwiegende Bedeutung auch hatte, wenn er zur Zeit des Ereignisses erst einen Anspruch darauf hatte, Gesellschafter zu werden. Das Reichsgericht erklärte, daß das im vorliegenden Falle, wo den Anlaß zur Geltendmachung des Übernahmeanspruchs eine Verfehlung des Beklagten gab, die mit seiner Treupflicht nicht unmittelbar zusammenhing, außer Zweifel stehe. Waren aber die Voraussetzungen des Übernahmeverlangens nach § 142 HGB. an sich gegeben, so war der Kläger auch berechtigt, aus dem Gesichtspunkte des Verzuges nicht nur als Entschädigung für die verspätete Erfüllung, sondern auch als Erfüllung unmittelbar die Einräumung der Rechtsstellung zu fordern, die er bei rechtzeitiger Erfüllung erlangt haben würde. (II, 27. November 1940.) Amtl. Sammlg. S. 260--271.

Vgl. Bd. 76 S. 147; Bd. 86 S. 143; Bd. 108 S. 59; Bd. 112 S. 283; Bd. 142 S. 19; Bd. 143 S. 274; JW. 1938 S. 2212.

40. Firmenrecht. Verwechslungsgefahr. (HGB. § 19; UnlWG. § 16; BGB. § 12.) Eine Gesellschaft mit beschränkter Haftung, welche die Firma „Gebr. Kn., Nieten- und Schraubenfabrik G. m. b. H." führte, stellte in ihrer Fabrik u. a. Präzisionsschrauben her. Einer ihrer Angestellten, der den Namen Kn. führte, schied aus ihren Diensten aus und gründete eine Kommanditgesellschaft, welche die Firma „Kommanditgesellschaft Kn. und Co., Fabrik für Präzisionsschrauben" führte. Die G. m. b. H. hatte ihren Sitz in A., die Kommanditgesellschaft in K.; beide Orte waren nicht weit voneinander entfernt. Die G. m. b. H. klagte gegen die Kommanditgesellschaft und ihren persönlich haftenden Gesellschafter mit dem Antrag, die Kommanditgesellschaft zu verurteilen, die Führung ihrer Firma oder der Unternehmenbezeichnung „Fabrik für Präzisionsschrauben" bei Strafvermeidung zu unterlassen und in die Löschung der Firma im Handelsregister einzuwilligen, ferner die Kommanditgesellschaft und den geschäftsführenden Gesellschafter zum Ersatz allen Schadens zu verurteilen, der ihr (der Klägerin) durch den Gebrauch der Firma und der Unternehmenbezeichnung entstanden sei und etwa noch entstehen werde. Das Landgericht wies die Klage ab. Im Berufungsverfahren erklärte sich die Kommanditgesellschaft auf einen Hilfsantrag der Klägerin bereit, ihre Firma nur in der Form zu führen: „K.er Präzisionsschrauben Arnold Kn. und Co. KommGes.“. In diesem Sinne wurde sie verurteilt; alle Kosten wurden der Klägerin auferlegt. Die Revision der Klägerin hatte keinen Erfolg. Das Berufungsgericht hatte Verwechslungsgefahr angenommen. Der Name Kn. kam zwar in der Umgebung der beiden Fabriken öfter vor, war aber im übrigen nicht häufig; die Firma der in besonderem Ansehen stehenden Klägerin und jene der Beklagten seien so wenig voneinander verschieden, daß die Unterschiede von der Allgemeinheit nicht beachtet und nicht genügend bemerkt würden; der Umstand, daß die Fabriken ihren Sitz an verschiedenen Orten hatten, sei von keiner wesentlichen Bedeutung, weil beide Orte in demselben gleichartigen Industriegebiet lägen und auch leicht der Gedanke aufkommen könne, daß es sich bei der beklagten Gesellschaft um eine Zweigstelle der Klägerin handle; die Verwechslungsgefahr sei um so größer, als beide Fabriken dieselbe Ware herstellten. Immerhin hielt das Berufungsgericht die Forderung der Klägerin für zu weitgehend. Der Zweitbeklagte habe den ihm zustehenden Familiennamen ordnungsgemäß in die Firma der von ihm begründeten Kommanditgesellschaft aufgenommen; dafür, daß er das in unlauterer Absicht getan habe, fehle es an genügendem Anhalt; bei billiger Abwägung der Belange beider Parteien sei die Klägerin genügend gesichert, wenn die beklagte Gesellschaft ihre Firma und ihre Geschäftsbezeichnung mit einem so unterscheidungskräftigen Zusatz versehe, daß eine Verwechslungsgefahr so gut wie ausgeschlossen sei. Diesen Ausführungen trat das Reichsgericht im wesentlichen bei. Dem Zweitbeklagten konnte nicht das Recht bestritten werden, nach seinem Austritt aus seiner bisherigen Stellung bei der Klägerin sich auf dem Gebiete selbständig zu machen, das ihm durch seine bisherige Tätigkeit und die Überlieferung in seiner Familie vertraut war. Die Klägerin hatte auf dem Gebiete der Erzeugung von Präzisionsschrauben kein Sonderrecht; weder Patente noch Warenzeichenrechte standen ihr zur Seite; sie hatte auch keinen Ausstattungsbesitz behauptet. Der Name „Kn.Schrauben" war keine Sonderbezeichnung für ihre Präzisionsschrauben um irgendeiner Sondergestaltung ihrer Erzeugnisse willen, sondern ein Sammelname für alle von ihr erzeugten Schrauben. Der Zweitbeklagte war verpflichtet, als einziger persönlich haftender Gesellschafter der Kommanditgesellschaft seinen Geschlechtsnamen in die Firma der Gesellschaft aufzunehmen; die Beifügung des Vornamens ist nicht vorgeschrieben. Der Name Kn. war nicht der Klägerin eigentümlich, mochte sie auch die bekannteste Kn.Firma sein. Wenn die Klägerin diesen Namen, der gar nicht mit der Wahrheit ihres Gesellschafterbestandes übereinstimmte, fortführen wollte, mußte sie es in den Kauf nehmen, daß sich aus dem rechtmäßigen Vorhandensein dieses Namens und seinem Auftreten im Geschäftsleben Unzuträglichkeiten ergaben. Sie konnte nicht verlangen, daß der wirkliche Namensträger es unterließ, von einer wettbewerblichen Betätigung abzusehen oder zur Vermeidung von Verwechslungen eine Unternehmensform zu wählen, die es ihm ermöglichte, die Aufnahme seines Namens in die Firma zu unterlassen. Dafür, daß die Beklagten im geschäftlichen Verkehr den Namen Kn. als ihr Firmenschlagwort ausgegeben und verbreitet hätten, fehlte jeder Nachweis; belanglos war es, wenn die Beklagten im täglichen Verkehr sich lediglich als Kn. bezeichneten, da dies zu Verwechslungen nur dann führen konnte, wo die Unterscheidung keine Rolle spielte. Die Abnehmer der beiden Parteien waren Großabnehmer, denen die Unterscheidung der beiden Firmen, zumal bei verschiedenem Sitz, durchaus geläufig war. Nach den Gepflogenheiten des Handelsstandes genügt sogar für Firmen am gleichen Orte, die denselben Familiennamen führen, die Beifügung eines anderen Vornamens als genügende Unterscheidung. Auch im Kostenpunkt konnte die Revision keinen Erfolg haben. Der ursprüngliche Antrag der Klägerin enthielt eine Zuvielforderung, gegen die sich die Beklagten wehren durften. Ob sie dabei zu weit gingen, indem sie jede Änderung ihrer Firma zunächst ablehnten, war für die Kostenfrage belanglos; als die Klägerin sich zu dem Hilfsantrage

verstand, hatten sie diesem sofort zugestimmt. (II, 28. November 1940.) Amtl. Sammlg. S. 271—288.
Vgl. Bd. 116 S. 209.

41. **Bergbau. Wasserlieferung an Gemeinden. Unzulässige Rechtsausübung. Klageänderung.** (PrABG. §§ 144, 148; ZPO. §§ 253, 268.) Gelegentlich der Enteignung eines Sees, der den Grubenbau einer Berggewerkschaft gefährdete, wurde durch einen gemeinsamen Beschluß des Oberbergamts und des zuständigen Bezirksausschusses der Gewerkschaft auferlegt, einer Gemeinde, deren Grundwasserverhältnisse durch den Bergbau beeinträchtigt wurden, das Trink- und Wirtschaftswasser in ausreichender Menge und zweckdienlicher Beschaffenheit zu liefern, soweit es für sie nach der im Einvernehmen mit dem Oberbergamt zu treffenden Entscheidung der Landespolizeibehörde erforderlich wäre. Die Gewerkschaft kam dieser Verpflichtung in der Weise nach, daß sie aus ihrem Wasserwerk das Wasser in einer behördlich genehmigten Rohrleitung mit mehreren Schöpfstellen in den Ort führte, während die Gemeinde zwecks Weiterführung des Wassers in die Häuser ein Leitungsnetz anlegte; am Eingange des Ortes war ein Wassermesser aufgestellt. Die zu liefernde Menge wurde im Jahr 1893 auf 70 cbm täglich festgesetzt, im Jahr 1924 auf 90 cbm und im Jahr 1939 auf 110 cbm erhöht. Im Jahr 1926 beschloß die Gemeinde, den über die festgesetzte Menge hinausgehenden Mehrverbrauch mit 20 *Rpf* für den Kubikmeter zu entschädigen. Da sich in den Jahren 1926 bis 1932 die in der Gemeinde verbrauchte Wassermenge unter der Pflichtleistung hielt, trat dieser Beschluß nicht in Wirksamkeit. Auch in der Folgezeit wurde die Mehrlieferung nicht vergütet. Die Gewerkschaft erhob Klage auf Zahlung der ihr nach dem Beschluß zukommenden Beträge; im zweiten Rechtsgange beantragte sie außerdem die Feststellung, daß die Gemeinde verpflichtet sei, für die Zukunft ein vom Gericht festzusetzendes Entgelt für die Mehrlieferung zu zahlen. In zwei Rechtszügen wurde die Klage abgewiesen. Das Reichsgericht verwies die Sache zurück. Der Beschluß vom Jahr 1926 entsprach allerdings nicht den Formvorschriften der maßgebenden Landgemeindeordnung, da es sich um kein Geschäft der laufenden Verwaltung handelte; die Festlegung von Preisen für Dauerlieferungen kehrt

in der Verwaltung einer Landgemeinde nicht regelmäßig wieder und ist auch nicht von unwesentlicher Bedeutung für deren Geldwesen. Für den Einwand unzulässiger Rechtsausübung war kein Platz. Es verstößt auch nicht gegen die guten Sitten und gesundes Volksempfinden, wenn eine öffentlichrechtliche Körperschaft, selbst bei jahrelanger Durchführung eines Vertrags, sich auf die Nichteinhaltung der zu ihrem Schutz erlassenen Formvorschriften beruft, es sei denn, daß besondere Umstände vorliegen. Eines näheren Eingehens auf diese Ausführungen bedurfte es nicht, da die Abweisung des von der Klägerin erhobenen Anspruchs aus anderen Gründen nicht gerechtfertigt war. Das Berufungsgericht hatte nicht geprüft, ob der Anspruch der Klägerin auf Bezahlung der Mehrlieferung, wenn auch nicht zu einem bestimmten Preis, so doch mit einem angemessenen Betrag, sich nicht schon ohne besondere Vereinbarung aus dem durch den Enteignungsbeschluß begründeten und durch die Parteien ausgestalteten Rechtsverhältnis über die Lieferung des Wassers ergab. Die der Klägerin auferlegte Verpflichtung war ein Ausgleich des Schadens, der der Beklagten durch den Bergbau der Klägerin entstanden war und noch weiter zu entstehen drohte. Dem stand nicht entgegen, daß die Schadensregelung nicht im Rechtswege getroffen, sondern von den Enteignungsbehörden festgelegt worden war. Die Enteignungsbehörden können dem Bergbauunternehmer auch zugunsten dritter, von der Enteignung nicht unmittelbar betroffener Personen bestimmte Auflagen machen, die mit den infolge der Enteignung zu treffenden Maßnahmen im Zusammenhang stehen. Lassen die Beteiligten eine solche Regelung rechtskräftig werden, so ist sie ebenso bindend wie eine innerhalb des Verfahrens geschlossene Vereinbarung. Umfang und Inhalt der Verpflichtung ergeben sich, soweit sie nicht ausdrücklich festgelegt sind, zunächst aus der rechtlichen Grundlage, auf der sie beruht, nämlich der Schadenersatzpflicht, weiter aber auch aus der Ausgestaltung des Rechtsverhältnisses durch die tatsächliche Handhabung der Beteiligten in der späteren Zeit. Ein unbeschränkter Anspruch auf Lieferung von Wasser konnte aus der im Berggesetz begründeten Schadenersatzpflicht der Klägerin selbst dann nicht hergeleitet werden, wenn die früheren Grundwasserverhältnisse eine solche übermäßige Verwendung zugelassen hätten. Der Enteignungsbeschluß trug allen zulässigen Ansprüchen auf künftige Erweiterung der Wasserlieferungspflicht dadurch Rechnung, daß er die Höchstmenge des zu liefernden Wassers nicht ein für allemal festlegte, sondern die Bestimmung dieser Grenze der Landespolizeibehörde überließ. Deren Entscheidungen waren für die Parteien bindend. Diese Höchstgrenze konnte aber nicht die Bedeutung haben, daß die Klägerin Mehrlieferungen schlechthin ablehnen durfte. Der Wasserverbrauch in landwirtschaftlichen Betrieben ist sehr ungleich; auch muß aus Gründen des Feuerschutzes die Möglichkeit bestehen, die Wasserentnahme plötzlich stark ansteigen zu lassen. Da anderseits die Klägerin nach dem Sinn und Zweck der ihr gemachten Auflage als eines Schadensausgleichs nur die von der Landespolizeibehörde festgesetzten Mengen unentgeltlich zu liefern hatte, folgte aus dem Wesen des Lieferungsverhältnisses, wie es durch den Enteigungsbeschluß begründet und durch die ständige Handhabung inhaltlich näher bestimmt war, die Verpflichtung der Beklagten, die über die Pflichtmenge hinaus bezogenen Wassermengen zu bezahlen. Wenn auch der Beschluß vom Jahr 1926 keinen selbständigen Verpflichtungsgrund bildete, weil er den für die Gemeinde damals geltenden Formvorschriften nicht entsprach, lag doch in ihm eine Bestätigung dafür, daß die Verpflichtung zur Bezahlung des Mehrverbrauchs Inhalt des zwischen den Parteien bestehenden Lieferverhältnisses war. Dem stand nicht entgegen, daß der Preis für den Mehrverbrauch nicht sofort festgelegt wurde. Diese Festlegung konnte späterer Einigung vorbehalten bleiben; kam eine solche nicht zustande, so war äußerstenfalls der angemessene Preis durch die Gerichte zu bestimmen. — Im zweiten Rechtsgange hatte die Klägerin Feststellung beantragt, daß die Beklagte verpflichtet sei, soweit sie künftig über die auf Grund des Enteignungsbeschlusses festgesetzte Pflichtmenge Wasser aus der Zuleitung der Klägerin entnehme, ein vom Gericht festzusetzendes angemessenes Entgelt zu zahlen. Das Berufungsgericht hatte in diesem Antrag eine unzulässige Klageänderung erblickt. Das Reichsgericht trat dieser Auffassung entgegen. Der Anspruch ging bei diesem Antrag wie bei dem ursprünglichen Antrag auf Bezahlung der über die Pflichtmenge hinaus bezogenen Wassermenge; auch die tatsächliche

Grundlage für beide Ansprüche war die gleiche. Der Antrag genügte auch den Anforderungen des § 253 ZPO. Die Rechtsprechung hat Anträge, welche die Festsetzung eines Betrages dem Ermessen des Gerichts oder eines Sachverständigen anheimstellen, stets zugelassen, wenn das sonstige Vorbringen genügende Grundlagen für die Anwendung des Ermessens bot. Im vorliegenden Fall ergab sich die Höchstgrenze des Anspruchs der Klägerin aus dem Zahlungsantrag und ihren späteren Ausführungen; der angemessene Preis konnte auch durch Einholung von Auskünften bei anderen Unternehmungen, die eine ähnliche Verpflichtung zu erfüllen hatten, ermittelt werden. Die Voraussetzungen für eine Feststellungsklage waren gegeben, da die beklagte Gemeinde sich immer wieder darauf berief, daß die Klägerin ihr und ihren Mitgliedern alles Wasser, dessen sie bedürfte, unentgeltlich liefern müsse; unter diesen Umständen konnte ein rechtliches Interesse der Klägerin, klare Verhältnisse über ihren Anspruch auf Bezahlung des Mehrverbrauchs zu erhalten, nicht verneint werden. (III, 15. November 1940.)

Amtl. Sammlg. S. 289—300.

42. **Milchwirtschaft. Auftrag. Geschäftsführung. Rechtsweg.** (BGB. §§ 662, 670; GVG. § 13; MilchwirtschVO. vom 22. Februar 1935, 17. April 1936, 29. Juli 1938.) Eine Gesellschaft mit beschränkter Haftung betrieb eine Trockenmilchfabrik; eine Anzahl Landwirte waren ihr zufolge Zuweisung des Milchwirtschaftsverbandes der Provinz ablieferungspflichtig. Im November 1937 wurde über sie das gerichtliche Vergleichsverfahren eröffnet. Der Milchwirtschaftsverband gab daraufhin das Einzugsgebiet frei, so daß es den Lieferanten der G. m. b. H. freistand, auf eigene Gefahr an sie weiter zu liefern, wenn sie sich nicht einem anderen Betrieb als Lieferanten zuweisen ließen. Alle suchten um neue Zuweisung nach; ein Teil wurde einer Aktiengesellschaft überwiesen. Hinsichtlich der rückständigen Forderungen der Lieferanten gegen die G. m. b. H. erklärte deren Geschäftsführer im November 1937 dem Verbandsvorsitzenden, sie werde dafür aufkommen und, falls die Aktiengesellschaft die Auszahlung übernehme, Ersatz leisten. Die Verhandlungen des Verbandsvorsitzenden mit der Aktiengesellschaft führten zu einer Vereinbarung, wonach die Aktiengesellschaft die Zahlung der Rückstände

übernahm; zum Ausgleich wurde sie vom Verbandsvorsitzenden ermächtigt, einen Reichspfennig für jedes Kilogramm von der Lieferantenmilch in Abzug zu bringen; die Zustimmung der Generaldirektion der Aktiengesellschaft wurde vorbehalten. Auf Grund dieser Vereinbarung zahlte die Aktiengesellschaft an die Lieferanten der G .m. b. H. mehr als 30000 *RM*. Die Hälfte davon bestritt sie aus der ihr vom Verbande bewilligten Ausgleichsgabe; für den Rest blieb sie unbefriedigt, weil die G. m. b. H. nach Aufhebung des Vergleichsverfahrens ihren Betrieb wieder aufgenommen und die Lieferanten wieder zugewiesen erhalten hatte. Der Vergleichsverwalter hatte sich im November 1938 verpflichtet, der Aktiengesellschaft die auf die Rückstände geleisteten Zahlungen zu erstatten. Die Aktiengesellschaft verlangte Ersatz vom Milchwirtschaftsverband. Nachdem dessen Rechte und Pflichten zufolge der Verordnung über den Zusammenschluß der deutschen Milch- und Fettwirtschaft vom 29. Juli 1938 auf die Hauptvereinigung der deutschen Milch- und Fettwirtschaft übergegangen waren, klagte diese gegen die G. m. b. H. mit dem Antrag, daß diese ihr den Erstattungsanspruch der Aktiengesellschaft von der Hand halte. Das Landgericht wies die Klage wegen Unzulässigkeit des Rechtswegs ab; das Berufungsgericht und das Reichsgericht erklärten den Rechtsweg für zulässig. Allerdings hatte der Milchwirtschaftsverband in der vorliegenden Sache Anordnungen hoheitlicher Art getroffen, wie sie ihm nach den milchwirtschaftlichen Verordnungen vom 17. April 1936 und 29. Juli 1938 zustanden; dahin gehörten die Freigabe des Einzugsgebietes der Beklagten, die Stillegung ihres Betriebs, die Zuweisungen von Lieferanten an die Aktiengesellschaft und die Genehmigung der Erhebung der Ausgleichsabgabe. Wenn er aber der Aktiengesellschaft zumutete, die rückständigen Schulden der G. m. b. H. wegzufertigen, konnte er das nicht durch eine obrigkeitliche Anordnung tun, war vielmehr auf eine Vereinbarung mit der Aktiengesellschaft angewiesen. Daß es sich bei der Vereinbarung nicht um eine obrigkeitliche Anordnung, sondern um einen bürgerlichrechtlichen Vertrag handelte, ergab sich schon aus dem Vorbehalt, in dem die Wirksamkeit von der Zustimmung der Generaldirektion abhängig gemacht wurde. Das wäre mit einer obrigkeitlichen Anordnung schlechthin unverträglich gewesen. Der Vertrag war als ein Auftrag anzusehen; daraus ergab sich die Verpflichtung des Verbandes, der Aktiengesellschaft zu ersetzen, was sie zur Ausführung des Auftrags aufwendete. Daran änderte es nichts, daß der Verband, um dieser Pflicht nachzukommen, seine obrigkeitlichen Befugnisse zur Hilfe nahm, indem er der Aktiengesellschaft erlaubte, von der Ausgleichsabgabe einen Reichspfennig für jedes Kilogramm der Lieferantenmilch abzuziehen. Das betraf nur die Art, wie er sich seiner Verpflichtung entledigte, nicht die Verpflichtung selbst. Auch das Rechtsverhältnis zwischen der Klägerin und der Beklagten war bürgerlichen Rechts, mochte es nach den Regeln des Auftrags, der Geschäftsführung ohne Auftrag, der ungerechtfertigten Bereicherung, des Schuldanerkenntnisses oder nach sonst einem bürgerlichrechtlichen Gesichtspunkte zu beurteilen sein. Der Rechtsweg war also zulässig, wenn nicht für Streitigkeiten der vorliegenden Art die Marktschiedsgerichte als besondere Gerichte bestellt sind. Nach der Verordnung vom 26. Februar 1935 können diese angerufen werden bei allen Streitigkeiten zwischen Mitgliedern der Zusammenschlüsse, die aus Maßnahmen der Zusammenschlüsse mit unmittelbarer Wirkung gegen beide Parteien entstehen. Das Landgericht hatte angenommen, daß die Schiedsgerichte, wenn sie für Streitigkeiten zwischen Mitgliedern der Zusammenschlüsse zuständig seien, erst recht zuständig sein müßten, wenn eine Partei der Zusammenschluß selbst sei. Das erklärte das Reichsgericht für einen Fehlschluß. Für Streitigkeiten zwischen dem Zusammenschluß und seinen Mitgliedern sind die Schiedsgerichte nur soweit zuständig, als dies ausdrücklich vorgesehen ist. Die Streitigkeit war auch nicht aus einer Maßnahme des Wirtschaftsverbandes entstanden, sondern aus der Vereinbarung, die zwischen dem Verbandsvorsitzenden und der Aktiengesellschaft getroffen worden war. (VI, 4. Dezember 1940.) Amtl. Sammlg. S. 300—307. Vgl. Bd. 156 S. 279; Bd. 157 S. 106, 115; Bd. 162 S. 37.

43. Vaterschaftsklage. Urteilszustellung. (ZPO. §§ 625, 640.) Eine Klage auf Feststellung der Vaterschaft wurde abgewiesen; das Urteil wurde im Parteibetriebe zugestellt. **Die Berufung des Klägers wurde mit der Begrün-**

dung verworfen, daß sie verspätet eingelegt worden sei. Die sofortige Beschwerde des Klägers hatte Erfolg. Das Berufungsgericht hatte angenommen, daß nur ein der Klage stattgebendes Urteil der amtlichen Zustellung bedürfe, da nur ein solches eine neue Rechtslage schaffe. § 625 ZPO., der gemäß § 640 ZPO. entsprechend anzuwenden ist, verlangt allerdings amtliche Zustellung nur für Urteile, welche die Scheidung oder Nichtigkeit der Ehe aussprechen. Das liegt daran, daß nur durch solche Urteile der bestehende Rechtszustand — nämlich der, daß die durch die Personenstandsregister ausgewiesene Ehe von jedermann als bestehend behandelt werden muß — eine Änderung erfährt. Das gleiche trifft für die Ehelichkeitsanfechtung zu, da bis zur erfolgreichen Anfechtung jeder die Ehelichkeit eines von einer Ehefrau geborenen Kindes anerkennen muß; die Zustellung von Amts wegen ist nur notwendig, wenn die Ehelichkeit verneint wird. Bei der Abstammungsfeststellungsklage besteht aber bis zur Entscheidung kein allgemein anzuerkennender Rechtszustand; dieser wird erst durch das Urteil geschaffen, und zwar sowohl durch ein solches, das die Abstammung feststellt, als durch ein solches, das sie verneint. Beide müssen von Amts wegen zugestellt werden. Zweifel könnten bei Urteilen entstehen, welche die Klage abweisen, weil die Abstammung nicht geklärt ist; es würde aber zu unnötigen Schwierigkeiten führen, wenn für solche Urteile die Notwendigkeit amtlicher Zustellung anerkannt würde. (IV, 9. Dezember 1940.) Amtl. Sammlg. S. 307—309.

44. Reichsautobahn. Unternehmer. Mittelbare Verpflichtung. Typische Bedingungen. (BGB. § 278.) Der Fuhrunternehmer F. wurde von der Reichsautobahn mit der Ausführung von Fahrten beschäftigt. Auf seinen Fahrzeugen beförderte er auch, ohne dazu verpflichtet zu sein, Arbeiter des für die Reichsautobahn tätigen Gärtners R. zu ihrer Arbeitsstätte. Ursprünglich erhielt er dafür keine Vergütung; später wurde ihm eine solche nach den jeweils aufgewendeten Lastkraftwagenstunden ausbezahlt. Ein mit solchen Arbeitern beladener Wagen stieß durch eine Nachlässigkeit seines Führers mit einem Wagen der städtischen Kleinbahn zusammen. Die Stadt wurde auf Grund des Reichshaftpflichtgesetzes sowohl von den verletzten Arbeitern als von den für sie eingetretenen Trägern der Reichsversicherung auf Ersatz der durch den Unfall entstandenen Schäden in Anspruch genommen und verlangte mit Klage von der Reichsautobahn Erstattung der von ihr geleisteten und noch künftig zu leistenden Zahlungen. Die Klage wurde damit begründet, daß die Reichsautobahn als Halterin des Kraftwagens anzusehen sei und daß sie auch die Pflicht gehabt habe, die Arbeiter sicher an ihre Arbeitsstätte zu befördern; das Verschulden des Wagenführers falle auch ihr zu Last. Die Beklagte bestritt, Halterin des Kraftwagens gewesen zu sein oder irgendetwas mit der Beförderung der Arbeiter zu ihrer Arbeitsstätte zu tun gehabt zu haben. Das Landgericht gab der Klage vollkommen, das Berufungsgericht zum Teil statt; das Reichsgericht wies sie ab. Als Halterin des Wagens, der den Zusammenstoß herbeiführte, hatte auch das Berufungsgericht die Beklagte nicht angesehen; es hatte aber angenommen, daß sie den Arbeitern gegenüber eine Verpflichtung zur sicheren Beförderung an die Arbeitsstätte gehabt habe. Dafür lag kein Anhaltspunkt vor. Als die Fuhrunternehmer die Arbeiter auf ihren Lastwagen mitnahmen, handelte es sich zunächst nur um eine Gefälligkeit gegenüber den Arbeitern, denen ein Anspruch auf Beförderung zur Baustelle nicht zustand. Da aber auch die Beklagte einen Vorteil davon hatte, wenn die Arbeiter nicht durch einen langen Anmarschweg ermüdet auf der Baustelle erschienen, fand sie sich bereit, den Fuhrunternehmern für diese Beförderung eine Vergütung zu zahlen; sie übernahm damit aber nicht gegenüber den Arbeitgebern, mit denen allen sie im Vertragsverhältnis stand, die Verpflichtung, ihre Arbeiter zur Arbeitsstelle zu befördern. Den Verträgen waren sozialpolitische Bedingungen zugrunde gelegt, wonach die Arbeitgeber dazu beizutragen hatten, den Weg zur Arbeitsstelle für die Arbeiter so kurz und so bequem als möglich zu gestalten; die oberste Bauleitung behielt sich vor, bei Sammeltransporten die erforderlichen Abmachungen mit den Transportunternehmern zu vermitteln. Als typische Vereinbarungen, die in ganz Deutschland den Verträgen der Beklagten zugrunde gelegt wurden, unterlagen diese Bedingungen der freien Nachprüfung des Reichsgerichts. Hienach oblag die Verpflichtung, die Arbeiter zur Arbeitsstelle zu befördern, ihrem Arbeitgeber, also R. Daß ein bei der Beklagten angestellter Straßenmeister sich um die Unterbringung der Arbeiter auf dem Wagen bekümmert hatte, reichte nicht zur Begründung der Annahme, daß die Beklagte an Stelle des R. die Sorge für die Beförderung der Arbeiter als eigene Verpflichtung übernommen habe. Demzufolge war auch der Führer des Wagens weder Erfüllungsgehilfe der Beklagten noch von ihr zu der Verrichtung bestellt, in deren Ausführung er den Arbeitern Schaden zugefügt hatte. (VI, 11. Dezember 1940.)

Amtl. Sammlg. S. 309—314.
Vgl. Bd. 111 S. 278.

45. **Abstammung. Rechtskraft.** (ZPO. §§ 256, 322, 640.) In einem Rechtsstreit wurde festgestellt, daß der Beklagte der außereheliche Vater des Klägers sei. Das Urteil erlangte die Rechtskraft. Mehrere Jahre später klagte der damals Verurteilte auf Feststellung, daß er nicht der Vater des Beklagten sei. In allen Rechtszügen wurde die Klage mit der Begründung abgewiesen, daß das Gegenteil schon rechtskräftig feststehe. Das frühere Urteil stellte nicht nur die Zahlvaterschaft, sondern auch die blutmäßige Vaterschaft des damaligen Beklagten fest. Damit war die Frage der Vaterschaft zwischen den Parteien endgültig entschieden. Darauf, ob diese Frage auch für und gegen alle wirksam entschieden war, kam es nicht an, da der anhängige Rechtsstreit die gleiche Frage zwischen den Parteien erneut austragen sollte. (IV, 16. Dezember 1940.)

Amtl. Sammlg. S. 315—317.

46. **Ehescheidung. Bedingte Aufhebungsklage. Schuldantrag.** (EheG. §§ 42, 60, 61.) Der Scheidungsklage der Frau gegenüber beantragte der Mann, das überwiegende Verschulden der Frau auszusprechen, da diese an einer schweren, ansteckenden Krankheit leide und ihn darüber beim Abschluß der Ehe in Unkenntnis gelassen habe. Das Landgericht schied die Ehe aus Alleinverschulden des Mannes. Das Berufungsgericht verwies die Sache zurück. Der Rekurs der Frau hatte keine Wirkung. Das Ehegesetz hat in wesentlich neuer Gestaltung der Eheanfechtung einige Aufhebungsgründe, die vordem als Nichtigkeitsgründe die Ehe mit rückwirkender Kraft als nicht geschlossen erscheinen ließen, zu Aufhebungsgründen gemacht, welche die Ehe nur mit Wirkung für die Zukunft auflösen. Damit sind die Wirkungen eines erfolgreichen

Aufhebungsbegehrens der Scheidung im wesentlichen gleichgestellt. Auch in der Verschuldungsfrage hat das Gesetz die Aufhebung und die Scheidung eng einander angepaßt. Gegenüber einer Klage auf Scheidung ist eine Widerklage auf Aufhebung der Ehe auch in der bedingten Form zulässig, daß der Anspruch nur für den Fall geltend gemacht wird, daß der Klage stattgegeben werde. Hienach bestehen keine durchschlagenden Gründe, es dem Beklagten zu versagen, dasselbe Ziel durch einen schlichten Schuldantrag zu erreichen. Sowohl im Ziel wie im tatsächlichen Ergebnis laufen der Schuldantrag und die bedingte Widerklage auf dasselbe hinaus: sie erstreben nicht selbständig die Lösung der Ehe, sondern verfolgen nur für den Fall, daß die Ehe auf die Klage hin gelöst wird, das Ziel, daß die Wirkungen dieser Lösung — d. i. im wesentlichen die Schuldfrage — so entschieden werden, wie wenn das Lösungsbegehren auch von dem Beklagten gestellt worden wäre. Es wäre eine rein äußerlich am Wortlaut des Gesetzes haftende Auslegung, wollte man dem verklagten Ehegatten für sein Begehren zwar den Weg der bedingten Aufhebungsklage eröffnen, ihm aber wegen des Fehlens einer entsprechenden Gesetzesverweisung im § 42 EheG. den Schuldantrag entsprechend den §§ 60, 61 EheG. versagen. Folgerichtig muß für diesen Fall auch die in den genannten Vorschriften der Fassung nach nur für die Scheidung getroffene Regelung gelten, daß die Schuldfeststellung als Billigkeitsanspruch auch noch auf solche Umstände gestützt werden kann, die infolge Zeitablauf oder aus anderen Gründen nicht mehr geeignet sind, angriffsweise geltend gemacht zu werden. Eine andere Regelung würde zu Härten und Ungleichheiten führen, besonders dann, wenn der bei Abschluß der Ehe von dem anderen Eheteil getäuschte Ehegatte die ihm angetane Unbill verziehen und die Ehe fortgesetzt hat, dann aber von dem anderen Teil wegen eines vielleicht weniger schwer wiegenden Umstandes auf Aufhebung oder Scheidung verklagt wird. Das muß auch für das Geltungsgebiet des österreichischen Prozeßrechts gelten, wenn dort auch Bedenken gegen die Zulassung einer bedingten Widerklage sich ergeben können; der Weg des schlichten Schuldantrags wird dadurch nicht ausgeschlossen. (IV, 15. Januar 1941.)

Amtl. Sammlg. S. 317—320. Vgl. Bd. 163 S. 315.

47. **Zwangsversteigerung. Zuschlag an einen Geisteskranken.** (ÖstExO. §§ 186, 187.) Bei der Zwangsversteigerung eines Grundstücks erlegte der Meistbietende ein Sparbuch, das ihm sein Sohn für Fälle der Not zur Verfügung gestellt hatte. Da er das Meistbot nicht erlegen konnte, wurde das Grundstück neuerdings versteigert; hiebei wurde ein viel geringerer Erlös erzielt. Für den Ausfall wurde der frühere Meistbietende haftbar gemacht; der hierüber ergangene Beschluß wurde rechtskräftig. Bald nachher wurde der Meistbietende wegen Geistesschwäche entmündigt. Ein Jahr später starb er. Der Nachlaßpfleger und der Sohn klagten gegen eine Hypothekgläubigerin, die sich an das Sparbuch halten wollte, auf Feststellung, daß der Erblasser zur Zeit der ersten Versteigerung geistesschwach gewesen sei und daß darum seine Beteiligung an der Versteigerung, die Versteigerung des Grundstücks und die Erlegung des Sparbuchs ohne rechtliche Wirkung geblieben seien. In zwei Rechtszügen drang die Klage durch. Das Reichsgericht wies sie ab. Darüber, ob ein Meistbot gültig ist oder nicht, hat der Exekutionsrichter im Verfahren über die Erteilung des Zuschlags und, wenn gegen die Erteilung des Zuschlags Rekurs eingelegt worden ist, der Rekursrichter zu entscheiden; ist der Zuschlag rechtskräftig geworden, so ist eine solche Nachprüfung nicht mehr möglich. Im vorliegenden Falle war der Zuschlagbeschluß noch nicht rechtskräftig geworden, weil er weder dem Vormund des Erblassers noch dem Nachlaßpfleger zugestellt worden war. Die Kläger hatten also noch die Möglichkeit, Rekurs einzulegen. (VIII, 9. Dezember 1940.) Amtl. Sammlg. S. 321—323.
Vgl. Bd. 35 S. 390; Bd. 73 S. 194.

48. **Wehrleistung. Schadenersatz. Rechtsweg. Revisionssumme.** (WeimVerf. Art. 131; WehrLeistG. §§ 26, 27; ReichsLeistG. §§ 26, 27; GVG. §§ 13, 71; ZPO. §§ 546; 547; VO. vom 1. September 1939 § 7). Der Eigentümer eines Lastkraftwagens wurde im September 1938 zu Leistungen für die Wehrmacht in Anspruch genommen. Hiebei stieß er mit einem Kraftwagen der Wehrmacht zusammen; sein Wagen wurde beschädigt. Das Reich ersetzte den Sachschaden. Er verlangte auch Erstattung des ihm während der Ausbesserung des Wagens entstandenen Verdienstausfalls und klagte zunächst 1000 *RM*

ein. In zwei Rechtszügen wurde die Klage wegen Unzulässigkeit des Rechtswegs abgewiesen. Das Reichsgericht verwies die Sache zurück. Die Klage war nicht nur auf die Vorschriften des Wehrleistungsgesetzes gestützt, sondern auch darauf, daß der Zusammenstoß durch ein Verschulden des Fahrers des anderen Wagens verursacht worden sei. Das Berufungsgericht hatte angenommen, daß ein Entschädigungsverlangen nach den Vorschriften der § 839 BGB. oder §§ 7, 17 KraftfahrzG. in Verbindung mit § 131 WeimVerf. und §§ 26, 27 WehrLeistG. ausgeschlossen sei. Daraus ergab sich, daß es die Klage nur insoweit als unzulässig behandelt hatte, als diese auf das Wehrleistungsgesetz gestützt war, daß es sie aber wegen Nichtbestehen des Anspruchs als unbegründet abgewiesen hatte, soweit der Anspruch aus den anderen Vorschriften hergeleitet wurde. Da der Klageanspruch die Revisionssumme nicht erreichte, kam für die Nachprüfung des Reichsgerichts nur die Abweisung der Klage wegen Unzulässigkeit des Rechtswegs oder wegen Mangel des Nachweises einer Amtspflichtverletzung in Betracht; soweit der Anspruch auf die Vorschriften des Kraftfahrzeuggesetzes gestützt war, konnte eine Nachprüfung nicht stattfinden (§ 71 Nr. 1, 2 GVG.). Die Annahme des Berufungsgerichts, daß für einen Anspruch aus dem Wehrleistungsgesetz der Rechtsweg nicht offenstehe, traf zu. Damit war aber der Klage der Boden nicht entzogen. Nach § 26 Abs. 2 WehrLeistG. wird für Verluste, Beschädigungen, außergewöhnliche Belästigungen und Haftpflichtschäden, die infolge oder gelegentlich der Leistung ohne grobes Verschulden des Leistungspflichtigen oder seines Beauftragten entstehen und für die ein Ersatz von dritter Seite nicht zu erlangen ist, eine angemessene Entschädigung gewährt. In erster Reihe wird also der Leistungspflichtige, dem ein Schaden zugefügt worden ist, auf den Anspruch gegen den Schädiger verwiesen. Das muß auch gelten, wenn der Schaden durch einen Angehörigen der Wehrmacht verursacht worden ist und der Entschädigungsanspruch sich gegen das Reich wendet. Daran hat auch das Reichsleistungsgesetz vom 1. September 1939 nichts geändert; dieses ist für die Entscheidung maßgebend, da die Zulässigkeit des Rechtswegs nach dem zur Zeit der Entscheidung geltenden Rechte beurteilt werden muß. Die Tatsache, daß die beiden Gesetze dem Wehrleistungspflichtigen einen von ihnen geschaffenen Anspruch auf Entschädigung geben, für diesen aber den Rechtsweg ausschließen, läßt dessen Zulässigkeit für Ansprüche anderer Art unberührt. Es kam auch nicht in Frage, daß der erhobene Anspruch nur zu dem Zweck der Eröffnung des Rechtswegs in das Gewand eines Schadenersatzanspruchs aus Art. 131 WeimVerf. gekleidet worden wäre, da der Sachvortrag des Klägers den ganzen Tatbestand des § 839 BGB. in schlüssiger Weise ergab, insbesondere auch Tatsachen anführte, aus denen das Vorliegen schuldhaften und amtspflichtverletzenden Verhaltens des Führers des Wagens der Wehrmacht abgeleitet wurde. — Das Reich hatte Widerklage auf Zahlung des Sachschadens erhoben, der durch den Zusammenstoß an dem Wagen der Wehrmacht verursacht worden war. Diese Klage war in zwei Rechtszügen als unbegründet abgewiesen worden; die Revision des Reichs wurde als unzulässig verworfen. Das Berufungsgericht hatte den Sachverhalt dahin gewürdigt, daß die Schädigung des Reichs durch den Kläger, wenn er dafür verantwortlich wäre, einen Haftpflichtschaden darstellte, der gelegentlich der Wehrleistung entstanden war; nach § 26 Abs. 2 WehrLeistG. könne der Kläger die Beseitigung dieser Haftpflicht vom Reich verlangen und diese bestehe hier darin, daß dieses unterlasse, die Haftpflicht geltend zu machen. Nicht näher geprüft hatte das Berufungsgericht die Frage, aus welchem Rechtsgrunde überhaupt ein Ersatzanspruch des Reichs gegen den Kläger gegeben sein könnte. Die Klarstellung der rechtlichen Natur dieses Anspruchs war im Rechtsgange der Revision schon deshalb nicht zu entbehren, weil die Revisionssumme von 10000 *RM* auch bei der Widerklage nicht erreicht wurde; eine Zusammenrechnung der Ansprüche der Klage und der Widerklage war nicht zulässig, ergab übrigens auch die Revisionssumme nicht. Die Revision des Beklagten wurde nicht schon dadurch zulässig, daß jene des Klägers in gewissem Umfange zulässig war; ihre Zulässigkeit konnte also nach § 547 Nr. 2 ZPO. nur darauf beruhen, daß der Anspruch zu denen gehörte, für die nach § 71 Abs. 2 GVG. die ausschließliche Zuständigkeit der Landgerichte vorgeschrieben ist. Das traf nicht zu. Das Berufungsgericht hatte nicht geprüft, ob nach den Umständen des Falles die Betätigung des Klägers in Erfüllung seiner Wehrleistungspflicht als Ausübung öffentlicher Gewalt und er somit als Beamter im Sinne des § 839 BGB. anzusehen war. Allerdings regelt diese Vorschrift nur das Verhältnis zwischen dem Beamten und einem geschädigten Dritten, nicht aber das Innenverhältnis zwischen dem Beamten und der öffentlichen Körperschaft, in deren Diensten er tätig gewesen ist; die Haftung für einen Schaden, den er dieser zugefügt hat, muß aus den für dieses Verhältnis maßgebenden Rechtssätzen abgeleitet werden. Für Beamte sind solche Rechtssätze im Beamtengesetz, für Soldaten in rechtsähnlichen Normen enthalten. Für Wehrleistungspflichtige, die außerhalb solcher Dauerverhältnisse für Belange des Reiches vorübergehend in Anspruch genommen werden, scheint ein entsprechender Rechtssatz zu fehlen. Die Pflicht des Wehrleistungspflichtigen schließt aber, wie alle Rechtspflichten, die Forderung sorgsamer Erfüllung und daher die Vermeidung aller Verhaltungsweisen ein, durch die Personen oder Sachen zu Schaden kommen können und dem Reich daher ein Ersatzanspruch erwachsen kann. Wird dieser Pflicht schuldhaft zuwidergehandelt, so wandelt sie sich in eine Verpflichtung zum Ausgleich des dem Reich dadurch verursachten Schadens. Dieser Anspruch wurde vom Reich mit der Widerklage verfolgt. Ob dafür der Rechtsweg offenstand, konnte dahingestellt bleiben; jedenfalls waren für seine Verfolgung nicht ausschließlich die Landgerichte zuständig, so daß hierauf die Zulässigkeit der Revision nicht gestützt werden konnte. Sie ließ sich auch nicht als Anschlußrevision halten. Anschlußrevisionen setzen die Möglichkeit einer in der Verfahrenslage begründeten Beziehung zu dem selbständigen Rechtsmittel des Gegners sowie einen Willen voraus, der diese Abhängigkeit umfaßt. Dem stand schon entgegen, daß das Reich seine Revision schon vor jener des Klägers eingelegt hatte und auch sonst in keiner Weise den Willen kundtat, seine Revision in Abhängigkeit von jener des Klägers gesetzt zu sehen. (III, 13. Dezember 1940.) Amtl. Sammlg. S. 323—336.

Vgl. Bd. 110 S. 160; Bd. 130 S. 401; Bd. 140 S. 415; Bd. 146 S. 244; Bd. 159 S. 247; Bd. 162 S. 230; Bd. 164 S. 1; JW. 1936 S. 815.

49. Ansteckende Erkrankung. Beweis des ersten Anscheins. (BGB. §§ 276, 611, 823; ZPO. §§ 286, 287.) Ein sechsjähriges Mädchen wurde wegen einer Halserkrankung in ein städtisches Krankenhaus eingeliefert und dort an Rachen- und Nasenpolypen operiert. Einige Tage nach der Operation erkrankte sie an Scharlach; in der Folge stellte sich eine doppelseitige Mittelohrentzündung ein, die zur Ertaubung beider Ohren führte. Sie behauptete, durch einen Knaben, der mit ihr in der gleichen Abteilung gelegen hatte, angesteckt worden zu sein. Dieser war mehrere Wochen vorher wegen Scharlacherkrankung in eine andere Abteilung eingeliefert worden, hatte ebenfalls eine Mittelohrentzündung bekommen und war nach deren Operation in die Abteilung verlegt worden, in die am folgenden Tage das Mädchen eingeliefert wurde; aus dieser wurde er nach vier Tagen wieder entfernt, da seine Erkrankung eine besondere Pflege nötig machte. Die Klage auf Feststellung der Verpflichtung der Stadt zum Ersatz allen aus der Scharlachansteckung entstandenen und noch entstehenden Schadens sowie auf ein angemessenes Schmerzensgeld drang in zwei Rechtszügen durch. Das Reichsgericht verwies die Sache zurück, soweit auf Zahlung von Schmerzensgeld erkannt und die Verpflichtung zum Ersatz nicht vermögensrechtlichen Schadens festgestellt worden war; im übrigen wurde die Revision verworfen. Die Untergerichte hatten die Klageansprüche sowohl aus dem Rechtsgrunde der Vertragsverletzung als aus jenem der unerlaubten Handlung als begründet angesehen. Der Nachweis des ursächlichen Zusammenhangs war nach den Grundsätzen über den Beweis des ersten Anscheins für geführt erachtet worden, nachdem mehrere Sachverständige erklärt hatten, daß bei dem Knaben zur Zeit seiner Einlieferung in die Abteilung, in der sich die Klägerin befand, noch Ansteckungsgefahr bestanden habe. Eine ernstliche Möglichkeit, daß die Klägerin schon bei ihrer Einlieferung in das Krankenhaus von anderer Seite mit Scharlach angesteckt worden war, war nicht dargetan worden. Bei dieser Sachlage war die Heranziehung der Grundsätze über den Beweis des ersten Anscheins nicht zu beanstanden. Zwar ist der Verlauf einer Krankheit nicht bloß von dem Vorhandensein bestimmter äußerer Entstehungsursachen, sondern auch von der kör-

perlichen Veranlagung und dem augenblicklichen körper-
lichen Zustande des Kranken abhängig, also von Ursachen,
die dem Arzt nicht immer hinreichend erkennbar sind und
deren Einfluß er nicht immer genau beurteilen kann. Ein
außergewöhnlicher Verlauf einer Krankheit rechtfertigt
also nicht ohne weiteres die Annahme eines Kunstfehlers
oder eines Verschuldens des Arztes, so daß dieser die
Gründe und Ursachen für den regelwidrigen Verlauf der
Krankheit angeben müßte. Indessen sind die ansteckenden Krankheiten Massenerscheinungen, die in leichteren
Fällen, und namentlich in ihren Anfangserscheinungen,
im allgemeinen einen gleichmäßigen und eigenartigen (typischen) Verlauf zu nehmen pflegen. Vor allem ist die
Ansteckung, die Übertragung des Krankheitserregers, nach
gefestigter wissenschaftlicher Erfahrung von dem Vorhandensein gewisser äußerer Umstände wie gewisser körperlicher Zustände (Ansteckungsbereitschaft) des Menschen abhängig, so daß das Vorhandensein bestimmter Umstände
die Ansteckung gerade auf einem bestimmten Wege nahelegt, das Fehlen dieser Umstände anderseits sie in der
Regel ausschließt. Das Berufungsgericht hatte geprüft,
ob die Klägerin schon mit einer Ansteckung in das Krankenhaus gekommen war; wenn es diese Möglichkeit verneinte, lag das auf dem Gebiete der tatrichterlichen Würdigung, die im Revisionsverfahren nicht nachgeprüft werden konnte, da beachtliche Verfahrensrügen in dieser Hinsicht nicht erhoben waren. Um die Annahme der Ansteckung durch einen bestimmten Kranken auszuschalten,
genügt nicht schon die theoretische Möglichkeit, daß irgendeine Mittelsperson irgendwie und irgendwo mit einem
anderen Kranken in Berührung gekommen ist und die
Krankheitskeime weitergegeben hat; es müssen bestimmte
Umstände dargetan werden, daß nicht bloß eine Möglichkeit, sondern eine gewisse Wahrscheinlichkeit für eine solche
mittelbare Beziehung des Erkrankten mit anderen Kranken bestand. Wenn das Berufungsgericht die Möglichkeit
einer Ansteckung der Klägerin durch eine solche Mittelsperson gegenüber der nach dem gewöhnlichen Verlauf der
Dinge viel näher liegenden Ansteckung durch den in ihre
Abteilung gebrachten Knaben als so fernliegend ansah,
daß sie nicht ernstlich als Krankheitsursache in Betracht
gezogen werden könne, lag das wiederum auf tatsächlichem

Gebiet und entsprach nur dem von der Rechtsprechung aufgestellten Grundsatz, daß ein anderer Ablauf in dem Bereich einer ernstlich in Betracht zu ziehenden Möglichkeit liegen müsse. Mit Recht war bei dieser Sachlage ein Verstoß der behandelnden Ärzte gegen ihre Pflicht bejaht worden. Es ist Pflicht jedes Arztes, durch äußerste Vorsicht seinen Kranken vor jeder Ansteckung und den damit verbundenen Gefahren zu bewahren, namentlich wenn dieser wegen seiner Erkrankung für die Ansteckung mit einer anderen Krankheit besonders empfindlich ist. Nach den Gutachten war eine Ansteckung der Klägerin, die durch ihre Operation in ihrer Widerstandskraft geschwächt war, nicht ausgeschlossen; bei Scharlacherkrankungen kommen trotz äußerer Heilung in zahlreichen Fällen Ansteckungen auch noch nach Ablauf der üblichen Absonderungszeit von 6 Wochen vor und besteht namentlich bei einer Erschwerung wie der durch Mittelohrentzündung die Ansteckungsgefahr darüber hinaus fort. Das mußte auch den behandelnden Ärzten der beklagten Stadt bekannt sein; sie durften sich deshalb nicht über die nach den Gutachten bestehende Gepflogenheit hinwegsetzen, Kinder, die infolge Scharlacherkrankung eine Mittelohrentzündung haben, während der Dauer dieser Erkrankung auch nach der äußeren Abheilung des Scharlachs nicht in ein Zimmer mit Kindern zu legen, die nicht an Scharlach leiden. Die teilweise Aufhebung des Urteils erfolgte wegen der Ausführungen des Berufungsgerichts über die Schadenersatzpflicht der Beklagten aus unerlaubter Handlung. Die Gründe sind nicht veröffentlicht. (III, 13. Dezember 1940.)

Amtl. Sammlg. S. 336—342.

Vgl. Bd. 78 S. 432; Bd. 128 S. 121; Bd. 150 S. 283; JW. 1933 S. 2701; 1935 S. 3540.

50. Geistliche. Ruhegehalt. Revision. (DBeamtG. § 174; GVG. § 71; ZPO. § 547; RG. vom 26. Juni 1935 über das Beschlußverfahren in Rechtsangelegenheiten der Evangelischen Kirche § 1.) Ein Pfarrer der Evangelischen Landeskirche in Preußen trat im Jahr 1902 in den Ruhestand. Er wandte sich dem Studium der Philologie zu und wurde als Studienrat angestellt. Im Jahr 1925 wurde er aus dieser Stellung in den Ruhestand versetzt. Seitdem bezog er sowohl als früherer Pfarrer als auch als früherer Studienrat einen Ruhegehalt. Im Jahr 1936 eröffnete ihm die Finanzabteilung beim Evangelischen Konsistorium, daß bei der Berechnung seines Ruhegehalts als Pfarrer die Vorschriften des Preußischen Gesetzes über die Pensionierung der unmittelbaren Staatsbeamten vom 27. März 1872 anzuwenden gewesen wären und daß er demgemäß in der Zeit vom 1. Januar 1933 bis zum 30. Juni 1936 insgesamt rund 6000 *RM* zuviel erhalten habe; diesen Betrag müsse er zurückzahlen. In der Folgezeit wurden ihm entsprechende Abzüge gemacht. Seine Beschwerde gegen die Kürzungen wiesen die kirchlichen Behörden zurück. Er klagte auf Auszahlung der einbehaltenen Beträge für die Zeit vom 1. Juli 1936 bis zum 31. Dezember 1937. Das Berufungsgericht gab der Klage statt. Das Reichsgericht erkannte auf Aussetzung des Verfahrens. Der Wert des Beschwerdegegenstandes lag unter der Revisionssumme. Die Zulässigkeit der Revision hing davon ab, ob für die Ansprüche von Geistlichen der öffentlichrechtlichen Religionsgesellschaften auf ihre Bezüge an Gehalt oder Ruhegehalt die Landgerichte ausschließlich zuständig sind. Eine ausdrückliche Vorschrift hierüber fehlt. Die Geistlichen sind auch, obwohl sie den Treueid auf den Führer zu leisten haben, nicht Beamte im Sinne des Deutschen Beamtengesetzes, da sie weder unmittelbar noch mittelbar staatliche Aufgaben zu erfüllen haben. Die Regelung ihrer Besoldung ist von der staatlichen Gesetzgebung den einzelnen Religionsgesellschaften überlassen worden; sie kann in den Kirchen verschieden sein und braucht sich nicht an das Besoldungsrecht des Reiches anzuschließen. Die Ausdehnung der Vorschriften über die ausschließliche Zuständigkeit der Landgerichte auf die Gehaltsansprüche der Geistlichen ergibt sich aber aus bestimmten Vorschriften des Deutschen Beamtengesetzes. § 174 ordnet die Geltung der Vorschriften des Abschnitt IX über den Rechtsweg wegen vermögensrechtlicher Ansprüche für die öffentlichrechtlichen Religionsgesellschaften und deren Verbände an. Zu den hienach anwendbaren Vorschriften gehört § 142, nach dem vermögensrechtliche Ansprüche der Beamten und Ruhestandsbeamten durch Klage vor den Verwaltungsgerichten geltend zu machen sind. Das Inkrafttreten dieser Vorschrift ist durch § 162 bis zur Errichtung des Reichsverwaltungsgerichts aufgeschoben; im übrigen sind aber die Vorschriften des Abschnitts IX auch für die Geistlichen in Kraft getreten. Es ergibt sich also als Wille des Gesetzes, daß die vermögensrechtlichen Ansprüche der Geistlichen aus dem Dienstverhältnis mit den entsprechenden Ansprüchen der Beamten in der gerichtlichen Verfolgung eine übereinstimmende Behandlung erfahren sollen. Daraus folgt die ausschließliche Zuständigkeit der Landgerichte für diese Ansprüche. — Das Versorgungswesen der Geistlichen der Evangelischen Kirche der Altpreußischen Union war früher durch die Ruhegehaltsordnung vom 26. Mai 1909 und das Preußische Gesetz über die Pfarrerbesoldung vom gleichen Tage geregelt. Danach hatten die Geistlichen im Ruhestand ein Ruhegehalt zu beanspruchen. Die Mittel hiefür waren einer Ruhegehaltskasse für evangelische Geistliche zu entnehmen, die als selbständiger Fonds verwaltet wurde. Das Vermögen dieser Kasse ging durch die Geldentwertung großenteils verloren; die Fortsetzung ihrer Leistungen konnte die Kirche nur mit Hilfe erheblicher Staatszuschüsse ermöglichen. Hienach konnte es fraglich werden, ob der Kläger überhaupt in der Lage war, einen rechtlichen Anspruch auf Ruhegehalt gegen die Beklagte geltend zu machen. Das Reichsgericht bejahte die Frage. Wenn sich auch die Kirche zur Erfüllung der Ruhegehaltsansprüche ihrer Geistlichen der Ruhegehaltskasse bediente und diese als selbständiger Fonds verwaltet wurde, ließ das doch die Tatsache unberührt, daß es die Kirche selbst war, welche die Altersversorgung ihrer Geistlichen in dieser Weise regelte. Die damit anerkannte Verpflichtung kam durch die Unmöglichkeit, sie in der bisherigen Weise durchzuführen, nicht in Wegfall. Der Preußische Staat machte auch seine Zuschüsse davon abhängig, daß sie zur Fortzahlung der Gehalts- und Ruhegehaltsbezüge der Geistlichen verwendet wurden. Die beklagte Kirche hatte ihre Anordnung gegenüber dem Kläger auf die §§ 28, 28a des Preußischen Pensionsgesetzes in der Fassung vom 12. Juli 1923 und den § 29 des Preußischen Gesetzes über die Einstellung des Personalabbaus gestützt. Nach diesen Bestimmungen ist neben einem Ruhegehalt aus einem nach der ersten Versetzung in den Ruhestand angetretenen neuen Amt das alte Ruhegehalt nur bis zur Erreichung des Betrages auszuzahlen, der sich für die gesamte Dienstzeit aus dem der Festsetzung des alten Ruhegehalts zugrunde gelegten

Diensteinkommen ergibt. Der Bescheid erging unter der Herrschaft des Gesetzes über das Beschlußverfahren in Rechtsangelegenheiten der Evangelischen Kirche vom 26. Juni 1935. Dort ist in § 1 vorgeschrieben: Hängt die Entscheidung eines bürgerlichen Rechtsstreites davon ab, ob seit dem 1. Mai 1933 in der Evangelischen Landeskirche oder in der Deutschen Evangelischen Kirche getroffene Maßnahmen gültig sind oder nicht, so hat das Gericht das Verfahren bis zur Entschließung der Beschlußstelle in Rechtsangelegenheiten der Evangelischen Kirche auszusetzen. So lag die Sache im vorliegenden Falle. Das Verfahren war also auszusetzen. (III, 29. November 1940.)

Amtl. Sammlg. S. 343—352.

Vgl. Bd. 79 S. 1; Bd. 114 S. 220; Bd. 152 S. 1.

51. **Ehescheidung. Ekelerregende Krankheit.** (EheG. §§ 52, 54.) Eine Frau mußte sich wegen einer bösartigen Mastdarmgeschwulst einer Operation unterziehen; der ganze Unterteil des Mastdarms wurde entfernt und ein künstlicher Darmausgang an der linken Bauchseite geschaffen. Ihr Mann hob einige Monate später die häusliche Gemeinschaft mit ihr auf und klagte auf Scheidung. In zwei Rechtszügen wurde die Klage abgewiesen. Das Reichsgericht verwies die Sache zurück. Der körperliche Zustand der Beklagten wich von der natürlichen körperlichen Beschaffenheit so sehr ab, daß er als ein schwerer Krankheitszustand anzusehen war. Nicht nur durch ihr äußeres Erscheinungsbild, sondern auch durch die Wirkung auf den Geruchsinn rief dieser Zustand bei einem normal empfindenden Menschen Ekel hervor. Die Entscheidung über die Klage hing also nur davon ab, ob dem Scheidungsbegehren des Klägers die sittliche Rechtfertigung abzusprechen war. Diese unter Berücksichtigung aller Umstände, besonders auch der Belange des Klägers, vorzunehmende Prüfung stand in erster Reihe dem Tatrichter zu. (IV, 16. Dezember 1940.)

Amtl. Sammlg. S. 352—354.

Vgl. Bd. 160 S. 240.

52. **Ehescheidung. Unfruchtbarkeit. Frist.** (EheG. §§ 53, 54, 58.) Der Klage auf Aufhebung oder Scheidung der Ehe wegen Unfruchtbarkeit der Frau setzte diese den Einwand entgegen, daß sie bei Eingehung der Ehe nicht unfruchtbar gewesen sei, dieser Zustand vielmehr seine Ursache

darin habe, daß in den ersten Jahren ihrer Ehe beim Geschlechtsverkehr empfängnisverhütende Mittel angewendet worden seien. In zwei Rechtszügen wurde die Klage abgewiesen. Das Reichsgericht verwies die Sache zurück. Die Abweisung der Aufhebungsklage wurde gebilligt, weil das Vorbringen der Frau, daß sie erst während der Ehe unfruchtbar geworden sei, glaubwürdig erschien. Die Scheidungsklage war abgewiesen worden, weil die Unfruchtbarkeit der Beklagten durch einen ärztlichen Eingriff beseitigt werden könne. Das Reichsgericht trat dieser Auffassung insofern bei, als es erklärte, daß nur dauernde Unfruchtbarkeit als Scheidungsgrund angesehen werden könne. Damit ist aber noch nicht gesagt, daß die Unfruchtbarkeit stets dann, wenn die Möglichkeit besteht, sie durch ärztlichen Eingriff zu beseitigen, als vorübergehend anzusehen sei. Die bloße Möglichkeit, daß dies der Fall sein werde, genügt hiefür nicht; es bedarf der Feststellung, daß nach den Erfahrungen der ärztlichen Wissenschaft bei regelmäßigem Verlauf des ärztlichen Eingriffs auf Erfolg zu rechnen ist. Die Beklagte hatte sich zur Vornahme des Eingriffs bereit erklärt. Es mußte aber auch geprüft werden, ob dem Kläger zuzumuten war, sich auf weitere Versuche zur Behebung der Unfruchtbarkeit der Beklagten einzulassen. Solche Versuche waren seit einer Reihe von Jahren schon gemacht worden, ohne daß ein Erfolg eingetreten war. Sowohl für die Beurteilung der Frage, ob dem Kläger noch ein weiterer Versuch in dieser Richtung und damit ein weiteres Zuwarten angesonnen werden konnte, als auch unter dem für die Scheidungsvorschrift des § 53 EheG. im Vordergrund stehenden bevölkerungspolitischen Gesichtspunkt konnte ferner an dem Vorbringen des Klägers nicht vorübergegangen werden, daß die Parteien, die beide 35 Jahre alt waren, immer weniger Aussicht hatten, Kinder aufzuziehen, und daß ihm aus diesem Grunde nicht angesonnen werden könne, nach nunmehr zehnjährigem Bestand der Ehe seinen Wunsch nach Kindern länger zurückzustellen. Ergab sich bei Berücksichtigung dieser Gesichtspunkte, daß dem Kläger nicht zugemutet werden konnte, sich auf weitere Versuche zur Behebung der Unfruchtbarkeit noch einzulassen, so mußte die Unfruchtbarkeit trotz der Möglichkeit, sie durch einen Eingriff zu beseitigen, und trotz der Bereitschaft der Beklagten, ihn an sich ausführen zu lassen, als eine dauernde im Sinne des Gesetzes angesehen werden. Bei der Prüfung der sittlichen Rechtfertigung des Scheidungsbegehrens war zu dem Vorbringen der Beklagten Stellung zu nehmen, daß ihre Unfruchtbarkeit auf die in der ersten Zeit der Ehe einverständlich angewendeten empfängnisverhütenden Mittel zurückzuführen sei. Beweispflichtig war insoweit die Beklagte. Gegenüber der von der Revision vorgetragenen Ansicht, daß die Beklagte es in der Hand habe, durch Verzögerung des Rechtsstreits den Ablauf der Fristen des § 58 EheG. herbeizuführen, verwies das Reichsgericht darauf hin, daß diese Fristen schon durch die Klageerhebung gewahrt waren. (IV, 16. Dezember 1940.)

Amtl. Sammlg. S. 354—358.

Vgl. Bd. 164 S. 290.

53. **Ehebruch. Unterhaltsanspruch des Kindes.** (ABGB. §§ 138, 158, 159, 870, 871, 1432.) Eine Frau beging, während ihr Mann zum Heeresdienst eingerückt war, Ehebruch. Sowohl sie als ihr Liebhaber betrachteten ein Kind, das die Frau in der Folge gebar, als aus diesem Verkehr hervorgegangen. Der Ehemann der Frau setzte die Ehe mit ihr fort, ohne die Ehelichkeit des Kindes anzufechten. Nachdem der Erzeuger des Kindes gestorben war, fand sich in seinem Nachlaß ein Sparbuch mit einer Einlage von 2500 *RM*, das auf den Namen des Kindes lautete. Im Verlassenschaftsabhandlungstermin erschien die Mutter des Kindes und erklärte, daß sie als Ersatz für die Verpflegung und Erziehung des Kindes einen Betrag beanspruche, der ungefähr der im Sprachbuch angelegten Summe gleichkam. Die Erben erkannten die Forderung an und erklärten sich bereit, der Frau das Sparbuch zu überlassen; sie widerriefen aber ihr Anerkenntnis sowie die Vereinbarung mit der Begründung, daß die Frau ihnen verschwiegen habe, daß sie zur Zeit der Geburt des Kindes verheiratet war und daß das Kind als ehelich in die Matrik eingetragen war. Der Klage der Frau setzten sie den weiteren Einwand entgegen, daß der Versorgungsanspruch nur von dem Kinde selbst erhoben werden könne. In allen Rechtszügen drang die Klage durch. Das Landgericht hatte auf Grund der Aussage der Klägerin und verschiedener Zeugenaussagen festgestellt, daß das Kind **aus dem Ehebruch der Klägerin mit dem Erblasser der Be-**

klagten stammte. Der Einwand der Revision, daß das Kind als aus der Ehe der Klägerin stammend anerkannt werden müsse, traf insofern zu, als es sich um den Familienstand des Kindes handelte; dieses hatte in jeder Beziehung die Stellung eines ehelichen Kindes der Klägerin mit allen Wirkungen, die sich in familienrechtlichen, erbrechtlichen und sonstigen rechtlichen Beziehungen hieraus ergaben. Das schloß jedoch nicht aus, daß die Frage, ob der Erblasser der wirkliche Erzeuger des Kindes war, in einem anderen Zusammenhang von Bedeutung war und daher von den Gerichten zum Gegenstand der Prüfung und Beweiserhebung gemacht werden konnte. Wenn ein Kind im Ehebruch erzeugt ist und der Ehemann die Ehelichkeit nicht anficht, wird dadurch allerdings die Klagbarkeit des Unterhaltsanspruchs gegen den Erzeuger ausgeschlossen; es besteht aber für ihn eine auf der Tatsache der Erzeugung beruhende natürliche Pflicht, für das Kind zu sorgen und anderen, die das an seiner Stelle tun, ihre Aufwendungen zu ersetzen. Eine vertragsmäßige Anerkennung dieser Verpflichtung ist gültig und wirksam und begründet deren Klagbarkeit. Von einer listigen Irreführung der Beklagten durch die Klägerin konnte keine Rede sein. Die Beklagten hätten den Nachweis führen müssen, daß die Klägerin bewußt und absichtlich ihnen etwas Wesentliches verschwiegen hatte, und daß gerade die verschwiegenen Tatsachen für ihren Entschluß, die Forderung der Klägerin anzuerkennen, ursächlich waren. Diesen Nachweis hatten die Untergerichte nicht als erbracht angesehen; im Gegenteil hatten sie es sogar für erwiesen erachtet, daß die Klägerin sich nicht absichtlich als zur Zeit der Geburt des Kindes ledig ausgegeben hatte. Der Einwand der mangelnden Klagebefugnis der Klägerin scheiterte schon daran, daß die Klägerin jahrelang den ganzen Aufwand für den Unterhalt des Kindes getragen hatte und daß ihr darum der Anspruch auf Ersatz dieses Aufwandes gegen die Beklagten zustand. (VIII, 8. Januar 1941.) Amtl. Sammlg. S. 358—365.

54. **Staatshaftung. Dienstreise. Beamteneigener Kraftwagen.** (WeimVerf. Art. 131; BGB. § 839.) Der Telegraphenbauführer R., der in einem beamteneigenen Wagen von seinem Wohnort an einen anderen Ort fuhr, um dort ein Dienstgeschäft vorzunehmen, stieß auf der

Fahrt mit einem anderen Kraftwagen zusammen. Ein Mann, der als Fahrgast sich in diesem Wagen befand, wurde erheblich verletzt. Er klagte gegen die Reichspost auf Schmerzensgeld, Schadenersatz und Feststellung der Verpflichtung zum Ersatz weiter noch entstehenden Schadens. In zwei Rechtszügen drang die Klage durch. Das Reichsgericht verwies die Sache zurück. Die frühere Auffassung, nach der die mit dem Betriebe der Reichspost verbundenen Geschäfte grundsätzlich dem bürgerlichrechtlichen Kreise zugerechnet wurden, ist seit längerer Zeit aufgegeben. Das hoheitsrechtliche Wesen der Tätigkeit, welche die Reichspost auf dem Gebiete des deutschen Fernmeldewesens entfaltet, ergibt sich übrigens schon unmittelbar aus dem Gesetz über Fernmeldeanlagen vom 14. Januar 1928. Nach diesem Gesetz übt der Reichspostminister neben dem Reichswehrminister das ausschließlich dem Reiche zustehende Recht aus, Fernmeldeanlagen zu errichten und zu betreiben. Die Tätigkeit der Reichspost auf diesem Gebiet ist also Ausfluß der Staatsgewalt. Demgemäß war auch das Dienstgeschäft, das R. auszuüben hatte, hoheitsrechtlicher Natur. Die Reichspost hatte dies nicht bestritten, aber geltend gemacht, daß die Zurücklegung des Weges vom Wohnsitz des Beamten bis zur Arbeitsstelle nicht in den Umkreis der öffentlichen Gewaltausübung falle, zumal er den Weg in einem ihm gehörigen Kraftwagen zurückgelegt habe; sie hatte versucht, die Fahrt des Beamten dem Falle gleichzustellen, daß sich ein Beamter von seiner Wohnung aus zu seiner Dienststelle begibt. Eine solche Gleichstellung erklärte das Reichsgericht für unzulässig. Der Beamte, der sich zu seiner Dienststelle begibt, tritt erst dort seinen Dienst an; wenn aber ein Beamter außerhalb seines Amtssitzes ein Dienstgeschäft vorzunehmen hat, ist auch die Fahrt dorthin dienstlich. Allerdings ist nicht jede dienstliche Kraftwagenfahrt schon Ausübung öffentlicher Gewalt; sie fällt, wie auch die sonstigen dienstlichen Handlungen des Beamten, entweder in den bürgerlichrechtlichen oder in den öffentlichrechtlichen Geschäftskreis der Behörde. Ob das eine oder das andere zutrifft, muß sich im Einzelfalle nach Wesen und Zweck der Fahrt bestimmen. Die Fahrt ist zweifellos hoheitsrechtlich, wenn sie unmittelbar der Verwirklichung hoheitsrechtlicher Ziele dient (bei der Reichspost z. B. der Briefbeförderung).

Wenn ein Beamter eine Dienstreise vorzunehmen hat, bleibt es ihm für die Regel überlassen, in welcher Weise er sie ausführen will; seine Pflicht ist nur, rechtzeitig zur Stelle zu sein. Die Auswahl und Benutzung der Beförderungsmittel steht dann in einem solchen Abstand von dem hoheitsrechtlichen Hauptgeschäft, daß ein innerer Zusammenhang zwischen beiden fehlt. Benutzt etwa ein Richter zur Wahrnehmung eines Ortstermins einen gemieteten oder eigenen Kraftwagen, so liegt das jenseits der Ausübung öffentlichrechtlicher Gewalt. Der Richter ist dann bloßer Verkehrsteilnehmer, nicht anders, als wenn er die Eisenbahn oder Straßenbahn benutzen würde. Anders ist es aber, wenn eine Hoheitsverwaltung eigene Kraftfahrzeuge in den Dienst ihrer Aufgaben stellt. Regelmäßig wird das um der hoheitsrechtlichen Aufgaben willen geschehen. Die Kraftfahrzeuge haben den Zweck, die Erledigung dieser Art von Geschäften zu fördern und die Beamtenschaft dafür beweglicher und daher auch geeigneter zu machen. Die Beamten sind dann regelmäßig dienstlich gehalten, wenn nicht sogar verpflichtet, sich zur Erledigung ihrer Amtsgeschäfte der Dienstwagen zu bedienen. Ihre Benutzung ist eine dienstliche Angelegenheit und steht, wenn der Zweck der Einzelfahrt, sei es auch nur mittelbar, der Ausübung eines hoheitsrechtlichen Geschäftes dient, damit in einem engen, von der Behörde selbst geschaffenen und gewollten Zusammenhang. Die Kraftwagen der Telegraphenbauämter bilden ein notwendiges Zubehörstück, das ihnen für ihre Amtstätigkeit beigegeben ist; die dienstliche Benutzung dieser Wagen läßt sich von dem Amtsbetriebe dieser Behörden nicht trennen. Die Fahrt des N. wäre also zweifellos in den Bereich der Ausübung öffentlicher Gewalt gefallen, wenn der Wagen behördeneigen gewesen wäre. Daß der Wagen beamteneigen war, begründete keinen Unterschied. Als beamteneigen gelten solche Kraftwagen, die den Beamten aus dienstlichem Anlaß zugewiesen, in der Regel aus Mitteln der Reichspost beschafft, betrieben und unterhalten werden, nach Ablauf ihrer Tilgungszeit aber in das Eigentum des Beamten übergehen. Der Beamte unterliegt bei der Verwendung des Wagens zum persönlichen Gebrauch bis zur Tilgung des Kaufpreises gewissen Beschränkungen; in der rechsgeschäftlichen Verfügung über

den Wagen bleibt er auch nach Tilgung des Kaufpreises solange beschränkt, als die Reichspost die dienstliche Benutzung des Wagens beansprucht. Der Beamte muß sich verpflichten, alle dienstlich erforderlichen Fahrten, soweit irgend möglich, mit dem ihm zugewiesenen Wagen auszuführen. Diese Regelung zeigt unverkennbar, daß der beamteneigene Wagen lediglich einen behördeneigenen ersetzen soll. Sie ist in dieser Form getroffen worden, um eine möglichst schonende Behandlung der Wagen zu erreichen. Der beamteneigene Wagen muß, selbst wenn der Beamte das Eigentum an ihm nach Tilgung des Kaufpreises erworben hat, dem behördeneigenen Wagen gleichgestellt werden; er ist aus demselben Grunde wie dieser in den Dienst der Behörde gestellt und muß von dem Beamten bei seinen Dienstreisen, soweit das angängig ist, verwendet werden. Das Berufungsgericht hatte festgestellt, daß die R. die ihm dem Kläger gegenüber obliegende Amtspflicht zum vorsichtigen, den Verkehrsregeln entsprechenden Verfahren schuldhaft verletzt hatte. Eine Inanspruchnahme der Reichspost nach Art. 131 Weim. Verf. und § 839 BGB. wäre allerdings ausgeschlossen gewesen, wenn der Kläger auf andere Weise Ersatz seines Schadens zu erlangen vermocht hätte. Der Umstand, daß R. in seiner Eigenschaft als Halter des Wagens dem Kläger nach § 7 KraftFahrzG. ersatzpflichtig war, kam dabei nicht in Betracht. Würde die Haftung des R. aus seiner Amtspflichtverletzung (§ 839 BGB.) nicht auf die Reichspost übergegangen sein, wie es nach Art. 131 WeimVerf. der Fall war, so hätte sich R. von ihr nicht durch Berufung auf seine daneben bestehende Haftung nach § 7 KraftFahrzG. befreien können; es ist undenkbar, daß die Haftung eines Beamten für eine von ihm begangene Amtspflichtverletzung um deswillen nicht eintreten sollte, weil er etwas weit geringeres getan, nämlich nur eine Gefährdung gesetzt und damit die Möglichkeit für die Schadensentstehung geschaffen hat. Zu prüfen war aber, ob der Kläger nicht von dem Halter des Wagens, in dem er gefahren war, Ersatz seines Schadens verlangen konnte. (III, 10. Januar 1941.) Amtl. Sammlg. S. 365—374.
Vgl. Bd. 129 S. 149; Bd. 156 S. 401; Bd. 158 S. 83; Bd. 164 S. 273; JW. 1938 S. 1562.

55. **Ausländisches Urteil. Vollstreckungsgegenklage. Zuständigkeit.** (ZPO. §§ 276, 722, 723, 767, 802; RG. über das deutsch-schweizerische Abkommen über die gegenseitige Anerkennung und Vollstreckung von gerichtlichen Entscheidungen und Schiedssprüchen vom 28. Juli 1930; VO. zur Ausführung dieses Abkommens vom 23. August 1930.) Eine Aktiengesellschaft in der Schweiz erwirkte bei dem Bezirksgericht Zürich ein Urteil gegen einen in Deutschland wohnenden Deutschen. Nachdem es rechskräftig geworden war, beantragte sie beim Landgericht Köln, das Urteil für vollstreckbar zu erklären. Das Landgericht gab die Sache zuständigkeitshalber an das Amtsgericht Köln. Dieses wies den Antrag ab. Das Landgericht Köln erklärte als Berufungsgericht das Urteil für vollstreckbar. Während des Berufungsverfahrens wurde beim Landgericht Köln eine Klage eingereicht, die sich als Vollstreckungsgegenklage bezeichnete und auf die §§ 826 BGB. sowie § 767 ZPO. gestützt wurde; in ihr wurde verlangt, die Zwangsvollstreckung aus dem schweizer Urteil für unzulässig zu erklären. Zur Begründung wurde vorgebracht, daß das Bezirksgericht Zürich für die Entscheidung der Sache nicht zuständig gewesen sei, daß zwischen den Parteien ein Vergleich geschlossen worden sei, durch den der Klageanspruch seine Befriedigung gefunden habe, und daß das Urteil durch unwahre Behauptungen erschlichen worden sei. In zwei Rechtszügen wurde die Klage abgewiesen. Das Reichsgericht verwies die Sache an das Amtsgericht Köln. Soweit die Klage auf § 767 ZPO. gestützt war, hatten schon die Untergerichte ihre Zuständigkeit verneint. Auch das Reichsgericht entschied in diesem Sinne. Bildet den Gegenstand der Vollstreckungsgegenklage die Frage, ob die im ausländischen Urteil ausgesprochene, durch Urteil eines inländischen Gerichts erteilte Vollstreckbarkeit wegen Wegfall ihrer Voraussetzung entfallen soll, so kann als Prozeßgericht nur das Gericht in Frage kommen, bei dem der Streit über die Gewährung der Vollstreckbarkeit im ersten Rechtsgang auszutragen war. Das war nach dem deutsch-schweizerischen Abkommen über die gegenseitige Anerkennung und Vollstreckung von gerichtlichen Entscheidungen vom 28. Juli 1930 das Amtsgericht Köln. Es stand nicht entgegen, daß es sich um einen Streitwert handelte, der an sich weit über die Zuständigkeit der Amtsgerichte hin-

ausging. Die sachliche wie die örtliche Zuständigkeit dieses Gerichts war eine ausschließliche. Mit dieser Klagebegründung war aber eine zweite aus § 826 BGB. verbunden. Die Verbindung war nicht in der Weise erfolgt, daß neben einen Hauptantrag ein Hilfsantrag gestellt wurde; vielmehr war ein einheitlicher Klageantrag (Unzulässigkeit der Zwangsvollstreckung) mit verschiedenen tatsächlichen Behauptungen begründet. Die beiden Begründungen waren dem Gericht gehäuft und als gleichwertig unterbreitet worden, wenngleich ihm die Wahl freigestellt war, aus welcher der beiden tatsächlich und rechtlich voneinander unabhängigen Begründungen es der Klage stattgeben wollte. Es waren also beide Gründe zu prüfen. Für den zweiten Teil der Klage war nach allgemeinen Vorschriften die Zuständigkeit des Landgerichts Köln gegeben. Nachdem das Reichsgericht einem im Revisionsverfahren gestellten Hilfsantrag des Klägers entsprechend die Sache an das Amtsgericht Köln verwiesen hatte, blieb es dem Kläger überlassen, ob er bei der vor dem Amtsgericht fortzusetzenden Verhandlung über die Vollstreckungsgegenklage den Klagegrund der unerlaubten Handlung als solchen ohne Rücksicht darauf vorbringen wollte, daß für eine selbständige Klage aus diesem Rechtsgrunde die sachliche Zuständigkeit des Amtsgerichts Köln fehlen würde. (VII, 21. Januar 1941.) Amtl. Sammlg. S. 374—384. Vgl. Bd. 13 S. 349; Bd. 95 S. 287; Bd. 148 S. 270; JW. 1886 S. 195.

56. Grundstückkauf. Preisbildung. Rechtskraft. (BGB. §§ 134, 812; ZPO. § 322; PreisbildG. §§ 1, 2, 6; PreisstopVO. §§ 1, 2; Runderlaß vom 6. Oktober 1937 Nr. 155/37.) Am 26. Juli 1937 wurde ein Haus um 164 500 *RM* verkauft und aufgelassen; der Käufer wurde als Eigentümer im Grundbuch eingetragen. Von dem Kaufpreis wurden 90 000 *RM* durch Übernahme einer Hypothek beglichen, 47 500 *RM* bezahlt. Auf Zahlung des Restes von 27 000 *RM* wurde Klage erhoben. Der Käufer verweigerte die Zahlung wegen angeblicher Mängel des Hauses, wurde aber in zwei Rechtszügen verurteilt; das Urteil erlangte die Rechtskraft. Nach Zahlung der Summe beantragte der Käufer bei der Preisprüfungsstelle, den volkswirtschaftlich gerechtfertigten Kaufpreis festzustellen. Diese erklärte einen Kaufpreis von 116 500 *RM* für gerechtfertigt. Der Käufer klagte nun auf Rückzahlung von zunächst 6500 RM, weil der Kauf in Höhe des Überpreises von 48 000 RM nichtig sei. In zwei Rechtszügen drang die Klage durch. Das Reichsgericht wies sie ab. Das Urteil auf Zahlung von 27 000 RM stand der neuen Klage nicht entgegen. Die Rechtskraft reicht nur soweit, als über den durch die Klage erhobenen Anspruch entschieden ist. Damals war nur ein Teil des Kaufpreises eingeklagt; diesen Teil konnte der Kläger keinesfalls zurückfordern. Die neue Klage betraf aber einen anderen Teil des Kaufpreises. Auf diesen Teil erstreckte sich die Rechtskraft des früheren Urteils nicht. Ein solcher Einfluß konnte auch nicht daraus entnommen werden, daß damals die eingeklagte Summe als Restbetrag bezeichnet war und daß sie nur auf Grund der Annahme der Gültigkeit des Vertrags zugesprochen werden konnte. Nirgends war aber mit Rechtskraftwirkung zwischen den Parteien festgestellt worden, daß der Kaufvertrag gültig sei. Demgemäß waren die Gerichte nicht gehindert, jetzt über diese Frage anders zu urteilen, als im früheren Rechtsstreit geschehen war. Die Klage war aber nicht hinlänglich begründet. Der Kläger erstrebte die Herabsetzung des vereinbarten Kaufpreises auf die angeblich volkswirtschaftlich gerechtfertigte Höhe. Das Verlangen wurde nicht auf eine Anordnung der Preisprüfungsstelle gestützt, sondern nur auf eine von dieser erteilte Auskunft. Zur Zeit des Vertragsschlusses waren das Preisbildungsgesetz vom 29. Oktober 1936 und die Preisstopverordnung vom 26. November 1936 in Kraft, nicht aber der Runderlaß des Reichskommissars für die Preisbildung vom 6. Oktober 1937. Dieser Runderlaß gibt die für die Preisbildungsüberwachung bei Grundstücken zu beachtenden Gesichtspunkte und die in Betracht kommenden Maßnahmen an. Er besagt, daß nicht nur jede weitere ungerechtfertigte Erhöhung der Bodenpreise verhindert, sondern auch dort, wo die Bodenpreise schon am 17. Oktober 1936 als überhöht anzusehen waren, versucht werden müsse, sie nach Möglichkeit wieder auf eine volkswirtschaftlich gerechtfertigte Höhe zurückzuführen. Dem Erlaß kam keine rückwirkende Kraft zu; insbesondere ist nirgends seine Erstreckung auf Grundstückkäufe angeordnet, die schon zum Eigentumsübergang geführt haben. Aus dem Preisbildungsgesetz und der

Preisstopverordnung ergab sich aber nicht, daß bei allen nach ihrem Inkrafttreten abgeschlossenen Grundstückveräußerungen Preise über das volkswirtschaftlich gerechtfertigte Maß hinaus verboten seien. Erst nach dem Runderlaß vom 6. Oktober 1937 konnten solche Käufe unter diesem Gesichtspunkt beanstandet werden. Jede Grundstückveräußerung hat nunmehr ein Prüfungsverfahren zu durchlaufen. Ohne Einschaltung dieses Prüfungsverfahrens ist eine Herabsetzung eines vereinbarten Kaufpreises nicht möglich. Eine nachträgliche Vernichtung schon durchgeführter Veräußerungen liegt nicht im Sinne der im Rundschreiben gegebenen Anweisungen. (V, 19. Dezember 1940.) Amtl. Sammlg. S. 385—393.

Vgl. Bd. 158 S. 40; JW. 1936 S. 3047.

57. **Klageerweiterung. Revision.** (ÖstEisenbahnHaftPflG. § 3; überlVO. §§ 6, 7.) Auf Grund eines Eisenbahnunfalls wurde Klage auf Ersatz der Heilungskosten, Schmerzensgeld und Zahlung einer Rente sowie auf Feststellung erhoben, daß die Reichsbahn verpflichtet sei, allen weiteren durch den Unfall entstandenen Schaden zu ersetzen. Den Streitgegenstand bezifferte die Klägerin auf 10 000 *RM*, wovon auf das Feststellungsbegehren rund 2940 *RM* entfallen sollten. Das Landgericht erkannte durch Zwischenurteil den Anspruch der Klägerin als dem Grunde nach zu Recht bestehend an, das Berufungsgericht nur zu drei Vierteln; in den Gründen der Entscheidung des Berufungsgerichts war hervorgehoben, daß durch das Urteil nicht über das Feststellungsbegehren entschieden sei. Nach Zustellung des Urteils, aber vor Ablauf der Revisionsfrist, teilte die Klägerin dem Landgericht mit, daß sie ihre Forderung für Heilungskosten von 460 *RM* auf 6000 *RM* erhöhe; sie beantragte neben der Feststellung der Schadenersatzpflicht der Reichsbahn deren Verurteilung zum Ersatz von drei Vierteln der verlangten Beträge. Die Revision der Beklagten wurde als unzulässig zurückgewiesen. Die Frage ihrer Zulässigkeit war für das Reichsgericht nicht schon dadurch bindend gelöst, daß beide Untergerichte sie bejaht und die Revision daher dem Reichsgericht zur Entscheidung vorgelegt hatten. Der Wert des Beschwerdegegenstandes in Revisionssachen kann immer nur mit Beziehung auf den Anspruch oder den für das weitere Rechtsmittelverfahren strittig gebliebenen Teil des An-

spruchs festgesetzt werden, der bereits dem Berufungsgericht zur Entscheidung vorgelegen hat. Der Beschwerdegegenstand kann begrifflich niemals größer sein als der Streitgegenstand des Berufungsverfahrens. Die Untergerichte hatten bisher nur über den Anspruch der Klägerin auf Ersatz der Heilungskosten, Schmerzensgeld und Zahlung einer Rente entschieden; der Wert dieser Ansprüche erreichte nicht 10000 *RM*. Die Zulässigkeit der Revision hing also davon ab, ob für die Ansprüche die Landgerichte ohne Rücksicht auf den Wert des Streitgegenstandes zuständig waren. Nach dem österreichischen Eisenbahnhaftpflichtgesetz sind Klagen auf Grund dieses Gesetzes beim Handelsgericht anzubringen. Handelsgerichte sind nach der Sprache der österreichischen Jurisdiktionsnorm die Landgerichte. Es handelte sich also um eine Rechtsstreitigkeit, die im ersten Rechtszug ausschließlich vor die Landgerichte gehörte. Die Zulässigkeit der Revision für solche Klagen setzt aber weiter voraus, daß die ausschließliche Zuständigkeit der Landgerichte ausdrücklich angeordnet ist. Das traf nicht zu. (VIII, 18. Januar 1941.)

Amtl. Sammlg. S. 393—398.

58. **Doppelehe. Zwischenstaatliches Recht.** (ÖstABGB. §§ 4, 37, 62, 111; EgzBGB. Art. 17.) Eine in Österreich im Jahr 1898 eingegangene Ehe wurde im Jahr 1911 einverständlich von Tisch und Bett geschieden. Um eine nach österreichischem Recht unzulässige neue Ehe eingehen zu können, ließ sich der Ehemann von einem ungarischen Staatsangehörigen an Kindes Statt annehmen und erwarb dadurch die ungarische Staatsangehörigkeit; er wurde auch aus dem österreichischen Staatsverband entlassen. Auf Grund einer Klage vor einem ungarischen Gericht wurde seine Ehe gelöst. Er ging eine neue Ehe mit einer Ungarin ein. Nachdem er im Jahr 1937 gestorben war, machten beide Frauen Versorgungsansprüche aus seiner Beamtenstellung geltend. Da die Verwaltungsbehörden die Ansprüche ablehnten, klagte die frühere Ehefrau auf Feststellung, daß im Gebiete des österreichischen Rechts sowohl das ungarische Trennungsurteil als auch die zweite Ehe des Mannes keine Rechtswirkung hätten. Das Berufungsgericht wies die Klage ab. Die Revision der Klägerin hatte teilweise Erfolg. Der Wechsel der Staatsangehörigkeit des Mannes war für die Klägerin ohne

Rechtswirkung geblieben, weil ihre Ehe von Tisch und Bett geschieden war. Die Bedeutung des ungarischen Urteils für die Klägerin war nach der Rechtslage zur Zeit des Ausspruchs zu beurteilen. Die Ehe ist ein persönliches Rechtsverhältnis, das zwar auf einem Vertrag beruht, bei dem aber das Wesen des Vertrages stark zurücktritt. Der Inhalt der Ehe ist der Parteiwillkür vollständig entrückt; hiefür gibt es keine Vertragsfreiheit. Maßgebend ist grundsätzlich das Recht des Staates, dem der Ehemann angehört (Personalstatut). Hieraus erwachsen keine Schwierigkeiten, wenn beide Ehegatten dem gleichen Personalstatut unterstehen. Ist aber die Frau Inländerin geblieben, so tritt der Anwendung des für den Mann geltenden Rechts der Umstand entgegen, daß der Staat besonderen Wert darauf legt, für seine Staatsangehörigen die von ihm gesetzten Normen über die Scheidung und Trennung der Ehe anzuwenden. Er allein bestimmt, ob und unter welchen Umständen das seine Staatsangehörigen bindende Eheband zu lösen ist und die öffentlichen Belange eine Befreiung von den Pflichten aus der Ehe gestatten. Einer Scheidung oder Trennung durch ein ausländisches Gericht ist daher die Anerkennung jedenfalls dann zu versagen, wenn die Voraussetzungen des inländischen Rechts nicht erfüllt sind. Der Trennung der Ehe dem Bande nach stand nach österreichischem Recht das Hindernis des § 111 ABGB. entgegen, da beide Ehegatten katholisch waren. Das ungarische Urteil war also für den Bestand der Ehe der Klägerin ohne Bedeutung. Ihre Ehe verlor auch ihre Geltung nicht durch die Eingehung der in Ungarn geschlossenen und für Ungarn gültigen Ehe. Der Staat muß eine und dieselbe Ehe immer vom gleichen Recht aus beurteilen, und zwar, wenn es sich um das eheliche Schicksal eines eigenen Staatsbürgers handelt, nach den von ihm aufgestellten Normen. Da aber beide Ehegatten im Zeitpunkt der Schließung der zweiten Ehe Ungarn waren, mußte die Gültigkeit dieser Ehe nach ungarischem Recht beurteilt werden, und zwar einschließlich der Vorfrage, ob die erste Ehe des Mannes dem Bande nach damals noch bestand. Nach ungarischem Recht war sie wirksam aufgelöst, so daß der zweiten Ehe in Ungarn das Ehehindernis des bestehenden Ehebandes nicht entgegenstand. Beide Ehen waren von verschiedenen rechtlichen Standpunkten aus zu betrachten. Dem Klagebegehren, die Rechtsunwirksamkeit der zweiten Ehe festzustellen, konnte also nicht entsprochen werden. Demzufolge konnte auch nicht festgestellt werden, daß die Klägerin die alleinige Witwe ihres verstorbenen Mannes war, sondern nur, daß sie dessen Witwe war. Die Beschränkung der Wirkungen dieser Feststellung auf das Geltungsgebiet des österreichischen Rechts hatte zu unterbleiben. Ein von einem deutschen Gericht gefälltes Erkenntnis bindet alle deutschen Gerichte und hat im ganzen Deutschen Reiche Geltung. (IV, 24. Januar 1941.)

Amtl. Sammlg. S. 398—404.

Gesetzesregister.

16. **Reichsgesetz vom 26. Juli 1935 über das Beschlußverfahren in Rechtsangelegenheiten der Evangelischen Kirche: 1** 50.
17. **Reichsleistungsgesetz vom 1. September 1939: 26** 48; **27** 48.
18. **Reichsversicherungsordnung (RVO.): 368** 13; **368d** 13; **369b** 13; **1542** 34.
19. **Unlauterer Wettbewerbsgesetz (UnlWG.): 16** 40.
20. **Vierjahresplangesetz (VierJahrPlG.): 2** 33.
21. **Wehrleistungsgesetz (WehrLeistG.): 26** 48; **27** 48.
22. **Weimarer Verfassung (WeimVerf.): 131** 13, 38, 48, 54.
23. **Sonstige Reichsgesetze und -verordnungen:** 2, 3, 6, 13, 14, 16, 20, 23, 25, 31, 37, 48, 50, 55, 56, 57.
24. **Landesgesetze und -verordnungen:** 20, 32, 35, 41, 57.
25. **Österreichisches Allgemeines Bürgerliches Gesetzbuch (Öst.-ABGB.): 4** 58; **37** 58; **62** 58; **111** 58; **138** 53; **158** 53; **159** 53; **163** 27; **164** 27; **380** 16; **424** 16; **425** 16; **426** 16; **431** 16; **432** 16; **433** 16; **773** 29; **783** 29; **820** 29; **821** 29; **870** 53; **871** 53; **988** 20; **989** 20; **1432** 53.
26. **Österreichische Zivilprozeßordnung (ÖstZPO.): 182** 9; **471** 14; **482** 3; **483** 3; **502** 37; **507** 14; **513** 14; **514** 14; **519** 14; **526** 16; **528** 14.
27. **Österreichisches Eisenbahnhaftpflichtgesetz (ÖstEisBahn.-HaftpflG.):** 57.
28. **Österreichisches Exekutionsordnung (ÖstExO.): 186** 47; **187** 47.
29. **Österreichische Jurisdiktionsnorm (ÖstJurN.): 55** 37.

Seitenzahlen der amtlichen Sammlung.

1 1—22; **2** 22—24; **3** 24—26; **4** 26—32; **5** 32—35; **6** 35—41;
7 41—54; **8** 54—62; **9** 62—64; **10** 65—68; **11** 68—85; **12** 85—91;
13 91—107; **14** 107—109; **15** 109—116; **16** 117—121;
17 121—125; **18** 125—128; **19** 129—136; **20** 136—140;
21 140—146; **22** 146—148; **23** 149—154; **24** 155—160;
25 161—165; **26** 166—185; **27** 187—192; **28** 193—207;
29 207—209; **30** 209—219; **31** 219—223; **32** 223—225;
33 226—236; **34** 236—242; **35** 242—248; **36** 248—254;
37 255—256; **38** 257—260; **39** 260—271; **40** 271—288;
41 289—300; **42** 300—307; **43** 307—309; **44** 309—314;
45 315—317; **46** 317—320; **47** 321—323; **48** 323—336;
49 336—342; **50** 343—352; **51** 352—354; **52** 354—358;
53 358—365; **54** 365—374; **55** 374—384; **56** 385—393;
57 393—398; **58** 398—404.

Sachregister.

FSC
www.fsc.org
MIX
Papier | Fördert
gute Waldnutzung
FSC® C083411

Kolibri 360 GmbH
Ferdinand-Jühlke-Straße 7
99095 Erfurt, Deutschland
produktsicherheit@kolibri360.de